The Handbook of Infrared and Raman Spectra of Inorganic Compounds and Organic Salts (a 4-volume set)

VOLUME 4

INFRARED SPECTRA OF INORGANIC COMPOUNDS

(3800-45cm⁻¹)

The Handbook of Infrared and Raman Spectra
of Inorganic Compounds and Organic Salts
(a 4-volume set)

Volume 4
INFRARED SPECTRA
OF INORGANIC COMPOUNDS

$(3800\text{-}45\,cm^{-1})$

Richard A. Nyquist and Ronald O. Kagel

The Dow Chemical Company
Midland, Michigan

ACADEMIC PRESS, INC.

San Diego London Boston
New York Sydney Tokyo Toronto

ACADEMIC PRESS, INC.
525B Street, Suite 1900, San Diego, CA 92101-4495, USA
1300 Boylston Street, Chestnut Hill, Massachusetts 02167
http://www.apnet.com

ACADEMIC PRESS LIMITED
24–28 Oval Road, London NW1 7DX, UK
http://www.hbuk.co.uk/ap/

Library of Congress Cataloging-in-Publication Data

Nyquist, Richard A.
   Infrared spectra of inorganic compounds (3800-45 cm⁻¹) /
Richard A. Nyquist, Ronald O. Kagel.
      p.   cm.
   Includes bibliographical references.
   ISBN 0-12-523450-3
   1. Infrared spectra—Tables.   2. Inorganic compounds—
Spectra—Tables.   I. Kagel, Ronald O.   II. Title.
QC457.N93   1971
544'63—dc20   76-180797
                                                    CIP

Printed in the United States of America
96   97   98   99   EB   9   8   7   6   5   4   3   2   1

CONTENTS

Infrared Spectra of Inorganic Compounds

PREFACE

Infrared spectroscopy has proved to be a most useful tool in structural determination and identification of organic compounds. This method of analysis has the advantage of being rapid and unambiguous, it can easily be applied to small quantities of sample, and is accessible to most organic chemists because of the availability of low-cost infrared spectrometers. For these reasons, over the last 30 years large numbers of standard spectra of organic compounds have been collected and catalogued, and are presently available for reference.

In contrast, relatively few infrared spectra of inorganic compounds are available for reference, and no present collection covers both the mid- and far-infrared regions from 3800 to 45 cm^{-1}. The present work is intended to help fill this void by presenting a comprehensive compendium of reference spectra of inorganic compounds.

The spectra included in this volume were obtained, we trust, with sufficient care so that sample preparation difficulties that are often encountered with inorganic materials, such as reaction between sample and window support material, were effectively avoided. Spectra-structure correlation charts and many anion frequency assignments are presented as aids in the identification of unknown inorganic samples.

Richard A. Nyquist
Ronald O. Kagel

ABOUT THE AUTHORS

Richard A. Nyquist retired from Dow Chemical Company, U.S.A. at the end of 1994 with the rank of Research Scientist. He received his B.A. degree in chemistry from Augustana College, Rock Island, IL, his M.S. degree from Oklahoma State University, and his doctorate from Utrecht University, The Netherlands. He joined Dow Chemical Company in 1953. His career has been mainly in the field of vibrational spectroscopy. He utilizes IR and Raman spectroscopy for solving chemical problems, for the elucidation of molecular structure, and for qualitative and quantitative analysis. Nyquist is the author or coauthor of more than 160 scientific articles including books, book chapters, and patents. He is a member of the American Chemical Society and the Society for Applied Spectroscopy. In 1985 he received the Williams-Wright Award from the Coblenty Society for his contributions to industrial IR spectroscopy. Nyquist was subsequently named an honorary member of the Coblenty Society for his contributions to vibrational spectroscopy, and in 1989 he was a national tour speaker for the Society of Applied Spectroscopy. The Association of Analytical Chemists honored him with the ANACHEM Award in 1993 for his contributions to analytical chemistry. Nyquist is listed in *Who's Who in Science and Engineering 1992–93*, *Who's Who in America 1992*, and *Who's Who in the World*. Dow Chemical Company has honored Nyquist with the V.A. Stenger Award in 1981 and the Walter Graf European Award in 1994 for excellence in analytical chemistry. Nyquist has also served as a member of ASTM, and he received the ASTM Award of Appreciation for his contributions to the Practice of Qualitative Infrared Analysis. He is currently president of Nyquist Associates.

Ronald O. Kagel retired from Dow Chemical Company in 1993 as an Environmental Consultant in Dow U.S.A. Manufacturing. He received his B.S. degree in chemistry from the University of Wisconsin and his doctorate from the University of Minnesota in physical chemistry. His career has been in the fields of vibrational spectroscopy and environmental analyses and regulations. Kagel is author or coauthor of more than 100 reports and publications including two books and several book chapters. He is a member of the Coblentz Society, the American Chemical Society, the New York Academy of Sciences, the American Institute of Chemists, the Association for the Advancement of Science, Alpha Chi Sigma, Phi Lambda Upsilon, and Sigma Xi. Kagel was elected a Fellow of the American Institute of Chemists in 1986. He founded, served as Chairman of the Board of Directors of, and was elected an honorary member of the Coalition for Responsible Waste Incineration. He is listed in *Who's Who in the Midwest*, *Who's Who in Science and Technology*, *American Men and Women of Science*, *World Environmental Directory*, *Men of Achievement* (International Biography Center, Cambridge, England), and *Personalities of America*. Kagel received the V. A. Stenger Award from Dow Chemical Company in 1974; the Award for Invaluable Contributions, Michigan Analytical Association of Chemists, 1981; and several Certificates of Appreciation from the American Chemical Society. He was recognized by the U.S. EPA for his peer review contributions in 1977. Kagel served on ASTM D-19, ANSI ISO/TC, the National Research Council, SOCMA, CMA, Chlorine Institute, and CRWI. He is presently an environmental consultant to companies in the U.S., Canada, Switzerland, and Germany.

ACKNOWLEDGMENTS

The success of a venture such as this, spawned out of necessity to expand the problem-solving capabilities of a large industrial firm, indeed depends upon the patronage of management and the inspiration of many colleagues. We wish to thank those who have in any way contributed to this book.

The Dow Chemical Company is gratefully acknowledged for permission to prepare this work for publication and for the extensive use of their facilities during the course of its preparation. A scientific organization is, of course, only as good as its people regardless of the degree of sophistication of its equipment and instrumentation. In this respect, we are particularly fortunate to be endowed with a highly competent staff of technicians who prepared the samples and recorded the spectra. We sincerely acknowledge F. F. Stec, G. R. Ward, J. H. Strope, T. L. Reder, G. R. Spencer, and G. W. Huffman whose skill and patience are reflected throughout.

Over thirty years ago the first application of infrared spectroscopy to an industrial problem was being actively pursued by N. Wright in this laboratory. Our work in infrared spectroscopy was always encouraged and enthusiastically supported by him. We feel that this book is, in a sense, one of the fruits of his pioneering efforts. W. J. Potts, Jr. rendered invaluable advice and support during the course of this work. We wish to thank L. K. Frevel for his support in the use of both infrared and X-ray techniques as a means of solving analytical problems and also H. W. Rinn for helping us verify by X-ray diffraction many of the compounds used to generate this compendium. S. P. Klesney, who aided in checking the chemical nomenclature, is gratefully acknowledged, as is Mrs. N. H. Carney for typing the manuscript. Last, but certainly not least, we recognize and appreciate the understanding and cooperation of Mrs. Irene M. Cote Nyquist and Mrs. Lois J. Kagel during the preparation of this book.

The Handbook of Infrared and Raman Spectra of Inorganic Compounds
and Organic Salts (a 4-volume set)

VOLUME 4

INFRARED SPECTRA OF INORGANIC COMPOUNDS
($3800-45 cm^{-1}$)

INFRARED SPECTRA OF INORGANIC COMPOUNDS

INTRODUCTION

The utility of infrared spectroscopy to the organic chemist is perhaps unsurpassed within the framework of most modern laboratories. Experimental, theoretical, and empirical correlations between functional organic groups and the infrared spectrum have been thoroughly studied and reported. The vast body of literature devoted to the results of these studies provides a rather solid base for use by the analytical spectroscopist. Through the efforts of several authors this accumulated literature has been summarized and reviewed in several excellent books (*1–7*).

The application of infrared spectroscopy to the identification of inorganic compounds has been somewhat less successful. Many simple inorganic compounds such as the borides, silicides, nitrides, and oxides, do not absorb radiation in the region between 4000 and 600 cm^{-1} which, for many years, was the extent of the infrared region covered by most commercial spectrometers. Only within the last 10 years have instruments become available which include the region below 600 cm^{-1}, and it has been even more recent that instrumentation has been developed to cover the far-infrared region between 200 and 10 cm^{-1}. These are the regions in which most inorganic compounds absorb infrared radiation.

The region 4000–600 cm^{-1} has proved to be very useful for the identification of polyatomic anions of the type CO_3^{2-}, SO_4^{2-}, NO_3^{-}, etc. When standard spectra are available, a compound such as KNO_3 can easily be distinguished from $NaNO_3$ or $Ca(NO_3)_2$, but in the absence of standard spectra, specific identification of a cation–anion pair is usually not possible by infrared spectroscopy. The differences between the spectra of KNO_3 and $Ca(NO_3)_2$, for example, are largely due to two effects: (1) the extent to which the cation perturbs the internal vibrations of the anion and (2) changes in the crystal structure of the system. The latter is more pronounced in the far-infrared region than in the region 4000–600 cm^{-1}. These effects are usually not predictable.

In obtaining infrared spectra of inorganic solids, an experimental complication arises from possible chemical reaction (cation exchange) between the inorganic compound and the infrared window material or support medium. The literature contains many examples of standard spectra of inorganic compounds in which this type of chemical reaction has obviously taken place. Care has been exercised in the preparation of samples here so as to avoid this difficulty.

In the present compendium, spectra of inorganic compounds in the solid phase are presented. The majority of these compounds are (powdered) crystalline solids in which the crystallographic unit cell may contain several polyatomic ions or molecules. The internal modes of vibration of the polyatomic group generally occur in the region 4000–400 cm^{-1}; many of these have been extensively documented in the literature. Other optical

1

modes called lattice modes of vibration result from the motion of one polyatomic group relative to another within the unit cell. Lattice modes generally occur in the region 400–10 cm^{-1} and are characteristic of a specific crystal geometry. They can be used as fingerprints for an inorganic compound in much the same way as the internal modes of vibration of organic compounds are used in the region 4000–400 cm^{-1}. The purpose of this work is to present reference spectra and empirical spectra–structure correlations. We do not intend to cover the theoretical aspects of the solid state. For this the reader is referred to several excellent review articles and books (8–13).

EXPERIMENTAL

The mid-infrared spectra were scanned using a Beckman Model IR-9 and two Herscher-Dow prism grating spectrometers in the region 3800–400 cm^{-1} and a Perkin-Elmer Model 225 in the region 3800–200 cm^{-1}. Far-infrared spectra were scanned on a Beckman Model IR-11 in the region 600–45 cm^{-1}. Extensive descriptive material about the instrumentation is given in several books (14–16).

The samples were prepared as mulls, using as mulling agents Fluorolube for the region between 3800–1333 cm^{-1} and Nujol for the region between 1333–400 and 600–45 cm^{-1}, the technique hereinafter being referred to as a "split mull." In the mulling technique, finely ground particles are suspended in the mulling agent and the slurry is supported between two infrared transmitting windows. Samples were not subjected to prolonged grinding, but were treated in a routine manner, the grinding time seldom exceeding 10 minutes. Mechanical grinding devices were not employed. BaF$_2$ windows were used in the region 3800–1333 cm^{-1}, AgCl in the region 1333–400 cm^{-1}, and polyethylene in the region 600–45 cm^{-1}. These window materials are inert to reaction with respect to most inorganic compounds. Standard window materials such as potassium bromide, sodium chloride, cesium bromide, and cesium iodide were found to be highly prone to ion exchange with a number of inorganic compounds, and for this reason their use was avoided.

To illustrate the extent of ion exchange effects, pure samples of Pb(NO$_3$)$_2$ (verified by X-ray diffraction) were prepared as split mulls on sodium chloride, potassium bromide, cesium iodide, barium fluoride, and silver chloride plates. Spectra A and D in Fig. 1 are of pure Pb(NO$_3$)$_2$ and NaNO$_3$, respectively, scanned as split mulls between BaF$_2$ (3800–1333 cm^{-1}) and AgCl (1333–400 cm^{-1}) plates. Spectrum B is a freshly prepared Pb(NO$_3$)$_2$ split mull between NaCl plates, and spectrum C is the Nujol portion of that mull 2 hours after preparation, having been in intimate contact with the NaCl plates. The out-of-plane NO$_3^-$ deformation of NaNO$_3$ which occurs at 838 cm^{-1} is clearly present in spectra B and C. The band intensity increases with contact time (spectra B to C), indicating the continuing formation of NaNO$_3$ by ion exchange between Pb(NO$_3$)$_2$ and the NaCl plate. Similar reactions were observed between Pb(NO$_3$)$_2$ and KBr and CsI.

The potassium bromide pellet technique for preparing samples was strictly avoided. Anomalies in the infrared spectra of inorganic compounds prepared by this technique have been extensively studied. In addition to a possible cation exchange reaction with KBr, the material under investigation may also undergo changes in crystalline form as a result of the high mechanical pressures (10,000 psi) used in the pelleting process. Extreme caution should be exercised when applying the potassium bromide pellet technique to obtain infrared spectra of inorganic compounds.

ARRANGEMENT OF SPECTRA

The spectra are arranged to bring together compounds containing similar anions, in order to facilitate recognition of characteristic group frequencies. The arrangement is based on the position in the periodic table of the central atom in the anion. Where there is no central atom (e.g., CN⁻) the anions are arranged by lowest group; thus CN⁻ falls under C.

In grouping the anions by their central atom, these have been arranged in order of, first, increasing group number, then increasing atomic number within a group: B, Al, C, Si, N, P, O, S, F, Cl, Br. In subarrangement under a given central atom, for example, N, the anions are given in order of increasing number of N atoms in the anion: N^{3-}, N_2^{4-}, N_3^-, etc. The polyatomic anions are arranged in orders of decreasing ratio of N atoms to other atoms in the anion, such as $N_2O_2^{2-}$, NO_2, NO_3. Under a specific anion, individual compounds appear in the order of increasing atomic number of the cation within a given group. For the nitrates (NO_3^-) the order is NH_4NO_3; (Group I) $NaNO_3$, KNO_3 $\cdots CsNO_3$; (Group II) $Ca(NO_3)_2$ $\cdots Ba(NO_3)_2$; (Group III) $Al(NO_3)_2$; (Group IV) $Ga(NO_3)_3$; etc. Compounds containing the ammonium ion have been placed at the beginning of each such grouping.

Two indices are provided. The first of these contains compounds as they appear in the book in numerical sequence. The second index is arranged alphabetically by anion.

SPECTRA–STRUCTURE CORRELATIONS

Characteristic infrared frequencies and band intensities of the different anions are summarized in Table 1 and Figs. 2 and 3. Frequency assignments for the fundamental vibrations of complex anions taken from the literature are summarized in Table 2. The assignment of the anion fundamental vibration is based on point group symmetry; the anion usually belongs to the same point group in the solid phase, regardless of the space group of the unit cell. In Table 2, the notation and numbering system of the fundamental vibrations are taken from Herzberg (1). For a detailed discussion of group theory the reader is referred to any one of several excellent texts (18–21). A correlation chart for metal oxides is given in Fig. 4.

The following paragraphs are devoted to a general résumé of the more common or important structural information contained in Table 1 and a brief discussion of lattice vibrations. For detailed information on individual groups reference should be made to the publications listed in the bibliography.

Strong bands associated with OH stretching vibrations of water and hydroxyl groups occur between 3200 and 3700 cm⁻¹. The hydroxyl group is characterized by a strong sharp absorption band in the region 3650–3700 cm⁻¹. Water of hydration usually exhibits one strong sharp band near 3600 cm⁻¹ and one or more strong sharp bands near 3400 cm⁻¹. Water of hydration is easily distinguished from hydroxyl groups by the presence of the H–O–H bending motion which produces a medium band (often multicomponent) in the region 1600–1650 cm⁻¹. Free water has a strong *broad* absorption band centered in the region 3200–3400 cm⁻¹; the H–O–H bending motion generally occurs near 1650 cm⁻¹.

The O–H stretching vibration of $HAsO_4$, HCO_3^-, HSO_4^-, HPO_4^{2-}, etc., characteristically exhibits a strong to medium-strong broad multipeaked band often extending from 2000 to 3400 cm⁻¹. This type of absorption is a distinguishing feature of the very acidic protons of the salts and corresponding acids.

The NH_4^+ group is characterized by a strong broad absorption centered near 3250 cm⁻¹ resulting from NH_4^+ antisymmetric stretching vibrations and a strong absorption band near 1400 cm⁻¹ resulting from an NH_4^+ bending vibration.

Multiple bond stretching and metal–hydrogen stretching vibrations usually occur in the region 1500–2500 cm⁻¹. The groups CN⁻, SCN⁻, and OCN⁻ exhibit a strong absorption in the region 2000–2300 cm⁻¹. The CN⁻ group is characterized by one or more strong very sharp absorption bands in the region 2000–2200 cm⁻¹. A strong sharp band in the region 2050–2200 cm⁻¹ is characteristic of the SCN⁻ group. The absorption band characteristic of the OCN⁻ group is strong but somewhat broader than the CN⁻ or SCN⁻ bands and occurs near or above 2200 cm⁻¹.

The NO stretching vibration of the NO⁻ group gives rise to a strong band at 1940 cm⁻¹.

Inorganic compounds containing the HPO_3^{2-} and $H_2PO_2^-$ groups are readily characterized by medium to strong bands in the region 2300–2400 cm⁻¹ often showing submaxima; these bands arise from the P–H stretching vibrations.

Polyatomic anions of inorganic compounds show characteristic absorption bands in the region 1500 ≃ 300 cm⁻¹ which result from stretching and bending vibrations. These characteristic bands are summarized in Table 1 and Fig. 1. Table 2 gives the frequency assignments of the fundamentals for many different anions, as summarized from the literature.

For a detailed discussion of the anion (point group) symmetry the reader is referred to Nakamoto (22). In certain cases where there is high point group symmetry, T_d, for example, vibrations normally infrared inactive will often appear as a weak band; also doubly and triply degenerate vibrations split into two and three components, respectively. These effects result either from lowering of the point group symmetry or from factor group splitting as a result of different crystalline environments.

Lattice vibrations can occur as high as 600 cm⁻¹, but usually occur in the region below 300 cm⁻¹. These vibrations are unique for a specific crystalline compound, and are useful fingerprints for identification. Many inorganic compounds absorb only in this region of the spectrum, particularly ionic metal halides, nitrides, silicides, tellurides, and heavy metal oxides. Lattice frequencies can be correlated with crystal structure in an isomorphous series. For example, the lattice modes in the alkali halides which are cubic show a decrease in frequency with increasing mass or atomic radius of the cation or anion. Compare spectra* of the series of compounds:

NaF (599), NaCl (717), and NaBr (796)
KCl (718), KBr (797), and KI (831)
RbCl (719), RbBr (798), and RbI (832)
AgCl (741) and AgBr (811)
NaCl (717), KCl (718), and RbCl (719)
NaBr (796), KBr (797), and RbBr (798)

Different crystalline forms of the same compound also show different lattice vibrations as well as spectral differences at higher frequency in the region where the fundamental vibrations occur in the spectrum. For example, SrP_2O_7 exists in two crystal forms, the α-form which is orthorhombic and the β-form which is tetragonal (compare spectra 265 and 266). Other such cases appear in this compilation of infrared spectra.

The infrared spectra of crystalline polyatomic inorganic compounds usually show more bands than can be assigned to fundamental vibrations and sum or difference tones of the internal vibrations of the polyatomic ion. These other bands result from sum or difference tones of the lattice modes with the internal fundamental vibrations of the polyatomic ion.

REFERENCES

1. G. Herzberg, "Molecular Spectra and Molecular Structure." Part II. Van Nostrand-Reinhold, Princeton, New Jersey, 1945.
2. R. N. Jones and C. Sandorfy, in "Chemical Applications of Spectroscopy" (W. West, ed.), Vol. IX, Chapter IV. Wiley (Interscience), New York, 1956.
3. A. B. F. Duncan, in "Chemical Applications of Spectroscopy" (W. West, ed.), Vol. IX, Chapter III. Wiley (Interscience), New York, 1956.

*Spectrum number given in parentheses.

REFERENCES

4. L. J. Bellamy, "The Infrared Spectra of Complex Molecules." Wiley, New York, 1958.

5. C. N. R. Rao, "Chemical Applications of Infrared Spectroscopy." Academic Press, New York, 1964.

6. N. B. Colthup, L. H. Daly, and S. E. Wiberley, "Introduction to Infrared and Raman Spectroscopy." Academic Press, New York, 1964.

7. L. J. Bellamy, "Advances in Infrared Group Frequencies." Methuen, London, 1968.

8. R. S. Halford, *J. Chem. Phys.* 14, 8 (1946).

9. W. Vedder and D. F. Horning, *in* "Advances in Spectroscopy" (H. W. Thompson, ed.), Vol. II, p. 189. Wiley (Interscience), New York, 1961.

10. H. Jones, "Theory of Brillouin Zones and Electronic States in Crystals." North-Holland Publ., Amsterdam, 1962.

11. S. S. Mitra, *Solid State Phys.* 13, 1-80 (1962).

12. Born and Huang, "Dynamical Theory of Crystal Lattices." Oxford Univ. Press, London and New York, 1966.

13. L. M. Falicov, "Group Theory and its Physical Applications." Univ. of Chicago Press, Chicago, 1966.

14. G. R. Harrison, R. C. Lord, and J. R. Lorfbourow, "Practical Spectroscopy." Prentice-Hall, Englewood Cliffs, New Jersey, 1948.

15. W. Brugel, "Infrared Spectroscopy." Methuen, London, 1962.

16. W. J. Potts, "Chemical Infrared Spectroscopy." Wiley, New York, 1963.

17. K. E. Lawson, "Infrared Absorption of Inorganic Substances." Van Nostrand-Reinhold, Princeton, New Jersey, 1961.

18. G. Herzberg, "Molecular Spectra and Molecular Structure," Part I. Van Nostrand-Reinhold, Princeton, New Jersey, 1940.

19. E. B. Wilson, J. C. Decius, and P. C. Cross, "Molecular Vibrations." McGraw-Hill, New York, 1955.

20. J. Lecomte, *in* "Handbuch der Physik " (S. Flugge, ed.), Vol. 26, p. 244. Springer-Verlag, Berlin and New York, 1958.

21. E. P. Wigner, "Group Theory." Academic Press, New York, 1959.

22. K. Nakamoto, "Infrared Spectra of Inorganic and Coordination Compounds." Wiley, New York, 1963.

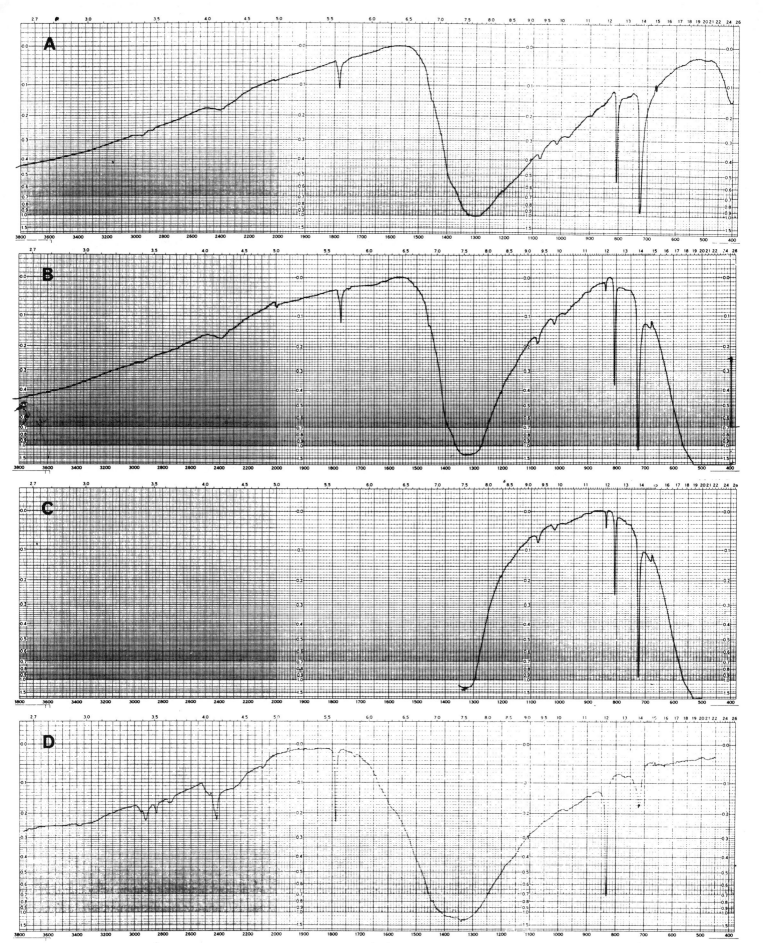

Fig. 1. Spectrum A: (Pb(NO$_3$)$_2$. Spectrum D: NaNO$_3$. A and D are scanned as split mulls between BaF$_2$ (3800-1333 cm^{-1}) and AgCl (1333-400 cm^{-1}) plates. Spectrum B: Pb(NO$_3$)$_2$ scanned as split mull between NaCl plates. Spectrum C: Nujol portion of B 2 hours after preparation, having been in contact with NaCl plates.

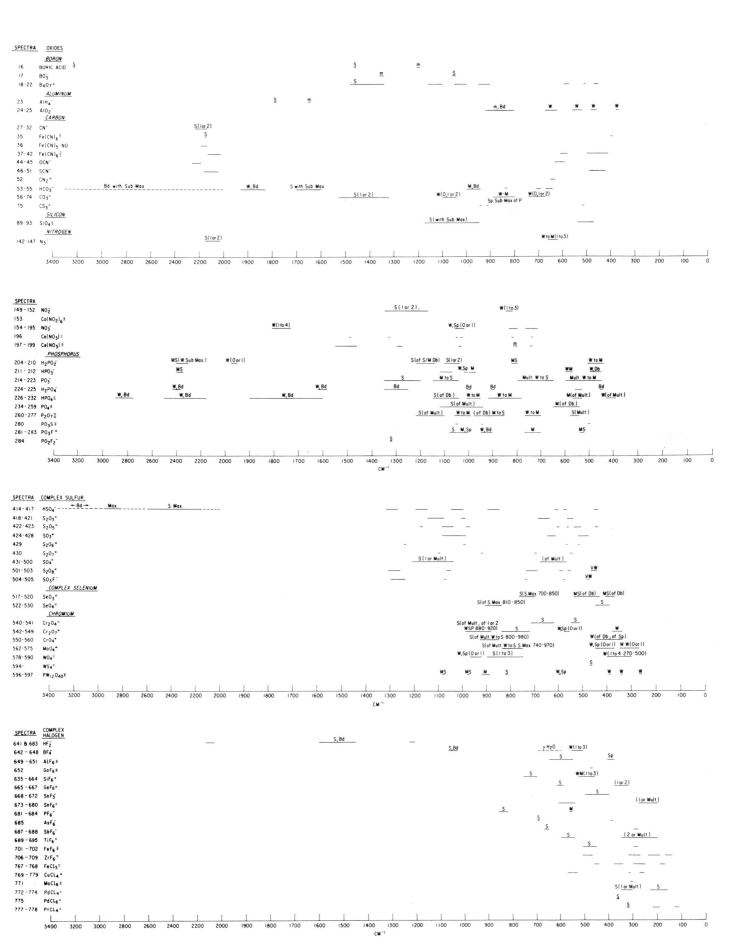

Fig. 2. Characteristic frequencies and band intensities.

7

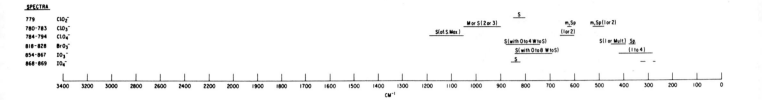

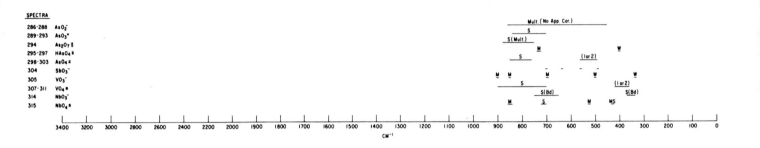

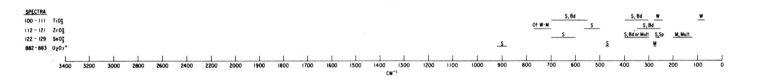

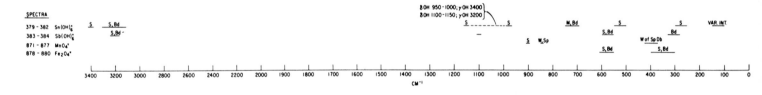

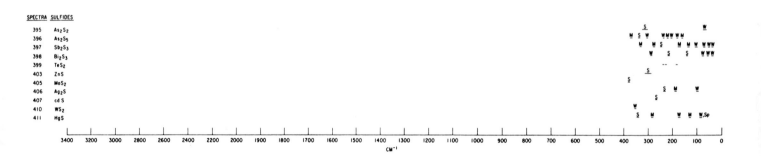

Fig. 3. Characteristic frequencies and band intensities.

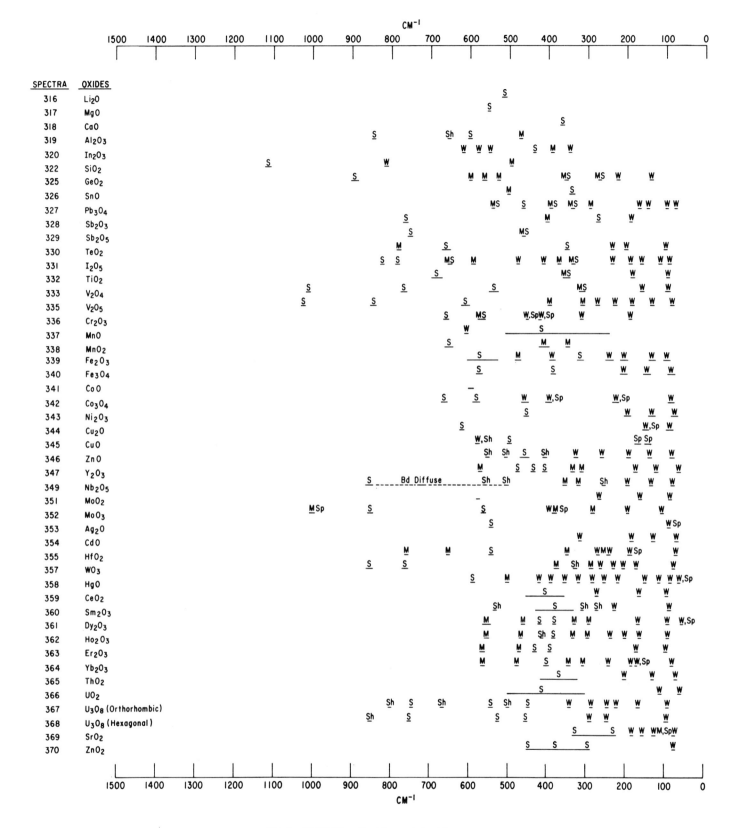

Fig. 4. Correlation chart for metal oxides.

9

Table 1. Characteristic Frequencies for Inorganic Ions

Element	Formula	Ion	Spectra	Characteristic absorption[†]
Boron		boride	1-15	60-100 bd
	$B_2O_7{}^{2-}$	tetraborate	18-22	1340-1480 stg (SMAX), 1100-1150 wk-bd (OD), 1000-1050 wk-bd (OD), 900-950 w, ~825 wk, 520-545 wk, 500-505 wk, 450-470 wk
Aluminum	$AlH_4{}^-$	tetrahydroaluminate	23	~1785 stg, ~1645 m, ~710 m bd
	$AlO_2{}^-$	aluminate	24-25	800-920 m bd, 620-670 wk, 515-560 wk, 450-480 wk, 370-380 wk
Carbon		carbide	26	No correlation
	CN^-	cyanide	27-33	2130-2230 stg (SD)
	$Fe(CN)_6{}^{3-}$	ferricyanide	35	~2140 and ~395
	$Fe(CN)_5NO^{2-}$	nitroferricyanide	36	2130-2170 wk-m(m), ~1929 stg
	$Fe(CN)_6{}^{4-}$	ferrocyanide	37-42	2020-2130 stg (SM), 580-610 wk, 410-500 wk
	OCN^-	cyanate	44-45	2180-2250 stg, 590-630 m
	SCN^-	thiocyanate	46-51	2040-2160 stg, 420-490 (OD) separated by 13-33 cm^{-1}
	$CN_2{}^{2-}$	cyanamide	52	~2000 stg, ~1935 stg, ~1300, ~1215, ~640, ~630
	$HCO_3{}^-$	bicarbonate	53-55	2000-3300 bd (SMAX), 1840-1930 wk bd, 1600-1700 stg (SMAX). 940-1000 m bd, 830-840 m, 690-710 m, 640-670 wk
	$CO_3{}^{2-}$	carbonate	56-74	1320-1530 stg (OD), 1040-1100 wk (OD). 800-890 wk-m. 670-745 (owk or WD)
	$CS_3{}^{2-}$	thiocarbonate	75	~931 stg, ~910 stg, ~518 wk
Silicon		silicide	76-86	60-110 bd
	$SiO_4{}^{4-}$	orthosilicate	89-93	860-1175 stg (SMAX), 470-540 stg
Titanium	$TiO_3{}^{2-}$	titanate (IV)	100-111	500-700 stg bd, 360-450. 200-400 bd (SMAX)
Zirconium	$ZrO_3{}^{2-}$	zirconate (IV)	112-121	700-770 (owk), 500-600 stg, 300-500 stg (SMAX), 230-240 wk
Tin	$SnO_3{}^{2-}$	stannate (IV)	122-129	600-700 stg bd, 300-450

Element	Name	Formula	Pages	Correlations
Nitrogen	nitride		130-141	No correlation
	azide	N_3^-	142-147	3150-3400 wk-m (OD), 2025-2150 stg, 620-660 wk-m (OD or M)
	nitrite	NO_2^-	149-152	1170-1350 stg (OD), 820-850 wk (OD or M)
	nitrate	NO_3^-	154-191	1730-1810 wk sp (SM), 1280-1520 stg (OD or M), 1020-1060 (owk) sp, 800-850 (W-M) sp, 715-770 wk-m* (OD)
	hexanitrocerate (IV)	$Ce(NO_3)_8$	197-199	1465-1550 stg, 1275-1300 stg, 1030-1045 m, 800-820 m, 740-750 m
Phosphorus	phosphide		200-203	No correlation
	hypophosphite	$H_2PO_2^-$	204-210	2300-2400 m-stg (SMAX 2200-2430), 1950-1975 (owk), 1140-1220 stg (OD), 1075-1102 wk sp (OD), 1035-1065 wk sp (OD), 800-825 m-stg, 440-510 wk-m
	orthophosphite	HPO_3^{2-}	211-212	2340-2400 m stg, 1070-1120 stg, 1005-1020 wk sp, 970-1000 m, 570-600 wk m, 450-500 wk (D)
	metaphosphate	PO_3^-	214-223	1200-1350 stg, 1040-1150 m-stg, 650-800 wk-stg (M), 450-600 w-m (M)
	orthophosphate (mono-basic)	$H_2PO_4^-$	224-225	~2700 wk bd, ~2400 wk bd, ~1700 bd, ~1250 bd, ~1100 bd, ~900 bd, 530-560, ~450
	orthophosphate (dibasic)	HPO_4^{2-}	226-232	2750-2900 wk bd, 2150-2500 wk bd, 1600-1900 wk bd, 1200-1410 w-m, 1040-1150 stg (OD), 950-1110 wk-m, 830-920 wk-m, 530-570 m (OM), 390-430 wk (OM)
	orthophosphate	PO_4^{3-}	234-259	940-1120 stg (OM), 540-650 m (OD) (AlPO₄ is an exception, see 240)
	pyrophosphate	$P_2O_7^{4-}$	260-277	1100-1220 stg (OM), 960-1060 wk-m (OD and O sp), 850-980 m, 705-770 w-m, 545-580 m stg (m 500-600)
	phosphorothioate	PO_2S^{3-}	280	~1050, ~945, ~500
	phosphorofluoridate	PO_3F^{2-}	281-283	1010-1080 stg, 1000-1020 m sp, 900-950 wk bd, 700-770 m, 525-540 m stg
	phosphorodifluoridate	$PO_2F_2^-$	284	~1315 stg, ~1150 stg, ~800 stg, ~500 stg
Arsenic	metaarsenite	AsO_2^-	286-288	450-860 (M with no apparent correlations)
	orthoarsenite	AsO_3^{2-}	289-293	700-840 stg
	pyroarsenate	$As_2O_7^{4-}$	294	750-880 stg (M), ~540 m, ~400 stg
	orthoarsenate dibasic	$HAsO_4^{2-}$	296-297	~840 stg, 720-740 m, ~400
	orthoarsenate	AsO_4^{3-}	298-303	770-850 stg (OM or SMAX)

11

Table 1. (continued)

Element	Formula	Ion	Spectra	Characteristic absorption[†]
Antimony	SbO_3^-	antimonate (V)	304	~700, ~635, ~560, ~490
Vanadium	VO_4^{3-}	orthovanadate	307–313	700–900 stg (O SMAX)
		oxides	316–368	Strong bands usually in region below 1300 cm⁻¹; as a rule of thumb, frequencies decrease progressing down thru each group in the periodic table of the elements.
Oxygen	OH^-	hydroxide	372–378	3750–2000 stg (sp, M, or bd)
	$Sn(OH)_6^{2-}$	hexahydroxostannate (IV)	379–382	3000–3400 stg bd, 2200–2300 wk bd, 950–1150 stg, 650–800 m bd, 500–550 stg, 250–300 stg
	$Sb(OH)_6^-$	hexahydroantimonate (V)	383–384	~3200 stg bd, ~1340 wk, 1075–1150 wk bd, ~720 and ~580, stg, ~450, 300–350 bd
	BiOX	bismuth oxyhalide (Cl, Br, and I)	388–390	480–530 wk, 240–375 stg bd, 70–150 m bd (bands decrease in frequency in the series Cl to I)
	UO_2X_2	uranyl halide (F and Cl)	393–394	850–1020 stg (M, SMAX), 380–470 bd. 250
Sulfur		sulfide	395–411	Bands below 400
	HSO_4^-	hydrogen sulfate (bisulfate)	414–417	3400–2000 bd (MAX near 2900; SMAX 2200–2600), 850–900, 605–620, 565–585, 450–480
	$S_2O_3^{2-}$	thiosulfate	418–421	1080–1150 stg (M or SMAX), 990–1010 stg, 640–690 m–s, 540–570 wk (o m)
	$S_2O_5^{2-}$	pyrosulfite	422–423	~1175 stg, 1040–1090 m, 970–990 stg, 650–660 m, 560–570 m, 510–540 m, 440–450 m
	SO_3^{2-}	sulfite	424–428	990–1090 stg (OM or SMAX), 615–660 m (o SMAX), 470–525 m (OD)
	$S_2O_6^{2-}$	dithionate	429	~1240 stg, ~995 m–stg, ~570 m–stg, ~520 m
	$S_2O_7^{2-}$	pyrosulfate	430	~1325 wk, ~1100 stg, ~920 m, ~700 wk, ~550 m
	SO_4^{2-}	sulfate	431–499	1040–1210 stg (OM or SMAX), (960–1030, often 1 or 2 wk sp bands), 570–680 m (OD or M)
	$S_2O_6^{2-}$	peroxydisulfate	501–503	1260–1310 stg, 1050–1070 m sp, 690–740 m, 580–600 wk m sp, ~560 m
	SO_3F^-	fluorosulfonate	504–505	1260–1300 stg, 1070–1080 m sp, ~740 m, ~580 m, ~480 wk
Selenium		selenide	506–516	Bands below 400
	SeO_3^{2-}	selenite	517–520	700–770 stg (SMAX 700–850), 430–540 m stg (OD), 360–410 (OD)
	SeO_4^{2-}	selenate	522–530	840–910 stg (o wk sh 810–850). 390–450 wk m
		telluride	531–538	No correlations

12

	Formula	Name	No.	Bands
Chromium	$Cr_2O_4{}^{2-}$	chromite	540-541	~620-720 stg, 515-550 m stg
	$Cr_2O_7{}^{2-}$	dichromate	542-549	880-990 stg (om, often 1 or 2 wk, sp bands 880-920). 720-840 stg, (555-580 o wk sp), 340-380 wk
	$CrO_4{}^{2-}$	chromate	550-560	850-930 stg (OM or SMAX 800-980)
Molybdenum	$MoO_4{}^{2-}$	molybdate(II)	562-575	750-835 stg (OM or SMAX 740-970), (370-450 owk sp), 308-350 wk, (268-315 owk)
Tungsten	$WO_4{}^{2-}$	tungstate	578-590	(920-970 owk sh), 750-900 stg (OM), 270-500 wk (OM with higher intensity band 300-400)
	$WS_4{}^{2-}$	tetrathiotungstate	594	~465
	$PW_{12}O_{40}{}^{3-}$	phosphotungstate	596-597	1080 n-stg, ~975 m-stg, 890-922 m, 810-820 stg, 590-600 wk sp, ~390 wk, ~340 wk, 260-270 wk
Halogen	F^-	fluoride	598-640	(See text)
	$HF_2{}^-$	hydrogen fluoride	641 (see 682)	2050-2122 m bd, ~1600 stg bd, 1205-1235 m stg
	$BF_4{}^-$	tetrafluoroborate	642-648	1000-1100 (MAX near 1050) stg. bd, (owk 760-780), 510-560 wk sp (O or M)
	$AlF_6{}^{3-}$	hexafluoroaluminate	650-651	550-650 stg, 380-410 sp
	$GaF_6{}^{3-}$	hexafluorogallate	652	~475 m
	$SiF_6{}^{2-}$	hexafluorosilicate	653-664	700-760 stg, 460-530 wk-m (o M)
	$GeF_6{}^{2-}$	hexafluorogermanate	665-667	590-620 stg, 320-380 m-stg (OD)
	$SnF_3{}^-$	trifluorostannate	668-671	450-490, 340-450
	$SnF_6{}^{2-}$	hexafluorostannate	672-679	540-610 stg, 200-280 (1 or more)
	$PF_6{}^-$	hexafluorophosphate	680-683	820-860 stg, 550-565 m, sp
	$AsF_6{}^-$	hexafluoroarsenate	684	~695 stg, 390 m
	$SbF_6{}^-$	hexafluoroantimonate(V)	686-687	650-670 stg, 280-300
	$TiF_6{}^{2-}$	hexafluorotitanate(IV)	688-694	540-600 stg, 200-350 (two or more)
	$FeF_6{}^{3-}$	hexafluoroferrate	700-702	460-510 stg, 280-300 m
	$ZrF_5{}^-$	pentafluorozirconate	703-704	450-500 stg, ~300
	$ZrF_6{}^{2-}$	hexafluorozirconate	705-708	440-500 stg, 270-320 m
	Cl^-	chloride	714-753	
	$SnCl_6{}^{2-}$	hexachlorostannate	764-765	300-325 stg
	$FeCl_5{}^{2-}$	pentachloroferrate(III)	766-767	440-460 m, 340-380 m, 250-300 stg, 170-195 m

Table 1. (continued)

Element	Formula	Ion	Spectra	Characteristic absorption†
	$CuCl_4^{2-} \cdot 2H_2O$	tetrachlorocuprate	768-769	530-570 wk-m (H_2O ?), 310-320 m, 255-275 wk, 100-110 wk. 95-100 wk, 58-65 wk
	$MoCl_6^{3-}$	hexachloromolybdate	770	~305 stg
	$PdCl_4^{2-}$	tetrachloropalladate	771-773	300-350 stg (M), 160-220 m-stg
	$PdCl_6^{2-}$	hexachloropalladate(IV)	774	~360 stg
	$PtCl_4^{2-}$	tetrachloroplatinate	776-777	315-325 stg, 190-220 m-stg, 110-130 m
	ClO_2^-	chlorite	779 (see 780)	800-850 stg (D)
	ClO_3^-	chlorate	780-783	900-1050 stg (D or M), 610-630 m sp, 475-525 m sp (OD)
	ClO_4^-	perchlorate	784-794	1050-1150 stg (O SMAX), 600-660 wk-m (OD)
	Br^-	bromide	794-815	
	BrO_3^-	bromate	818-827	740-850 stg (OD or SMAX), 390-450 wk-m, sp (OD), 350-380 m sp
	I^-	iodide	828-849	
	IO_3^-	iodate	853-866	690-830 stg (O SMAX or M), 300-420 (OD or M)
	IO_4^-	periodate	867-868	830-860 stg, 310-330 m, 260-270 wk-m
Manganese	MnO_3^{2-}	manganite	869	~635 m, ~550 stg, bd
	MnO_4^{2-}	manganate	870	800-900 stg (M)
	MnO_4^-	permanganate	871-876	870-950 stg (o M or SMAX), (o wk 830-840), 370-400 wk (OD)
Iron	$Fe_2O_4^{2-}$	ferrate(III)	877-879	550-610 stg, 400-450 m
Cobalt	CoO_2^-	cobaltite(III)	880	~660 m, ~570 stg
Uranium	$U_2O_7^{2-}$	uranate(VI)	881-882	880-900 stg, 470-480 m-stg, 270-280 wk

†(o = often); (owk) = often weak, but not always detected; (D) = doublet

wk = weak, m = medium, stg = strong, bd = broad, sp = sharp, (SD) = sometimes doublet, (OD) = often doublet, (M) = multiple,

(SM) = sometimes multiple), (SMAX) = with submaxima.

*Band at ~721 results from Nujol.

14

Table 2.   Fundamental Vibrations of Inorganic Ions

Element	Formula	Ion	Point group	Vibrations	Spectra	Ref.[†]
Aluminum	AlH_4^-	tetrahydroaluminate	T_d	$\nu_1(a_1)$ 1790, $\nu_2(e)$ 799, $\nu_3(f_2)$ 1740, $\nu_4(f_2)$ 764	23	(1)
Carbon	CN^-	cyanide		$\nu_{C \equiv N}$ (2080–2239)	27–33	(2)
	$Fe(CN)_6^{3-}$	ferricyanide	O_h	$\nu_6(f_{1u})$ 2105, $\nu_7(f_{1u})$ 511 w, 1460 vw, $\nu_8(f_{1u})$ 387 st.	35	(3)
	$Fe(CN)_5NO^{2-}$	nitroferricyanide	C_{4v}	IR a_1 2173, 2163, 1945, 653, 468, 408, a_2 – b_1 2157, – e2145, 663, 424, 417, 321 b_2 – e2145, Raman a_1 2174, 2162, 1947, 656, 493, 472, 408, 123 a_1 2157, 410 e2144, 422, 415, 164; 100	36	(4)
	$Fe(CN)_6^{4-}$	ferrocyanide	O_h	$\nu_6($ $)$ 2021 and 2033, $\nu_7(f_{1u})$ 585s, $\nu_8(f_{1u})$ 414m	37–42	(3)
	OCN^-	cyanate	$C_{\infty v}$	$\nu_1(\Sigma^+)$ 1292.6 and $2\nu_2(\Sigma^+)$ 1205.5 in Fermi Resonance, $\nu_2(\pi)$ 629.4, $\nu_3(\Sigma)$ 2169.6	44–45	(5)
	SCN^-	thiocyanate	$C_{\infty v}$	$\nu_1(\Sigma^+)$ 743, $\nu_2(\pi)$ 470, $\nu_3(\Sigma^+)$ 2066 (ν_2 often splits in the solid)	46–51	(6)
	CN_2^{2-}	cyanamide	$D_{\infty h}$	$\nu_1(\Sigma_g^+)$ 860, $\nu_2(\pi_u)$ 210, $\nu_3(\Sigma_g^+)$ 930	52	(7)
	CO_3^{2-}	carbonate	D_{3h}	$\nu_1(a_1')$ 1087, $\nu_2(a_2'')$ 874, $\nu_3(e)$ 1432, $\nu_4(e')$ 706	56–74	(8)
	CS_3^{2-}	thiocarbonate	D_{3h}	IR $\nu_1(a_1')$ 488–520, $\nu_2(a_2'')$ 325, $\nu_3(e)$ 920, $\nu_4(e')$ 475–520; Raman 516, 510, 420, 325	75	(9)
Silicon	SiO_4^{4-}	orthosilicate	T_d	$\nu_1(a_1)$ 800, $\nu_2(e)$ 500, $\nu_3(f_2)$ 1050, $\nu_4(f_2)$ 625	89–93	(10)
Titanium	TiO_3^{2-}	titanate (IV)	O_h	$\nu_1(f_{1u})$ ~540 be, $\nu_2(f_{1u})$ ~400 bd	100–111	(11)
Nitrogen	N_3^-	azide	$D_{\infty h}$	$\nu_1(\Sigma_g^+)$ 1344, $\nu_2(\pi_u)$ 645, $\nu_3(\Sigma_g^+)$ 2041	142–147	(12)
	NO_2^-	nitrite	C_{2v}	$\nu_1(a_1)$ 1320–1365, $\nu_2(a_1)$ 807–818, $\nu_3(b_1)$ 1221–1251	149–152	(13)
	NO_3^-	nitrate	D_{3h}	$\nu_1(a_1')$ 1018–1050, $\nu_2(a_2'')$ 807–850, $\nu_3(e)$ 1310–1405, $\nu_4(e')$ 697–716	154–191	(14)
Phosphorus	$H_2PO_2^-$	hypophosphite	C_{2v}		204–210	(15)
	HPO_3^{2-}	orthophosphite	C_{3v}	crystal $\nu_1(a_1)$ 2410, $\nu_2(a_1)$ 977, $\nu_3(a_1)$ 591, $\nu_4(e)$ 1110 and 1083, $\nu_5(e)$ 1021 and 1006, $\nu_6(e)$ 498 and 471, soln. $\nu_1(a_1)$ 2315, $\nu_2(a_1)$ 979, $\nu_3(a_1)$ 567, $\nu_4(e)$ 1085, $\nu_5(e)$ 1027, $\nu_6(e)$ 465	211–212	(15)
	$H_2PO_4^-$	orthophosphate monobasic	S_4	ν_{PO_4}(4b, 4e) 540 and 450	224–225	(16)
	HPO_4^{2-}	orthophosphate dibasic	C_{3v}	$\nu_1(a_1)$ 2900, $\nu_2(a_1)$ 988, $\nu_3(a_1)$ 862, $\nu_4(a_1)$ 537, $\nu_5(e)$ 1230, $\nu_6(e)$ 1076, $\nu_7(e)$ 537, $\nu_8(e)$ 394	226–232	(17)

15

Table 2. (continued)

Element	Formula	Ion	Spectra	Point Group	Vibrations	Ref.[†]
	PO_4^{3-}	orthophosphate	234–259	T_d	Raman $\nu_1(a_1)$ 935, $\nu_2(e)$ 420, $\nu_3(f_2)$ 1080, $\nu_4(f_2)$ 550	(18)
	$P_2O_7^{4-}$	pyrophosphate	260–277	D_{3h}	(a_1') 1212, 909, 477, (a_2'') 1165, 940, 553 (e'') 999, 573, 432, (e') 1124, 707, 615, 201 Divalent salts have lesser symmetry	(19)
Arsenic	PO_3S^{3-}	phosphorothioate	280	C_{3v}	$\nu_1(a_1)$ 960, $\nu_2(a_1)$ 611, $\nu_3(a_1)$ 480, $\nu_4(e)$ 1038, $\nu_5(e)$ 515, $\nu_6(e)$ 367	(20)
	$As_2O_7^{4-}$	pyroarsenate	294	D_{3d}	900–920, ~850, ~450, ~300	(21)
Vanadium	AsO_4^{3-}	orthoarsenate	298–303	T_d	$\nu_1(a_1)$ 813, $\nu_2(e)$ 342, $\nu_3(f_2)$ 813, $\nu_4(f_2)$ 402	(22)
	VO_4^{3-}	orthovanadate	307–313	T_d	$\nu_1(a_1)$ 870, $\nu_2(e)$ 345, $\nu_3(f_2)$ 825, $\nu_4(f_2)$ 480	(22)
Sulfur	$S_2O_3^{2-}$	thiosulfate	418–421	C_{3v}	$\nu_1(a_1)$ 995, $\nu_2(a_1)$ 669, $\nu_3(a_1)$ 446, $\nu_4(e)$ 1123, $\nu_5(e)$ 541, $\nu_6(e)$ 335	(23)
	SO_3^{2-}	sulfite	424–428	C_{3v}	$\nu_1(a_1)$ 1010, $\nu_2(a_1)$ 633, $\nu_3(e)$ 961, $\nu_4(e)$ 496	(24)
	SO_4^{2-}	sulfate	431–499	T_d	$\nu_1(a_1)$ 983, $\nu_2(e)$ 450, $\nu_3(f_2)$ 1105, $\nu_4(f_2)$ 983	(18)
Selenium	SeO_3^{2-}	selenite	517–520	C_{3v}	$\nu_1(a_1)$ 807, $\nu_2(a_1)$ 432, $\nu_3(e)$ 737, $\nu_4(e)$ 374	(25)
	SeO_4^{2-}	selenate	522–530	T_d	$\nu_1(a_1)$ 833, $\nu_2(e)$ 335, $\nu_3(f_2)$ 875, $\nu_4(f_2)$ 432	(18)
Chromium	$Cr_2O_7^{2-}$	dichromate	542–549	C_{2v}	(a_1, b_1, b_2) 924–966, (a_1) 900–910, (b_1) 880–892, (b_1) 760–780, (a_1) 550–570, (a_1, a_2, b_1, b_2) 365, (a_1) 220	(26)
Molybdenum	CrO_4^{2-}	chromate	550–560	T_d	$\nu_1(a_1)$ 847, $\nu_2(e)$ 348, $\nu_3(f_2)$ 884, $\nu_4(f_2)$ 368	(18)
	MoO_4^{2-}	molybdate (III)	562–575	T_d	$\nu_1(a_1)$ 940, $\nu_2(e)$ 220, $\nu_3(f_2)$ 895, $\nu_4(f_2)$ 365	(18)
Tungsten	WO_4^{2-}	tungstate	578–590	T_d	$\nu_1(a_1)$ 928, $\nu_2(e)$ 320, $\nu_3(f_2)$ 833, $\nu_4(f_2)$ 405	(18)
	WS_4^{2-}	tetrathiotungstate		T_d	$\nu_1(a_1)$ 487, $\nu_2(e)$ 179, $\nu_3(f_2)$ 440–465, ν_4	(27)
Fluorine	BF_4^-	tetrafluoroborate	642–648	T_d	$\nu_1(a_1)$ 769, $\nu_2(e)$ 353, $\nu_3(f_2)$ 984(B^{11}), 1016(B^{10}), $\nu_4(f_2)$ 524(B^{11}), 529(B^{10})	(28)
	SiF_6^{2-}	hexafluorosilicate	653–664	O_h	IR $\nu_3(f_{1u})$ 720, $\nu_4(f_{1u})$ 470 Raman $\nu_1(A_{1g})$ 656, $\nu_2(Eg)$ 510, $\nu_5(f_{2g})$ 402, [$\nu_6(f_{2u})$ 260 from combination tone]	(29)
	GeF_6^{2-}	hexafluorogermanate	665–667	O_h	IR $\nu_3(f_{1u})$ 600, $\nu_4(f_{1u})$ Raman $\nu_1(A_{1g})$ 627, $\nu_2(Eg)$ 454, $\nu_5(f_{2g})$ 318	(30)
	SnF_6^{2-}	hexafluorostannate	672–679	D_{3d}	$\nu_1(A_{1g})$ 572, $\nu_3(a_{2u}$ or $Eu)$ 555, $\nu_2(Eg)$ 460, $\nu_4(a_{2u}$ or $Eu)$ 256, $\nu_5(A_{2g}$ or $Eg)$ 247	(30)

Formula	Name	Point group		Frequencies	Range	Ref.
$PF_6{}^-$	hexafluorophosphate	O_h	IR	$\nu_3(f_{1u})$ 830, $\nu_4(f_{1u})$ 550, $[\nu_6(f_{2u})$ 317 from combination tone]	680–683	(29)
			Raman	$\nu_1(A_{1g})$ 735, $\nu_2(E_g)$ 563, $\nu_5(f_{2g})$ 462		
$TiF_6{}^{2-}$	hexafluorotitanate (III)	D_{3d}		$\nu_1(A_{1g})$ 608–613, $\nu_5(A_{2g}$ or $E_g)$ 275–281	688–694	(30)
$ZrF_6{}^{2-}$	hexafluorozirconate	D_{3d}		$\nu_1(A_{1g})$ 576–581, $\nu_3(a_{2u}$ or $E_u)$ 500, $\nu_4(A_{2u}$ or $E_u)$ 230, $\nu_5(A_{2g})$ or $E_g)$ 110–130	705–708	(30)
$SnCl_6{}^{2-}$	hexachlorostannate	O_h	Raman	$\nu_1(A_{1g})$ 311, $\nu_2(E_g)$ 229, $\nu_5(F_{2g})$ 158	764–765	(31)
$CuCl_4{}^{2-}$	tetrachlorocuprate	T_d		$\nu_1(a_1)$ 270–300, $\nu_2(e)$ 130–150, $\nu_3(f_2)$ 290–300, $\nu_4(f_2)$ 110–130	768–769	(32)
$PdCl_4{}^{2-}$	tetrachloropalladate	D_{4h}		(a_{2u}) 170, (e_u) 334 (Lattice 120, 111 and 95)	771–773	(33)
$PdCl_6{}^{2-}$	hexachloropalladate (IV)	O_h	Raman	$\nu_1(A_{1g})$ 317, $\nu_2(E_g)$ 292, $\nu_5(f_{2g})$ 164	774	(34)
$PtCl_4{}^{2-}$	tetrachloroplatinate	D_{4h}	Raman	$\nu_1(A_{1g})$ 335, $\nu_2(B_{1g})$ 164, $\nu_4(B_{2g})$ 304	776–777	(35)
$ClO_2{}^-$	chlorite	C_{2v}		$\nu_1(a_1)$ 790, $\nu_2(a_1)$ 400, $\nu_3(b_1)$ 840	778	(36)
$ClO_3{}^-$	chlorate	C_{3v}		$\nu_1(a_1)$ 910, $\nu_2(a_1)$ 617, $\nu_3(e)$ 960, $\nu_4(e)$ 493	779–782	(37)
$ClO_4{}^-$	perchlorate	T_d		$\nu_1(a_1)$ 935, $\nu_2(e)$ 460, $\nu_3(f_2)$ 1050–1170, $\nu_4(f_2)$ 630	783–793	(38)
$BrO_3{}^-$	bromate	C_{3v}		$\nu_1(a_1)$ 806, $\nu_2(a_1)$ 421, $\nu_3(e)$ 836, $\nu_4(e)$ 356	818–827	(39)
$IO_3{}^-$	iodate	C_{3v}		$\nu_1(a_1)$ 779, $\nu_2(a_1)$ 390, $\nu_3(e)$ 826, $\nu_4(e)$ 330	853–866	(40)
$IO_4{}^-$	periodate	T_d		$\nu_1(a_1)$ 791, $\nu_2(e)$ 256, $\nu_3(f_2)$ 853, $\nu_4(f_2)$ 325	867–868	(18)
$MnO_4{}^{2-}$	manganate			ν_3 ~840	870	(41)
$MnO_4{}^-$	permanganate	T_d		$\nu_1(a_1)$ 840, $\nu_2(e)$ 340–350, $\nu_3(f_2)$ ~900, $\nu_4(f_2)$ ~387	871–876	(42)

† Key to references:

1. L. A. Woodward and H. L. Roberts, *Trans. Faraday Soc.* **52**, 1458 (1956).
2. K. Nakamoto, "Infrared Spectra of Inorganic and Coordination Compounds," p. 73. Wiley, New York, 1963.
3. I. Nakagawa and T. Shimanouchi, *Spectrochim. Acta* **18**, 101 (1962).
4. R. K. Khanna, C. W. Brown, and L. H. Jones, *Inorg. Chem.* **8**, 2195 (1969).
4a. O. Zakharieva, Th. Woike, and S. Hauesuehl, *Spectrochem. Acta* **51A**, 447 (1995).
5. A. G. Maki and J. C. Decius, *J. Chem. Phys.* **31**, 772 (1959).
6. L. H. Jones, *J. Chem. Phys.* **25**, 1069 (1956).
7. A. Tramer, *C. R. Acad. Sci.* **249**, 2755 (1959).
8. S. Bhagavantan and T. Venkatarayuda, *Proc. Indian Acad. Sci., Sect. A* **9**, 224 (1939).
9. A. Mueller and M. Stockburger, *Z. Naturforsch. B* **20**, 1242 (1965); A. Mueller and B. Krebs, *Spectrochim. Acta* **22**, 1535 (1966); B. Krebs, A. Mueller, and G. Gattow, *Z. Naturforch. B* **20**, 1017 (1956); B. Krebs and A. Mueller, *Z. Naturforsch. A* **20**, 1664 (1965).

10. D. Fortum and J. O. Edwards, *J. Inorg. Nucl. Chem.* **2**, 264 (1956).

11. J. T. Last, *Phys. Rev.* **105**, 1740 (1957); A. F. Yatsenko, *Izv. Akad. Nauk SSSR, Ser. Fiz.* **22**, 1456 (1958).

12. K. Nakamoto, "Infrared Spectra of Inorganic and Coordination Compounds," p. 77. Wiley, New York, 1963.

13. R. E. Weston and T. F. Brodasky, *J. Chem. Phys.* **27**, 683 (1957).

14. K. Nakamoto, "Infrared Spectra of Inorganic and Coordination Compounds," p. 92. Wiley, New York, 1963.

15. M. Tsuboi, *J. Amer. Chem. Soc.* **79**, 1351 (1957).

16. J. A. A. Ketelaar, *Acta Cryst.* **7**, 691 (1954); H. Ratajczak and Z. Mielke, *J. Mol. Struct.* **1**, 397 (1967-1968).

17. E. E. Berry and C. B. Baddiel, *Spectrochim. Acta, Part A* **23**, 2089 (1967).

18. K. Nakamoto, "Infrared Spectra of Inorganic and Coordination Compounds," p. 107. Wiley, New York, 1963.

19. A. Hazel and S. D. Ross, *Spectrochim. Acta, Part A* **23**, 1583 (1967); **24**, 131 (1968).

20. E. Steger and K. Martin, *Z. Anorg. All. Chem.* **308**, 330 (1960).

21. R. Hubin and P. Tarte, *Spectrochim. Acta, Part A* **23**, 1815 (1967); W. Bues, K. Buhler, and P. Kuhnle, *Z. Anorg. Allg. Chem.* **325**, 8 (1963).

22. H. Siebert, *Z. Anorg. Allg. Chem.* **275**, 225 (1954).

23. H. Gerding and K. Eriks, *Rec. Trav. Chim. Pays Bas* **69**, 659 (1950); H. Siebert, *Z. Anorg. Allg. Chem.* **277**, 225 (1957).

24. J. C. Evans and J. H. Bernstein, *Can. J. Chem.* **33**, 1270 (1955); A. Simon and K. Waldmann, *Z. Phys. Chem. (Leipzig)* **204**, 235 (1955).

25. K. Nakamoto, "Infrared Spectra of Inorganic and Coordination Compounds," p. 92. Wiley, New York, 1963.

26. H. Stammerich, D. Bassi, O. Sala, and H. Siebert, *Spectrochim. Acta* **13**, 192 (1958).

27. A. Muller and B. Krebs, *Spectrochim. Acta, Part A* **23**, 2809 (1967).

28. K. Nakamoto, "Infrared Spectra of Inorganic and Coordination Compounds," p. 106. Wiley, New York, 1963.

29. H. F. Shurvell, *Can. Spectrosc.* **12**, 156 (1967).

30. P. A. W. Dean and D. F. Evans, *J. Chem. Soc., A*, p. 698 (1967).

31. L. A. Woodward and L. E. Anderson, *J. Chem. Soc., London* p. 1284 (1957).

32. J. S. Avery, C. D. Burbridge, and D. M. L. Goodgame, *Spectrochim. Acta, Part A* **24**, 1721 (1968).

33. J. Hiraisha and T. Shimmanouchi, *Spectrochim. Acta* **22**, 1483 (1966).

34. L. A. Woodward and J. A. Creighton, *Spectrochim. Acta* **17**, 594 (1961).

35. H. Stammerich and F. Forneris, *Spectrochim. Acta* **16**, 363 (1960).

36. J. P. Mathieu, *C. R. Acad. Sci.* **234**, 2272 (1952).

37. J. L. Hollenberg and D. A. Dows, *Spectrochim. Acta* **16**, 1155 (1960).

38. H. Colm, *J. Chem. Soc., London* p. 4282 (1952).

39. M. Rolla, *Gazz. Chim. Ital.* **69**, 779 (1939); C. Rocchiccioli, *C. R. Acad. Sci.* **249**, 236 (1959).

40. K. Nakamoto, "Infrared Spectra of Inorganic and Coordination Compounds," p. 87. Wiley, New York, 1963.

41. C. Rocchiccioli, *C. R. Acad. Sci.* **256**, 1707 (1963).

42. P. J. Hendra, *Spectrochim. Acta, Part A* **24**, 125 (1967).

NUMERICAL INDEX OF SPECTRA

40	Potassium copper(II) ferrocyanide	$K_2CuFe(CN)_6$
41	Lead ferrocyanide	$Pb_2Fe(CN)_6 \cdot xH_2O$
42	Iron(III) ferrocyanide	$Fe_4[Fe(CN)_6]_3$
43	Barium cyanoplatinate	$BaPt(CN)_4 \cdot 4H_2O$
44	Sodium cyanate	$NaOCN$
45	Silver cyanate	$AgOCN$
46	Potassium thiocyanate	$KSCN$
47	Lead thiocyanate	$Pb(SCN)_2$
48	Iron(II) thiocyanate	$Fe(SCN)_2 \cdot 3H_2O$
49	Copper(I) thiocyanate	$CuSCN$
50	Silver thiocyanate	$AgSCN$
51	Mercury(II) thiocyanate	$Hg(SCN)_2$
52	Lead cyanamide	$PbCN_2$
53	Ammonium bicarbonate	NH_4HCO_3
54	Sodium bicarbonate	$NaHCO_3$
55	Potassium bicarbonate	$KHCO_3$
56	Lithium carbonate	Li_2CO_3
57	Sodium carbonate	$Na_2CO_3(\cdot <1H_2O)$
58	Sodium carbonate	$Na_2CO_3 \cdot 10H_2O$
59	Potassium carbonate	$K_2CO_3 \cdot <1.5H_2O$
60	Cesium carbonate	$Cs_2CO_3 \cdot xH_2O$
61	Calcium carbonate (calcite)	$CaCO_3$
61a	Calcium carbonate (vaterite)	$CaCO_3$
62	Strontium carbonate	$SrCO_3$
63	Barium carbonate	$BaCO_3$
64	Lead carbonate	$PbCO_3$
65	Manganese carbonate	$MnCO_3(\cdot xH_2O$ or wet$)$
66	Cobalt carbonate (basic)	$CoCO_3 \cdot xH_2O$
67	Silver carbonate	Ag_2CO_3
68	Cadmium carbonate	$CdCO_3$
69	Lead carbonate (basic)	$2PbCO_3 \cdot Pb(OH)_2$
70	Bismuth carbonate (basic)	$Bi_2O_2CO_3$
71	Nickel(II) carbonate	$NiCO_3 \cdot xH_2O$
72	Copper(II) carbonate (basic) (malachite)	$CuCo_3 \cdot Cu(OH)_2$
73	Copper(II) carbonate (basic) (azurite)	$2CuCO_3 \cdot Cu(OH)_2$
74	Zinc carbonate (basic)	$2ZnCO_3 \cdot 3Zn(OH)_2$
75	Barium thiocarbonate	$BaCS_3$
76	Magnesium silicide	Mg_2Si
77	Calcium silicide	$CaSi_2$
78	Boron silicide	B_6Si
79	Titanium silicide	$TiSi_2$
80	Vanadium silicide	VSi_2
81	Manganese silicide	$MnSi_2$
82	Zirconium silicide	$ZrSi_2$
83	Niobium silicide	$NbSi_2$
84	Molybdenum silicide	$MoSi$
85	Molybdenum silicide	$MoSi_2$
86	Tungsten silicide	WSi
87	Lithium silicate	$Li_2Si_2O_5$
88	Lithium metasilicate	Li_2SiO_3
89	Magnesium orthosilicate	$Mg_2SiO_4(\cdot xH_2O$ or wet$)$
90	Lead orthosilicate	$PbSiO_4$
91	Cobalt orthosilicate	$CoSiO_4$

92	Copper(II) orthosilicate	$CuSiO_4 \cdot xH_2O$
93	Zinc orthosilicate	$ZnSiO_4 (\cdot xH_2O$ or wet)
94	Silica gel	$(SiO_2)_n \cdot xH_2O$
95	Kaolin clay	$Al_2O_3 \cdot 2SiO_2 \cdot 2H_2O$
96	Magnesium calcium aluminum silicate	$Al_2O_3 \cdot (Mg,Ca) \cdot 0.5SiO_2 \cdot xH_2O$
97	Barium zirconium silicate	$BaZrSiO_5$
98	Lithium zirconium silicate	Li_4ZrSiO_6
99	Magnesium aluminum silicate	$Mg_2Al_4Si_5O_{18} (\cdot H_2O$ or wet)
100	Lithium titanate(IV)	Li_2TiO_3
101	Calcium titanate(IV)	$CaTiO_3$
102	Strontium titanate(IV)	$SrTiO_3$
103	Barium titanate(IV)	$BaTiO_3$
104	Lead titanate(IV)	$PbTiO_3$
105	Bismuth titanate(IV)	$Bi_2(TiO_3)_3$
106	Cobalt titanate(IV)	$CoTiO_3$
107	Nickel(II) titanate(IV)	$NiTO_3$
108	Copper(II) titanate(IV)	$CuTiO_3$
109	Zinc titanate(IV)	$ZnTiO_3$
110	Cerium titanate(IV)	$Ce(TiO_3)_2$
111	Europium titanate(IV)	$Eu_2(TiO_3)_3$
112	Lithium zirconate(IV)	Li_2ZrO_3
113	Magnesium zirconate(IV)	$MgZrO_3$
114	Calcium zirconate(IV)	$CaZrO_3$
115	Strontium zirconate(IV)	$SrZrO_3$
116	Barium zirconate(IV)	$BaZrO_3$
117	Lead zirconate(IV)	$PbZrO_3$
118	Bismuth zirconate(IV)	$Bi_2(ZrO_3)_3$
119	Zinc zirconate(IV)	$ZnZrO_3$
120	Cadmium zirconate(IV)	$CdZrO_3$
121	Cerium zirconate(IV)	$Ce(ZrO_3)_2$
122	Magnesium stannate(IV)	$MgSnO_3$
123	Calcium stannate(IV)	$CaSnO_3$
124	Strontium stannate(IV)	$SrSnO_3$
125	Barium stannate(IV)	$BaSnO_3$
126	Lead stannate(IV)	$PbSnO_3$
127	Bismuth stannate(IV)	$Bi_2(SrO_3)_3 \cdot 3H_2O$
128	Iron(III) stannate(IV)	$Fe_2(SnO_3)_3 \cdot 3H_2O$
129	Cerium stannate(IV)	$Ce(SnO_3)_2$
130	Calcium nitride	Ca_3N_2
131	Barium nitride	Ba_3N_2
132	Boron nitride	BN
133	Aluminum nitride	AlN
134	Silicon nitride	Si_3N_4
135	Titanium nitride	Ti_3N_4
136	Vanadium nitride	VN
137	Chromium(III) nitride	CrN
138	Zirconium nitride	ZrN
139	Niobium nitride	NbN_2
140	Molybdenum nitride	Mo_2N
141	Tantalum nitride	TaN_2
142	Ammonium azide	NH_4N_3
143	Sodium azide	NaN_3
144	Potassium azide	KN_3

145	Rubidium azide	RbN_3
146	Cesium azide	CsN_3
147	Barium azide	$Ba(N_3)_2$
148	Sodium hyponitrite	$Na_2N_2O_2 \cdot xH_2O$
149	Sodium nitrite	$NaNO_2$
150	Barium nitrite	$Ba(NO_2)_2 \cdot xH_2O$
151	Lead nitrite	$Pb(NO_2)_2 \cdot xH_2O$
152	Silver nitrite	$AgNO_2$
153	Sodium hexanitrocobaltate(III)	$Na_3Co(NO_2)_6$
154	Ammonium nitrate	NH_4NO_3
155	Sodium nitrate	$NaNO_3$
156	Potassium nitrate	KNO_3
157	Rubidium nitrate	$RbNO_3$
158	Cesium nitrate	$CsNO_3$
159	Calcium nitrate	$Ca(NO_3)_2 \cdot 4H_2O$
160	Strontium nitrate	$Sr(NO_3)_2$
161	Barium nitrate	$Ba(NO_3)_2$
162	Aluminum nitrate	$Al(NO_3)_3 \cdot 9H_2O$
163	Gallium nitrate	$Ga(NO_3)_3 \cdot xH_2O$
164	Indium nitrate	$In(NO_3)_3 \cdot (4\frac{1}{2}?)H_2O$
165	Thallium nitrate	$Tl(NO_3)_3$
166	Lead nitrate	$Pb(NO_3)_2$
167	Lead nitrate	$Pb(NO_3)_2 \cdot xH_2O$
168	Bismuth nitrate	$Bi(NO_3)_3 \cdot 5H_2O$
169	Scandium nitrate	$Sc(NO_3)_3 \cdot xH_2O$
170	Chromium(III) nitrate	$Cr(NO_3)_3 \cdot 9H_2O$
171	Iron(III) nitrate	$Fe(NO_3)_3 \cdot xH_2O$
172	Cobalt nitrate	$Co(NO_3)_2 \cdot 6H_2O$
173	Nickel(II) nitrate	$Ni(NO_3)_2 \cdot 6H_2O$
174	Zinc nitrate	$Zn(NO_3)_2 \cdot 6H_2O$
175	Yttrium nitrate	$Y(NO_3)_3 \cdot 6H_2O$
176	Zirconium nitrate	$Zr(NO_3)_4 \cdot 5H_2O$
177	Silver nitrate	$AgNO_3$
178	Cadmium nitrate	$Cd(NO_3)_2 \cdot 4H_2O$
179	Lanthanum nitrate	$La(NO_3)_3 \cdot 6H_2O$
180	Mercury(I) nitrate	$HgNO_3 \cdot H_2O$
181	Cerium nitrate	$Ce(NO_3)_3 \cdot 6H_2O$
182	Neodymium nitrate	$Nd(NO_3)_3 \cdot 6H_2O$
183	Samarium nitrate	$Sm(NO_3)_3 \cdot 6H_2O$
184	Gadolinium nitrate	$Gd(NO_3)_3 \cdot 5H_2O$
185	Terbium nitrate	$Tb(NO_3)_3 \cdot 6H_2O$
186	Dysprosium nitrate	$Dy(NO_3)_3 \cdot xH_2O$
187	Holmium nitrate	$Ho(NO_3)_3 \cdot xH_2O$
188	Erbium nitrate	$Er(NO_3)_3 \cdot 6H_2O$
189	Thulium nitrate	$Tm(NO_3)_3 \cdot xH_2O$
190	Ytterbium nitrate	$Yb(NO_3)_3 \cdot 4H_2O$
191	Thorium nitrate	$Th(NO_3)_3 \cdot 4H_2O$
192	Bismuth subnitrate	$BiONO_3 \cdot H_2O$
193	Tellurium nitrate (basic)	$4TeO_2 \cdot N_2O_5 \cdot 1\frac{1}{2}H_2O$
194	Zirconyl nitrate	$ZrO(NO_3)_2 \cdot 2H_2O$
195	Uranyl nitrate	$UO_2(NO_3)_2 \cdot 6H_2O$
196	Ammonium pentanitratocerate(III)	$(NH_4)_2Ce(NO_3)_5 \cdot 4H_2O$
197	Ammonium hexanitratocerate(IV)	$(NH_4)_2Ce(NO_3)_6$

198	Potassium hexanitratocerate(IV)	$K_2Ce(NO_3)_6$
199	Magnesium hexanitratocerate(IV)	$MgCe(NO_3)_6 \cdot 24H_2O$
200	Antimony phosphide	SbP
201	Bismuth phosphide	BiP
202	Iron(II) phosphide	FeP ($\sim50\%$) and Fe_2P ($\sim50\%$)
203	Zinc phosphide	Zn_3P_2
204	Ammonium hypophosphite	$NH_4H_2PO_2$
205	Lithium hypophosphite	LiH_2PO_2
206	Sodium hypophosphite	$NaH_2PO_2 \cdot H_2O$
207	Potassium hypophosphite	KH_2PO_2
208	Calcium hypophosphite	$Ca(H_2PO_2)_2$
209	Manganese hypophosphite	$Mn(H_2PO_2)_2 \cdot H_2O$
210	Iron(III) hypophosphite	$Fe(H_2PO_2)_3$
211	Sodium orthophosphite	$Na_2HPO_3 \cdot 5H_2O$
212	Barium orthophosphite	$BaHPO_3(\cdot xH_2O$ or wet$)$
213	Metaphosphoric acid	$(HPO_3)_x$
214	Sodium metaphosphate	$(NaPO_3)_x$
215	Potassium metaphosphate	$(KPO_3)_x$
216	Beryllium metaphosphate	$[Be(PO_3)_2]_x$
217	Magnesium metaphosphate	$[Mg(PO_3)_2]_x$
218	Calcium metaphosphate	$[Ca(PO_3)_2]_x \cdot xH_2O$
219	Strontium metaphosphate	$[Sr(PO_3)_2]_x$
220	Barium metaphosphate	$[Ba(PO_3)_2]_x$
221	Aluminum metaphosphate	$[Al(PO_3)_3]_x$
222	Lead metaphosphate	$[Pb(PO_3)_2]_x$
223	Zinc metaphosphate	$[Zn(PO_3)_2]_x$
224	Ammonium orthophosphate (monobasic)	$NH_4H_2PO_4$
225	Potassium orthophosphate (monobasic)	KH_2PO_4
226	Ammonium orthophosphate (dibasic)	$(NH_4)_2HPO_4$
227	Potassium orthophosphate (dibasic)	$K_2HPO_4(\cdot xH_2O$ or wet$)$
228	Calcium orthophosphate (dibasic)	$CaHPO_4 \cdot H_2O$
229	Strontium orthophosphate (dibasic), α-form	$SrHPO_4$
230	Strontium orthophosphate (dibasic), β-form	$SrHPO_4$
231	Barium orthophosphate (dibasic)	$BaHPO_4$
232	Cobalt orthophosphate (dibasic)	$CoHPO_4$
233	Sodium pyrophosphate (dibasic)	$Na_2H_2P_2O_7$
234	Lithium orthophosphate	$Li_3PO_4 \cdot \frac{1}{2}H_2O$
235	Sodium orthophosphate	$Na_3PO_4 \cdot 12H_2O$
236	Magnesium orthophosphate	$Mg_3(PO_4)_2 \cdot 8H_2O$
237	Magnesium orthophosphate (basic)	$Mg_3(PO_4)_2 \cdot Mg(OH)_2$
238	Calcium orthophosphate	$Ca_3(PO_4)_2$
239	Boron orthophosphate (tetragonal)	BPO_4
240	Aluminum orthophosphate	$AlPO_4$ (wet)
241	Tin(II) orthophosphate	$Sn_3(PO_4)_2$ (wet)
242a	Lead orthophosphate (apatite structure)	$Pb_3(PO_4)_2$
242b	Lead orthophosphate	$Pb_3(PO_4)_2$
243	Antimony orthophosphate	$SbPO_4$
244	Bismuth orthophosphate	$BiPO_4$
245	Chromium(III) orthophosphate	$CrPO_4 \cdot 6H_2O$
246	Iron(II) orthophosphate	$Fe_3(PO_4)_2 \cdot 8H_2O$
247	Iron(III) orthophosphate	$FePO_4 \cdot 2H_2O$
248	Nickel orthophosphate	$Ni_3(PO_4)_2 \cdot 8H_2O$
249	Copper(II) orthophosphate	$Cu_3(PO_4)_2 \cdot 3H_2O$

250	Zinc orthophosphate (hopeite)	$Zn_3(PO_4)_2 \cdot 4H_2O$
251	Silver orthophosphate	Ag_3PO_4
252	Cadmium orthophosphate	$Cd_3PO_4 (\cdot xH_2O$ or wet$)$
253	Mercury(II) orthophosphate	$Hg_3(PO_4)_2$
254	Ammonium magnesium orthophosphate	$NH_4MgPO_4 \cdot H_2O$
255	Ammonium manganese orthophosphate	NH_4MnPO_4
256	Ammonium cobalt orthophosphate	$NH_4CoPO_4 \cdot H_2O$
257	Dilithium sodium orthophosphate	$Li_2NaPO_4 (\cdot xH_2O$ or wet$)$
258	Calcium nickel orthophosphate	$Ca_8Ni(PO_4)_6 \cdot xH_2O$
259	Lead copper(I) orthophosphate	$PbCuPO_4$
260	Sodium pyrophosphate	$Na_4P_2O_7$
261	Sodium pyrophosphate	$Na_4P_2O_7 \cdot 10H_2O$
262	Potassium pyrophosphate	$K_4P_2O_7 \cdot (3?)H_2O$
263	Magnesium pyrophosphate	$Mg_2P_2O_7 \cdot 3H_2O$
264	Calcium pyrophosphate (β-form)	$Ca_2P_2O_7$
265	Strontium pyrophosphate (α-form, orthorhombic)	$Sr_2P_2O_7$
266	Strontium pyrophosphate (β-form, tetragonal)	$Sr_2P_2O_7$
267	Sodium potassium pyrophosphate	$Na_2K_2P_2O_7$
268	Calcium pyrophosphate (γ-form)	$Ca_2P_2O_7 \cdot xH_2O$
269	Aluminum pyrophosphate	$Al_4(P_2O_7)_3 \cdot xH_2O$
270	Barium pyrophosphate (α-form)	$Ba_2P_2O_7$
271	Barium pyrophosphate	$Ba_2P_2O_7 \cdot xH_2O$
272	Tin pyrophosphate	$Sn_2P_2O_7$
273	Lead pyrophosphate	$Pb_2P_2O_7$
274	Cobalt pyrophosphate	$Co_2P_2O_7$
275	Nickel pyrophosphate	$Ni_2P_2O_7$
276	Copper(II) pyrophosphate	$Cu_2P_2O_7 \cdot xH_2O$
277	Zinc pyrophosphate	$ZnP_2O_7 \cdot xH_2O$ $(x = 3-5)$
278	Sodium tripolyphosphate	$Na_5P_3O_{10} (\cdot xH_2O)$
279	Potassium tripolyphosphate	$K_5P_3O_{10} (\cdot xH_2O$ or wet$)$
280	Sodium phosphorothioate	$Na_3PO_3S \cdot xH_2O$
281	Sodium phosphorofluoridate	Na_2PO_3F
282	Potassium phosphorofluoridate	$K_2PO_3F \cdot xH_2O$
283	Barium phosphorofluoridate	$BaPO_3F \cdot xH_2O$
284	Potassium phosphorodifluoridate	KPO_2F_2
285	Manganese arsenide	$MnAs$
286	Sodium metaarsenite	$NaAsO_2$
287	Lead metaarsenite	$Pb(AsO_2)_2$ (wet)
288	Zinc metaarsenite	$Zn(AsO_2)_2$
289	Antimony orthoarsenite	$SbAsO_3$
290	Iron(III) orthoarsenite (basic)	$2FeAsO_3 \cdot Fe_2O_3 \cdot (5H_2O?)$
291	Copper(II) orthoarsenite	$Cu(AsO_3)_2 \cdot xH_2O$
292	Silver orthoarsenite	Ag_3AsO_3
293	Mercury(I) orthoarsenite	Hg_3AsO_3
294	Lead pyroarsenate	$Pb_2As_2O_7$
295	Potassium orthoarsenate (monobasic)	KH_2AsO_4
296	Ammonium orthoarsenate (dibasic)	$(NH_4)_2HAsO_4$
297	Sodium orthoarsenate (dibasic)	$Na_2HAsO_4 \cdot 7H_2O$
298	Antimony orthoarsenate	$SbAsO_4$
299	Iron(II) orthoarsenate	$Fe_3(AsO_4)_2 \cdot 6H_2O$
300	Cobalt orthoarsenate	$Co(AsO_4) \cdot 8H_2O$
301	Copper(II) orthoarsenate	$Cu_3(AsO_4)_2 \cdot 4H_2O$
302	Zinc orthoarsenate	$Zn_3(AsO_4)_2 \cdot 8H_2O$

303	Mercury(II) orthoarsenate	$Hg_3(AsO_4)_2$
304	Lead antimonate	$Pb(SbO_3)_2$
305	Ammonium metavanadate	NH_4VO_3
306	Sodium pyrovanadate	$Na_4V_2O_7 \cdot xH_2O$
307	Sodium orthovanadate	$Na_3VO_4 \cdot 10H_2O$
308	Calcium orthovanadate	$Ca_3(VO_4)_2$
309	Lead orthovanadate	$Pb_3(VO_4)_2$
310	Iron(III) orthovanadate	$FeVO_4 \cdot 2H_2O$
311	Silver orthovanadate	Ag_3VO_4
312	Calcium nickel hydroxyorthovanadate	$(Ca,Ni)_5OH(VO_4)_3$
313	Calcium copper(II) hydroxy orthovanadate	$(Ca,Cu)_5OH(VO_4)_3$
314	Potassium metaniobate	$KNbO_3(\cdot xH_2O$ or wet$)$
315	Potassium orthoniobate	$K_3NbO_4 \cdot xH_2O$
316	Lithium oxide	Li_2O
317	Magnesium oxide	MgO
318	Calcium oxide	CaO
319	Aluminum oxide (alundum)	Al_2O_3
320	Indium sesquioxide	In_2O_3
321	Thallium(III) oxide	Tl_2O_3
322	Silicon dioxide (cristobalite)	SiO_2
323	Silicon dioxide	SiO_2
324	Vycor glass	Primarily SiO_2
325	Germanium dioxide	GeO_2
326	Tin(II) oxide	SnO
327	Lead oxide	Pb_3O_4
328	Antimony trioxide	Sb_2O_3 or Sb_4O_6
329	Antimony pentoxide	Sb_2O_5
330	Tellurium dioxide	TeO_2
331	Iodine pentoxide	I_2O_5
332	Titanium dioxide	TiO_2
333	Vanadium tetroxide	$V_2O_4(\cdot H_2O$ or wet$)$
334	Vanadium oxide	V_6O_{13}
335	Vanadium pentoxide	V_2O_5
336	Chromium(III) sesquioxide	Cr_2O_3
337	Manganese(II) oxide	MnO
338	Manganese dioxide	MnO_2
339	Iron(III) oxide (hematite)	Fe_2O_3
340	Iron oxide (magnetite)	Fe_3O_4
341	Cobalt(II) oxide	CoO
342	Cobalt oxide	Co_3O_4 (plus 20–30% CoO)
343	Nickel oxide	Ni_2O_3
344	Copper(I) oxide	Cu_2O
345	Copper(II) oxide	CuO
346	Zinc oxide	ZnO
347	Yttrium oxide	Y_2O_3
348	Niobium oxide	NbO_2
349	Niobium pentoxide	Nb_2O_5
350	Niobium pentoxide	$Nb_2O_5 \cdot xH_2O$
351	Molybdenum dioxide	MoO_2
352	Molybdenum trioxide (molybdite)	MoO_3
353	Silver oxide	Ag_2O
354	Cadmium oxide	CdO
355	Hafnium oxide	HfO_2

356	Tantalum pentoxide	Ta_2O_5
357	Tungsten trioxide (wolframite)	WO_3
358	Mercury(II) oxide	HgO
359	Cerium(IV) dioxide	CeO_2
360	Samarium oxide	Sm_2O_3
361	Dysprosium oxide	Dy_2O_3
362	Holmium oxide	Ho_2O_3
363	Erbium oxide	Er_2O_3
364	Ytterbium oxide	Yb_2O_3
365	Thorium dioxide	ThO_2
366	Uranium dioxide	UO_2
367	Uranium oxide (orthorhombic)	U_3O_8
368	Uranium oxide (hexagonal)	U_3O_8
369	Strontium peroxide	SrO_2
370	Zinc peroxide	ZnO_2
371	Ammonium hydroxide hydrochloride	NH_3ClOH
372	Lithium hydroxide	$LiOH \cdot H_2O$
373	Sodium hydroxide	$NaOH$
374	Magnesium hydroxide	$Mg(OH)_2$
375	Barium hydroxide	$Ba(OH)_2 \cdot 8H_2O$
376	α-Aluminum hydroxide (gibbsite)	$Al(OH)_3$
377	Nickel hydroxide	$Ni(OH)_2$
378	Lanthanum hydroxide	$La(OH)_3$ (plus 20–30% hexagonal form of La_2O_3)
379	Potassium hexahydroxostannate(IV)	$K_2Sn(OH)_6$
380	Copper(II) hexahydroxostannate(IV)	$CuSn(OH)_6$
381	Zinc hexahydroxostannate(IV)	$ZnSn(OH)_6$
382	Cadmium hexahydroxostannate(IV)	$CdSn(OH)_6$
383	Sodium hexahydroxoantimonate(V)	$NaSb(OH)_6$
384	Potassium hexahydroxoantimonate(V)	$KSb_6(OH)_6 \cdot xH_2O$
385	γ-Aluminum oxyhydroxide (boehmite)	$AlOOH$ (10–20% NH_4Cl impurity)
386	Iron(III) oxyhydroxide (akaganeite)	β-$FeOOH$ (plus small amount of $NaCl$)
387	Antimony oxide chloride	$Sb_4O_5Cl_2$
388	Bismuth oxychloride	$BiOCl$
389	Bismuth oxybromide	$BiOBr$
390	Bismuth oxyiodide	$BiOI$
391	Zirconyl chloride	$ZrOCl_2 \cdot 8H_2O$
392	Molybdenum oxydichloride	$MoO_2Cl_2 \cdot (2?)H_2O$
393	Uranyl fluoride	$UO_2F_2 \cdot xH_2O$
394	Uranyl chloride	$UO_2Cl_2 \cdot 3H_2O$
395	Arsenic disulfide	As_2S_2
396	Arsenic pentasulfide	As_2S_5
397	Antimony trisulfide	Sb_2S_3
398	Bismuth trisulfide	Bi_2S_3
399	Tellurium sulfide	TeS_2
400	Titanium sesquisulfide	Ti_2S_3
401	Nickel monosulfide	NiS
402	Copper sulfide	Cu_2S
403	Zinc sulfide (α and β)	ZnS
404	Niobium sulfide	NbS
405	Molybdenum sulfide	MoS_2
406	Silver sulfide	Ag_2S
407	Cadmium sulfide	CdS

408	Tantalum sulfide	TaS
409	Tantalum disulfide	TaS_2
410	Tungsten sulfide	WS_2
411	Mercury(II) sulfide	HgS
411a	Sulfur	S_8 (CS_2 soln. 3800–400 cm^{-1})
412	Lead sulfamate	$Pb(NH_2SO_3)_2$
413	Ammonium imidodisulfate	$(NH_4)_2S_2NHO_6$
414	Ammonium hydrogen sulfate	$(NH_4)HSO_4$
415	Sodium hydrogen sulfate	$NaHSO_4 \cdot H_2O$
416	Potassium hydrogen sulfate	$KHSO_4$
417	Rubidium hydrogen sulfate	$RbHSO_4$
418	Potassium thiosulfate	$K_2S_2O_3 \cdot \frac{1}{3}H_2O$
419	Magnesium thiosulfate	$MgS_2O_3 \cdot 6H_2O$
420	Barium thiosulfate	$BaS_2O_3 \cdot H_2O$
421	Lead thiosulfate	PbS_2O_3
422	Sodium pyrosulfite	$Na_2S_2O_5$
423	Potassium pyrosulfite	$K_2S_2O_5$
424	Sodium sulfite	Na_2SO_3
425	Magnesium sulfite	$MgSO_3 \cdot 6H_2O$
426	Strontium sulfite	$SrSO_3$ (wet)
427	Barium sulfite	$BaSO_3$ (wet)
428	Lead sulfite	$PbSO_3$
429	Potassium dithionate	$K_2S_2O_6$
430	Silver pyrosulfite	$Ag_2S_2O_7$
431	Ammonium sulfate	$(NH_4)_2SO_4$
432	Lithium sulfate	$Li_2SO_4 \cdot H_2O$
433	Sodium sulfate	Na_2SO_4
434	Potassium sulfate	K_2SO_4
435	Rubidium sulfate	Rb_2SO_4
436	Cesium sulfate	Cs_2SO_4
437	Beryllium sulfate	$BeSO_4 \cdot 4H_2O$
438	Magnesium sulfate	$MgSO_4 \cdot H_2O$
439	Magnesium sulfate	$MgSO_4 \cdot 7H_2O$
440	Calcium sulfate	$CaSO_4 \cdot \frac{1}{2}H_2O$
441	Calcium sulfate	$CaSO_4 \cdot 2H_2O$
442	Strontium sulfate	$SrSO_4$
443	Barium sulfate	$BaSO_4$
444	Aluminum sulfate	$Al_2(SO_4)_3 \cdot 18H_2O$
445	Gallium sulfate	$Ga_2(SO_4)_3 \cdot 18H_2O$
446	Indium sulfate	$In_2(SO_4)_3 \cdot 9H_2O$
447	Thallium sulfate	$Tl_2(SO_4)_3$
448	Lead sulfate (tribasic)	$3PbO \cdot PbSO_4 \cdot xH_2O$
449	Antimony sulfate	$Sb_2(SO_4)_3 \cdot xH_2O$
450	Bismuth sulfate	$Bi_2(SO_4)_3 \cdot xH_2O$
451	Vanadium sulfate	$VSO_4 \cdot 7H_2O$
452	Manganese(II) sulfate	$MnSO_4 \cdot 4H_2O$
453	Iron(II) sulfate	$FeSO_4 \cdot 7H_2O$
454	Iron(III) sulfate	$Fe_2(SO_4)_3 \cdot 9H_2O$
455	Cobalt(II) sulfate	$CoSO_4 \cdot 7H_2O$
456	Nickel sulfate	$NiSO_4 \cdot 6H_2O$
457	Copper(II) sulfate	$CuSO_4 \cdot 5H_2O$
458	Zinc sulfate	$ZnSO_4 \cdot 6H_2O$
459	Zinc sulfate	$ZnSO_4 \cdot 7H_2O$

460	Yttrium sulfate	$Y_2(SO_4)_3 \cdot 8H_2O$
461	Zirconium sulfate	$Zr(SO_4)_2 \cdot 4H_2O$
462	Silver sulfate	Ag_2SO_4
463	Cadmium sulfate	$CdSO_4 \cdot 7H_2O$
464	Mercury(I) sulfate	Hg_2SO_4
465	Mercury(II) sulfate	$HgSO_4$
466	Mercury(II) sulfate (basic)	$HgSO_4 \cdot 2H_2O$
467	Cerium(II) sulfate	$Ce_2(SO_4)_3 \cdot (5?)H_2O$
468	Cerium(IV) sulfate	$Ce(SO_4)_2 \cdot 4H_2O$
469	Praseodymium sulfate	$Pr_2(SO_4)_3 \cdot 8H_2O$
470	Neodymium sulfate	$Nd_2(SO_4)_3 \cdot 8H_2O$
471	Samarium sulfate	$Sm(SO_4)_3 \cdot 8H_2O$
472	Europium sulfate	$Eu(SO_4)_3 \cdot 8H_2O$
473	Gadolinium sulfate	$Gd_2(SO_4)_3 \cdot 8H_2O$
474	Dysprosium sulfate	$Dy_2(SO_4)_3 \cdot 8H_2O$
475	Holmium sulfate	$Ho_2(SO_4)_4 \cdot 8H_2O$
476	Erbium sulfate	$Er_2(SO_4)_3 \cdot 8H_2O$
477	Ytterbium sulfate	$Yb_2(SO_4)_3 \cdot 8H_2O$
478	Thorium sulfate	$Th(SO_4)_2 \cdot (8?)H_2O$
479	Uranium sulfate	$UO_2SO_4 \cdot 3H_2O$
480	Copper tetraamine sulfate	$Cu(NH_4)_4SO_4$
481	Ammonium sodium sulfate	NH_4NaSO_4
482	Ammonium sulfate antimony trifluoride complex	$(NH_4)_2SO_4 \cdot SbF_3$
483	Ammonium chromium sulfate	$NH_4Cr(SO_4)_2 \cdot 12H_2O$
484	Ammonium manganese sulfate	$(NH_4)_2MnSO_4 \cdot xH_2O$
485	Ammonium iron(II) sulfate	$(NH_4)_2Fe(SO_4)_2 \cdot 6H_2O$
486	Ammonium iron(III) sulfate	$(NH_4)Fe(SO_4)_2 \cdot 3H_2O$
487	Ammonium iron(III) sulfate	$NH_4Fe(SO_4)_2 \cdot xH_2O$
488	Ammonium cobalt sulfate	$(NH_4)_2Co(SO_4)_2 \cdot 6H_2O$
489	Ammonium copper(II) sulfate	$(NH_4)_2Cu(SO_4)_2 \cdot xH_2O$
490	Sodium iron(III) sulfate	$NaFe(SO_4)_2 \cdot 4H_2O$
491	Potassium magnesium sulfate	$K_2Mg(SO_4)_2 \cdot 6H_2O$
492	Potassium aluminum sulfate	$KAl(SO_4)_2 \cdot 4H_2O$
493	Potassium chromium sulfate	$KCr(SO_4)_2 \cdot 12H_2O$
494	Potassium iron(III) sulfate	$KFe(SO_4)_2 \cdot 24H_2O$
495	Potassium nickel sulfate	$K_2Ni(SO_4)_2 \cdot 6H_2O$
496	Potassium copper(II) sulfate	$K_2Cu(SO_4)_2 \cdot 6H_2O$
497	Potassium cadmium sulfate	$K_2Cd(SO_4)_2 \cdot 6H_2O$
498	Rubidium aluminum sulfate	$RbAl(SO_4)_2 \cdot 12H_2O$
499	Cesium aluminum sulfate	$CsAl(SO_4)_2 \cdot 12H_2O$
500	Potassium basic alum	$K_2Al_6(SO_4)_5(OH)_{10} \cdot 4H_2O$
501	Ammonium peroxydisulfate	$(NH_4)_2S_2O_8$
502	Sodium peroxydisulfate	$Na_2S_2O_8$
503	Potassium peroxydisulfate	$K_2S_2O_8$
504	Ammonium fluorosulfonate	NH_4SO_3F
505	Potassium fluorosulfonate	KSO_3F
506	Gallium monoselenide	$GaSe$
507	Tin(II) selenide	$SnSe$
508	Lead selenide	$PbSe$
509	Titanium diselenide	$TiSe_2$
510	Chromium selenide	Cr_2Se_3
511	Zinc selenide	$ZnSe$ (wet)
512	Zirconium diselenide	$ZrSe_2$

513	Niobium diselenide	$NbSe_2$
514	Molybdenum diselenide	$MoSe_2$
515	Tantalum diselenide	$TaSe_2$
516	Tungsten diselenide	WSe_2
517	Sodium selenite	Na_2SeO_3
518	Potassium selenite	K_2SeO_3 (wet)
519	Barium selenite	$BaSeO_3$ (wet)
520	Zinc selenite	$ZnSeO_3$
521	Copper selenite	$Cu(OH)SeO_3H \cdot H_2O$
522	Ammonium selenate	$(NH_4)_2SeO_4$
523	Sodium selenate	$Na_2SeO_4 \cdot xH_2O$
524	Potassium selenate	K_2SeO_4
525	Magnesium selenate	$MgSeO_4 \cdot 6H_2O$
526	Calcium selenate	$CaSeO_4 \cdot 2H_2O$
527	Iron(II) selenate	$FeSeO_4 \cdot xH_2O$
528	Nickel selenate	$NiSeO_4 \cdot 6H_2O$
529	Copper(II) selenate	$CuSeO_4 \cdot 5H_2O$
530	Silver selenate	Ag_2SeO_4
531	Tin(II) telluride	$SnTe$
532	Bismuth telluride	Bi_2Te_3
533	Titanium telluride	$TiTe_2$
534	Vanadium telluride	VTe
535	Chromium telluride	Cr_2Te_3
536	Zinc telluride	$ZnTe$
537	Molybdenum telluride	$MoTe_2$
538	Tungsten telluride	WTe_2
539	Telluric acid	H_6TeO_6
540	Copper(I) chromite	$Cu_2Cr_2O_4$
541	Copper(II) chromite	$CuCr_2O_4$
542	Ammonium dichromate	$(NH_4)_2Cr_2O_7$
543	Lithium dichromate	$Li_2Cr_2O_7 \cdot 2H_2O$
544	Sodium dichromate	$Na_2Cr_2O_7 \cdot 2H_2O$
545	Potassium dichromate	$K_2Cr_2O_7$
546	Rubidium dichromate	$Rb_2Cr_2O_7$
547	Calcium dichromate	$CaCr_2O_7 \cdot xH_2O$
548	Zinc dichromate (OH impurity)	$ZnCr_2O_7$
549	Silver dichromate	$Ag_2Cr_2O_7$
550	Ammonium chromate	$(NH_4)_2CrO_4 \cdot xH_2O$
551	Lithium chromate	$Li_2CrO_4 \cdot xH_2O$
552	Sodium chromate	$Na_2CrO_4 \cdot (10?)H_2O$
553	Potassium chromate	K_2CrO_4
554	Cesium chromate	Cs_2CrO_4
555	Magnesium chromate	$MgCrO_4 \cdot 7H_2O$
556	Calcium chromate	$CaCrO_4 \cdot 2H_2O$
557	Aluminum chromate	$Al_2(CrO_4)_3 \cdot xH_2O$
558	Lead chromate	$PbCrO_4$
559	Cadmium chromate (carbonate impurity)	$CdCrO_4$
560	Lithium sodium chromate	$LiNaCrO_4 \cdot xH_2O$
561	Potassium zinc chromate	$K_2CrO_4 \cdot 3ZnCrO_4 \cdot Zn(OH)_2$
562	Lithium molybdate(VI)	Li_2MoO_4 (wet)
563	Sodium molybdate(VI)	$Na_2MoO_4 \cdot 2H_2O$
564	Potassium molybdate(VI)	$K_2MoO_4 \cdot xH_2O$ or wet
565	Calcium molybdate(VI)	$CaMoO_4$ (wet)

618	Chromium(III) fluoride	$CrF_3 \cdot 3H_2O$
619	Manganese fluoride	$MnF_2 \cdot xH_2O$
620	Iron(II) fluoride	$FeF_2 \cdot 4H_2O$
621	Iron(III) fluoride	$FeF_3 \cdot 4\frac{1}{2}H_2O$
622	Cobalt(II) fluoride	$CoF_2 \cdot 4H_2O$
623	Cobalt(III) fluoride	$CoF_3 \cdot xH_2O$
624	Nickel fluoride	$NiF_2 \cdot xH_2O$
625	Copper(II) fluoride	$CuF_2 \cdot 2H_2O$
626	Zinc fluoride	$ZnF_2 \cdot 4H_2O$
627	Yttrium fluoride	YF_3
628	Zirconium fluoride	$ZrF_4 \cdot xH_2O$
629	Silver(II) fluoride	AgF_2
630	Cadmium fluoride	CdF_2
631	Lanthanum fluoride	LaF_3
632	Hafnium fluoride	$HfF_4 \cdot xH_2O$
633	Cerium(III) fluoride	CeF_3
634	Samarium fluoride	SmF_3
635	Gadolinium fluoride	GdF_3
636	Dysprosium fluoride	DyF_3
637	Holmium fluoride	HoF_3
638	Erbium fluoride	ErF_3
639	Ytterbium fluoride	YbF_3
640	Thorium fluoride	$ThF_4 \cdot 4H_2O$
641	Sodium hydrogen fluoride	$NaHF_2$
642	Ammonium tetrafluoroborate	NH_4BF_4
643	Lithium tetrafluoroborate	$LiBF_4 \cdot xH_2O$
644	Sodium tetrafluoroborate	$NaBF_4 (\cdot xH_2O)$
645	Potassium tetrafluoroborate	KBF_4
646	Calcium tetrafluoroborate	$Ca(BF_4)_2 \cdot xH_2O$
647	Nickel tetrafluoroborate	$Ni(BF_4)_2 \cdot xH_2O$
648	Zinc tetrafluoroborate	$Zn(BF_4) \cdot xH_2O$
649	Ammonium tetrafluoroaluminate	NH_4AlF_4
650	Ammonium hexafluoroaluminate	$(NH_4)_3AlF_6$
651	Potassium hexafluoroaluminate	K_3AlF_6
652	Ammonium hexafluorogallate	$(NH_4)_3GaF_6$
653	Ammonium hexafluorosilicate	$(NH_4)_2SiF_6$
654	Lithium hexafluorosilicate	$Li_2SiF_6 \cdot 2H_2O$
655	Sodium hexafluorosilicate	Na_2SiF_6
656	Potassium hexafluorosilicate	K_2SiF_6
657	Magnesium hexafluorosilicate	$MgSiF_6 \cdot 6H_2O$
658	Calcium hexafluorosilicate	$CaSiF_6 \cdot 2H_2O$
659	Barium hexafluorosilicate	$BaSiF_6 \cdot xH_2O$
660	Manganese hexafluorosilicate	$MnSiF_6 \cdot 6H_2O$
661	Cobalt hexafluorosilicate	$CoSiF_6 \cdot 6H_2O$
662	Nickel hexafluorosilicate	$NiSiF_6 \cdot 6H_2O$
663	Copper(II) hexafluorosilicate	$CuSiF_6 \cdot 6H_2O$
664	Zinc hexafluorosilicate	$ZnSiF_6 \cdot 6H_2O$
665	Ammonium hexafluorogermanate	$(NH_4)_2GeF_6$
666	Sodium hexafluorogermanate	$Na_2GeF_6 (wet)$
667	Barium hexafluorogermanate	$BaGeF_6 (wet)$
668	Ammonium trifluorostannate	NH_4SnF_3
669	Sodium trifluorostannate	$NaSnF_3$
670	Potassium trifluorostannate	$KSnF_3$

671	Iron(II) trifluorostannate	$Fe(SnF_3)_2 \cdot 7H_2O$
672	Zinc trifluorostannate	$Zn(SnF_3)_2 \cdot 7H_2O$
673	Lithium hexafluorostannate	$Li_2SnF_6 \cdot xH_2O$
674	Sodium hexafluorostannate	$Na_2SnF_6 \cdot xH_2O$
675	Potassium hexafluorostannate	$K_2SnF_6 \cdot H_2O$
676	Magnesium hexafluorostannate	$MgSnF_6 \cdot xH_2O$
677	Calcium hexafluorostannate	$CaSnF_6 \cdot xH_2O$
678	Cobalt hexafluorostannate	$CoSnF_6 \cdot 7H_2O$
679	Nickel hexafluorostannate	$NiSnF_6 \cdot 7H_2O$
680	Copper(II) hexafluorostannate	$CuSnF_6 \cdot 7H_2O$
681	Ammonium hexafluorophosphate	NH_4PF_6
682	Potassium hexafluorophosphate	KPF_6
683	Potassium hexafluorophosphate and KHF_2	KPF_6 and KHF_2
684	Cesium hexafluorophosphate	$CsPF_6$
685	Potassium hexafluoroarsenate	$KAsF_6$ (wet)
686	Ammonium tetrafluoroantimonate	NH_4SbF_4
687	Potassium hexafluoroantimonate(V)	$KSbF_6$ (wet)
688	Silver hexafluoroantimonate(V)	$AgSbF_6 \cdot xH_2O$
689	Ammonium hexafluorotitanate(IV)	$(NH_4)_2TiF_6$
690	Lithium hexafluorotitanate(IV)	$Li_2TiF_6 \cdot xH_2O$
691	Sodium hexafluorotitanate(IV)	Na_2TiF_6
692	Potassium hexafluorotitanate(IV)	K_2TiF_6
693	Calcium hexafluorotitanate(IV)	$CaTiF_6$
694	Barium hexafluorotitanate(IV)	$BaTiF_6 \cdot xH_2O$
695	Nickel hexafluorotitanate(IV)	$NiTiF_6 \cdot xH_2O$
696	Potassium hexafluorochromate	$K_3CrF_6 (\cdot H_2O$ or wet)
697	Potassium pentafluoromanganate	$K_2MnF_5 \cdot H_2O$
698	Potassium hexafluoromanganate(IV)	K_2MnF_6
699	Potassium hexafluoromanganate(III)	$K_3MnF_6 \cdot xH_2O$
700	Potassium pentafluoro(aquo)ferrate	$K_2FeF_5 \cdot H_2O$
701	Ammonium hexafluoroferrate	$(NH_4)_3FeF_6$
702	Sodium hexafluoroferrate	Na_3FeF_6 (wet)
703	Potassium tetrafluorozincate	$K_2ZnF_4 (\cdot H_2O$ or wet)
704	Sodium pentafluorozirconate	$NaZrF_5 \cdot xH_2O$
705	Potassium pentafluorozirconate	$KZrF_5 \cdot xH_2O$
706	Ammonium hexafluorozirconate	$(NH_4)_2ZrF_6$
707	Sodium hexafluorozirconate	Na_2ZrF_6
708	Potassium hexafluorozirconate	$K_2ZrF_6 (\cdot H_2O$ or wet)
709	Indium hexafluorozirconate	$In(ZrF_6)_3 \cdot xH_2O$
710	Potassium heptafluorozirconate	$K_3ZrF_7 \cdot xH_2O$
711	Potassium heptafluoroniobate(V)	K_2NbF_7
712	Potassium hexafluorotantalate	$KTaF_6$
713	Potassium heptafluorotantalate(IV)	K_3TaF_7
714	Sodium pentafluorouranate	$NaUF_5 \cdot xH_2O$
715	Ammonium chloride	NH_4Cl
716	Lithium chloride	$LiCl \cdot xH_2O$
717	Sodium chloride	$NaCl$
718	Potassium chloride	KCl
719	Rubidium chloride	$RbCl$
720	Cesium chloride	$CsCl$ (wet)
721	Magnesium chloride	$MgCl_2 \cdot 6H_2O$
722	Calcium chloride	$CaCl_2 \cdot 6H_2O$
723	β-Calcium chloride	$CaCl_2 \cdot 2H_2O$

724	Strontium chloride	$SrCl_2 \cdot 2H_2O$
725	Barium chloride	$BaCl_2 \cdot 2H_2O$
726	Aluminum chloride	$Al_2Cl_6 \cdot 6H_2O$
727	Indium chloride	$InCl$
728	Indium chloride	$InCl_3 \cdot xH_2O$
729	Thallium chloride	$TlCl$
730	Lead chloride	$PbCl_2$
731	Vanadium chloride	$VCl_3 \cdot xH_2O$
732	Chromium(III) chloride	$CrCl_3 \cdot xH_2O$
733	Chromium(III) chloride	$CrCl_3 \cdot 6H_2O$
734	Iron(II) chloride	$FeCl_2 \cdot 4H_2O$
735	Cobalt chloride	$CoCl_2 \cdot 6H_2O$
736	Nickel chloride	$NiCl_2 \cdot 6H_2O$
737	Zinc chloride	$ZnCl_2 \cdot xH_2O$
738	Yttrium chloride	$YCl_3 \cdot H_2O$
739	Niobium chloride	$NbCl_5 \cdot xH_2O$
740	Palladium chloride	$PdCl_2$
741	Silver chloride	$AgCl$
742	Cadmium chloride	$CdCl_2 \cdot 2H_2O$
743	Lanthanum chloride	$LaCl_3 \cdot 7H_2O$
744	Hafnium chloride	$HfCl_4 \cdot xH_2O$
745	Tantalum chloride	$TaCl_5 \cdot xH_2O$
746	Tungsten chloride	WCl_6
747	Mercury(I) chloride	Hg_2Cl_2
748	Mercury(II) chloride	$HgCl_2$
749	Cerium chloride	$CeCl_3 \cdot xH_2O$
750	Praseodymium chloride	$PrCl_3 \cdot 7H_2O$
751	Samarium chloride	$SmCl_3 \cdot 6H_2O$
752	Gadolinium chloride	$GdCl_3 \cdot 6H_2O$
753	Holmium chloride	$HoCl_3 \cdot xH_2O$
754	Thorium chloride	$ThCl_4 \cdot xH_2O$
755	Mercury amide chloride	NH_2HgCl
756	Diammine palladium dichloride	$\begin{array}{ccc} Cl & & NH_3 \\ & Pd & \\ NH_3 & & Cl \end{array}$
757	Hexammine cobalt(III) chloride	$Co(NH_3)_6Cl_3$
758	Ammonium magnesium chloride	$NH_4MgCl_3 \cdot xH_2O$
759	Potassium magnesium chloride	$KMgCl_3 \cdot xH_2O$
760	Potassium magnesium dichloride bromide	$KMgBrCl_2 \cdot xH_2O$
761	Sodium aluminum chloride	$NaAlCl_4 \cdot xH_2O$
762	Ammonium gallium chloride	NH_4GaCl_4
763	Ammonium trichlorostannate(II)	NH_4SnCl_3
764	Potassium trichlorostannate (II)	$K_4SnCl_3 \cdot xH_2O$
765	Ammonium hexachlorostannate(IV)	$(NH_4)_2SnCl_6$
766	Cobalt hexachlorostannate(IV)	$CoSnCl_2 \cdot xH_2O$
767	Ammonium pentachloroferrate(III)	$(NH_4)_2FeCl_5 \cdot H_2O$
768	Potassium pentachloroferrate(III)	$K_2FeCl_5 \cdot H_2O$
769	Ammonium tetrachlorocuprate	$(NH_4)_2(CuCl_4) \cdot 2H_2O$
770	Potassium tetrachlorocuprate	$K_2CuCl_4 \cdot 2H_2O$
771	Potassium hexachloromolybdate	K_3MoCl_6
772	Ammonium tetrachloropalladate	$(NH_4)_2PdCl_4$
773	Sodium tetrachloropalladate	$Na_2PdCl_4 \cdot xH_2O$
774	Potassium tetrachloropalladate	K_2PdCl_4

775	Potassium hexachloropalladate(IV)	K_2PdCl_6
776	Barium cadmium chloride	$BaCdCl_4 \cdot 4H_2O$
777	Ammonium tetrachloroplatinate	$(NH_4)_2PtCl_4$
778	Potassium tetrachloroplatinate	K_2PtCl_4
779	Sodium chlorite	$NaClO_2$
780	Sodium chlorate	$NaClO_3$
781	Potassium chlorate	$KClO_3$
782	Strontium chlorate	$Sr(ClO_3)_2$
783	Barium chlorate	$Ba(ClO_3)_2 \cdot H_2O$
784	Ammonium perchlorate	NH_4ClO_4
785	Lithium perchlorate	$LiClO_4 \cdot 3H_2O$
786	Sodium perchlorate	$NaClO_4 \cdot H_2O$
787	Rubidium perchlorate	$RbClO_4$
788	Cesium perchlorate	$CsClO_4$
789	Magnesium perchlorate	$Mg(ClO_4)_2 \cdot 6H_2O$
790	Barium perchlorate	$Ba(ClO_4)_2 \cdot xH_2O$
791	Barium perchlorate	$Ba(ClO_4)_2 \cdot 3H_2O$
792	Gallium perchlorate	$Ga(ClO_4)_3 \cdot 6H_2O$
793	Zinc perchlorate	$Zn(ClO_4)_2 \cdot 6H_2O$
794	Cerium perchlorate	$Ce(ClO_4)_3 \cdot xH_2O$
795	Ammonium bromide	NH_4Br
796	Sodium bromide	$NaBr$
797	Potassium bromide	KBr
798	Rubidium bromide	$RbBr$
799	Cesium bromide	$CsBr$
800	Strontium bromide	$SrBr_2 \cdot 6H_2O$
801	Barium bromide	$BaBr_2 \cdot 2H_2O$
802	Indium bromide	$InBr_3 \cdot xH_2O$
803	Tin bromide	$SnBr_4 \cdot xH_2O$
804	Lead bromide	$PbBr_2$
805	Arsenic(III) bromide	$AsBr_3$
806	Antimony bromide	$SbBr_3$
807	Bismuth bromide	$BiBr_3 \cdot xH_2O$
808	Tellurium bromide	$TeBr_4 \cdot xH_2O$
809	Iron(II) bromide	$FeBr_2 \cdot 6H_2O$
810	Zinc bromide	$ZnBr_2 \cdot xH_2O$
811	Silver bromide	$AgBr$
812	Cadmium bromide	$CdBr_2$
813	Lanthanum bromide	$LaBr_3 \cdot 7H_2O$
814	Mercury(I) bromide and 814(a) Mercury(II) bromide	Hg_2Br_2 and $HgBr_2$
815	Neodymium bromide	$NdBr_3 \cdot xH_2O$
816	Holmium bromide	$HoBr_3 \cdot xH_2O$
817	Ammonium cadmium bromide	$(NH_4)_2CdBr_4$
818	Lithium bromate	$LiBrO_3$
819	Sodium bromate	$NaBrO_3$
820	Potassium bromate	$KBrO_3$
821	Rubidium bromate	$RbBrO_3$
822	Cesium bromate	$CsBrO_3$
823	Magnesium bromate	$Mg(BrO_3)_2 \cdot 6H_2O$
824	Barium bromate	$Ba(BrO_3)_2 \cdot H_2O$
825	Aluminum bromate	$Al(BrO_3)_3 \cdot 9H_2O$
826	Lead bromate	$Pb(BrO_3)_2 \cdot H_2O$
827	Zinc bromate	$Zn(BrO_3)_2 \cdot 6H_2O$

828	Cadmium bromate	$Cd(BrO_3)_2 \cdot xH_2O$
829	Ammonium iodide	NH_4I
830	Lithium iodide	$LiI \cdot (3?)H_2O$
831	Potassium iodide	KI
832	Rubidium iodide	RbI
833	Cesium iodide	CsI
834	Barium iodide	$BaI_2 \cdot 2H_2O$
835	Thallium iodide	TlI
836	Germanium iodide	GeI_4
837	Tin(IV) iodide	SnI_4
838	Lead iodide	PbI_2
839	Arsenic iodide	AsI_3
840	Antimony iodide	SbI_3
841	Bismuth iodide	BiI_3
842	Nickel iodide	NiI_2
843	Copper iodide	Cu_2I_2
844	Zirconium iodide	$ZrI_4 \cdot xH_2O$
845	Niobium iodide	$NbI_5 \cdot xH_2O$
846	Palladium iodide	PdI_2
847	Silver iodide	AgI
848	Mercury(I) iodide	Hg_2I_2
849	Mercury(II) iodide	HgI_2
850	Ytterbium iodide	$YbI_3 \cdot xH_2O$
851	Potassium bismuth iodide	$K_4BiI_7 \cdot xH_2O$
852	Copper tetraiodomercurate(II)	Cu_2HgI_4
853	Potassium iodocadmate	$K_2CdI_4 \cdot xH_2O$
854	Ammonium iodate	NH_4IO_3
855	Lithium iodate	$LiIO_3$
856	Sodium iodate	$NaIO_3$
857	Sodium iodate	$NaIO_3 \cdot H_2O$
858	Rubidium iodate	$RbIO_3$
859	Cesium iodate	$CsIO_3 \cdot xH_2O$
860	Calcium iodate	$Ca(IO_3)_2 \cdot 6H_2O$
861	Strontium iodate	$Sr(IO_3)_2$
862	Barium iodate	$Ba(IO_3)_2 \cdot H_2O$
863	Lead iodate	$Pb(IO_3)_2$
864	Chromium(III) iodate	$Cr(IO_3)_3 \cdot xH_2O$
865	Nickel iodate	$Ni(IO_3)_2 \cdot xH_2O$
866	Silver iodate	$AgIO_3$
867	Cesium iodate	$CsIO_3 \cdot xH_2O$
868	Sodium periodate	$NaIO_4$
869	Potassium periodate	KIO_4
870	Lithium manganite	Li_2MnO_3
871	Barium manganate	$BaMnO_4$
872	Lithium permanganate	$LiMnO_4 \cdot 3H_2O$
873	Sodium permanganate	$NaMnO_4 \cdot 3H_2O$
874	Potassium permanganate	$KMnO_4$
875	Magnesium permanganate	$Mg(MnO_4)_2 \cdot 6H_2O$
876	Barium permanganate	$Ba(MnO_4)_2$
877	Zinc permanganate	$Zn(MnO_4)_2 \cdot 6H_2O$
878	Cobalt ferrate(III)	$CoFe_2O_4$
879	Nickel ferrate(III)	$NiFe_2O_4$
880	Copper(II) ferrate(III)	$CuFe_2O_4$

881	Lithium cobaltite(III)	$LiCoO_2$
882	Ammonium uranate(VI)	$(NH_4)_2U_2O_7 \cdot xH_2O$
883	Sodium uranate(VI)	$Na_2U_2O_7 \cdot H_2O$
884	Lead calcium uranate(VI) (wolfsendorfite)	$Pb_5CaU_{12}O_{42} \cdot 12H_2O$

Miscellaneous minerals

885	Albite	$Na_2O \cdot Al_2O_3 \cdot 6SiO_2$
886	Apatite	$CaF_2 \cdot 3Ca_3P_2O_8$
887	Dolomite	$CaCO_3 \cdot MgCO_3$
888	Hectorite	$Si_8(Mg_{5.33}Li_{0.67})O_{20}(OH)_4$
889	Microcline	$K_2O \cdot Al_2O_3 \cdot 6SiO_2$
890	Pyrite (fool's gold)	FeS_2
891	Serpentine	$3MgO \cdot 2SiO_2 \cdot 2H_2O$
892	Quartz	SiO_2

ALPHABETICAL INDEX OF SPECTRA

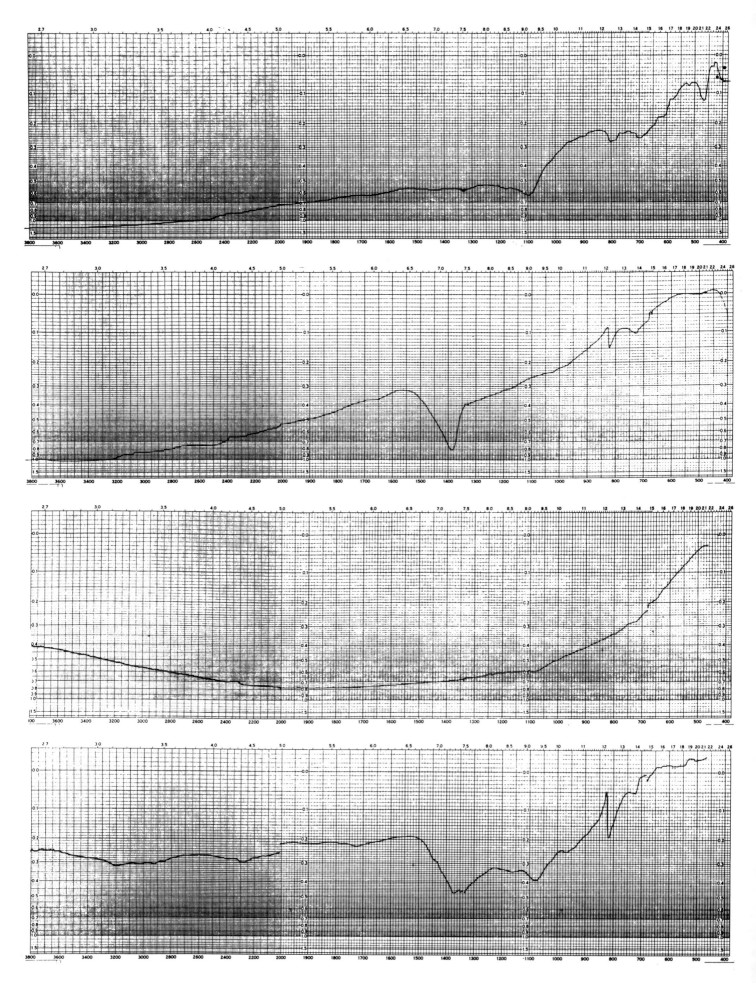

48

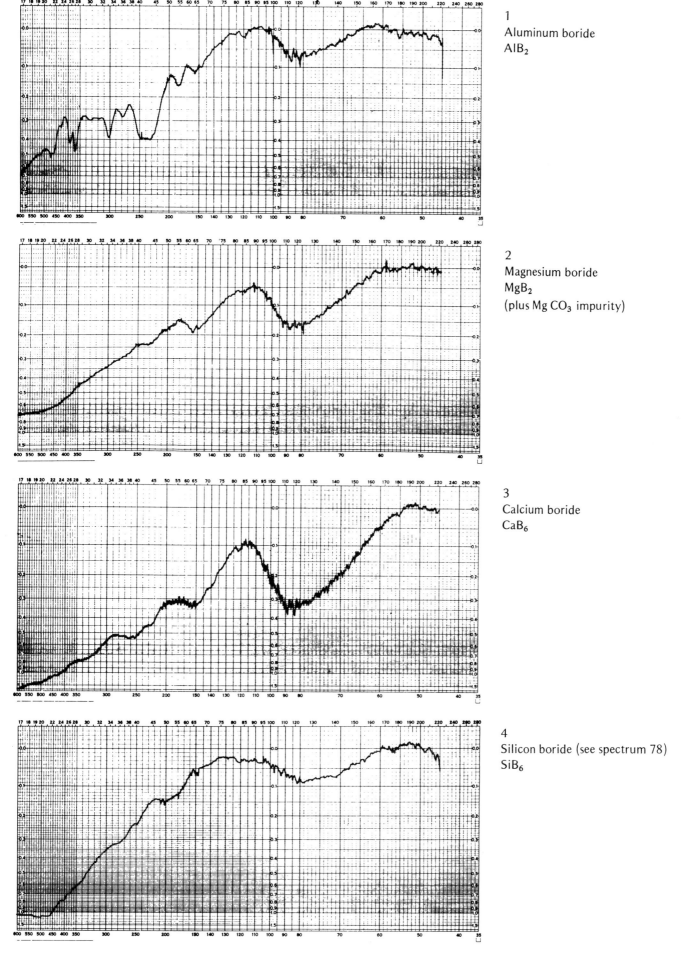

1
Aluminum boride
AlB_2

2
Magnesium boride
MgB_2
(plus $MgCO_3$ impurity)

3
Calcium boride
CaB_6

4
Silicon boride (see spectrum 78)
SiB_6

49

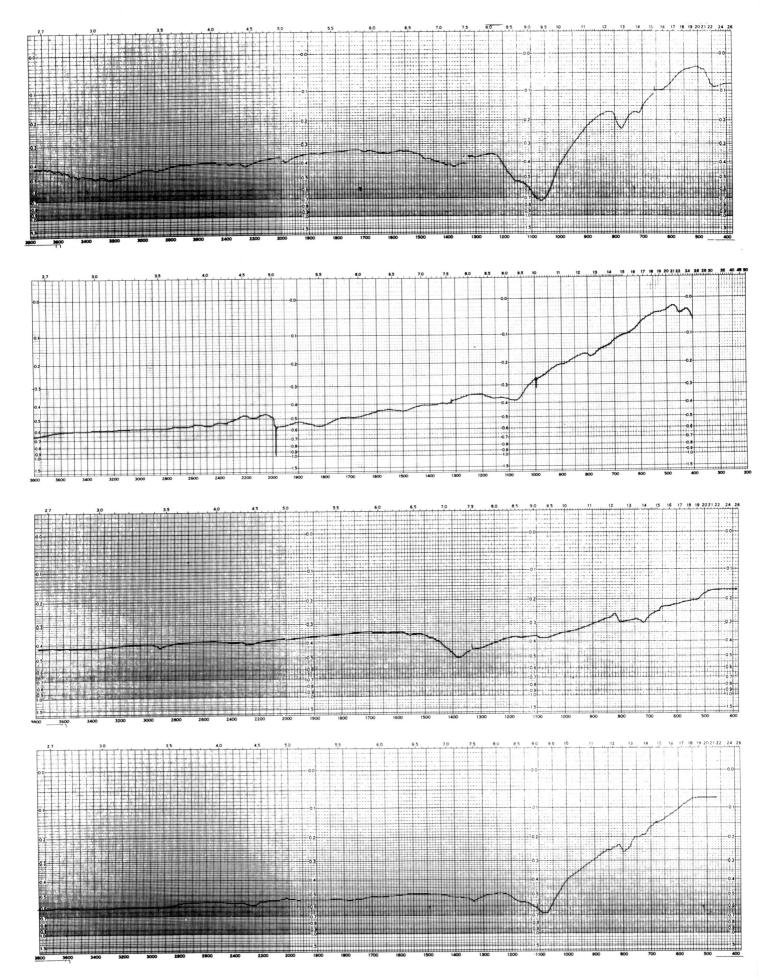

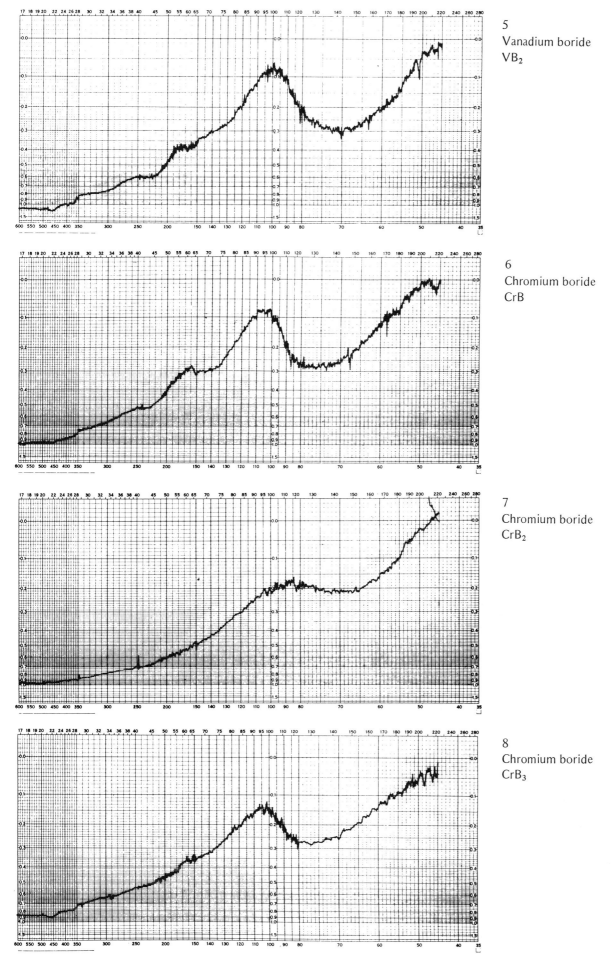

5
Vanadium boride
VB_2

6
Chromium boride
CrB

7
Chromium boride
CrB_2

8
Chromium boride
CrB_3

51

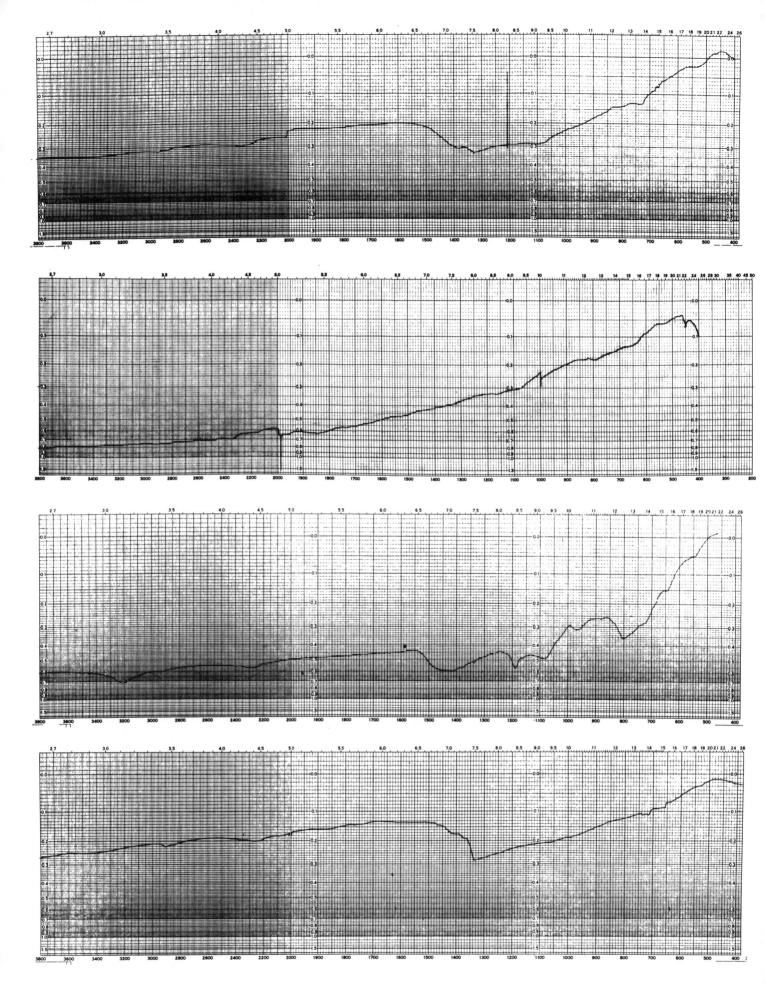

52

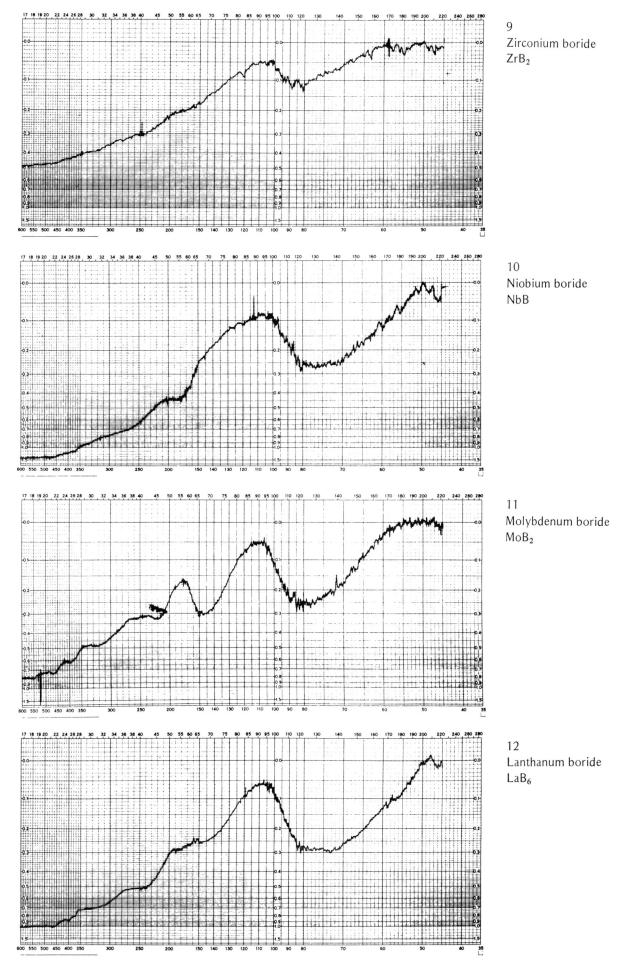

9
Zirconium boride
ZrB_2

10
Niobium boride
NbB

11
Molybdenum boride
MoB_2

12
Lanthanum boride
LaB_6

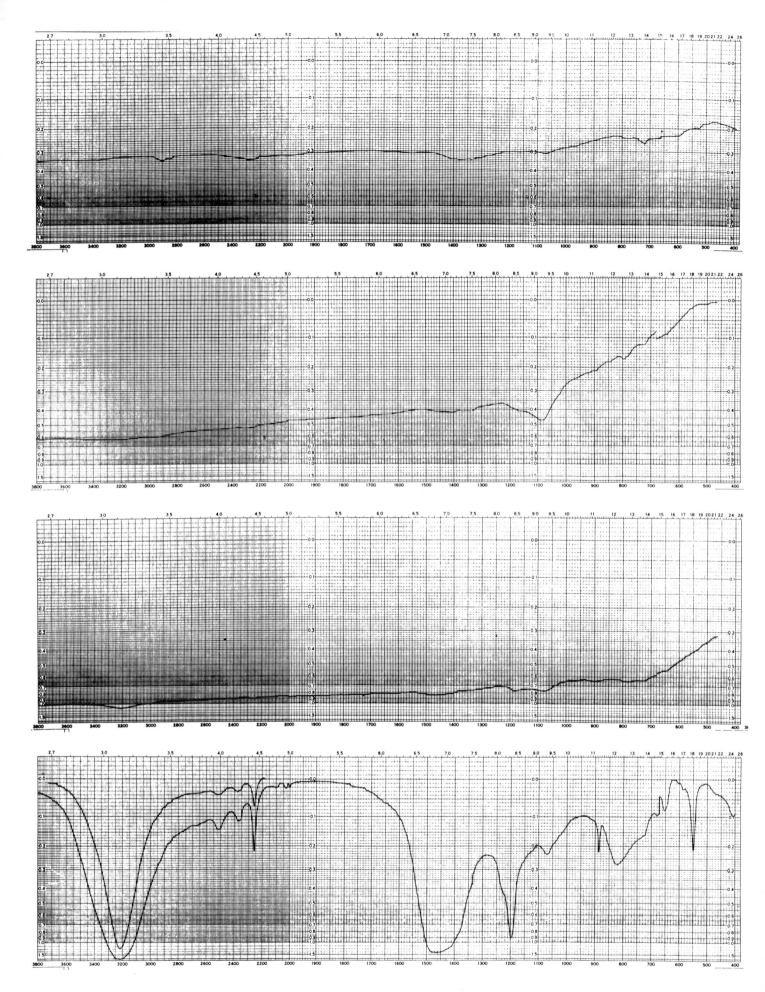

54

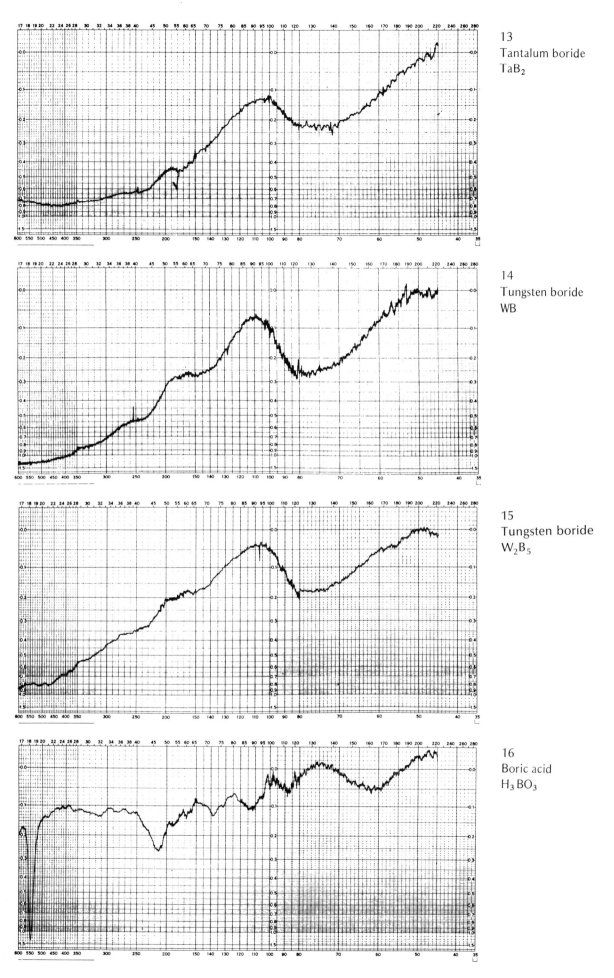

13
Tantalum boride
TaB_2

14
Tungsten boride
WB

15
Tungsten boride
W_2B_5

16
Boric acid
H_3BO_3

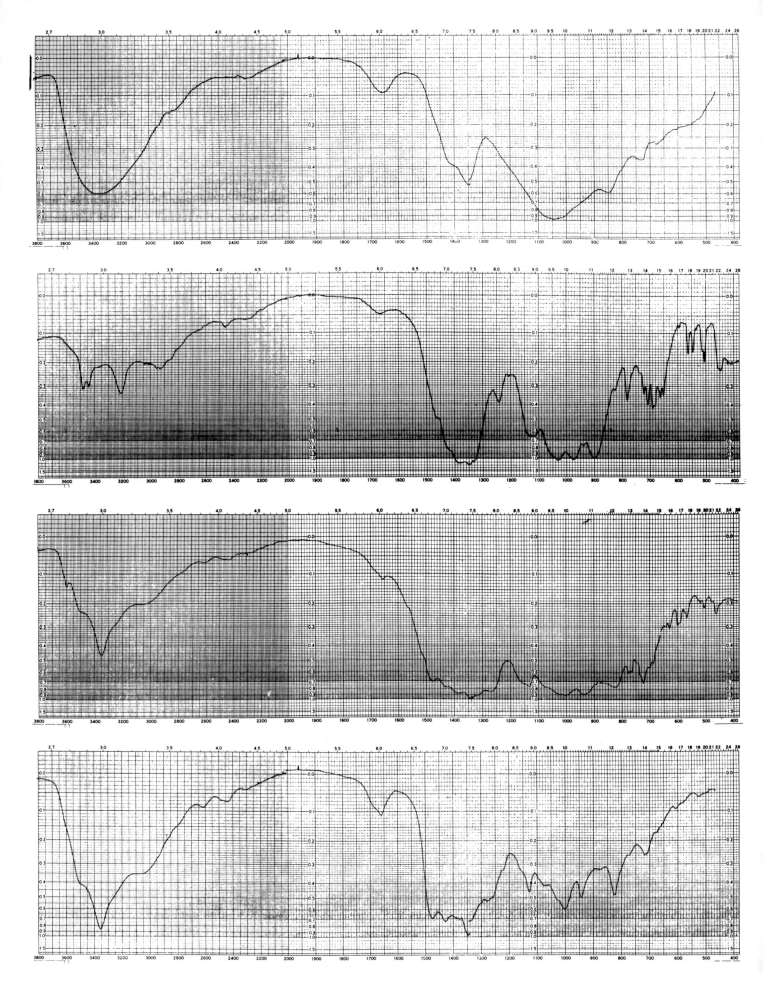

56

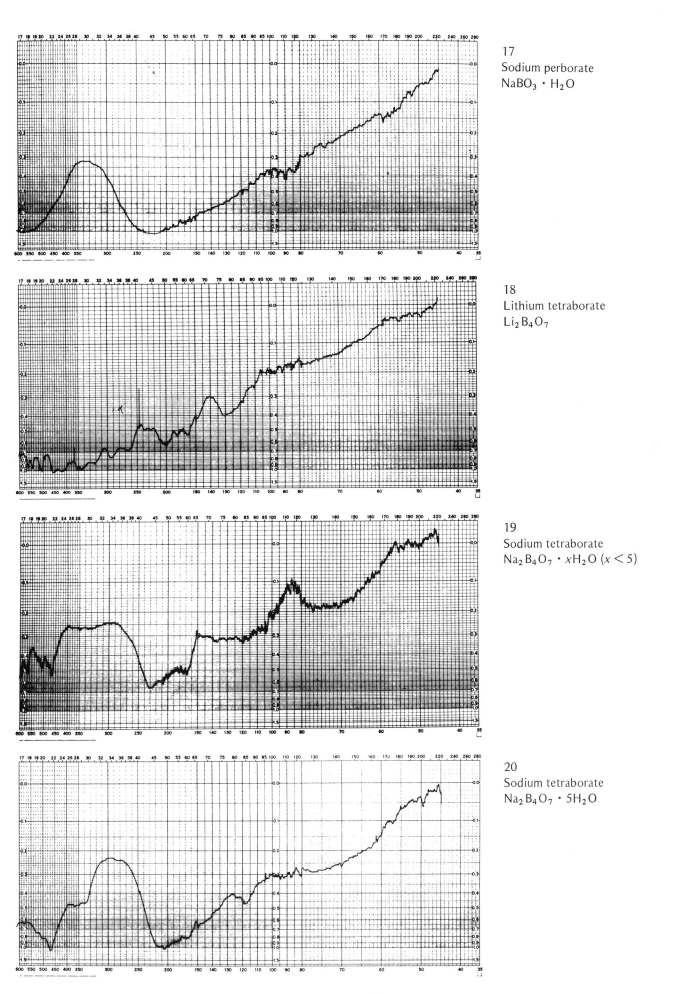

17
Sodium perborate
$NaBO_3 \cdot H_2O$

18
Lithium tetraborate
$Li_2B_4O_7$

19
Sodium tetraborate
$Na_2B_4O_7 \cdot xH_2O \; (x < 5)$

20
Sodium tetraborate
$Na_2B_4O_7 \cdot 5H_2O$

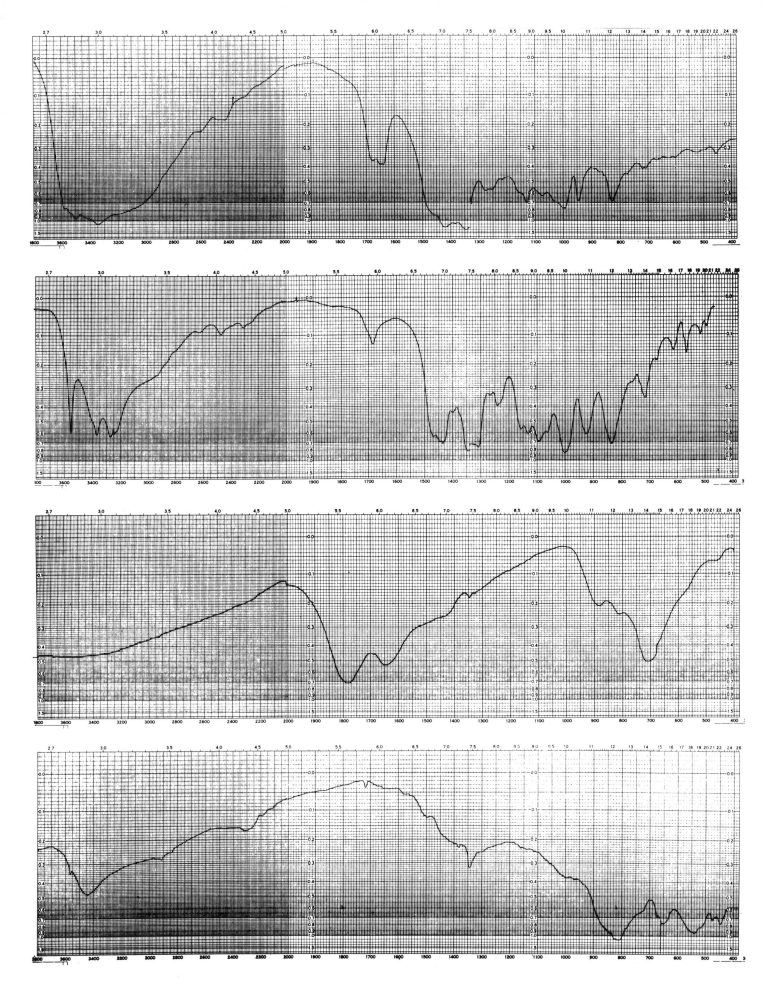

58

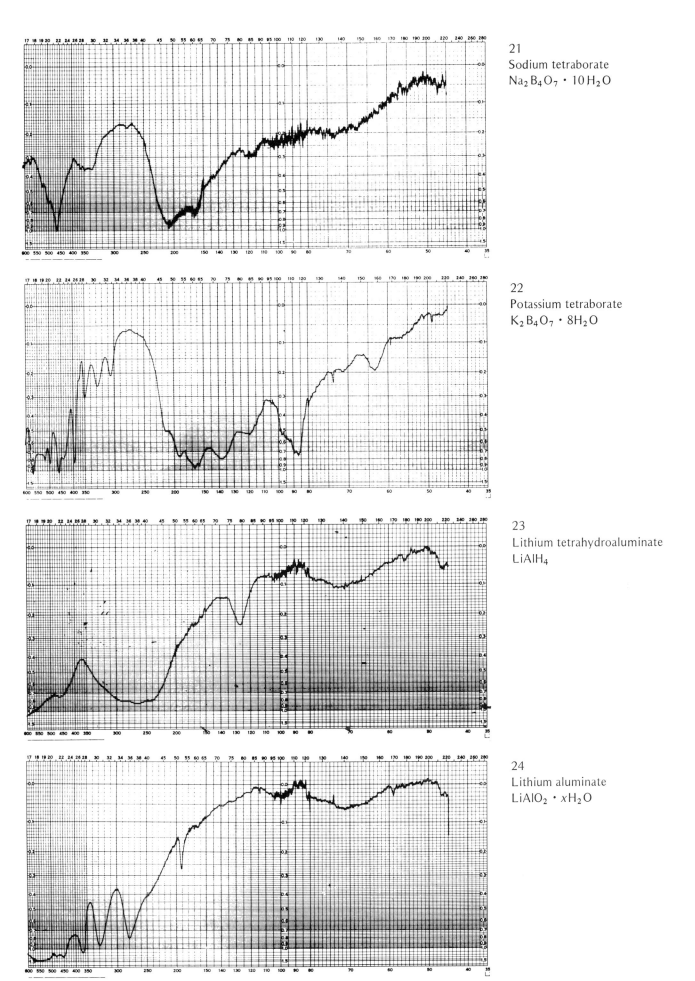

21
Sodium tetraborate
$Na_2B_4O_7 \cdot 10H_2O$

22
Potassium tetraborate
$K_2B_4O_7 \cdot 8H_2O$

23
Lithium tetrahydroaluminate
$LiAlH_4$

24
Lithium aluminate
$LiAlO_2 \cdot xH_2O$

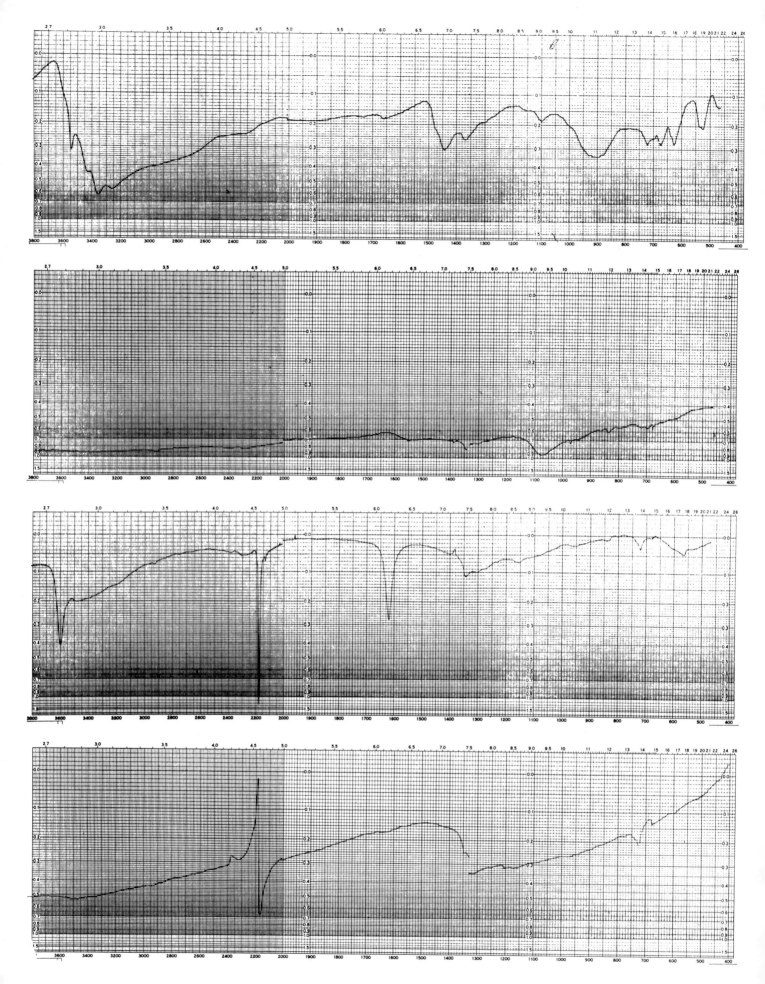

60

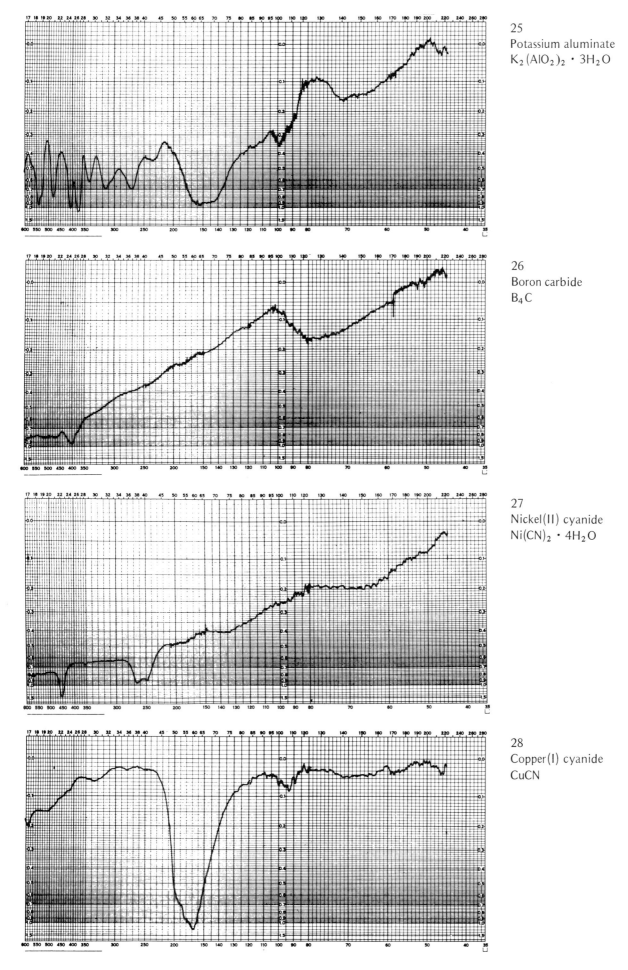

25
Potassium aluminate
$K_2(AlO_2)_2 \cdot 3H_2O$

26
Boron carbide
B_4C

27
Nickel(II) cyanide
$Ni(CN)_2 \cdot 4H_2O$

28
Copper(I) cyanide
CuCN

61

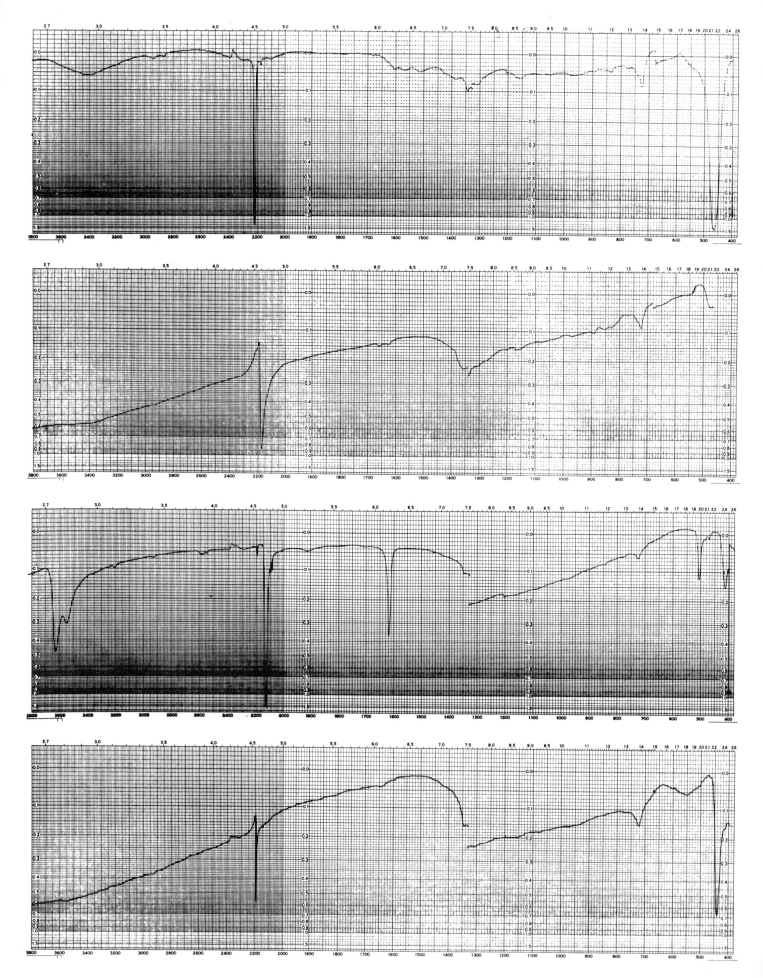

62

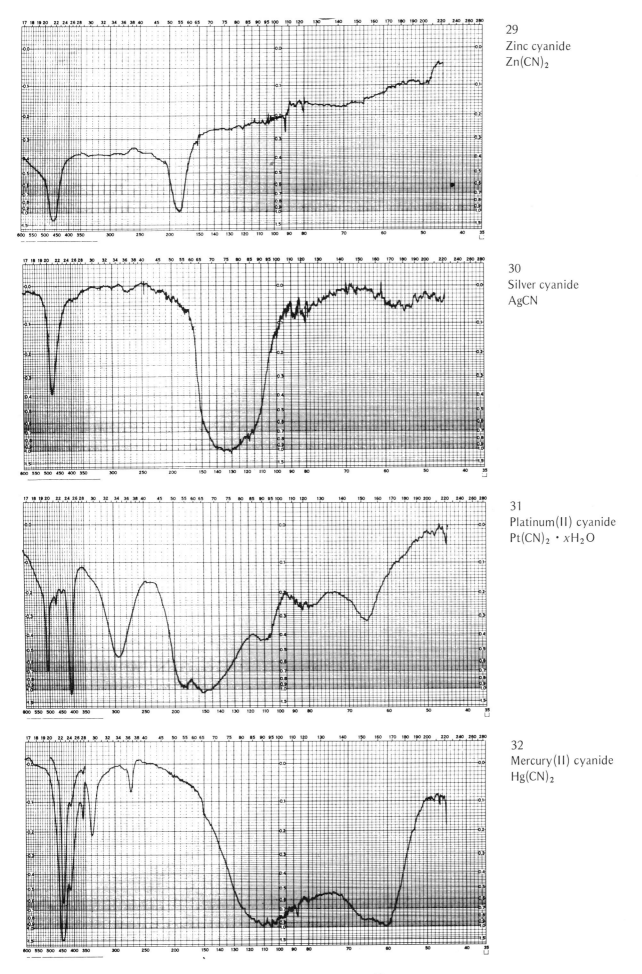

29
Zinc cyanide
$Zn(CN)_2$

30
Silver cyanide
AgCN

31
Platinum(II) cyanide
$Pt(CN)_2 \cdot x H_2O$

32
Mercury(II) cyanide
$Hg(CN)_2$

63

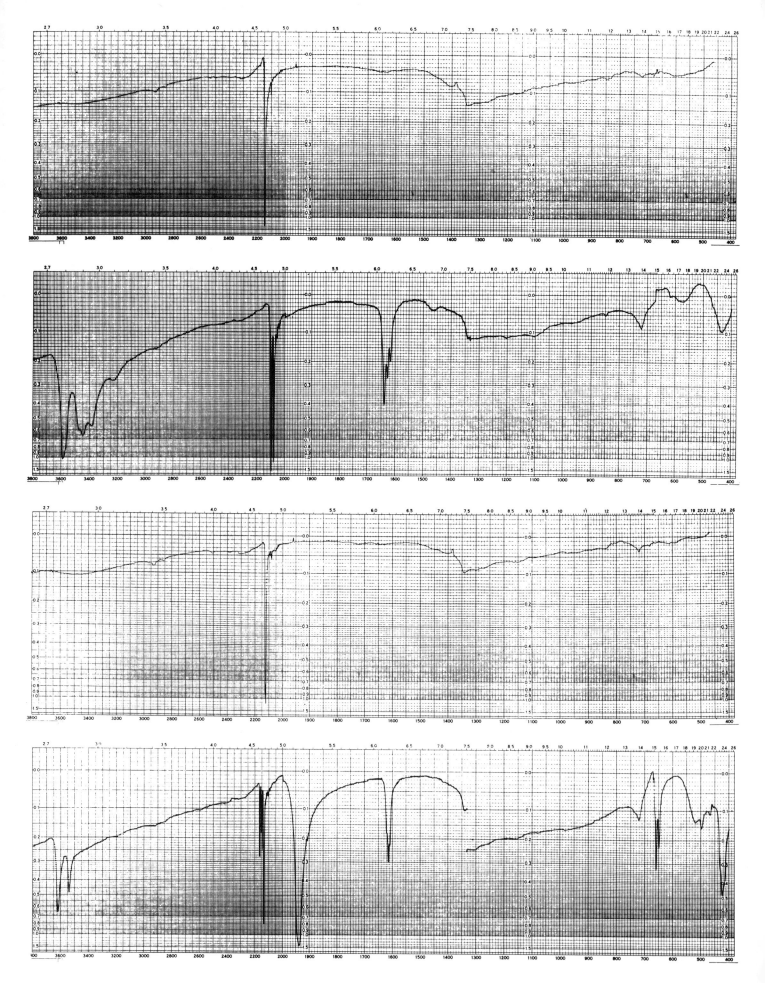

64

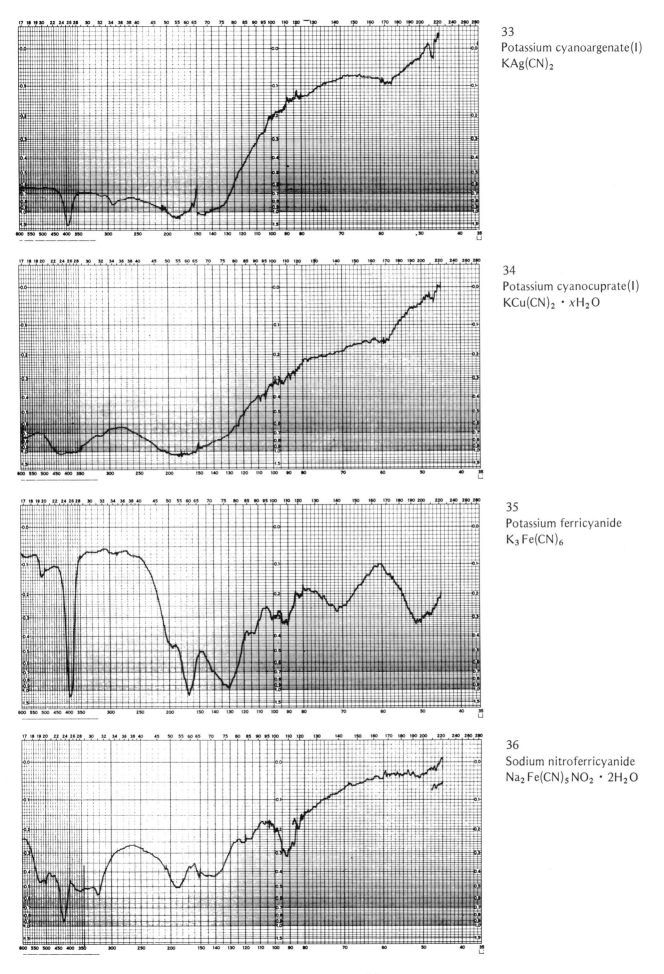

33
Potassium cyanoargenate(I)
KAg(CN)$_2$

34
Potassium cyanocuprate(I)
KCu(CN)$_2$ · xH$_2$O

35
Potassium ferricyanide
K$_3$Fe(CN)$_6$

36
Sodium nitroferricyanide
Na$_2$Fe(CN)$_5$NO$_2$ · 2H$_2$O

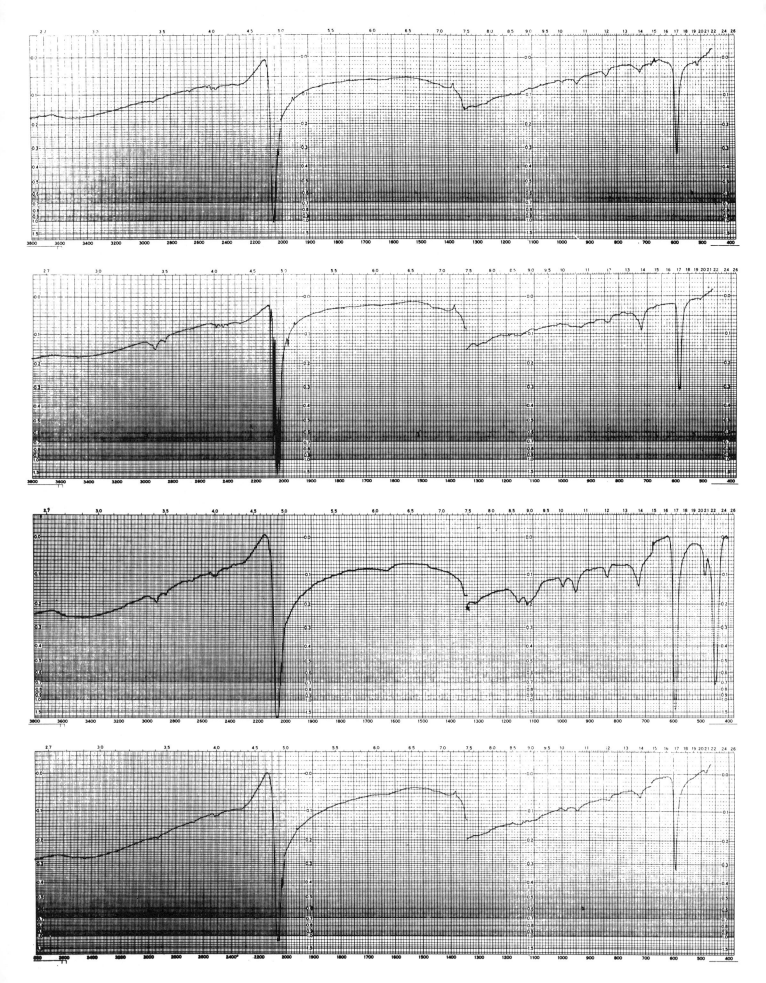

66

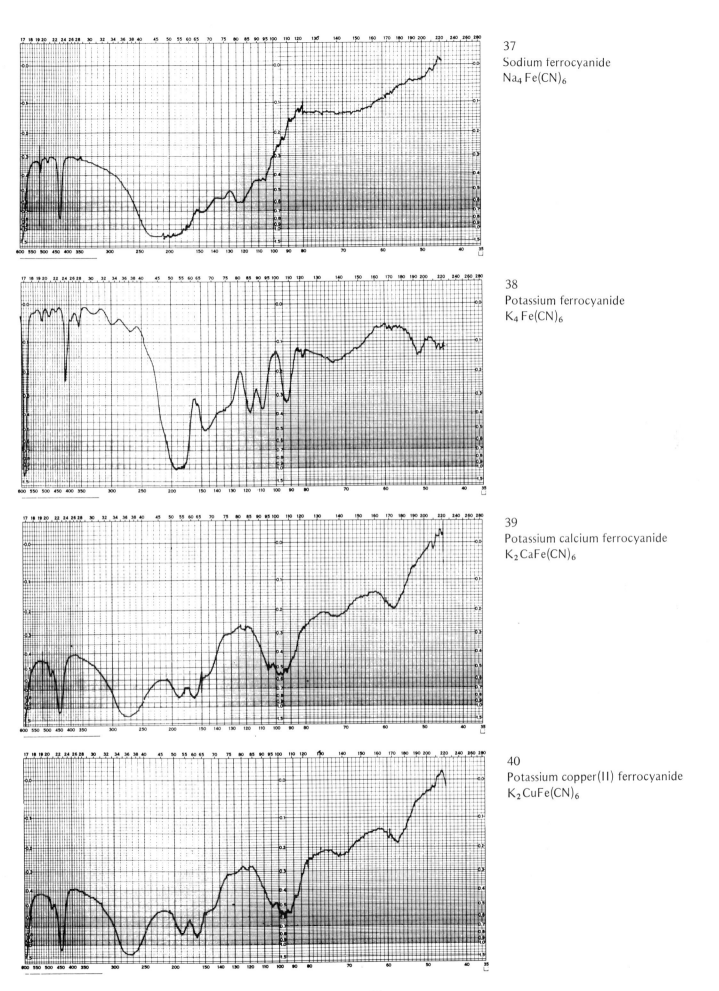

37
Sodium ferrocyanide
$Na_4Fe(CN)_6$

38
Potassium ferrocyanide
$K_4Fe(CN)_6$

39
Potassium calcium ferrocyanide
$K_2CaFe(CN)_6$

40
Potassium copper(II) ferrocyanide
$K_2CuFe(CN)_6$

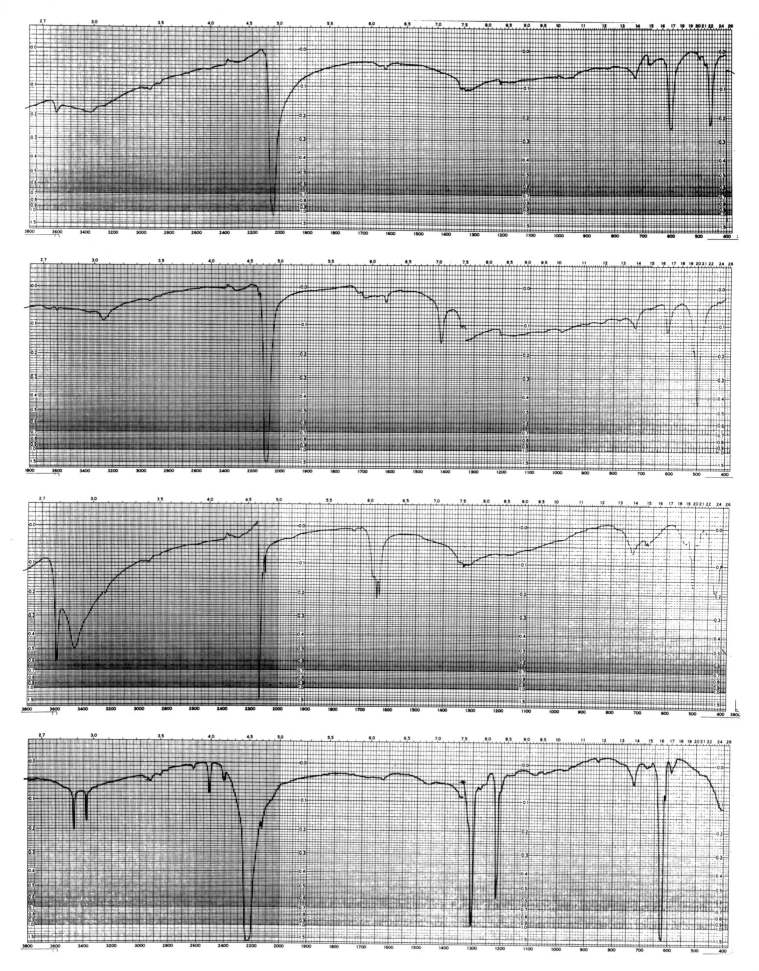

68

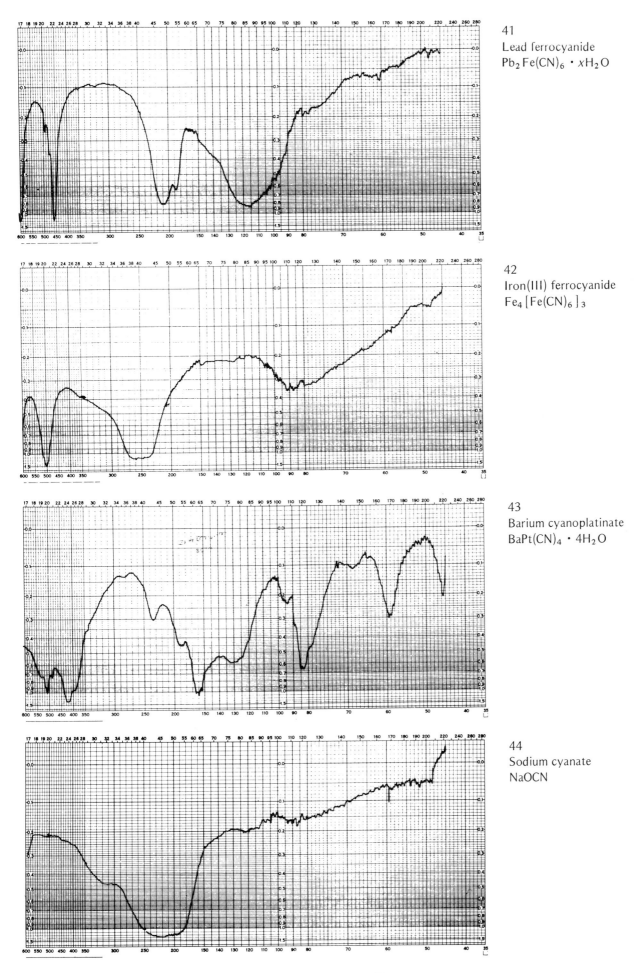

41
Lead ferrocyanide
$Pb_2Fe(CN)_6 \cdot xH_2O$

42
Iron(III) ferrocyanide
$Fe_4[Fe(CN)_6]_3$

43
Barium cyanoplatinate
$BaPt(CN)_4 \cdot 4H_2O$

44
Sodium cyanate
NaOCN

69

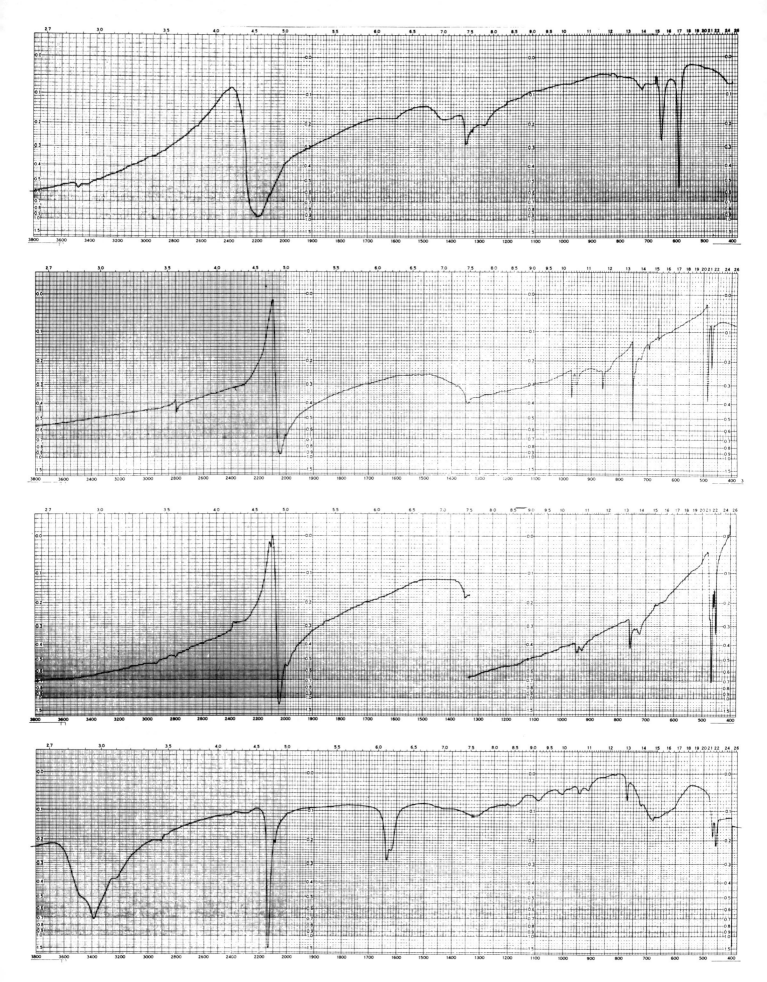

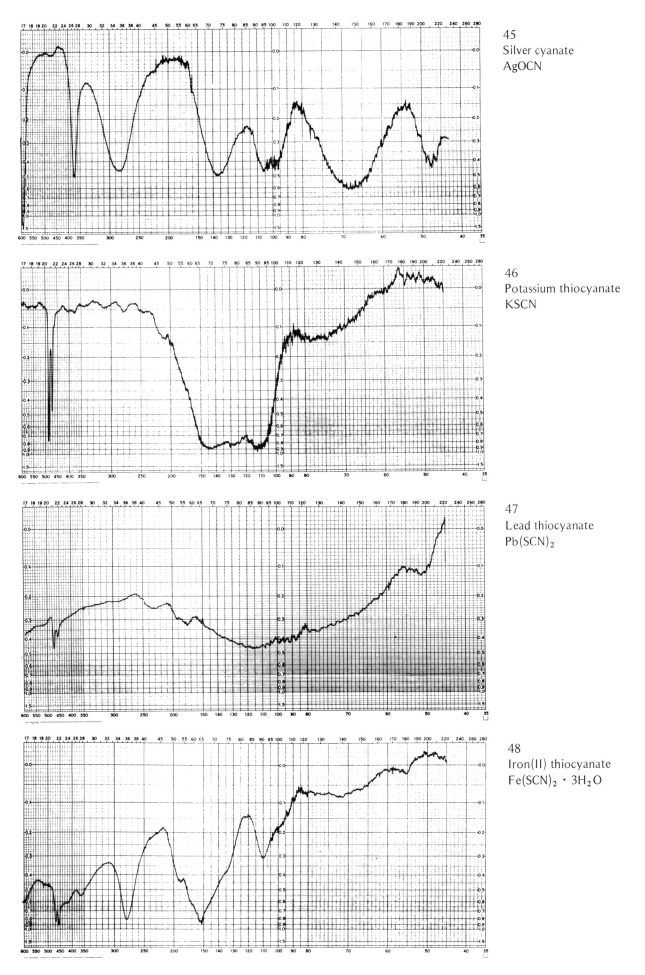

45
Silver cyanate
AgOCN

46
Potassium thiocyanate
KSCN

47
Lead thiocyanate
Pb(SCN)$_2$

48
Iron(II) thiocyanate
Fe(SCN)$_2$ · 3H$_2$O

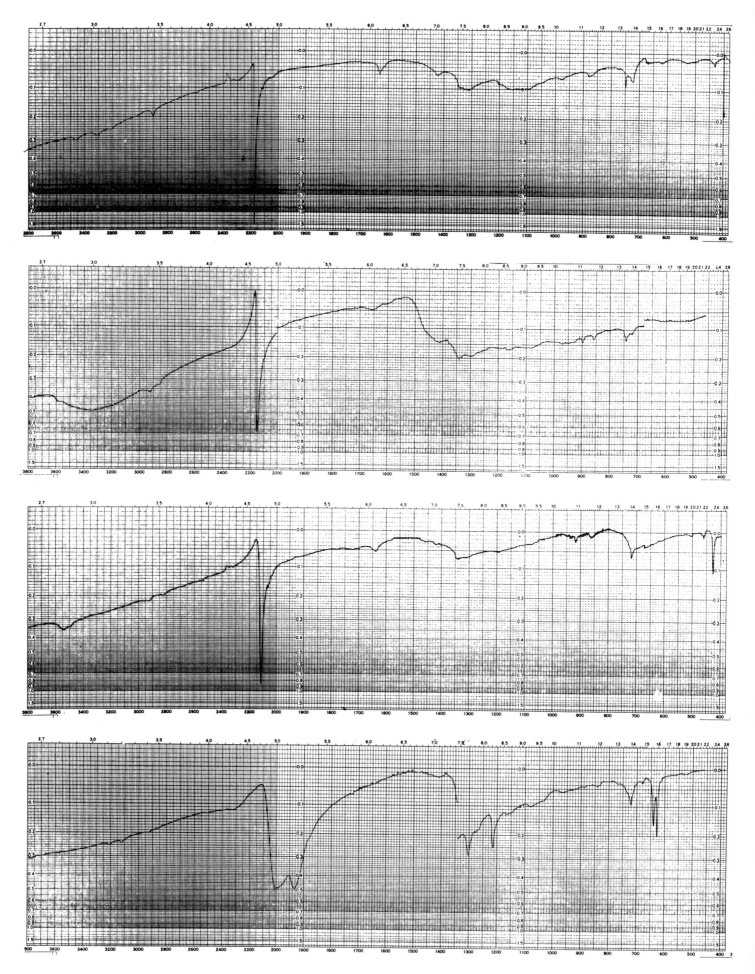

72

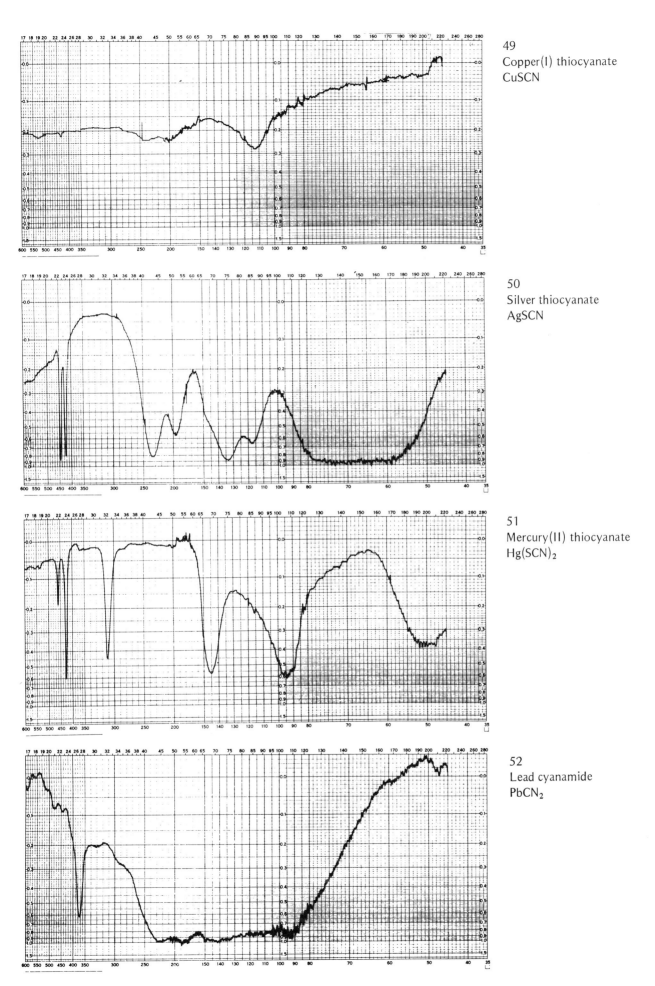

49
Copper(I) thiocyanate
CuSCN

50
Silver thiocyanate
AgSCN

51
Mercury(II) thiocyanate
Hg(SCN)$_2$

52
Lead cyanamide
PbCN$_2$

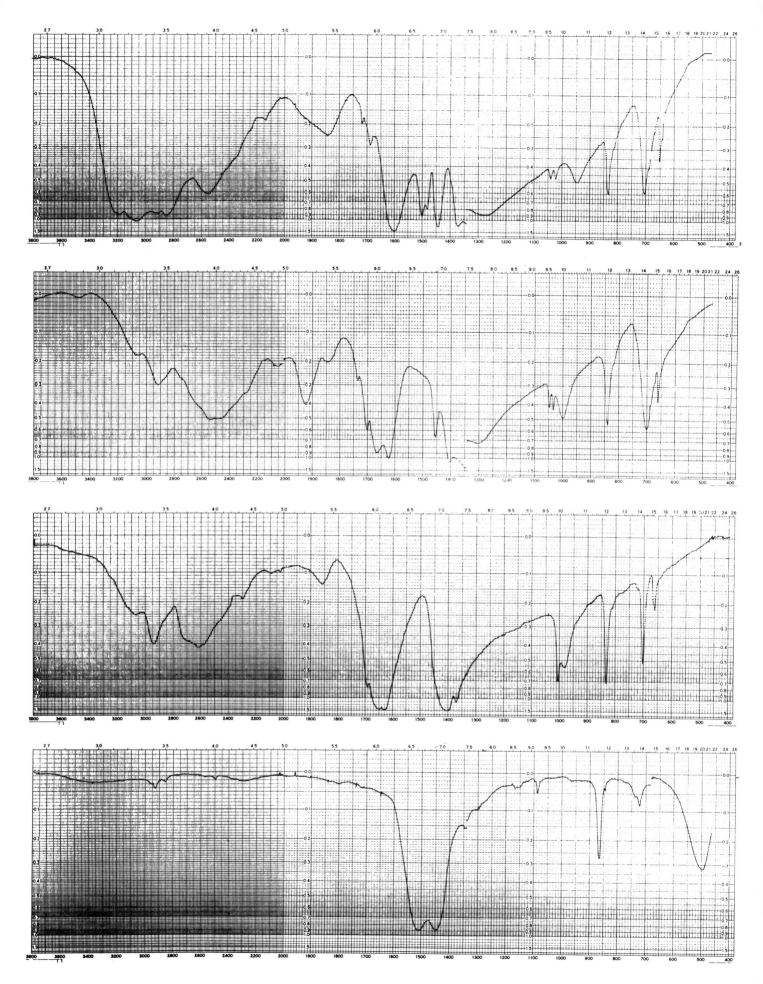

74

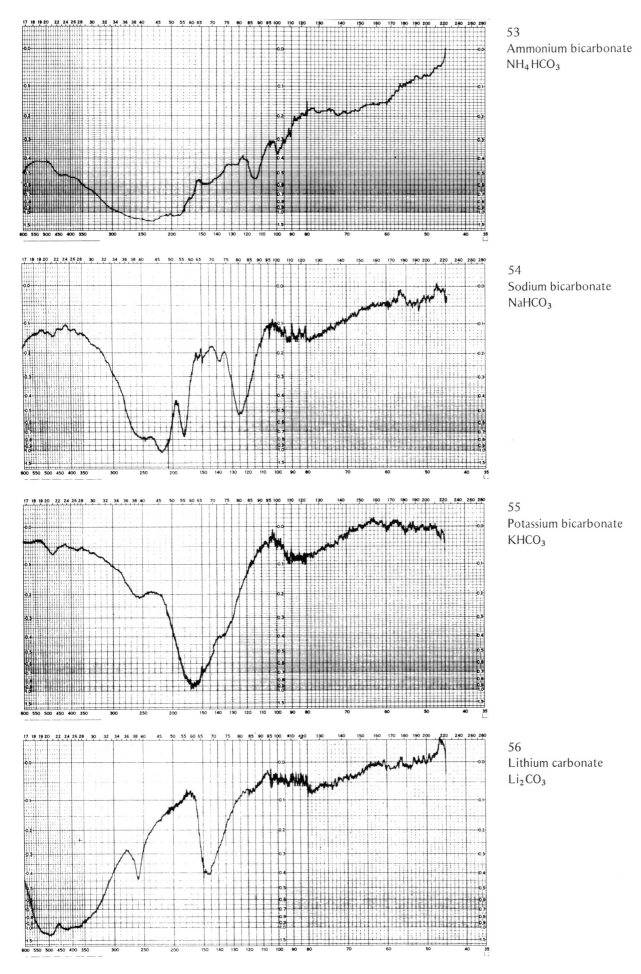

53
Ammonium bicarbonate
NH_4HCO_3

54
Sodium bicarbonate
$NaHCO_3$

55
Potassium bicarbonate
$KHCO_3$

56
Lithium carbonate
Li_2CO_3

75

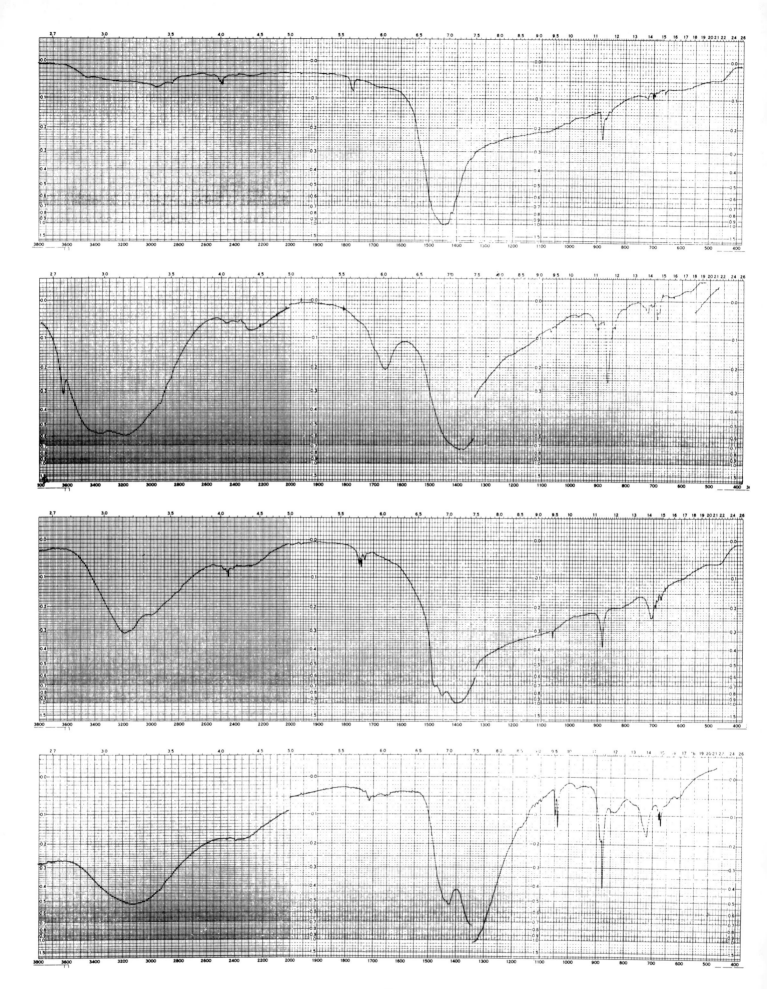

76

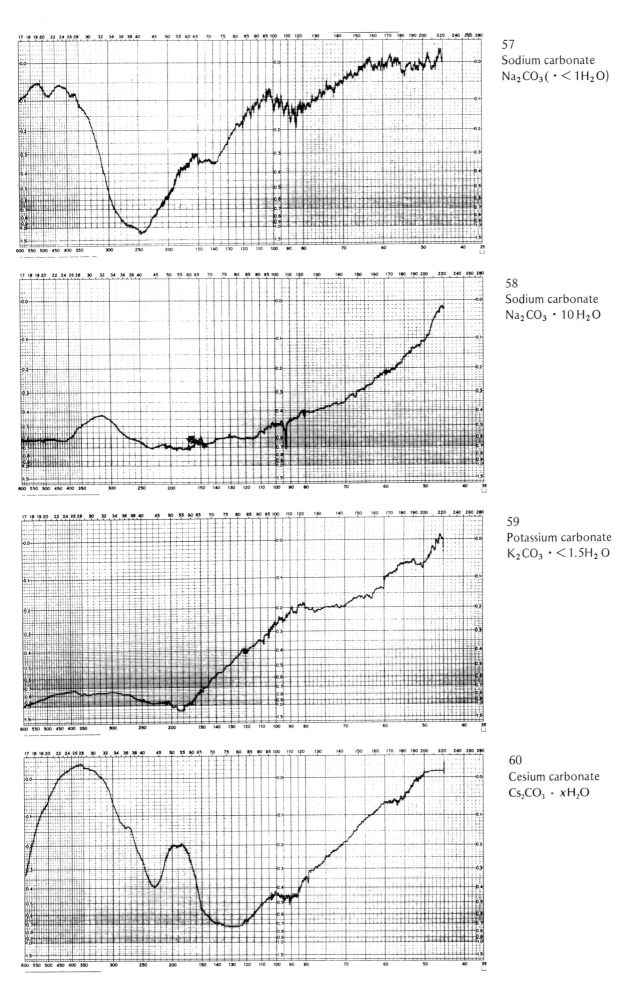

57
Sodium carbonate
$Na_2CO_3 (\cdot < 1H_2O)$

58
Sodium carbonate
$Na_2CO_3 \cdot 10 H_2O$

59
Potassium carbonate
$K_2CO_3 \cdot < 1.5H_2O$

60
Cesium carbonate
$Cs_2CO_3 \cdot xH_2O$

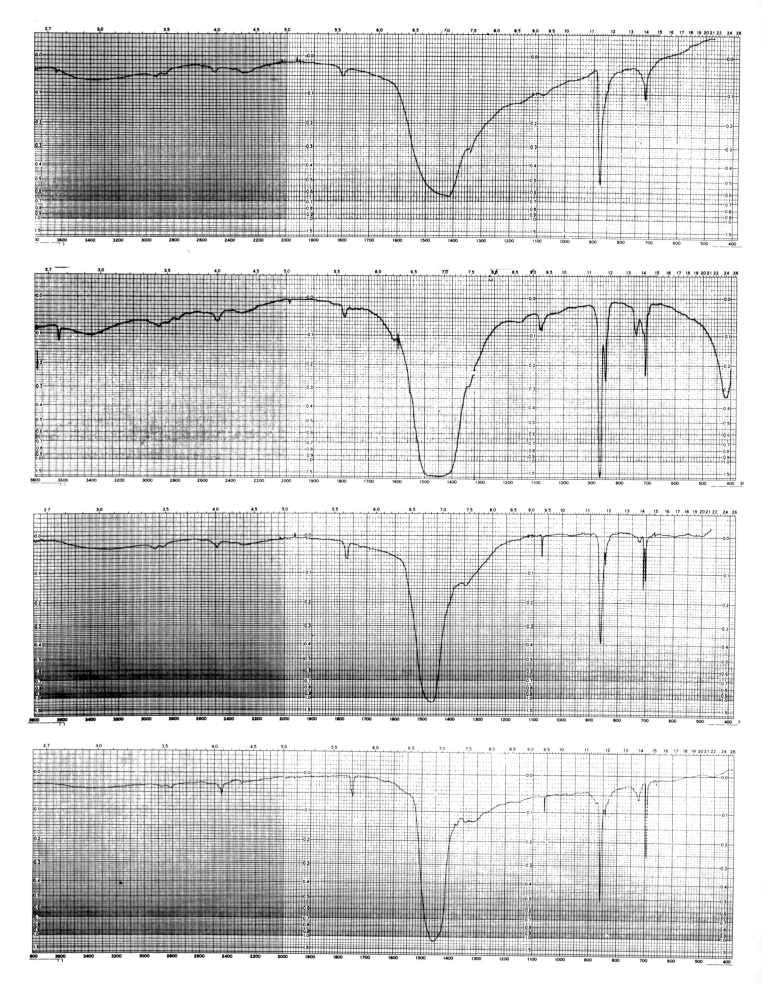

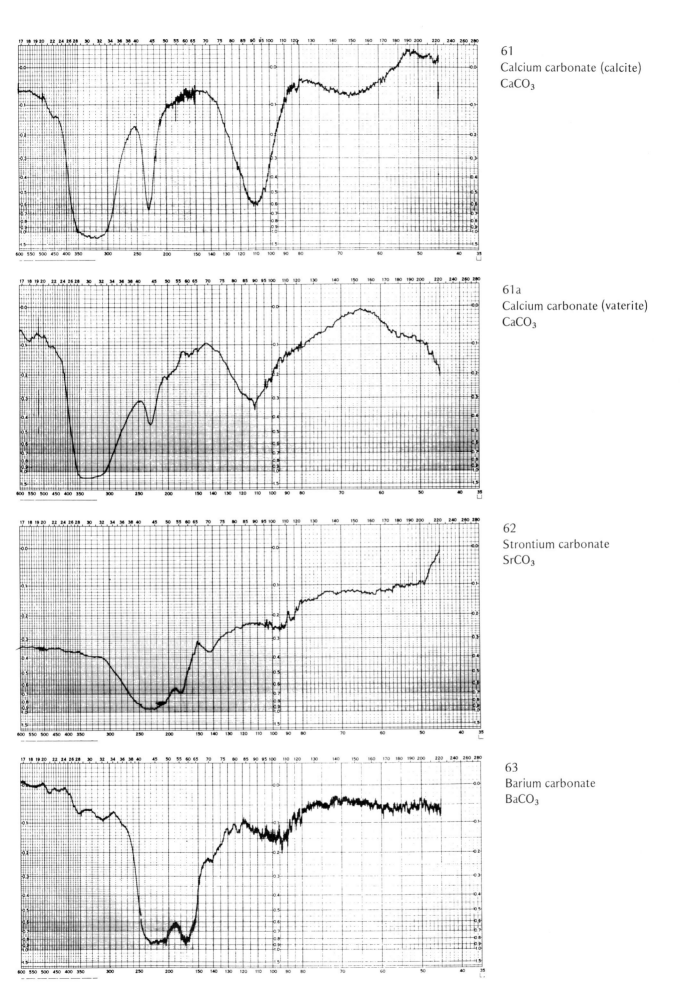

61
Calcium carbonate (calcite)
CaCO$_3$

61a
Calcium carbonate (vaterite)
CaCO$_3$

62
Strontium carbonate
SrCO$_3$

63
Barium carbonate
BaCO$_3$

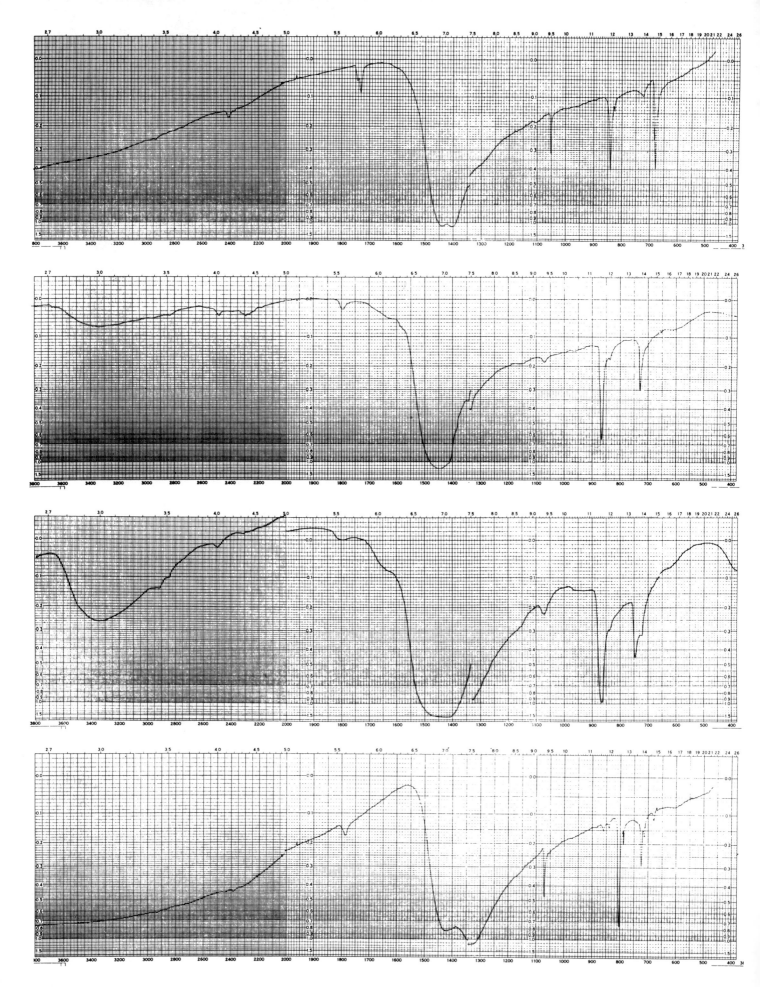

80

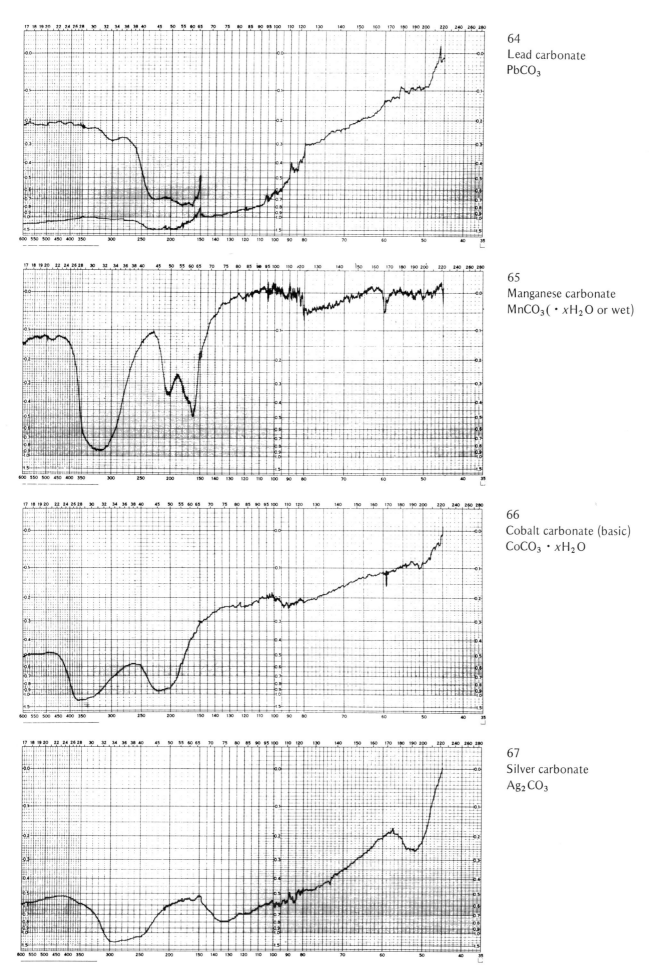

64
Lead carbonate
$PbCO_3$

65
Manganese carbonate
$MnCO_3 (\cdot xH_2O \text{ or wet})$

66
Cobalt carbonate (basic)
$CoCO_3 \cdot xH_2O$

67
Silver carbonate
Ag_2CO_3

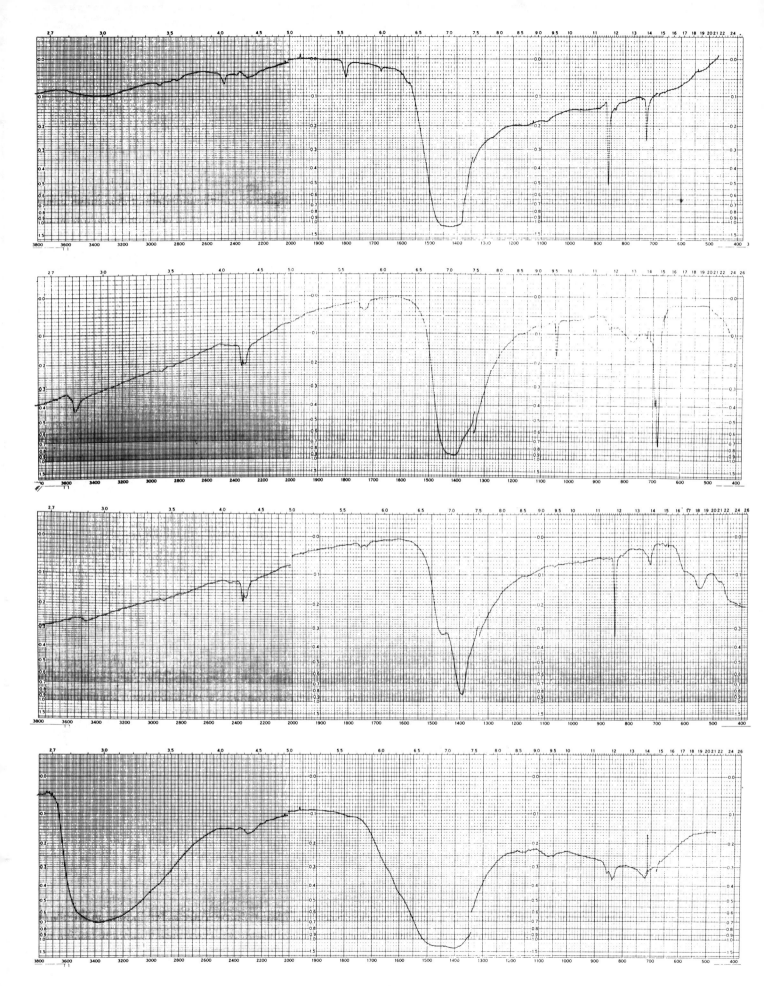

82

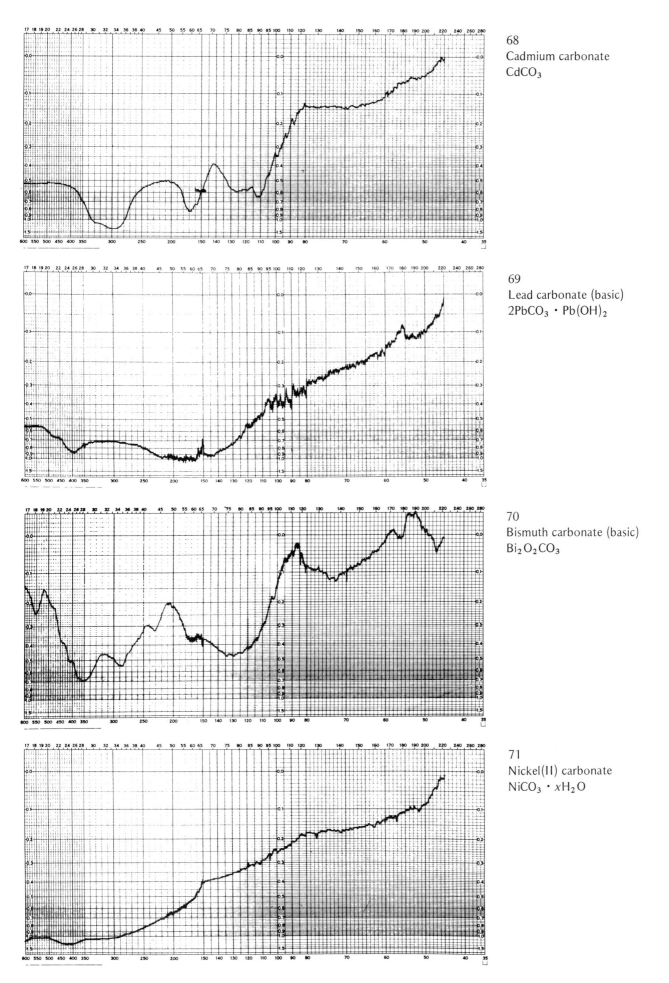

68
Cadmium carbonate
CdCO$_3$

69
Lead carbonate (basic)
2PbCO$_3$ · Pb(OH)$_2$

70
Bismuth carbonate (basic)
Bi$_2$O$_2$CO$_3$

71
Nickel(II) carbonate
NiCO$_3$ · xH$_2$O

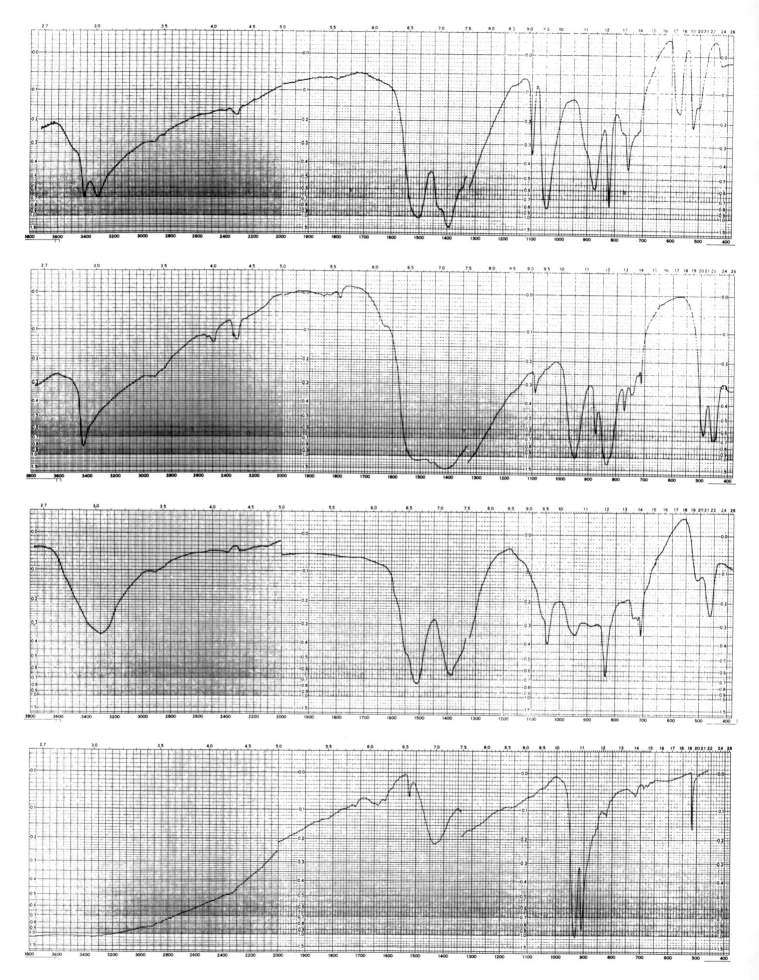

84

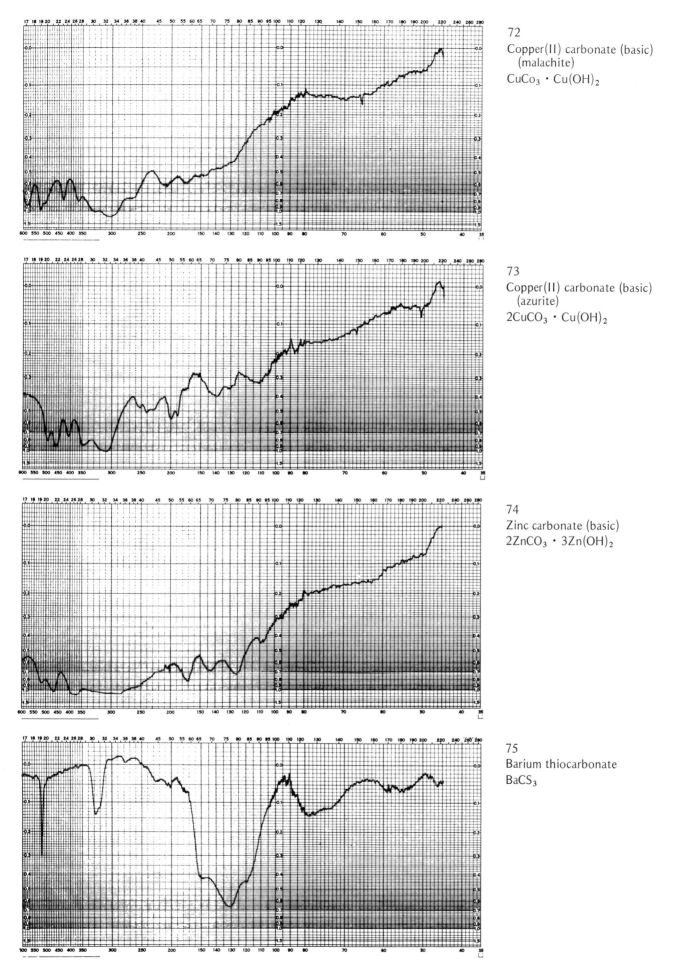

72
Copper(II) carbonate (basic)
(malachite)
$CuCo_3 \cdot Cu(OH)_2$

73
Copper(II) carbonate (basic)
(azurite)
$2CuCO_3 \cdot Cu(OH)_2$

74
Zinc carbonate (basic)
$2ZnCO_3 \cdot 3Zn(OH)_2$

75
Barium thiocarbonate
$BaCS_3$

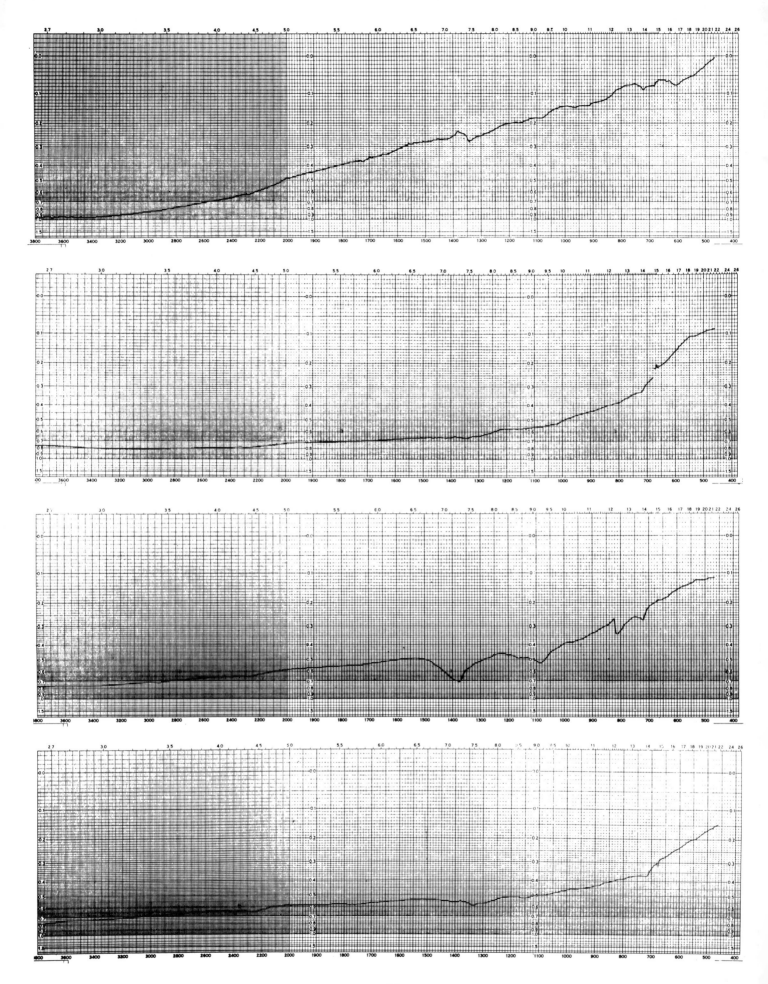

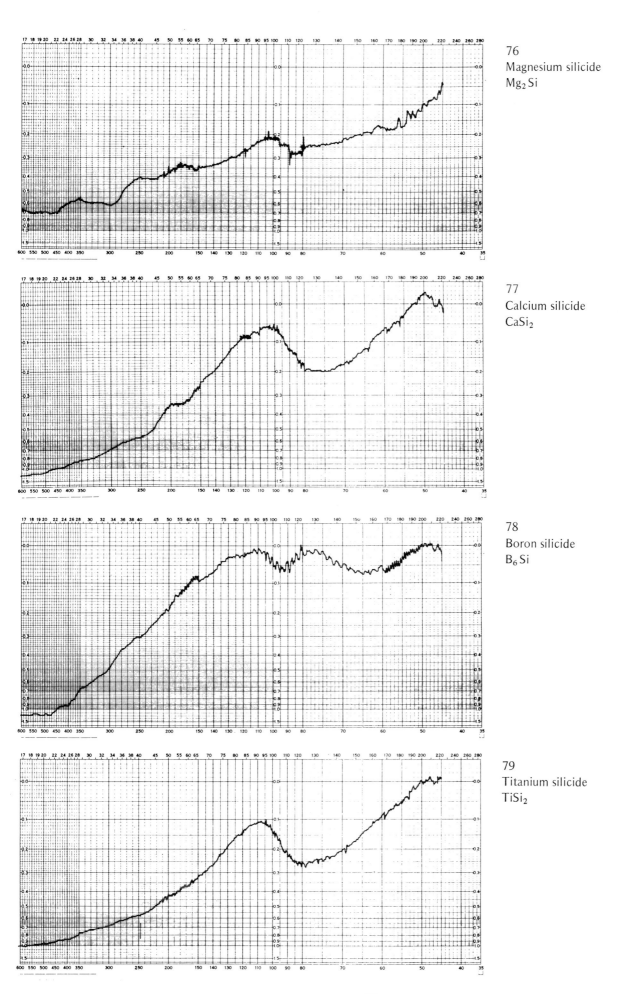

76
Magnesium silicide
Mg_2Si

77
Calcium silicide
$CaSi_2$

78
Boron silicide
B_6Si

79
Titanium silicide
$TiSi_2$

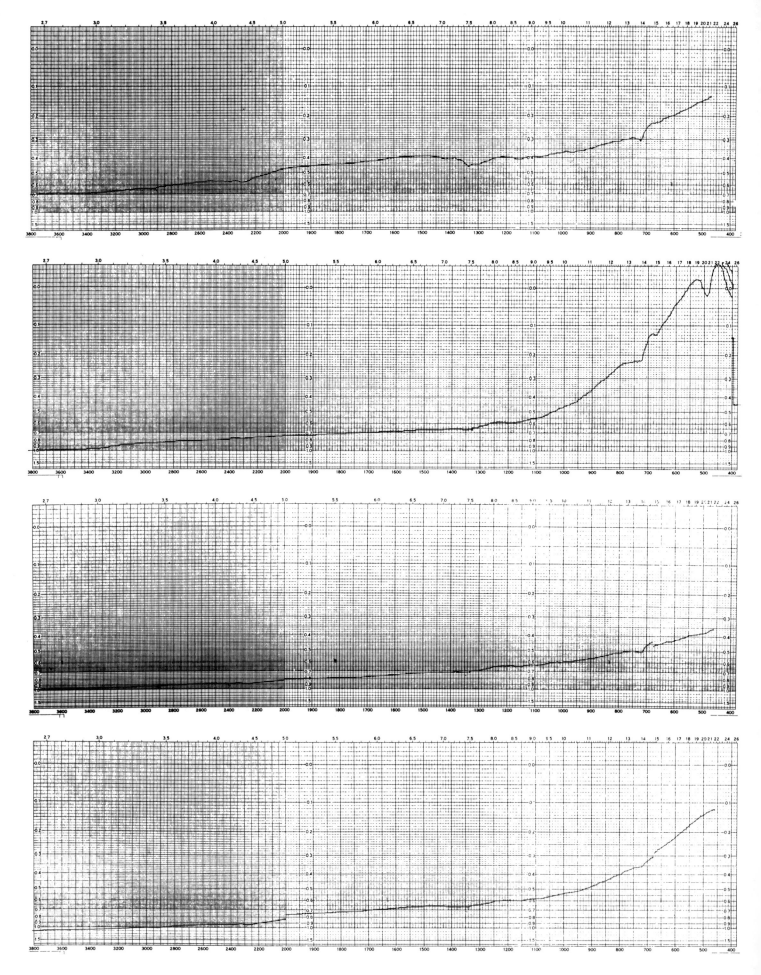

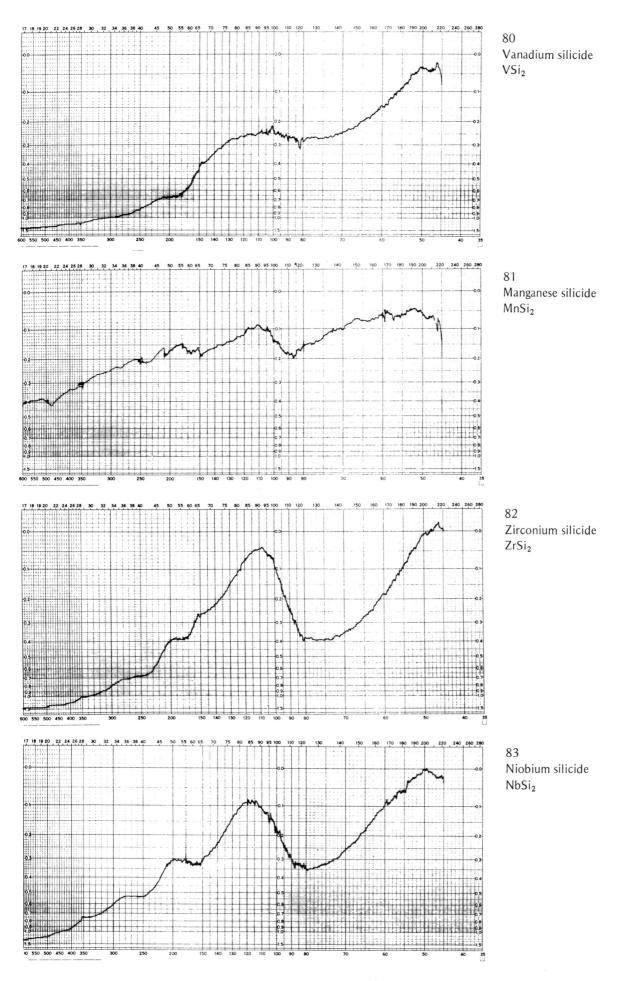

80
Vanadium silicide
VSi$_2$

81
Manganese silicide
MnSi$_2$

82
Zirconium silicide
ZrSi$_2$

83
Niobium silicide
NbSi$_2$

89

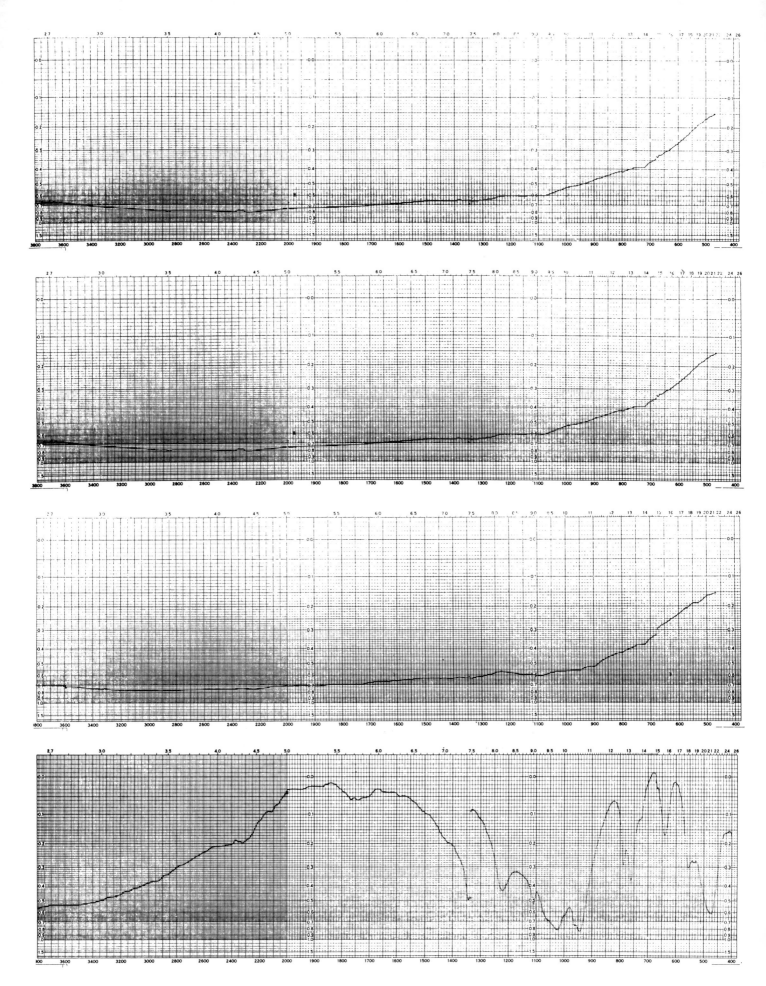

90

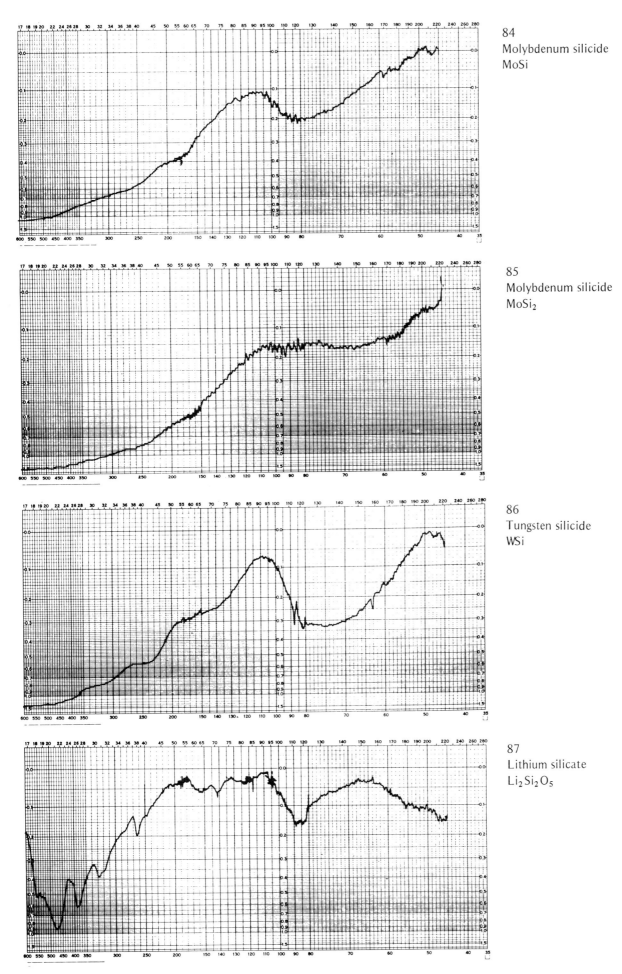

84
Molybdenum silicide
MoSi

85
Molybdenum silicide
$MoSi_2$

86
Tungsten silicide
WSi

87
Lithium silicate
$Li_2Si_2O_5$

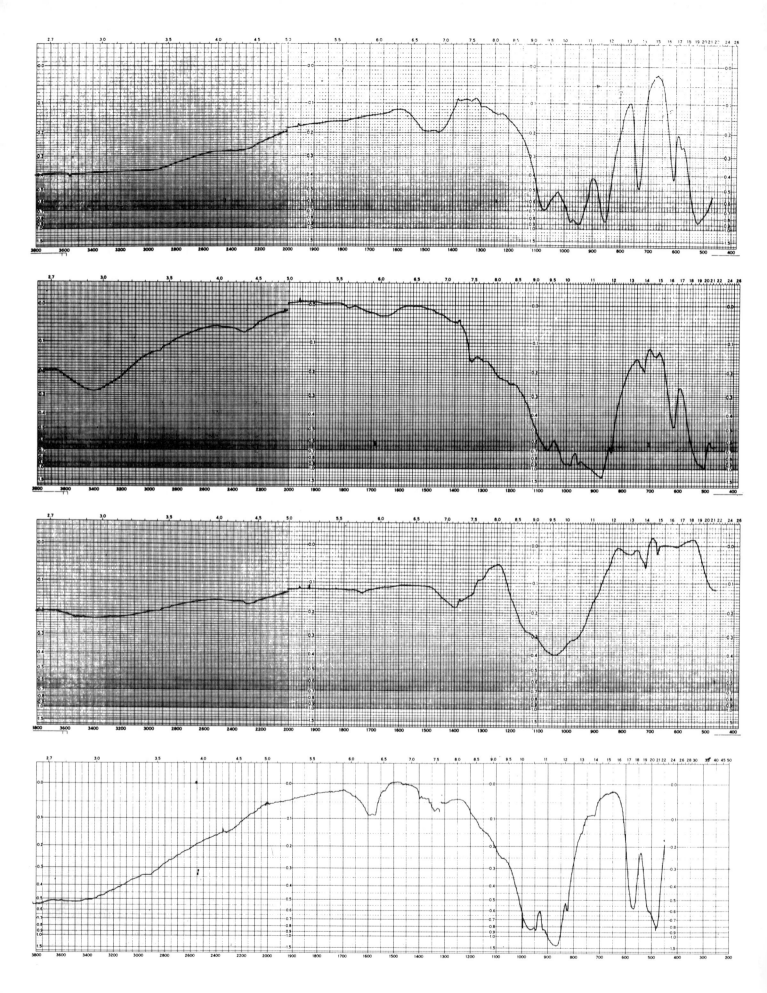

92

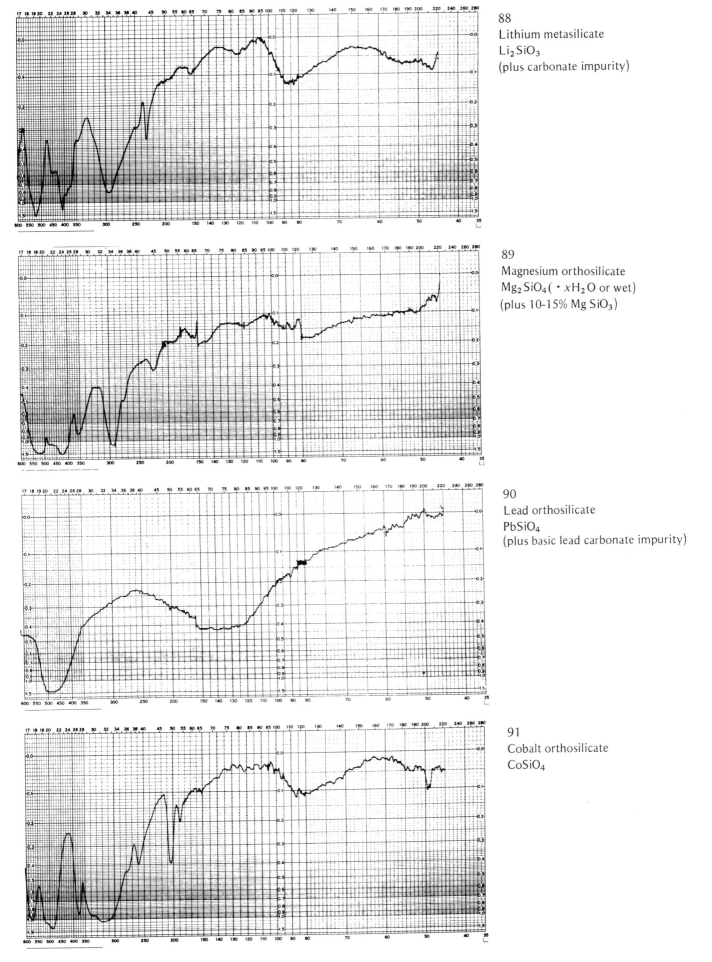

88
Lithium metasilicate
Li$_2$SiO$_3$
(plus carbonate impurity)

89
Magnesium orthosilicate
Mg$_2$SiO$_4$ ($\cdot$ xH$_2$O or wet)
(plus 10-15% Mg SiO$_3$)

90
Lead orthosilicate
PbSiO$_4$
(plus basic lead carbonate impurity)

91
Cobalt orthosilicate
CoSiO$_4$

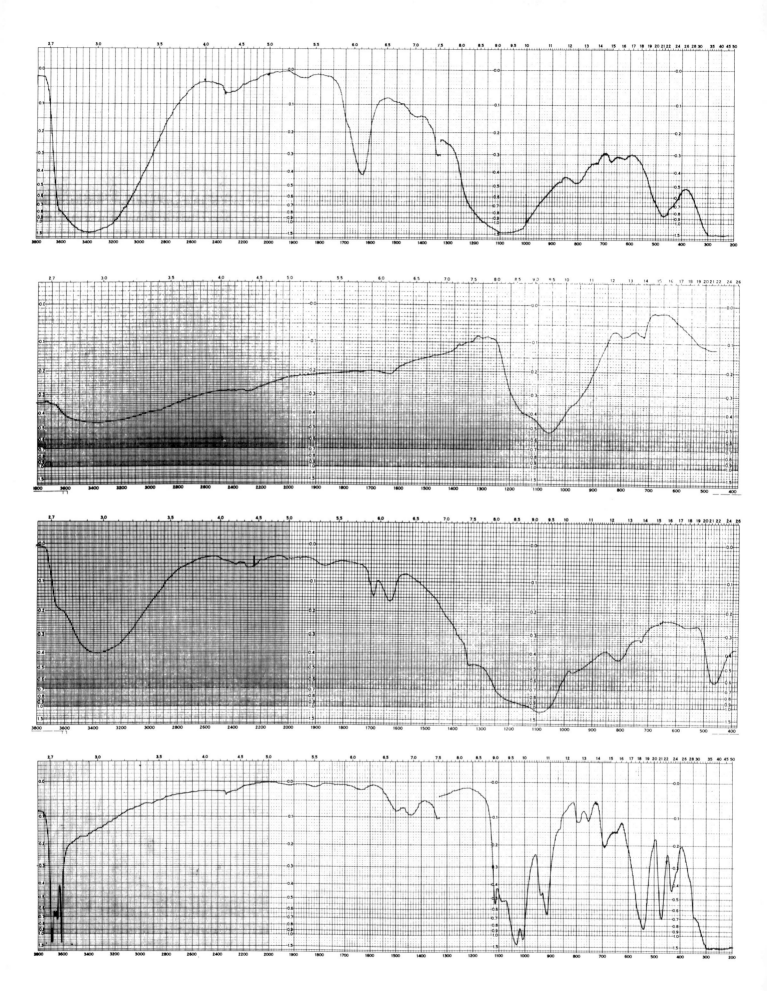

94

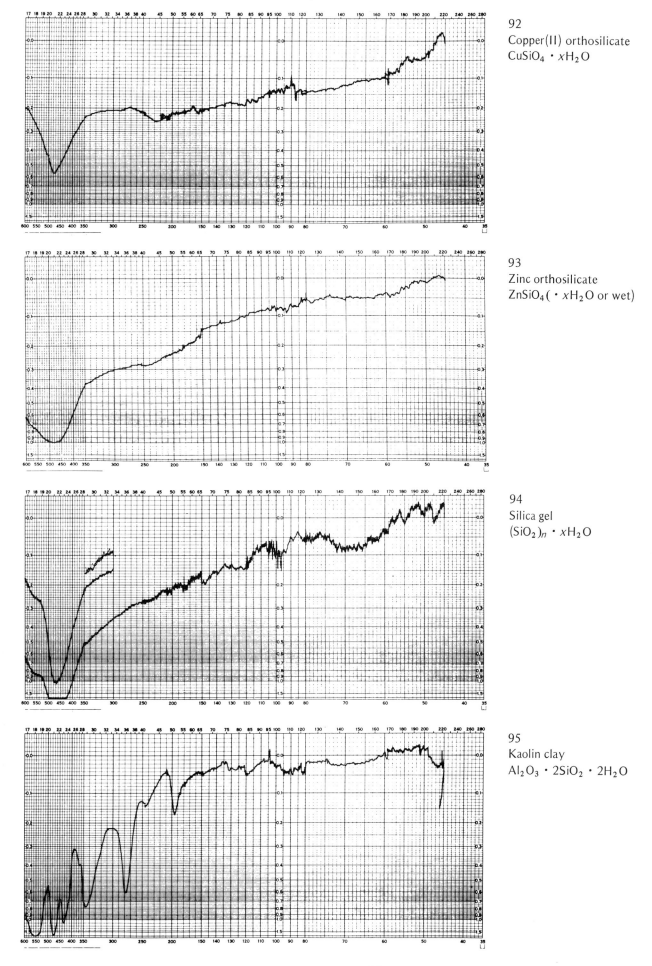

92
Copper(II) orthosilicate
$CuSiO_4 \cdot xH_2O$

93
Zinc orthosilicate
$ZnSiO_4 (\cdot xH_2O$ or wet)

94
Silica gel
$(SiO_2)_n \cdot xH_2O$

95
Kaolin clay
$Al_2O_3 \cdot 2SiO_2 \cdot 2H_2O$

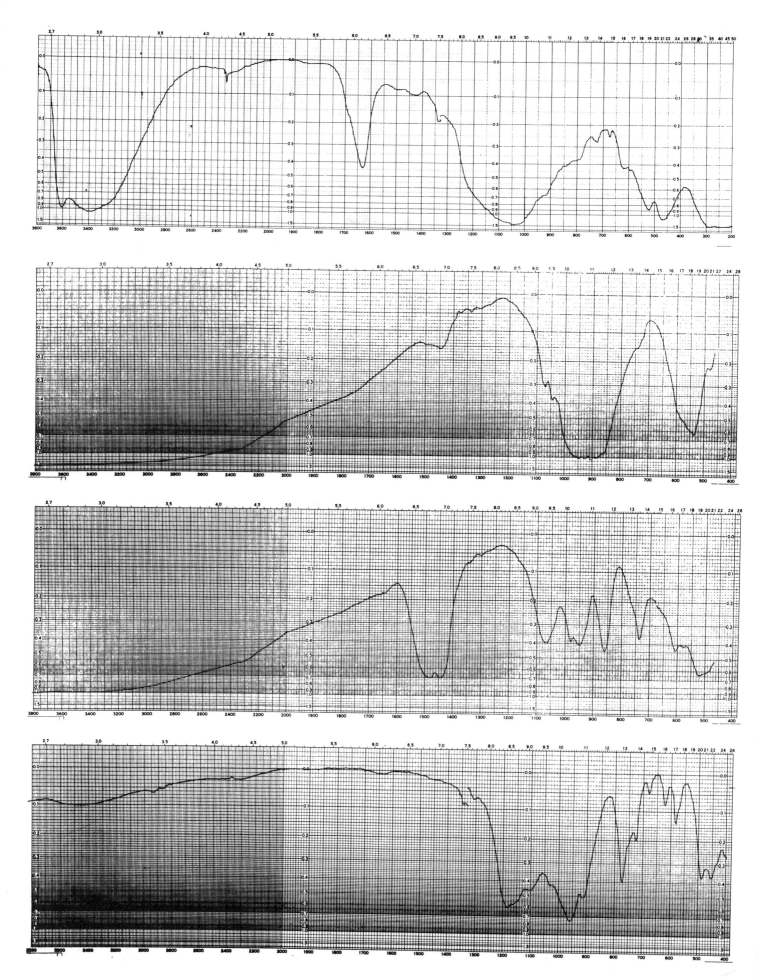

96

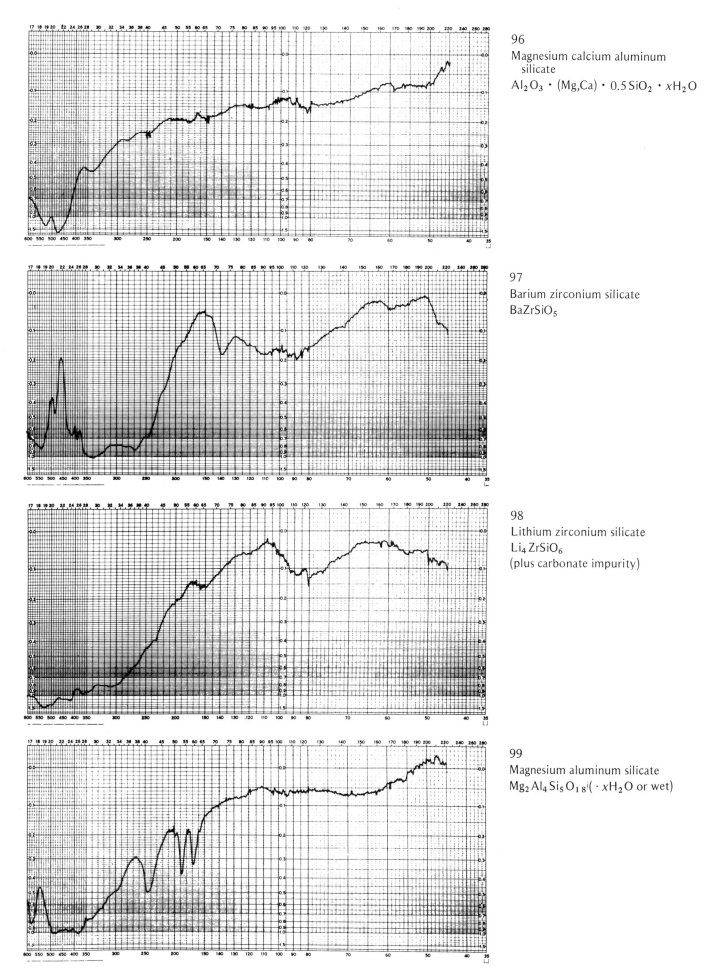

96
Magnesium calcium aluminum
  silicate
$Al_2O_3 \cdot (Mg,Ca) \cdot 0.5\,SiO_2 \cdot xH_2O$

97
Barium zirconium silicate
$BaZrSiO_5$

98
Lithium zirconium silicate
Li_4ZrSiO_6
(plus carbonate impurity)

99
Magnesium aluminum silicate
$Mg_2Al_4Si_5O_{18}(\cdot\,xH_2O$ or wet)

97

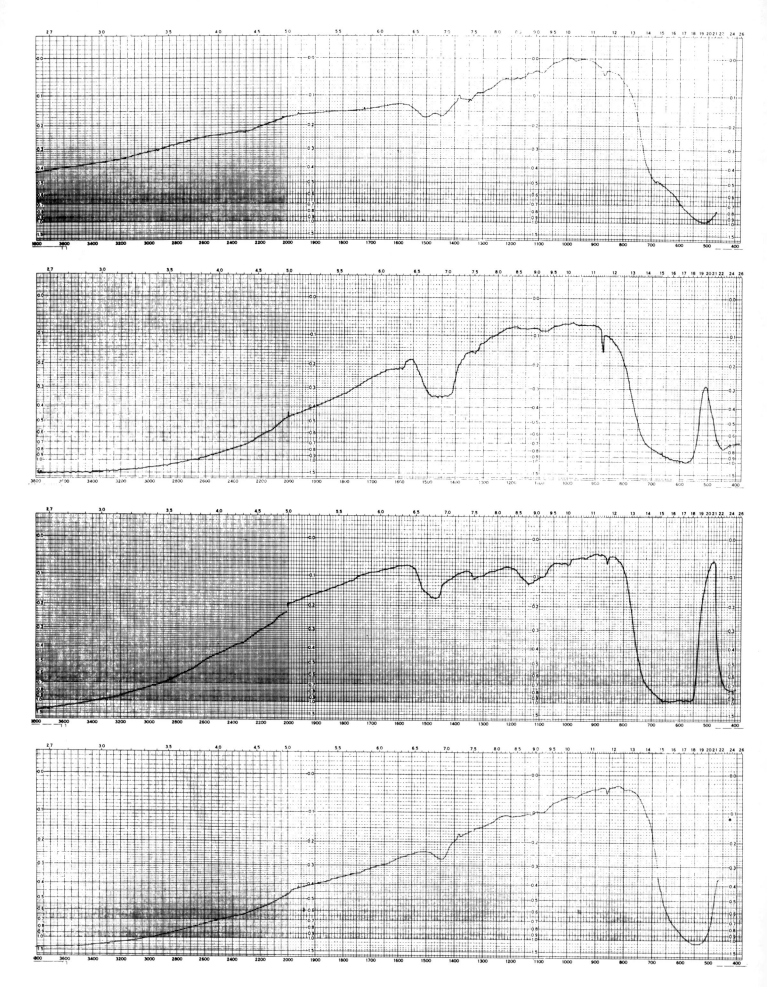

98

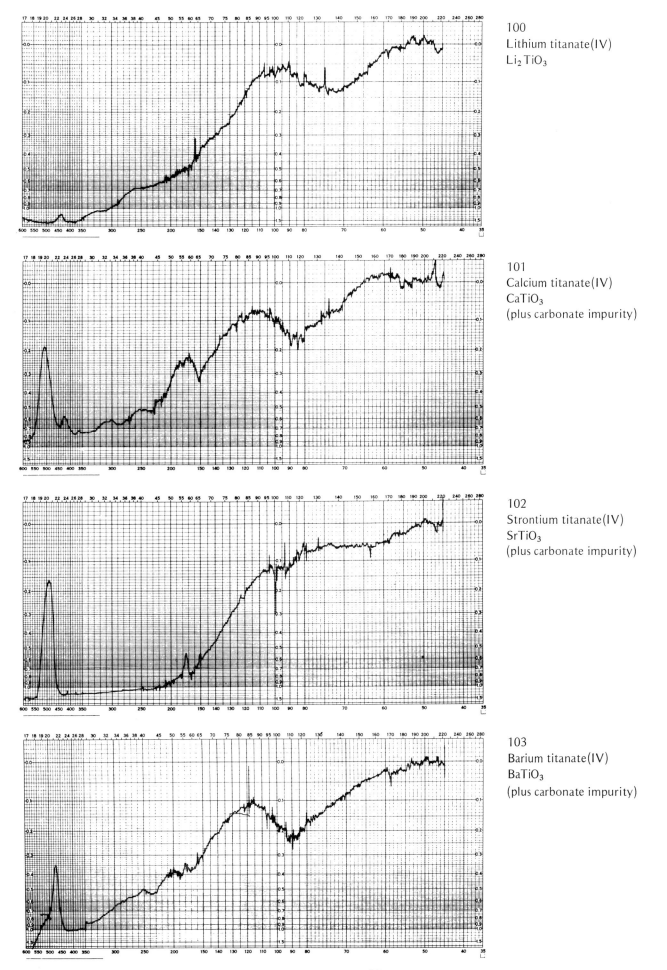

100
Lithium titanate(IV)
Li_2TiO_3

101
Calcium titanate(IV)
$CaTiO_3$
(plus carbonate impurity)

102
Strontium titanate(IV)
$SrTiO_3$
(plus carbonate impurity)

103
Barium titanate(IV)
$BaTiO_3$
(plus carbonate impurity)

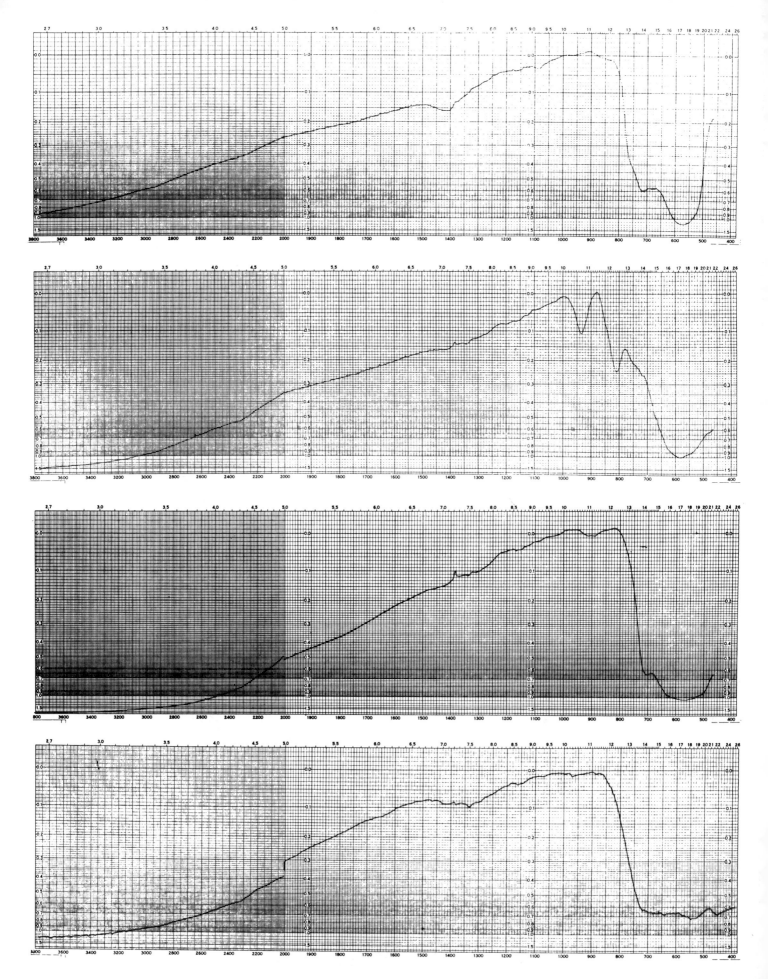

100

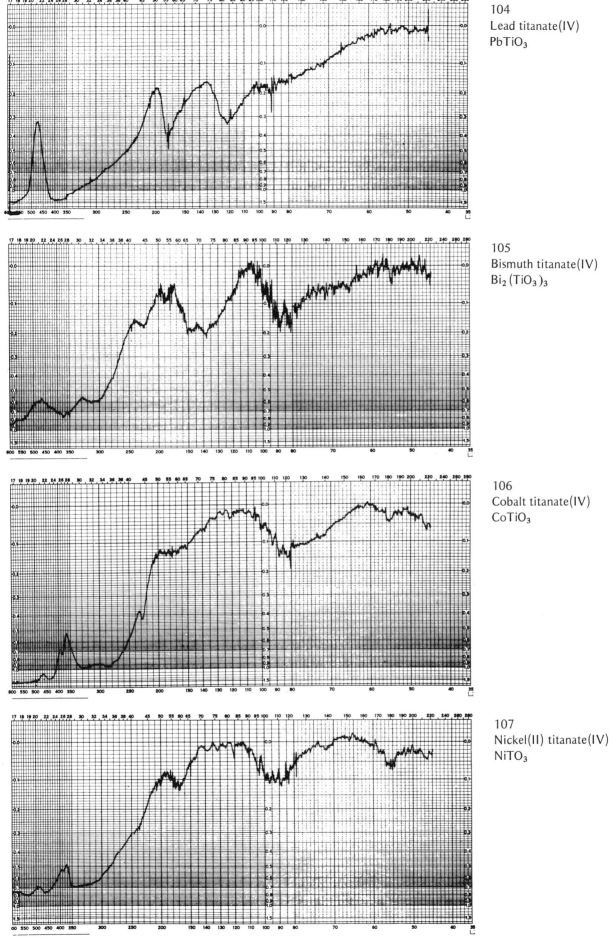

104
Lead titanate(IV)
PbTiO$_3$

105
Bismuth titanate(IV)
Bi$_2$(TiO$_3$)$_3$

106
Cobalt titanate(IV)
CoTiO$_3$

107
Nickel(II) titanate(IV)
NiTO$_3$

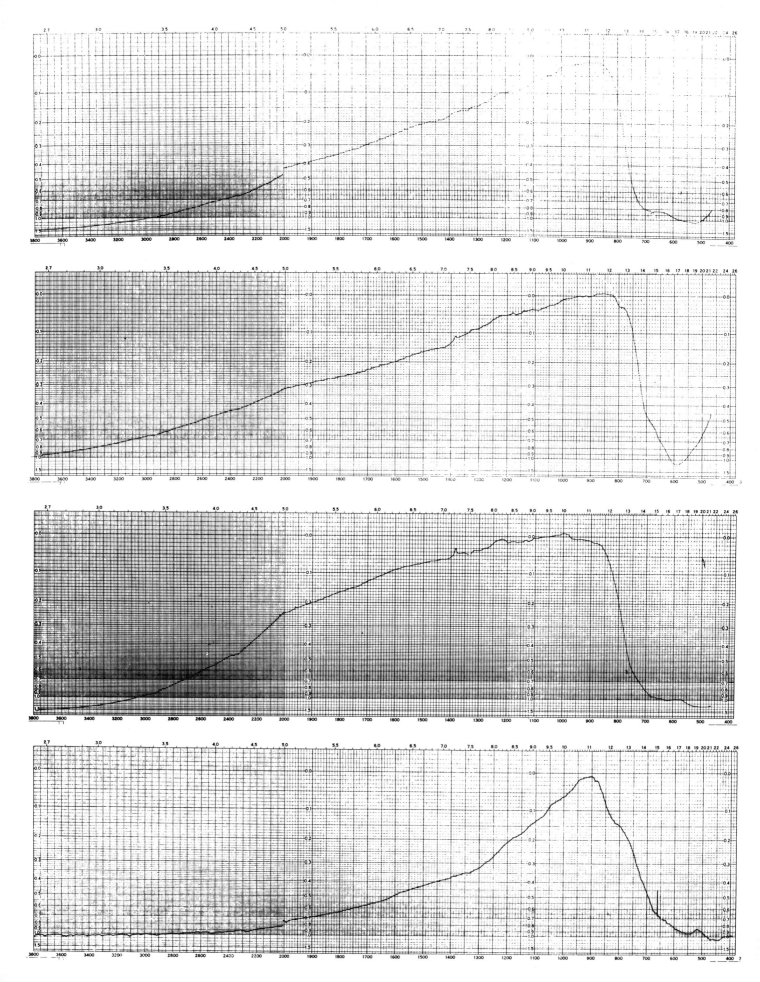

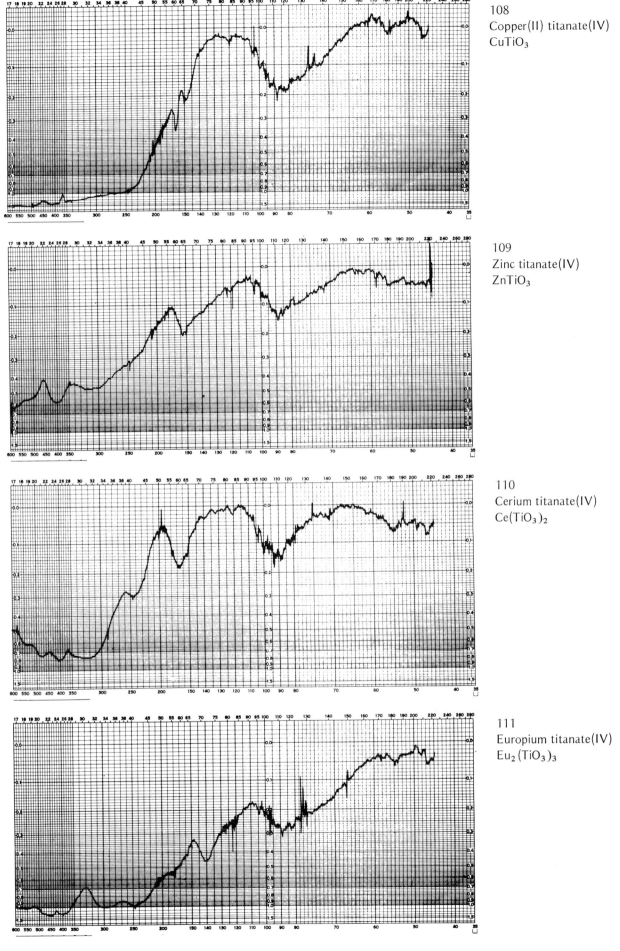

108
Copper(II) titanate(IV)
$CuTiO_3$

109
Zinc titanate(IV)
$ZnTiO_3$

110
Cerium titanate(IV)
$Ce(TiO_3)_2$

111
Europium titanate(IV)
$Eu_2(TiO_3)_3$

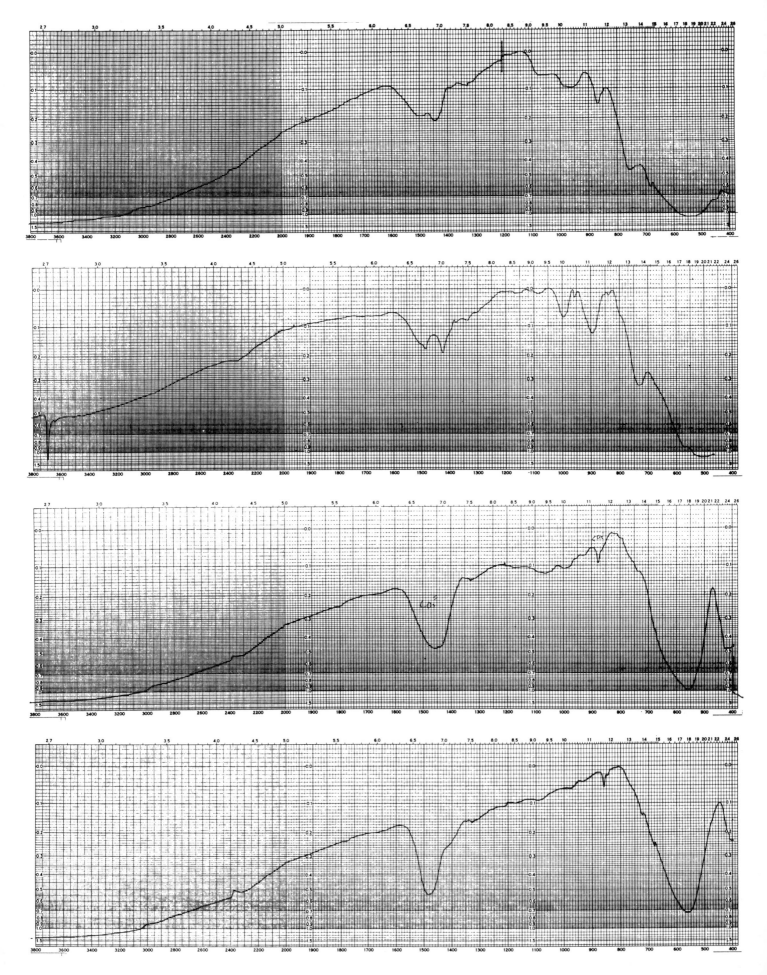

104

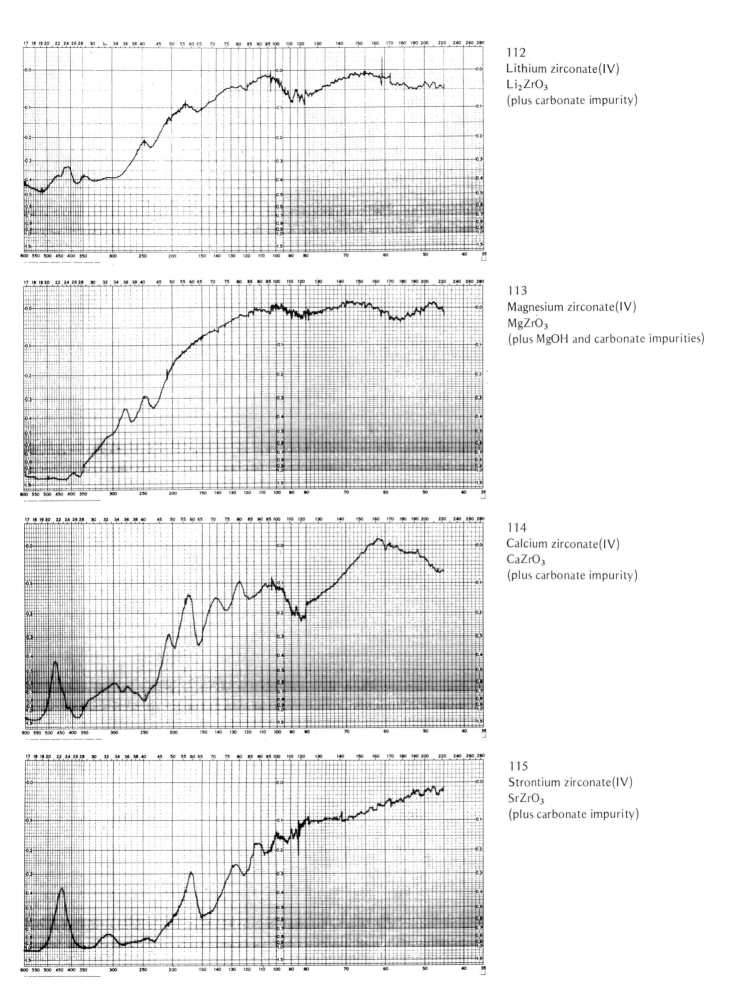

112
Lithium zirconate(IV)
Li$_2$ZrO$_3$
(plus carbonate impurity)

113
Magnesium zirconate(IV)
MgZrO$_3$
(plus MgOH and carbonate impurities)

114
Calcium zirconate(IV)
CaZrO$_3$
(plus carbonate impurity)

115
Strontium zirconate(IV)
SrZrO$_3$
(plus carbonate impurity)

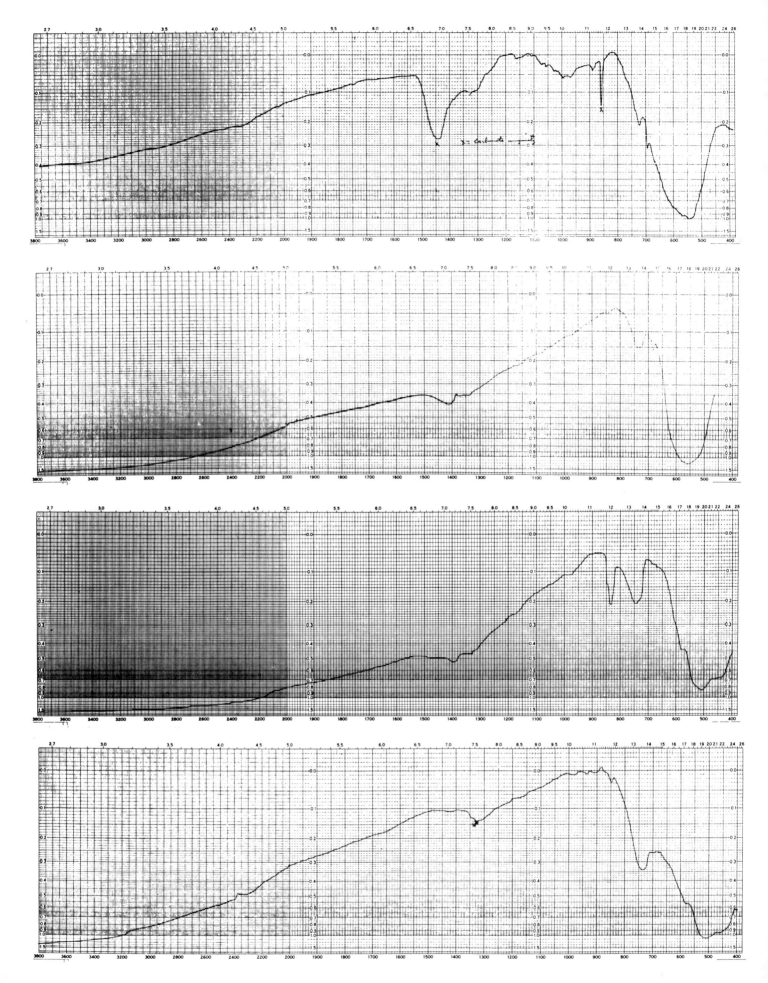

106

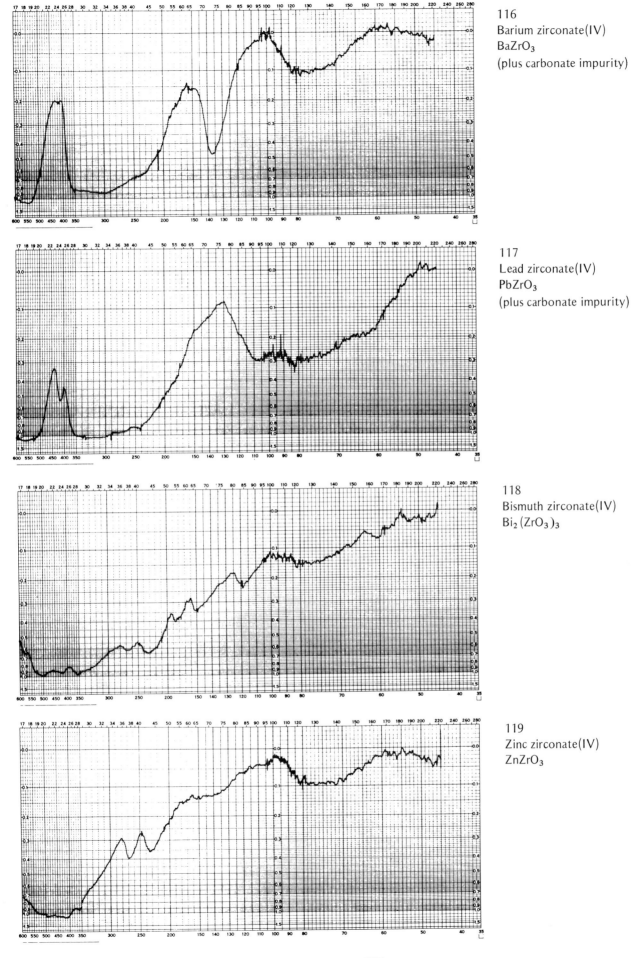

116
Barium zirconate(IV)
BaZrO₃
(plus carbonate impurity)

117
Lead zirconate(IV)
PbZrO₃
(plus carbonate impurity)

118
Bismuth zirconate(IV)
Bi₂(ZrO₃)₃

119
Zinc zirconate(IV)
ZnZrO₃

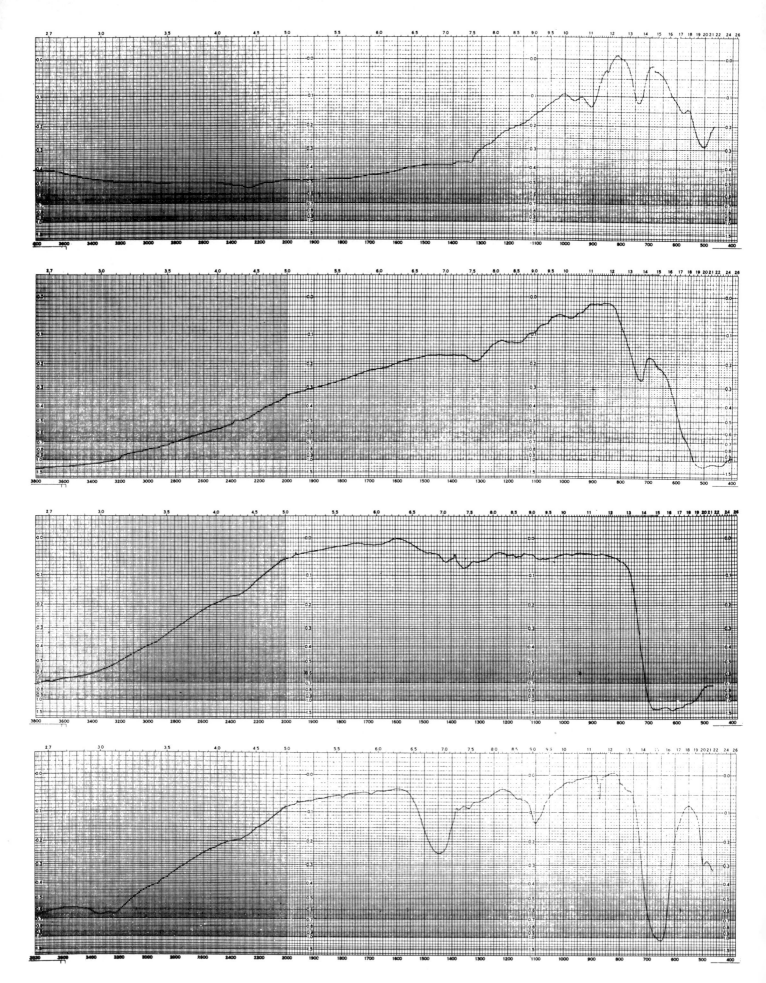

108

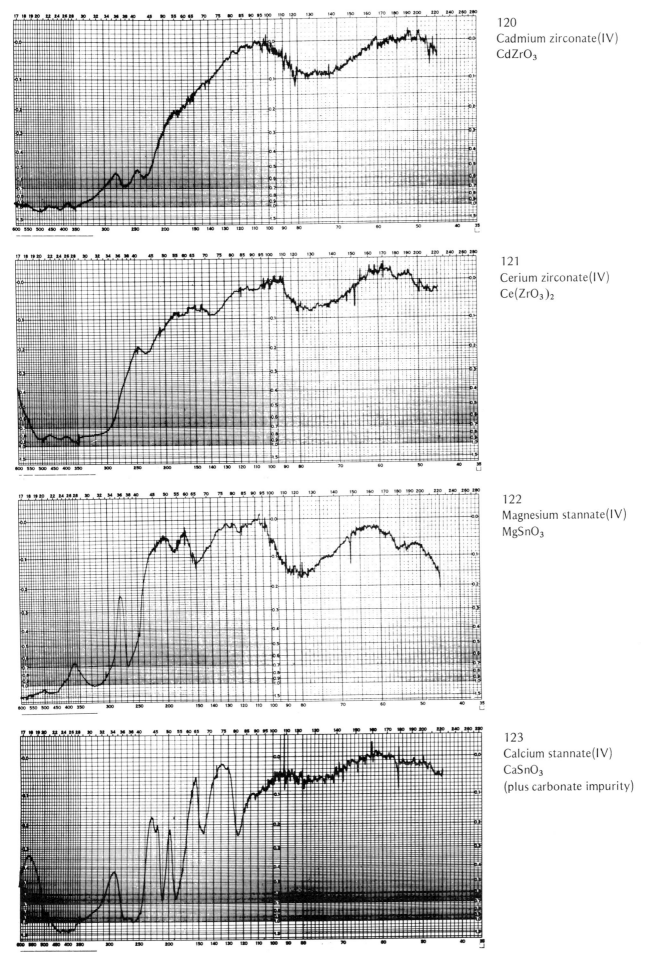

120
Cadmium zirconate(IV)
CdZrO₃

121
Cerium zirconate(IV)
Ce(ZrO₃)₂

122
Magnesium stannate(IV)
MgSnO₃

123
Calcium stannate(IV)
CaSnO₃
(plus carbonate impurity)

109

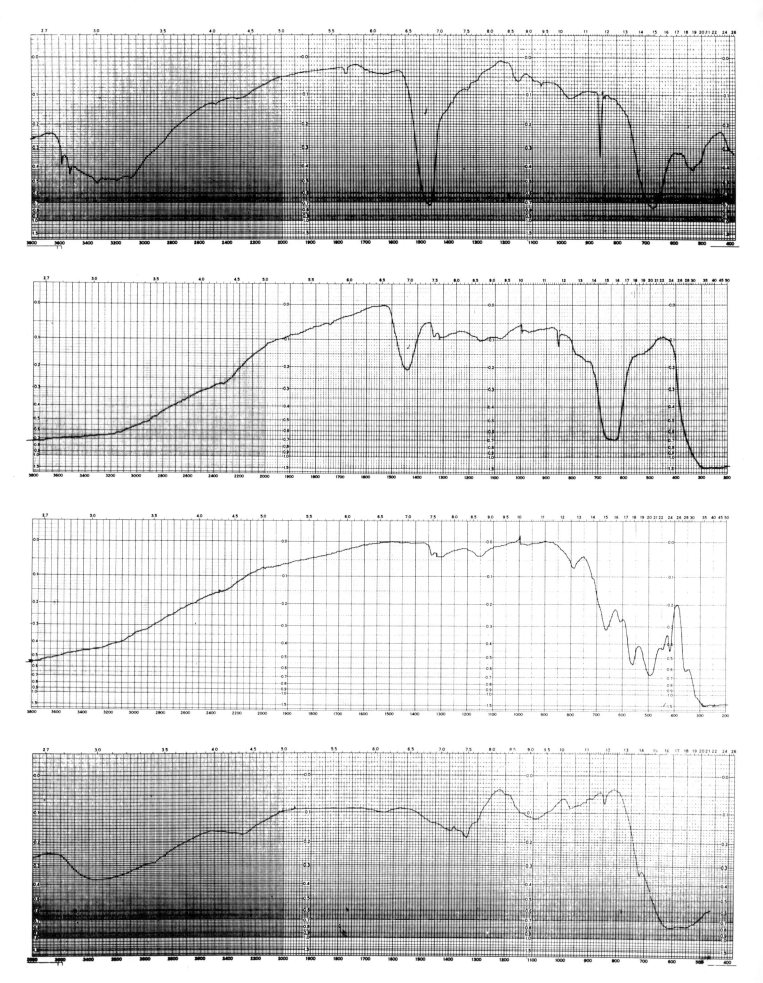

110

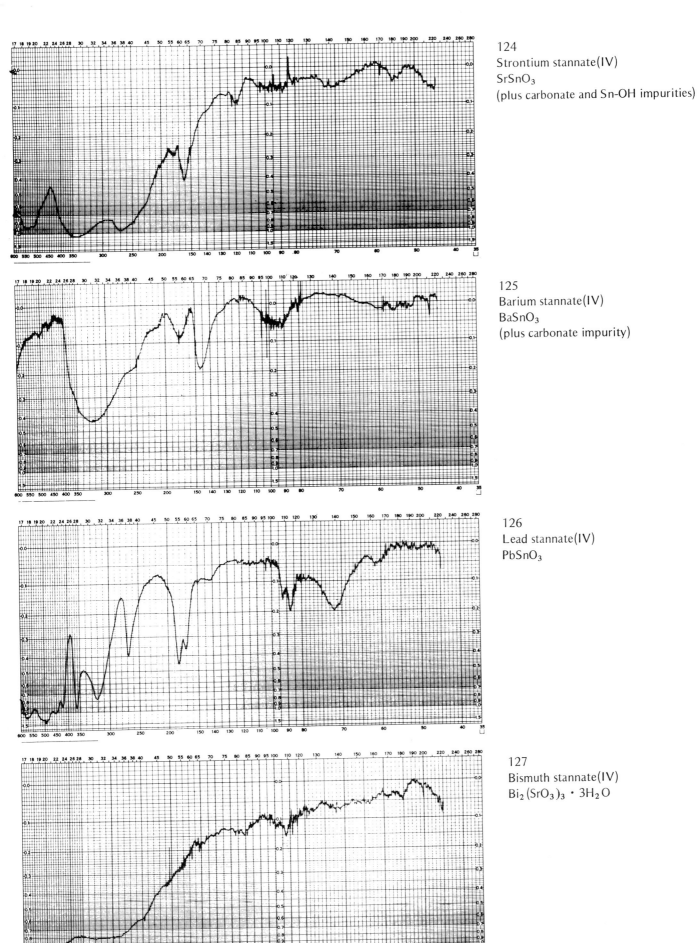

124
Strontium stannate(IV)
SrSnO₃
(plus carbonate and Sn-OH impurities)

125
Barium stannate(IV)
BaSnO₃
(plus carbonate impurity)

126
Lead stannate(IV)
PbSnO₃

127
Bismuth stannate(IV)
Bi₂(SrO₃)₃ · 3H₂O

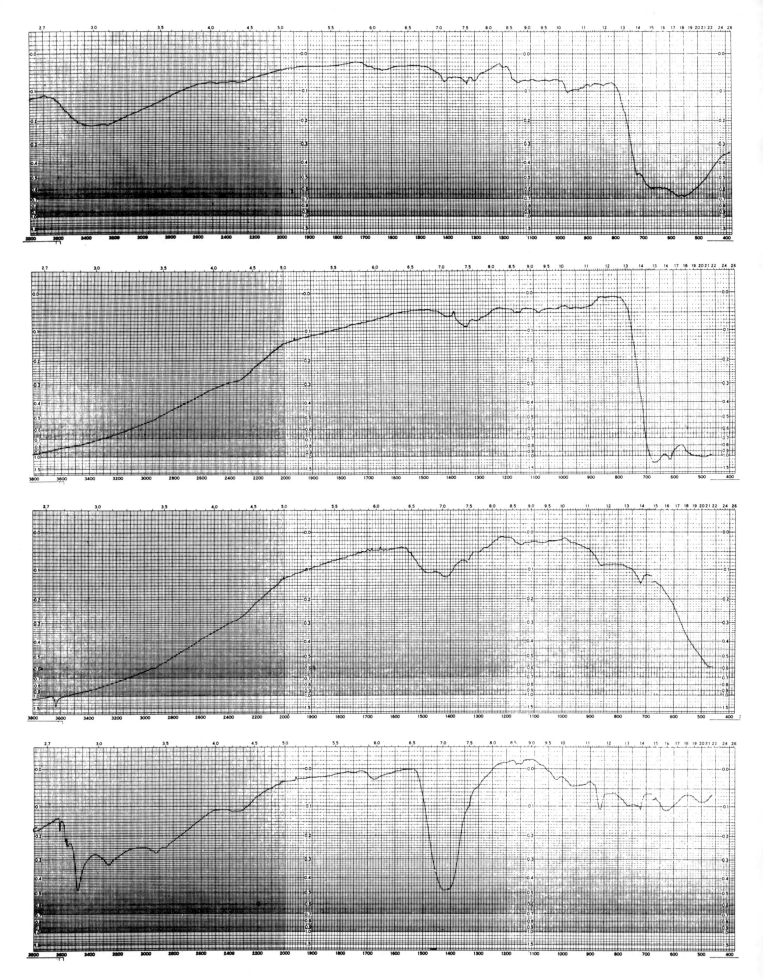

112

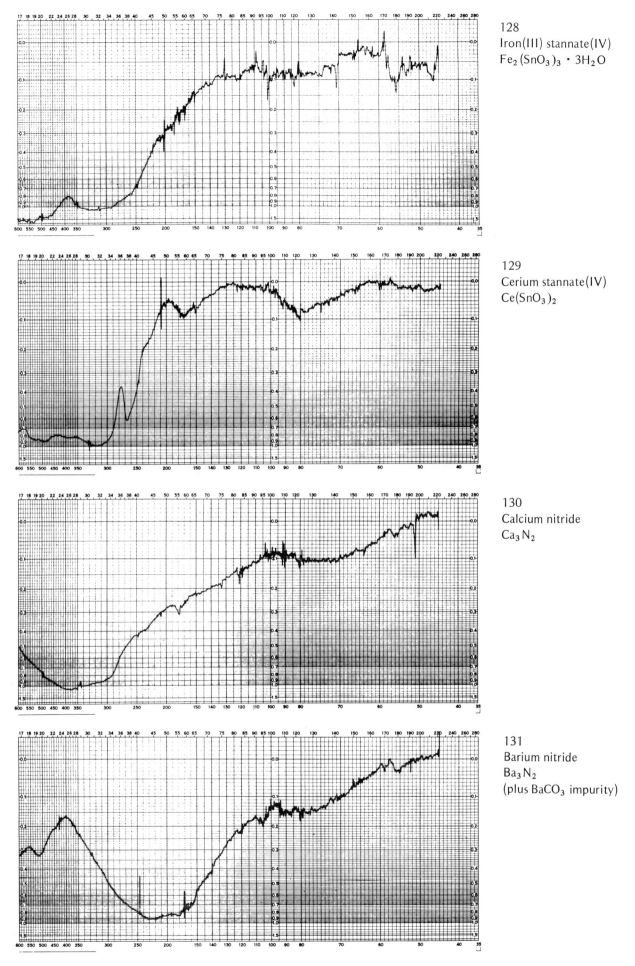

128
Iron(III) stannate(IV)
$Fe_2(SnO_3)_3 \cdot 3H_2O$

129
Cerium stannate(IV)
$Ce(SnO_3)_2$

130
Calcium nitride
Ca_3N_2

131
Barium nitride
Ba_3N_2
(plus $BaCO_3$ impurity)

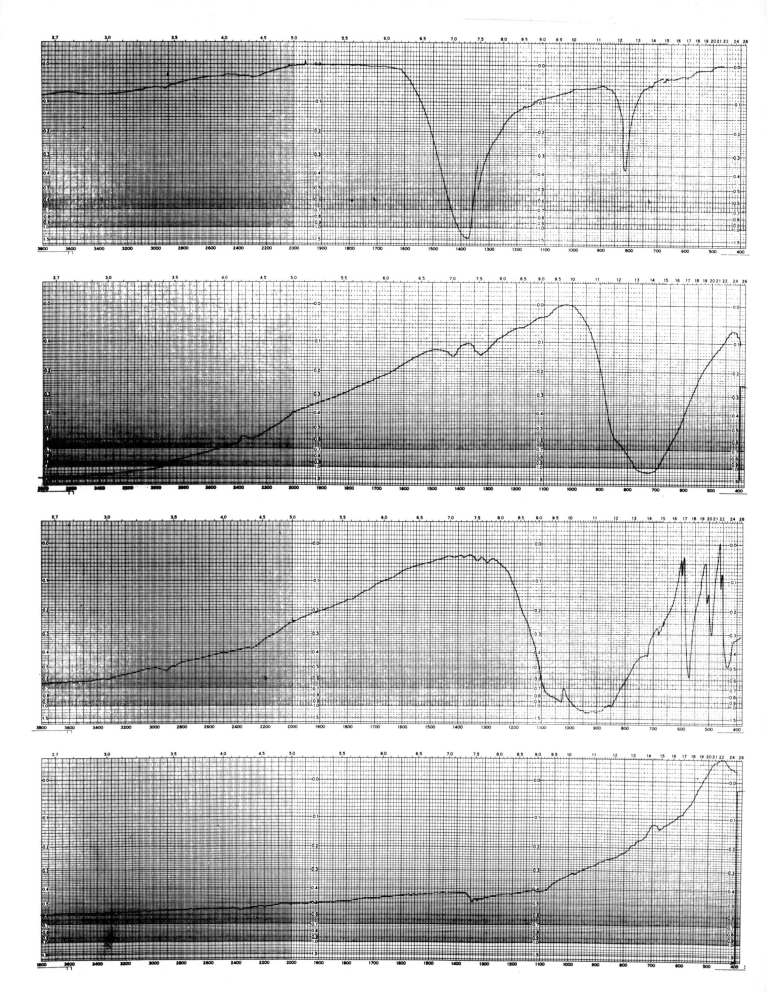

114

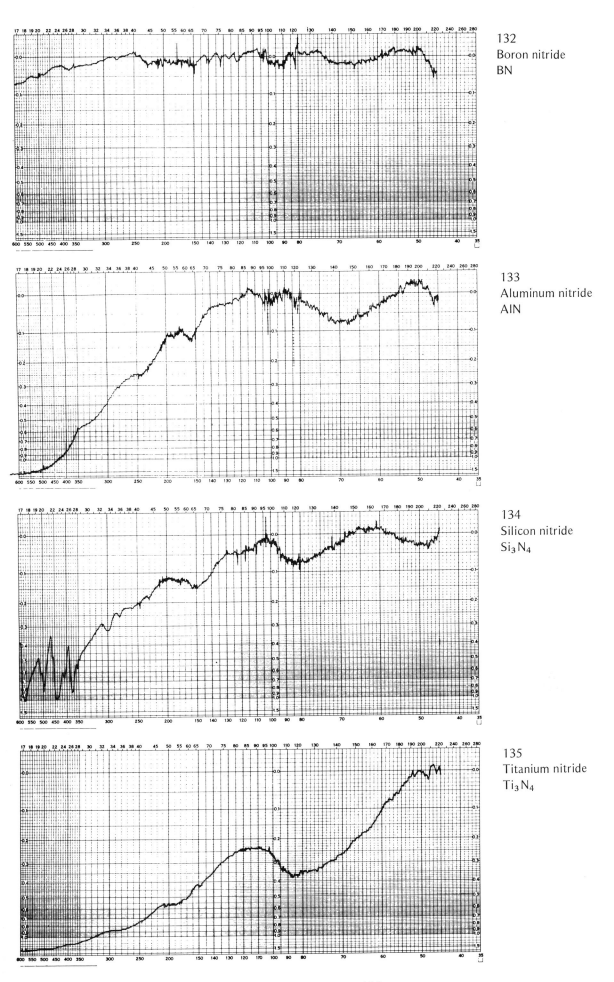

132
Boron nitride
BN

133
Aluminum nitride
AlN

134
Silicon nitride
Si_3N_4

135
Titanium nitride
Ti_3N_4

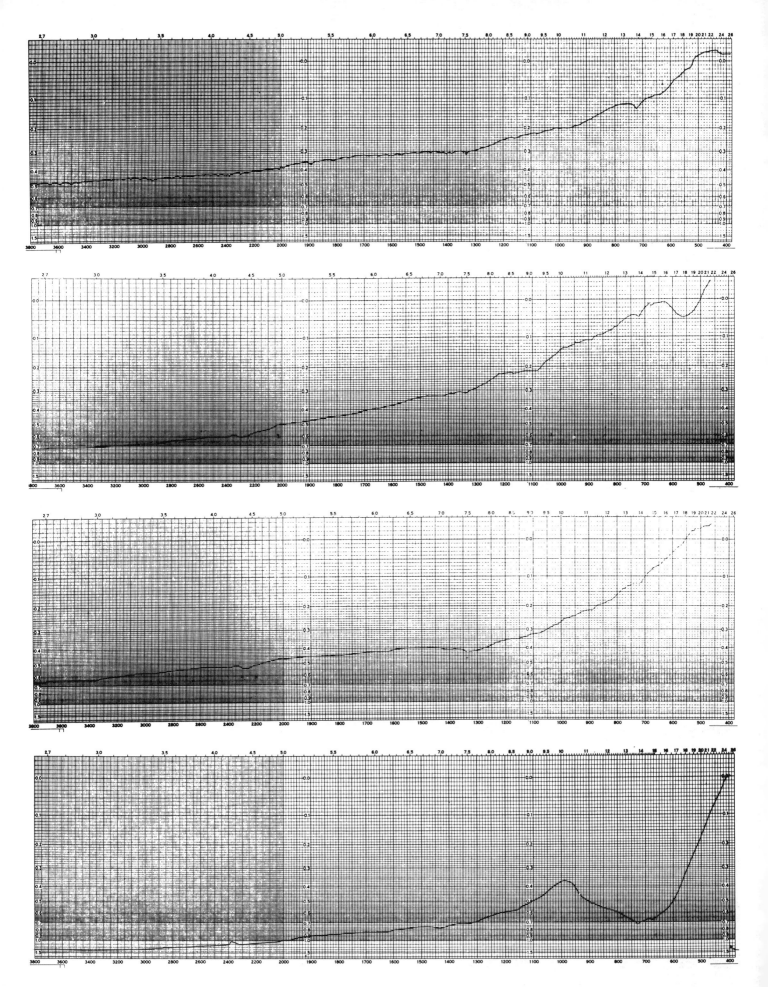

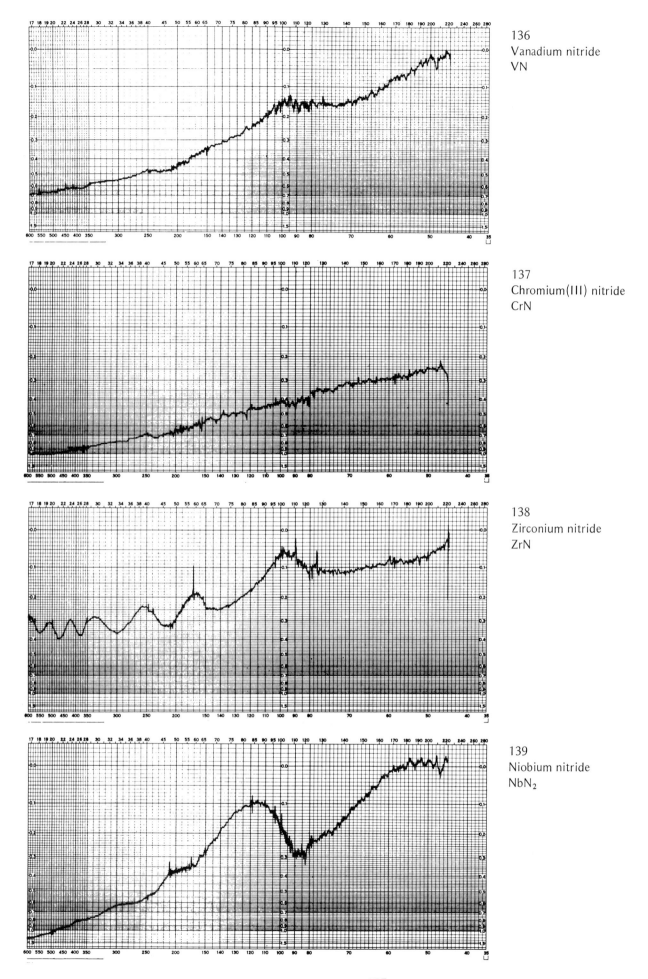

136
Vanadium nitride
VN

137
Chromium(III) nitride
CrN

138
Zirconium nitride
ZrN

139
Niobium nitride
NbN$_2$

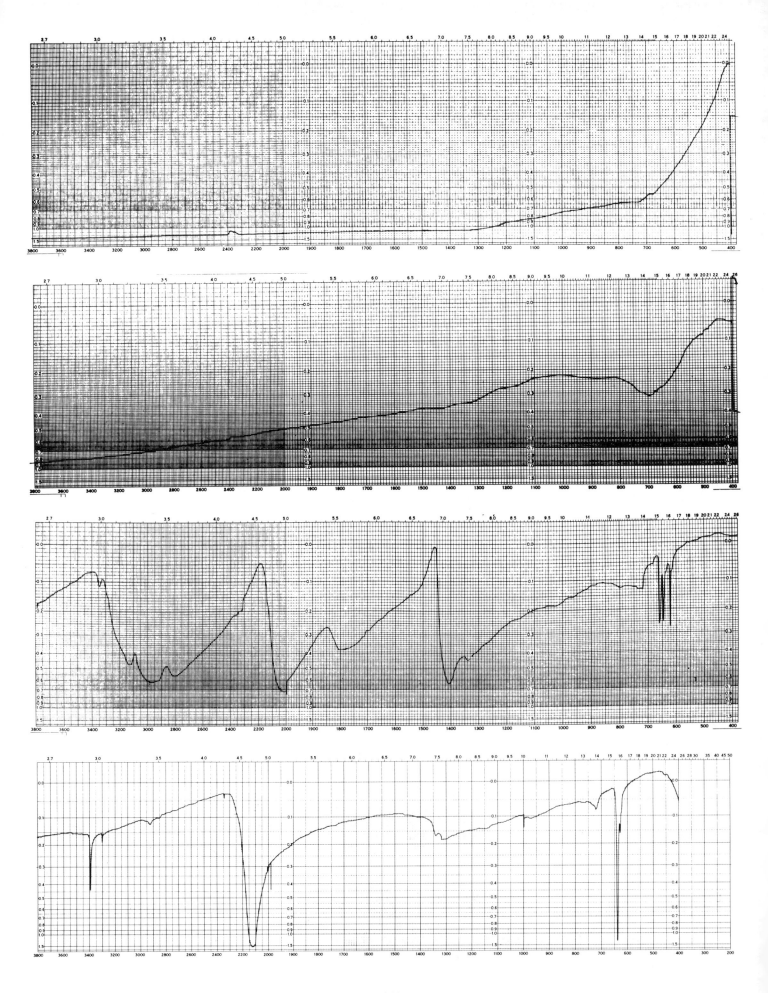

118

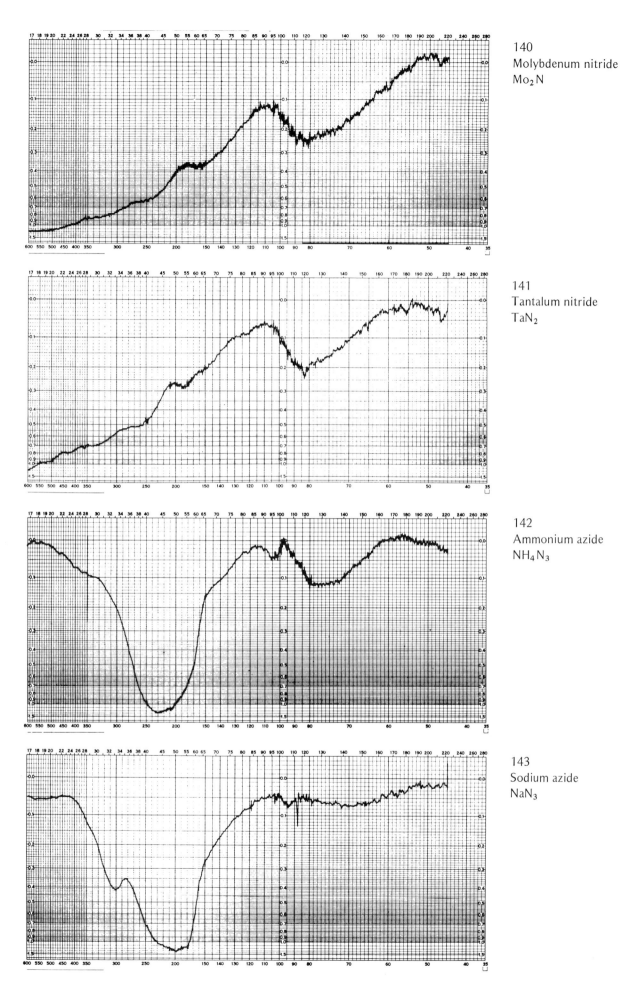

140
Molybdenum nitride
Mo_2N

141
Tantalum nitride
TaN_2

142
Ammonium azide
NH_4N_3

143
Sodium azide
NaN_3

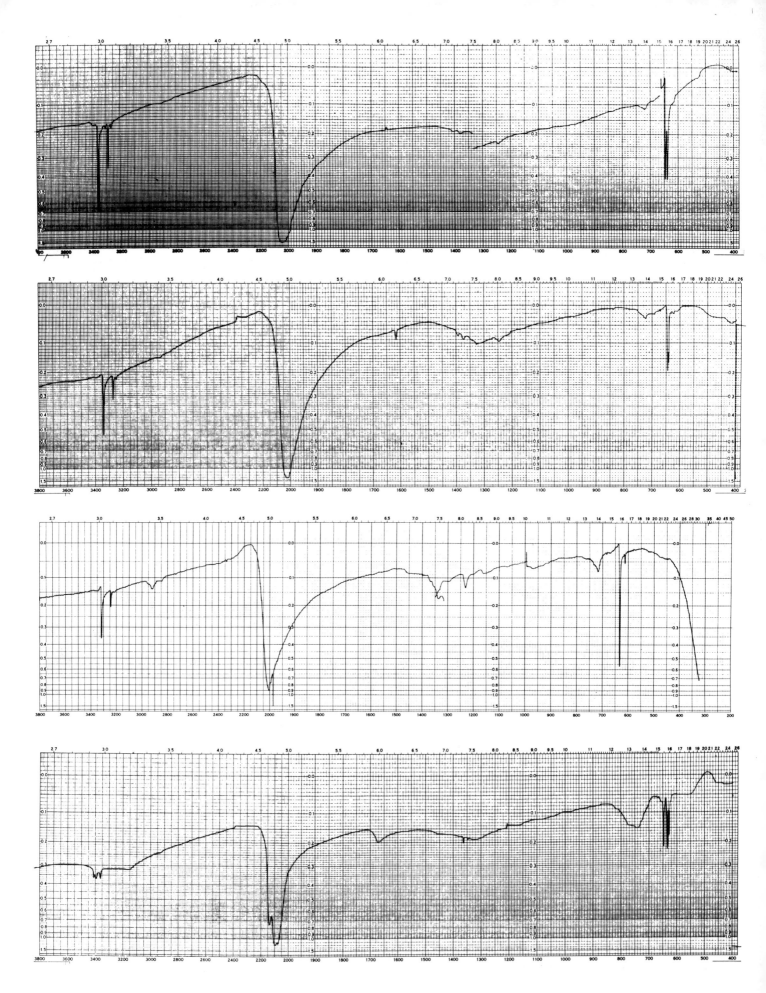

120

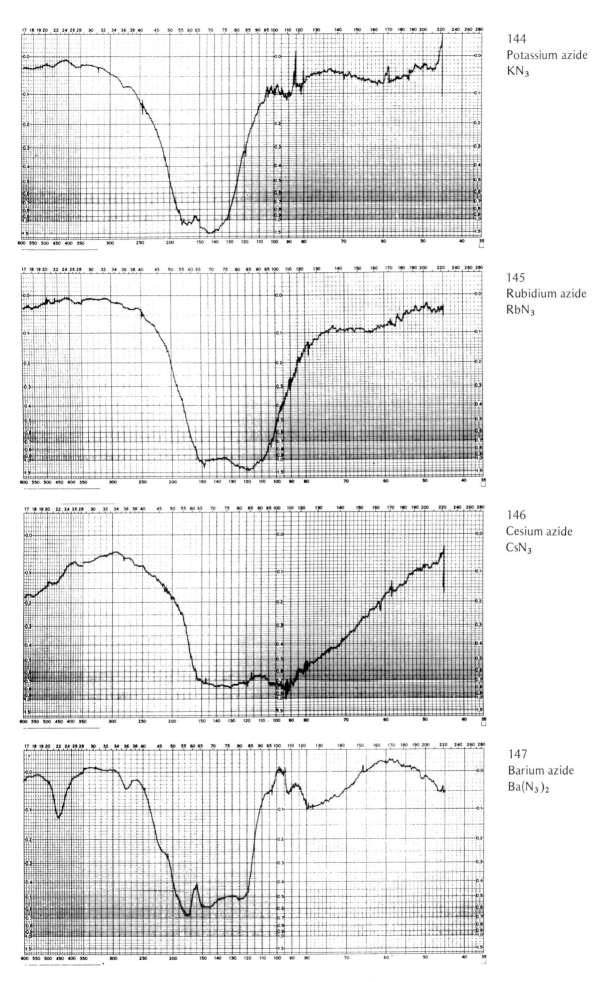

144
Potassium azide
KN$_3$

145
Rubidium azide
RbN$_3$

146
Cesium azide
CsN$_3$

147
Barium azide
Ba(N$_3$)$_2$

121

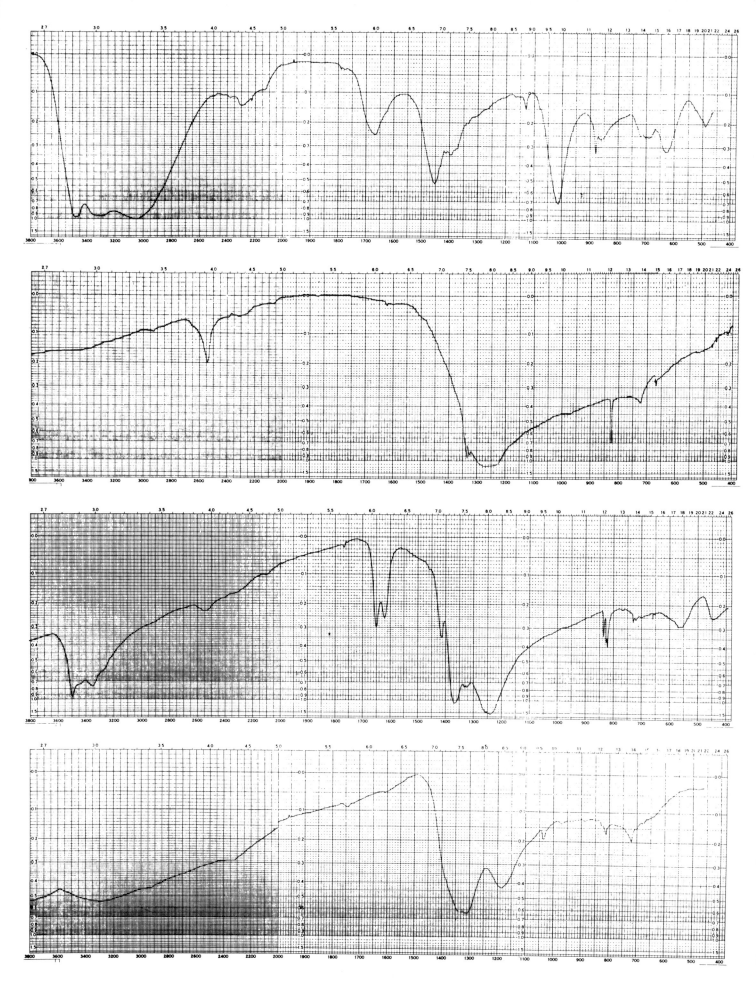

122

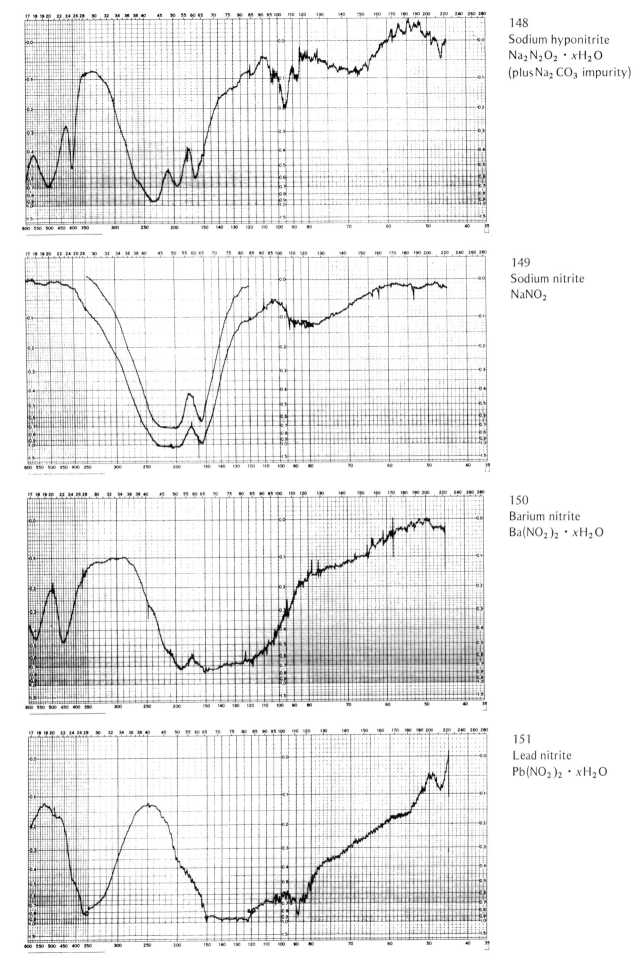

148
Sodium hyponitrite
Na$_2$N$_2$O$_2$ · xH$_2$O
(plus Na$_2$CO$_3$ impurity)

149
Sodium nitrite
NaNO$_2$

150
Barium nitrite
Ba(NO$_2$)$_2$ · xH$_2$O

151
Lead nitrite
Pb(NO$_2$)$_2$ · xH$_2$O

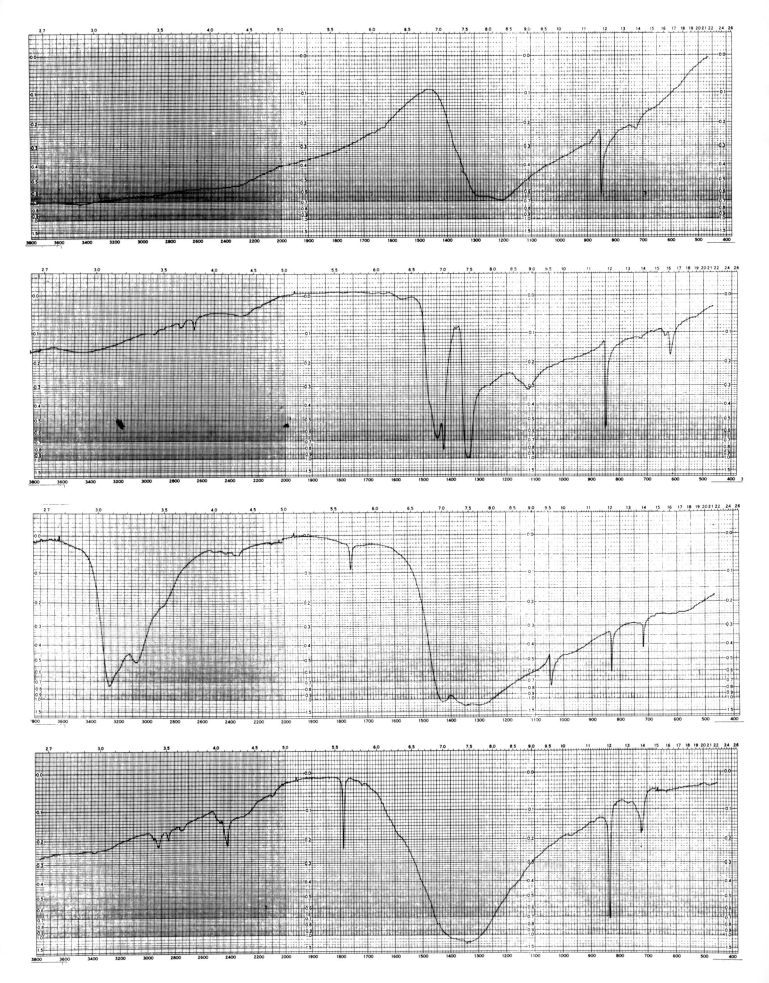

124

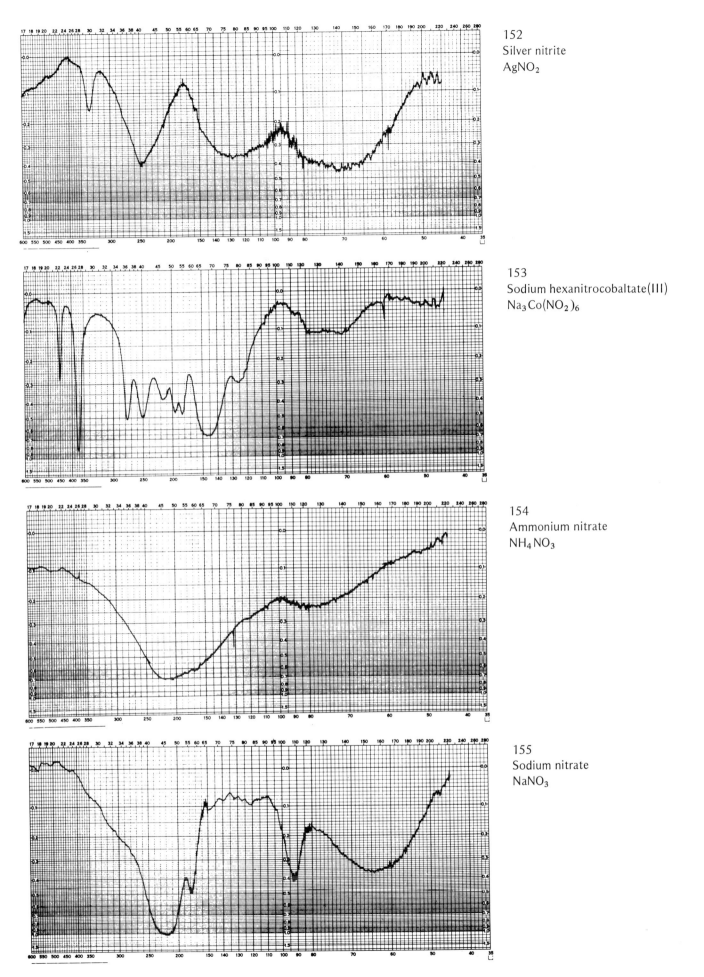

152
Silver nitrite
AgNO$_2$

153
Sodium hexanitrocobaltate(III)
Na$_3$Co(NO$_2$)$_6$

154
Ammonium nitrate
NH$_4$NO$_3$

155
Sodium nitrate
NaNO$_3$

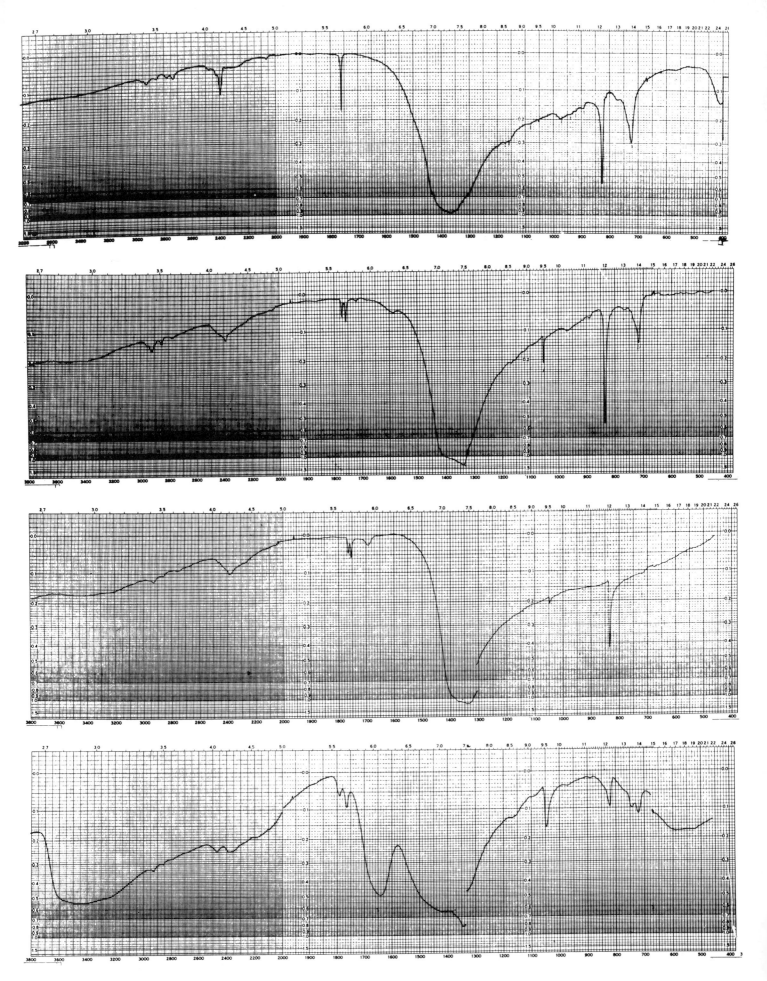

126

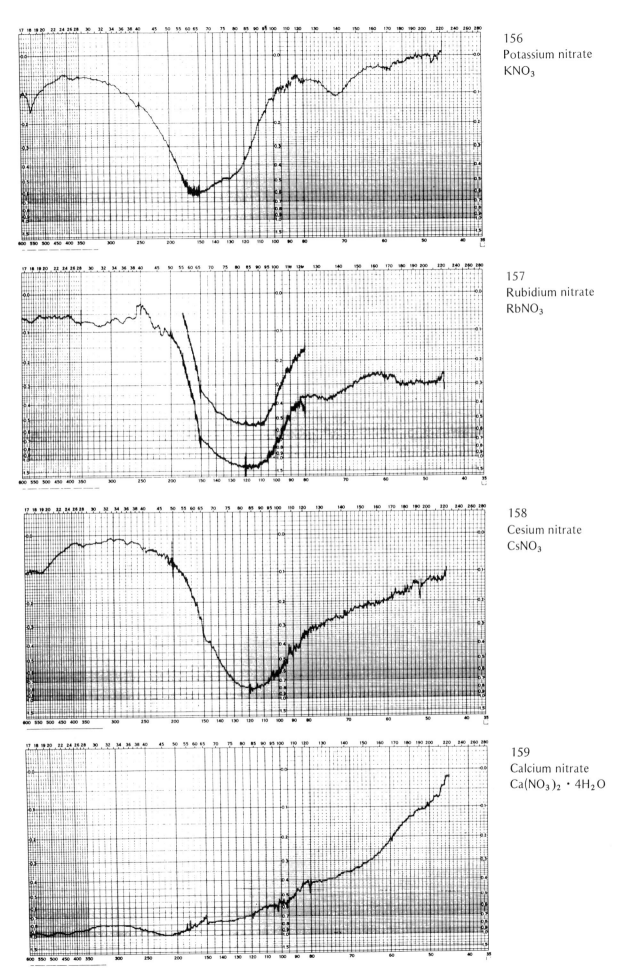

156
Potassium nitrate
KNO₃

157
Rubidium nitrate
RbNO₃

158
Cesium nitrate
CsNO₃

159
Calcium nitrate
Ca(NO₃)₂ · 4H₂O

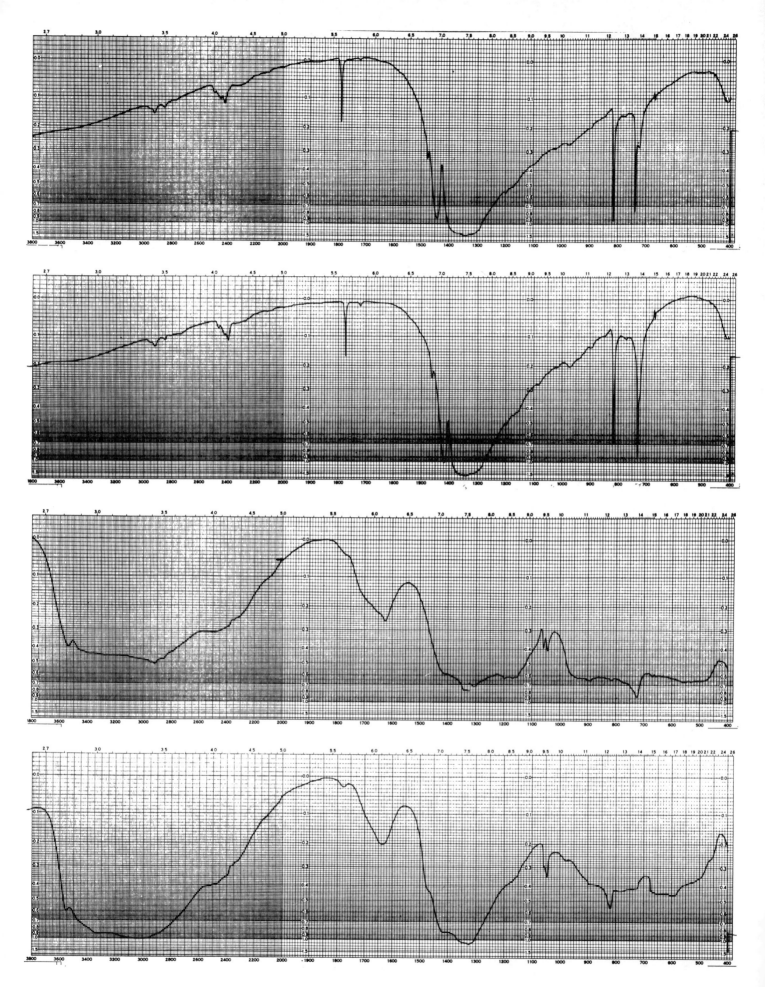

128

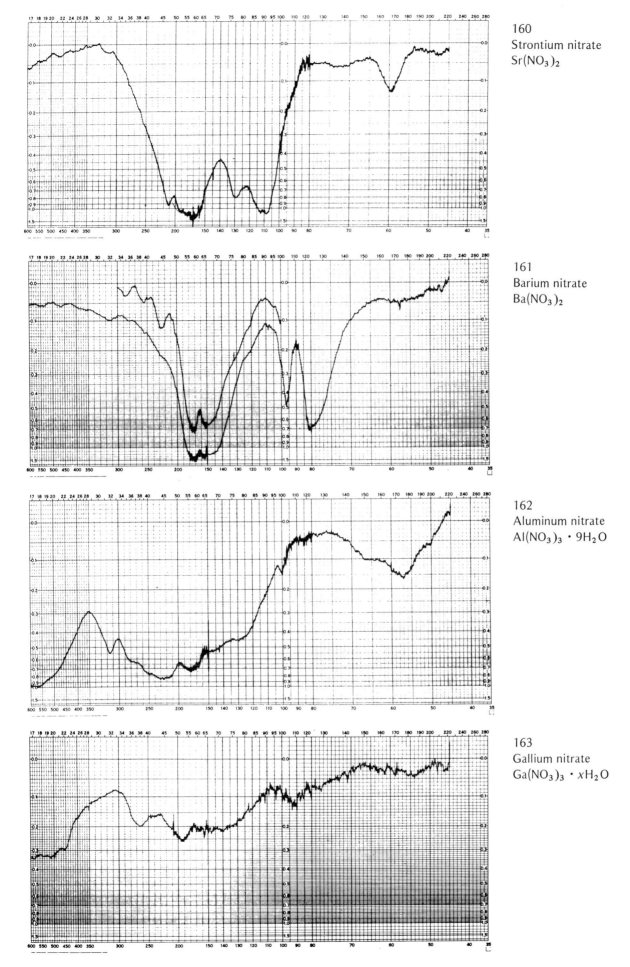

160
Strontium nitrate
$Sr(NO_3)_2$

161
Barium nitrate
$Ba(NO_3)_2$

162
Aluminum nitrate
$Al(NO_3)_3 \cdot 9H_2O$

163
Gallium nitrate
$Ga(NO_3)_3 \cdot xH_2O$

129

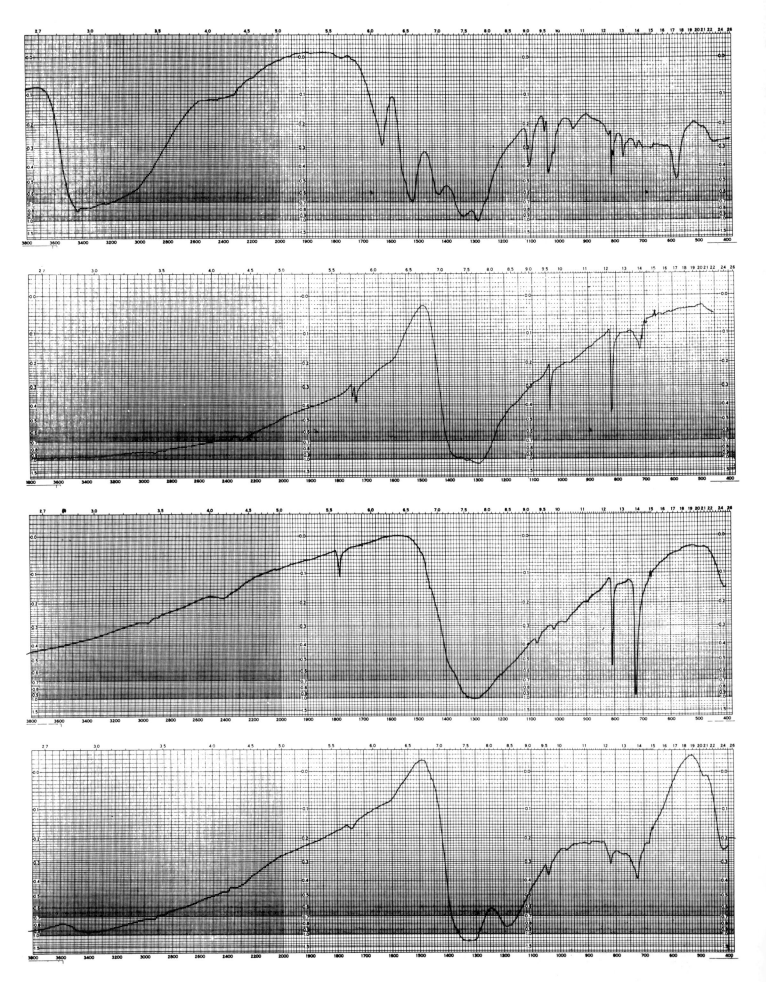

130

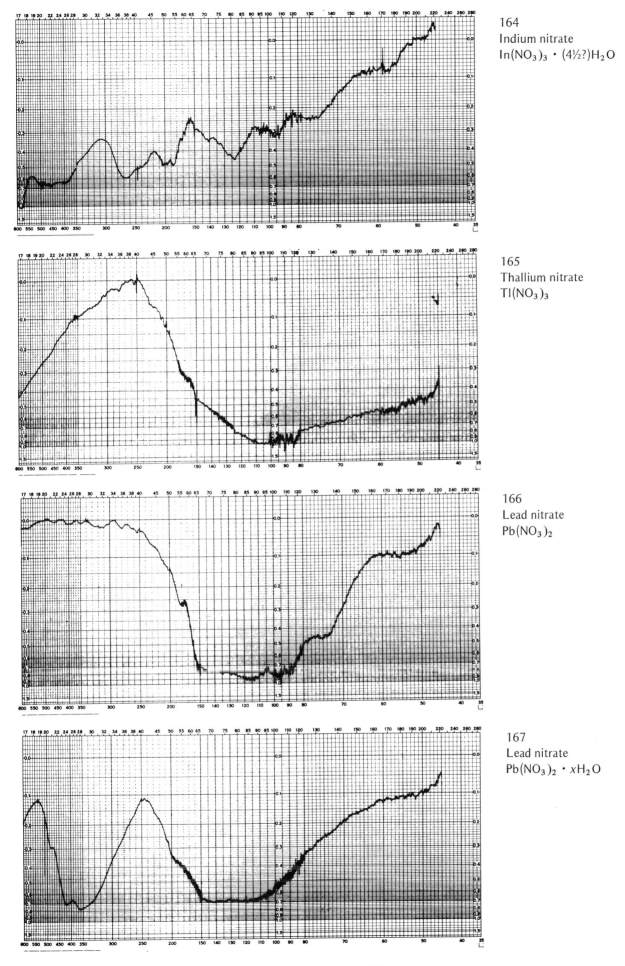

164
Indium nitrate
$In(NO_3)_3 \cdot (4\frac{1}{2}?)H_2O$

165
Thallium nitrate
$Tl(NO_3)_3$

166
Lead nitrate
$Pb(NO_3)_2$

167
Lead nitrate
$Pb(NO_3)_2 \cdot xH_2O$

131

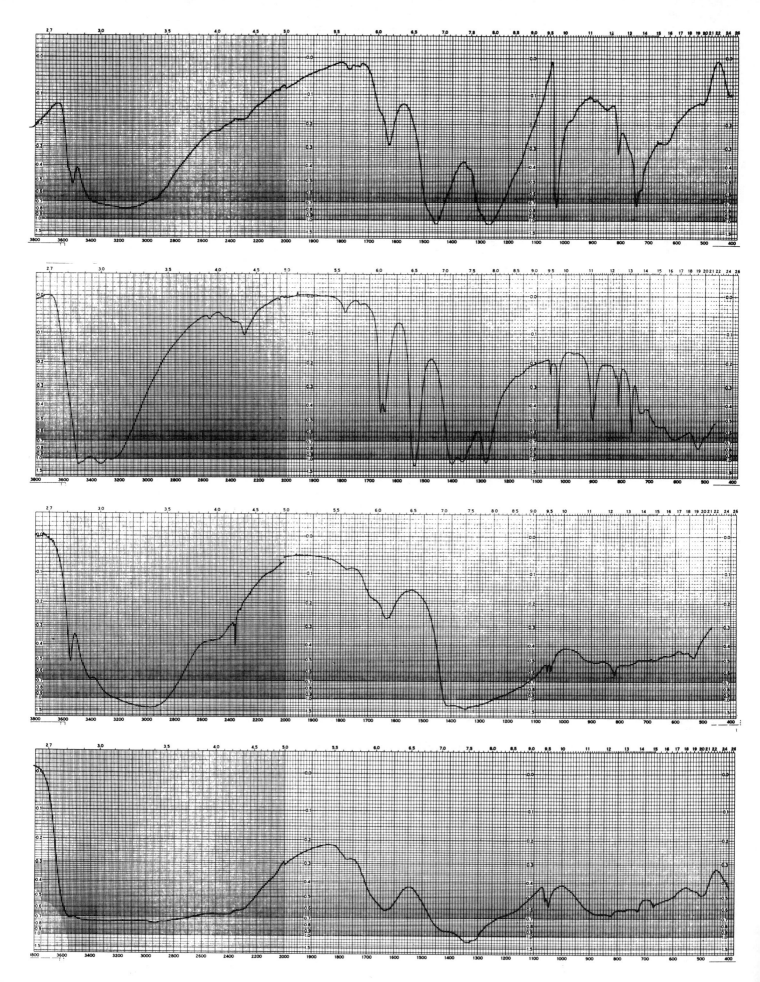

132

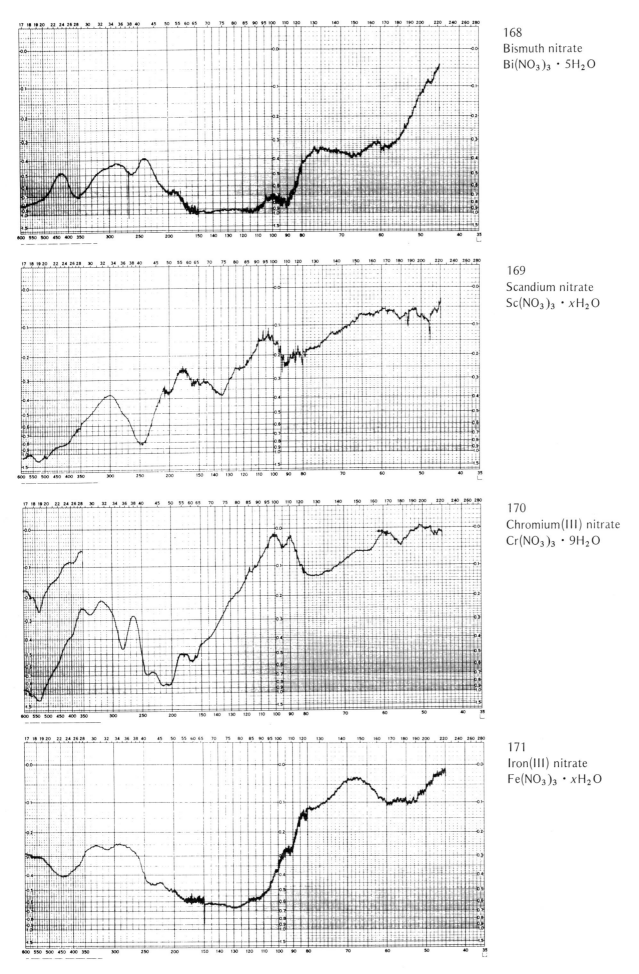

168
Bismuth nitrate
Bi(NO$_3$)$_3$ · 5H$_2$O

169
Scandium nitrate
Sc(NO$_3$)$_3$ · xH$_2$O

170
Chromium(III) nitrate
Cr(NO$_3$)$_3$ · 9H$_2$O

171
Iron(III) nitrate
Fe(NO$_3$)$_3$ · xH$_2$O

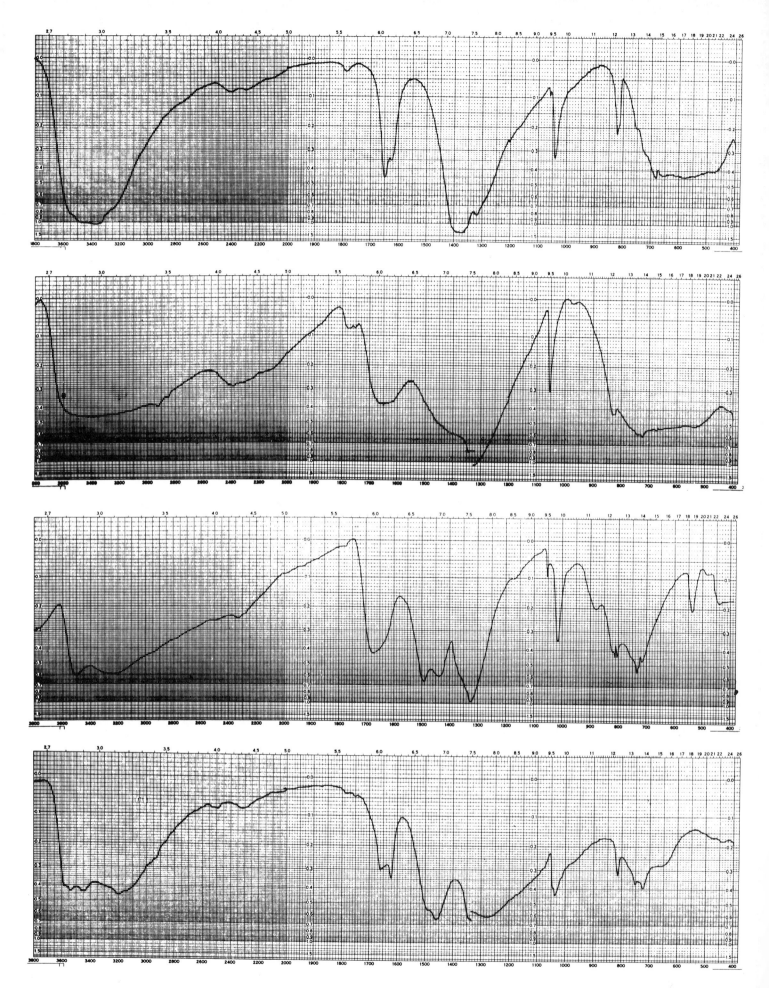

134

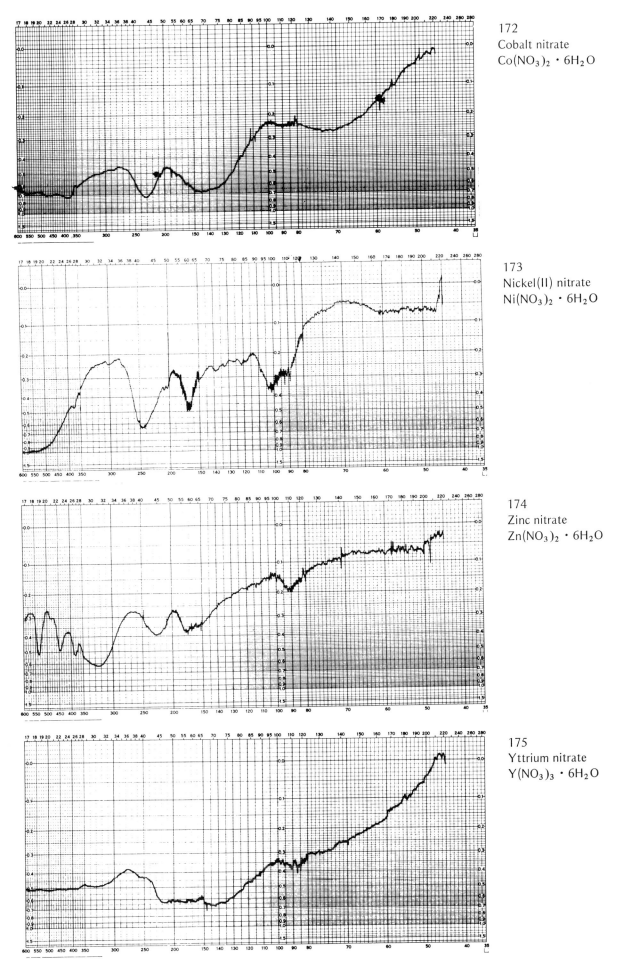

172
Cobalt nitrate
Co(NO$_3$)$_2$ · 6H$_2$O

173
Nickel(II) nitrate
Ni(NO$_3$)$_2$ · 6H$_2$O

174
Zinc nitrate
Zn(NO$_3$)$_2$ · 6H$_2$O

175
Yttrium nitrate
Y(NO$_3$)$_3$ · 6H$_2$O

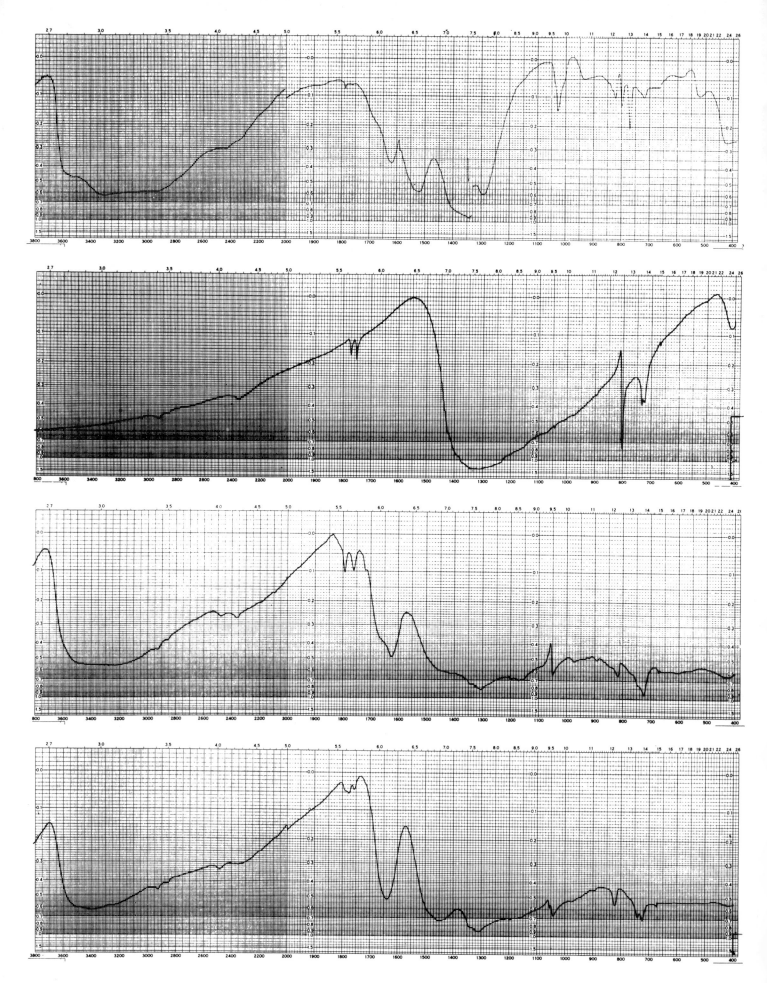

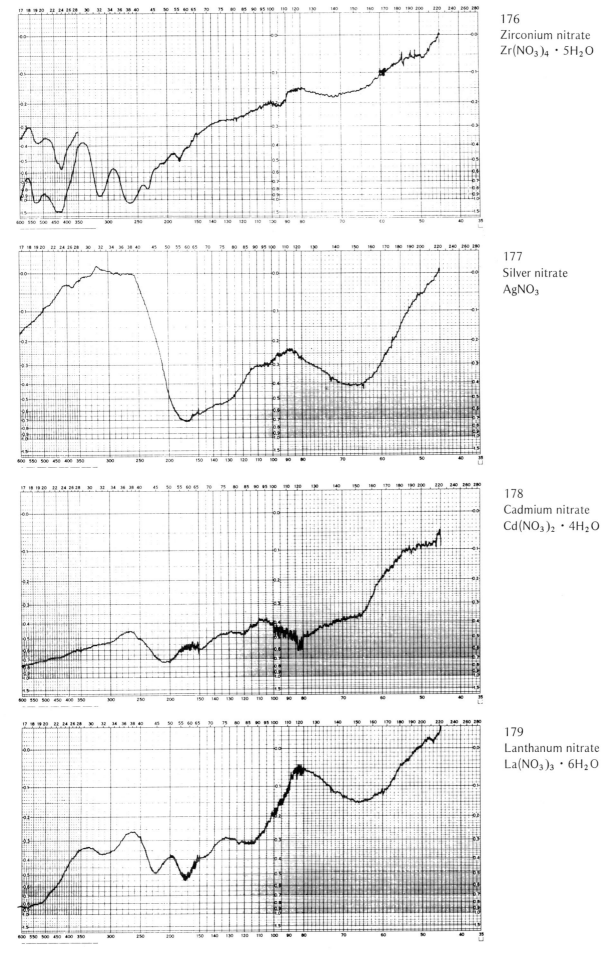

176
Zirconium nitrate
$Zr(NO_3)_4 \cdot 5H_2O$

177
Silver nitrate
$AgNO_3$

178
Cadmium nitrate
$Cd(NO_3)_2 \cdot 4H_2O$

179
Lanthanum nitrate
$La(NO_3)_3 \cdot 6H_2O$

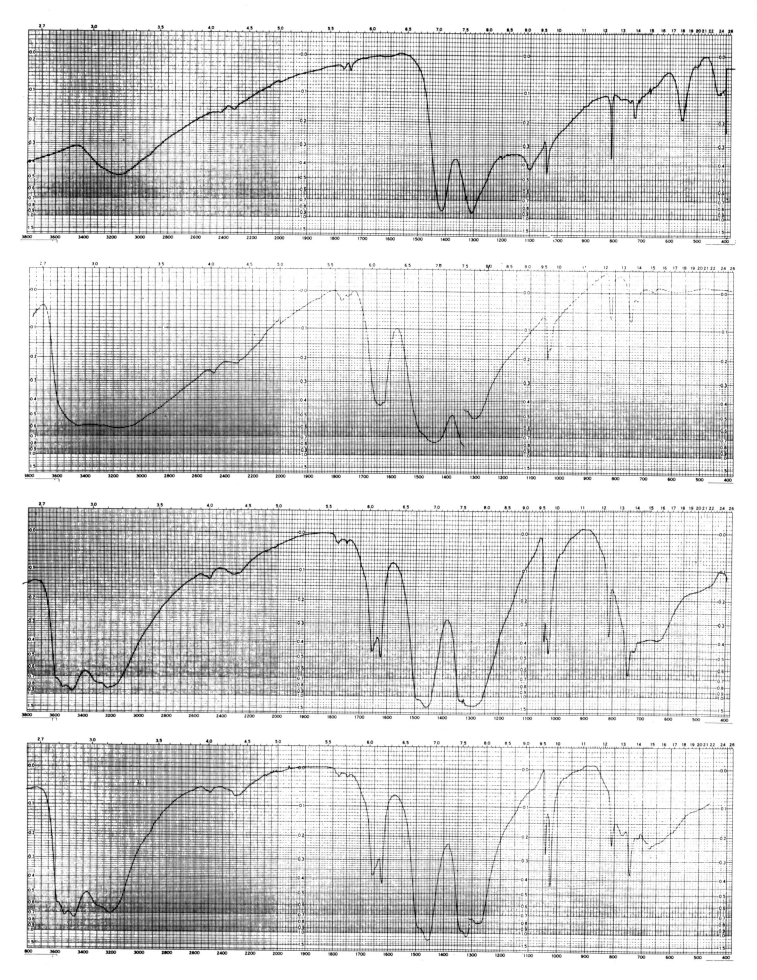

138

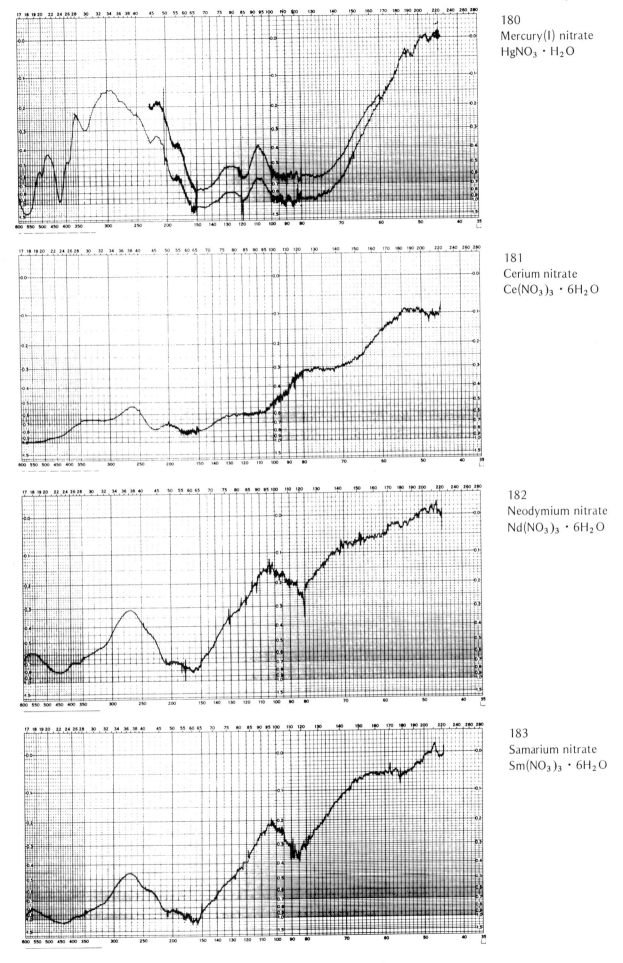

180
Mercury(I) nitrate
$HgNO_3 \cdot H_2O$

181
Cerium nitrate
$Ce(NO_3)_3 \cdot 6H_2O$

182
Neodymium nitrate
$Nd(NO_3)_3 \cdot 6H_2O$

183
Samarium nitrate
$Sm(NO_3)_3 \cdot 6H_2O$

139

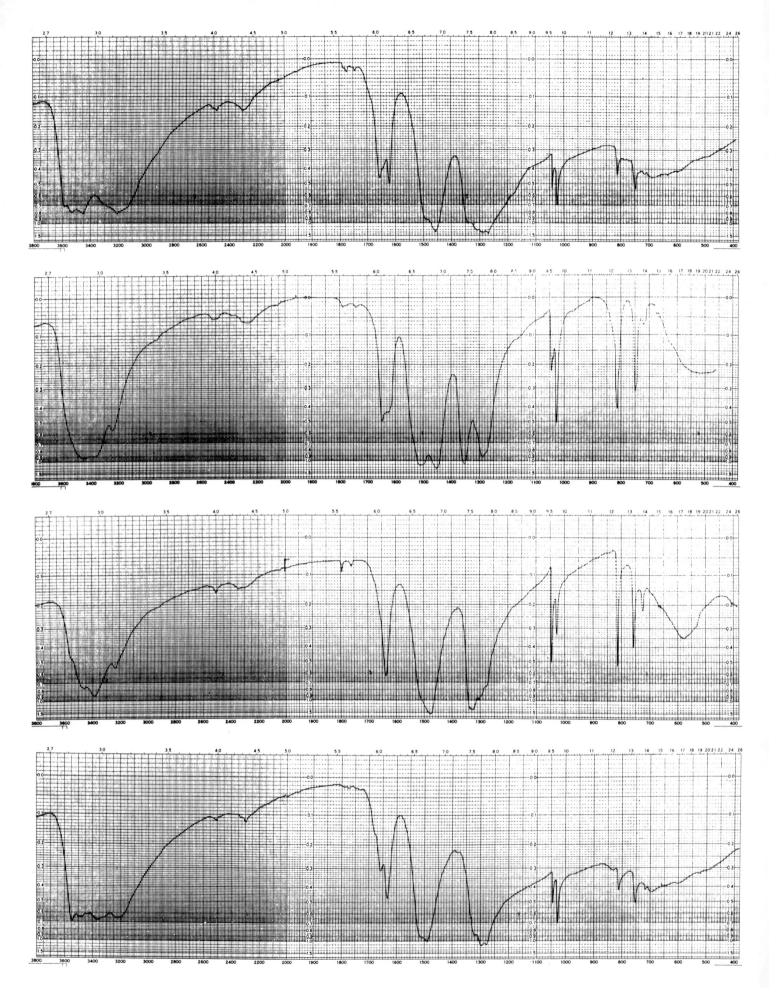

140

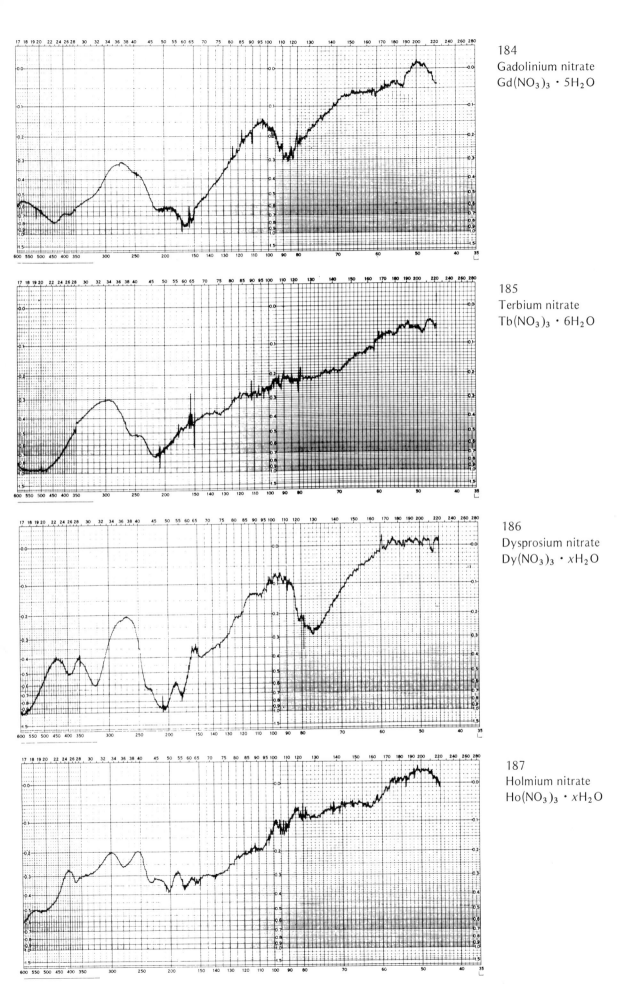

184
Gadolinium nitrate
$Gd(NO_3)_3 \cdot 5H_2O$

185
Terbium nitrate
$Tb(NO_3)_3 \cdot 6H_2O$

186
Dysprosium nitrate
$Dy(NO_3)_3 \cdot xH_2O$

187
Holmium nitrate
$Ho(NO_3)_3 \cdot xH_2O$

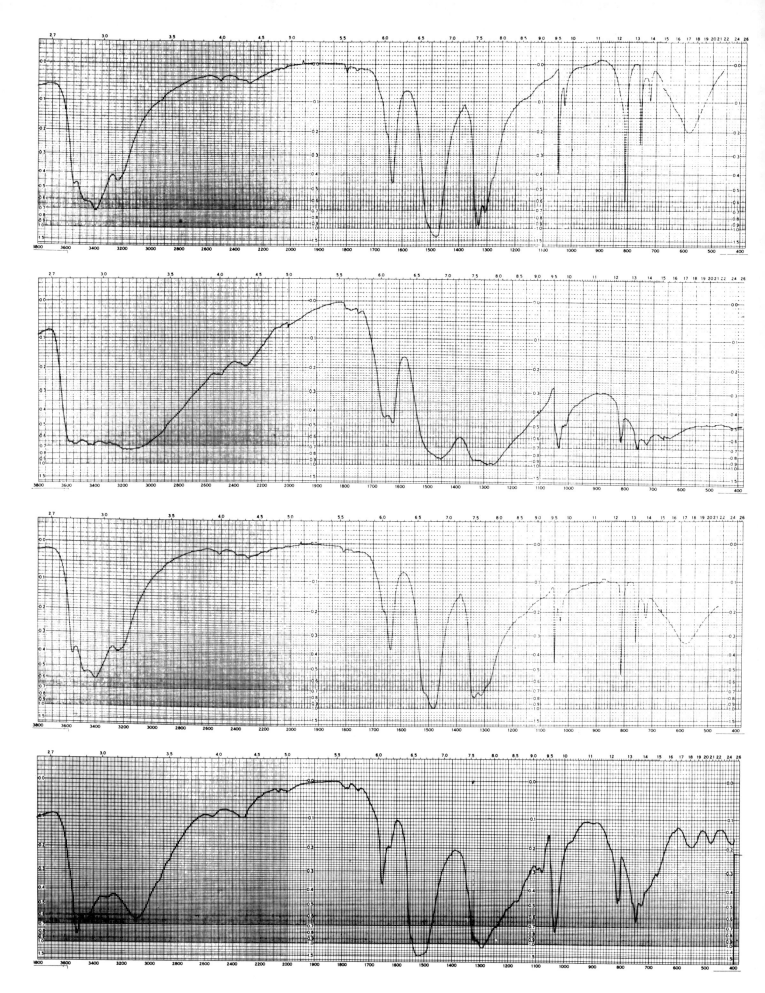

142

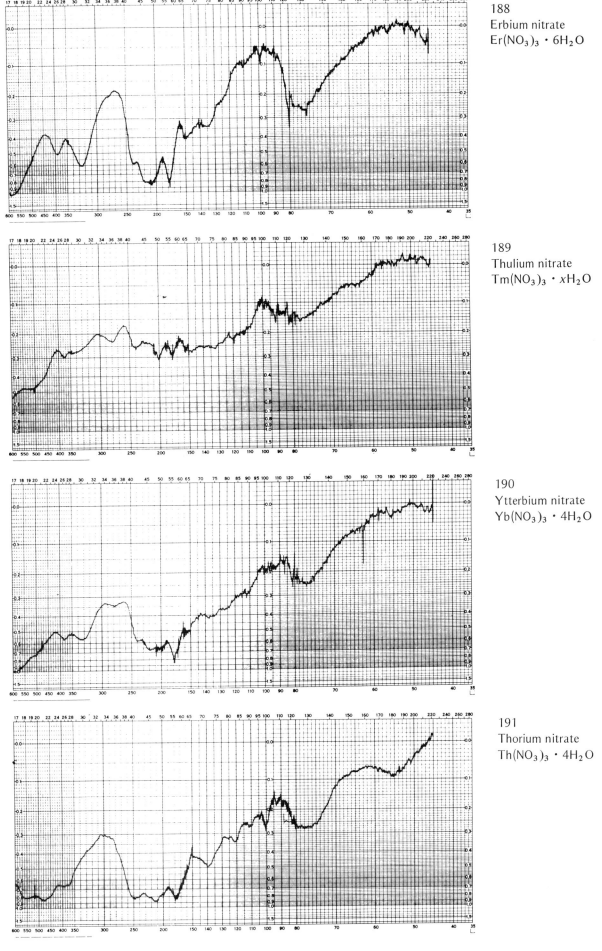

188
Erbium nitrate
$Er(NO_3)_3 \cdot 6H_2O$

189
Thulium nitrate
$Tm(NO_3)_3 \cdot xH_2O$

190
Ytterbium nitrate
$Yb(NO_3)_3 \cdot 4H_2O$

191
Thorium nitrate
$Th(NO_3)_3 \cdot 4H_2O$

143

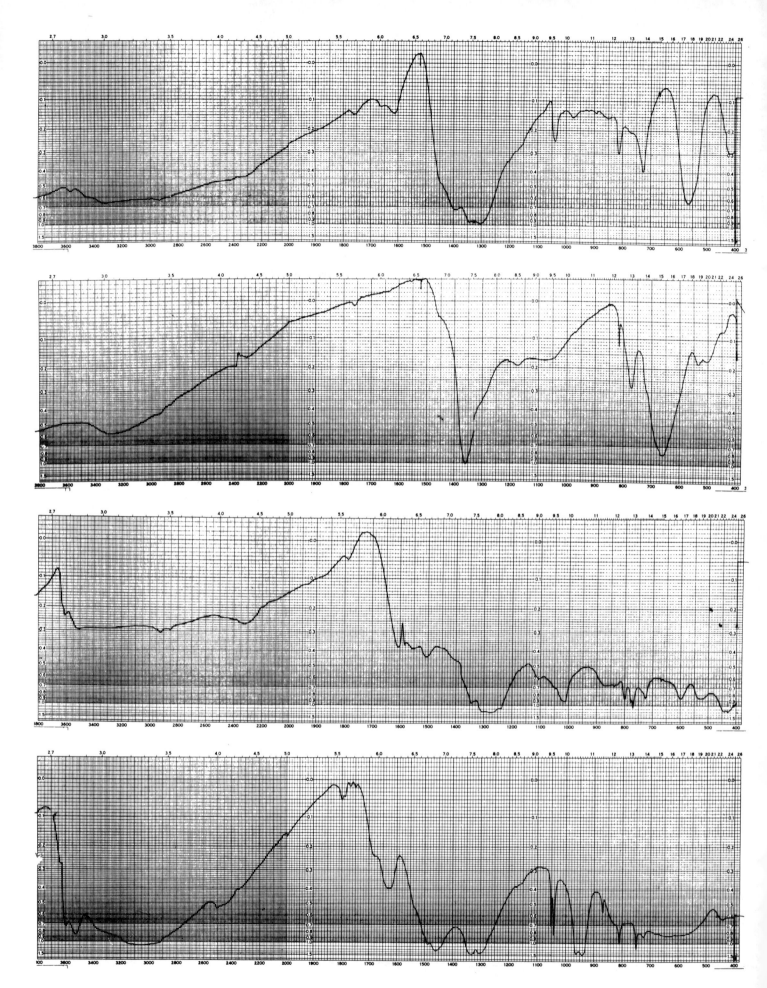

144

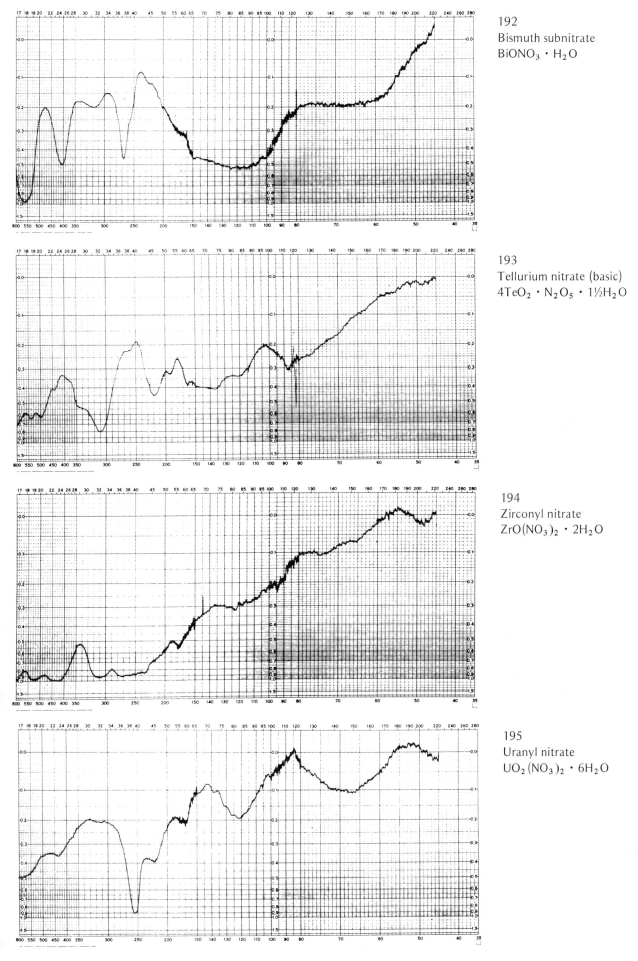

192
Bismuth subnitrate
$BiONO_3 \cdot H_2O$

193
Tellurium nitrate (basic)
$4TeO_2 \cdot N_2O_5 \cdot 1\frac{1}{2}H_2O$

194
Zirconyl nitrate
$ZrO(NO_3)_2 \cdot 2H_2O$

195
Uranyl nitrate
$UO_2(NO_3)_2 \cdot 6H_2O$

145

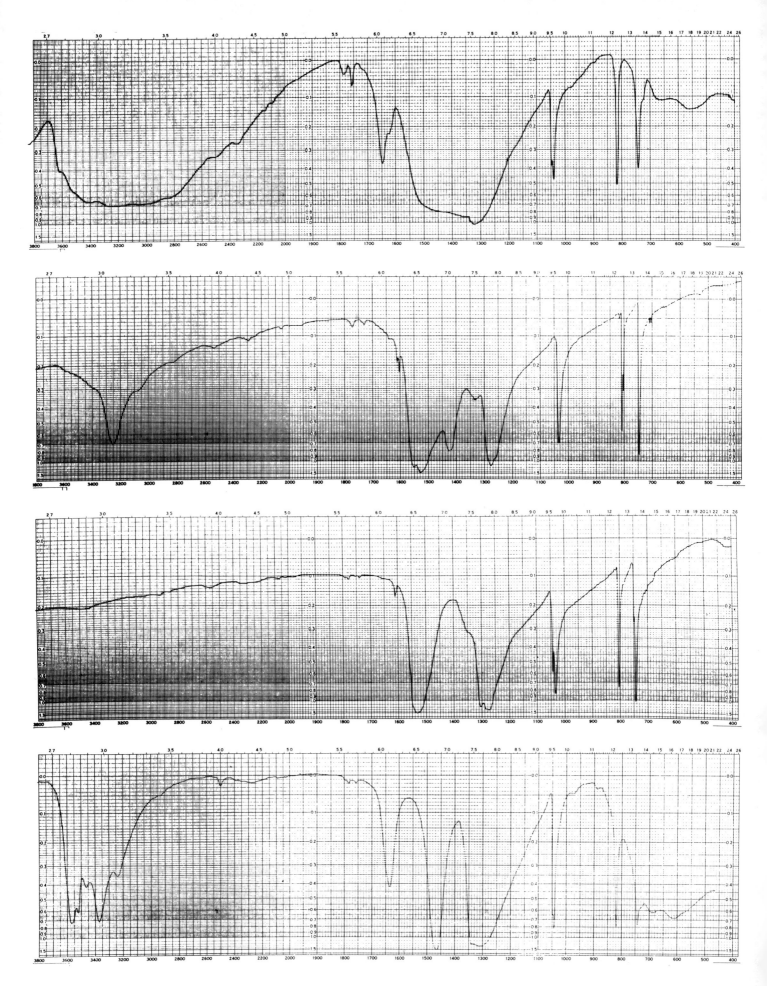

146

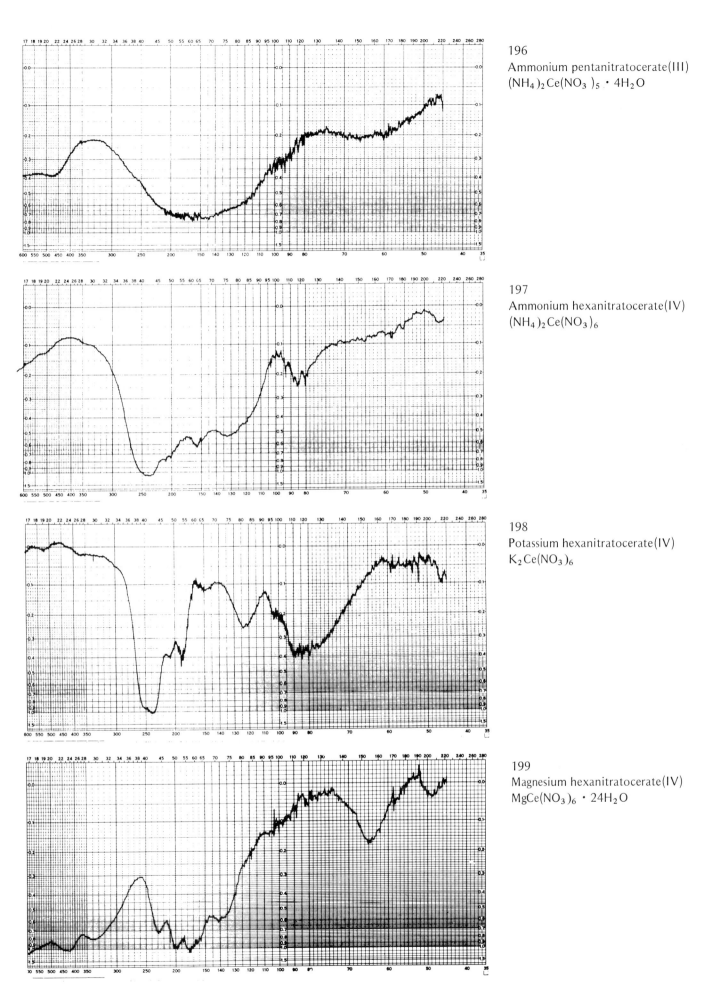

196
Ammonium pentanitratocerate(III)
$(NH_4)_2Ce(NO_3)_5 \cdot 4H_2O$

197
Ammonium hexanitratocerate(IV)
$(NH_4)_2Ce(NO_3)_6$

198
Potassium hexanitratocerate(IV)
$K_2Ce(NO_3)_6$

199
Magnesium hexanitratocerate(IV)
$MgCe(NO_3)_6 \cdot 24H_2O$

147

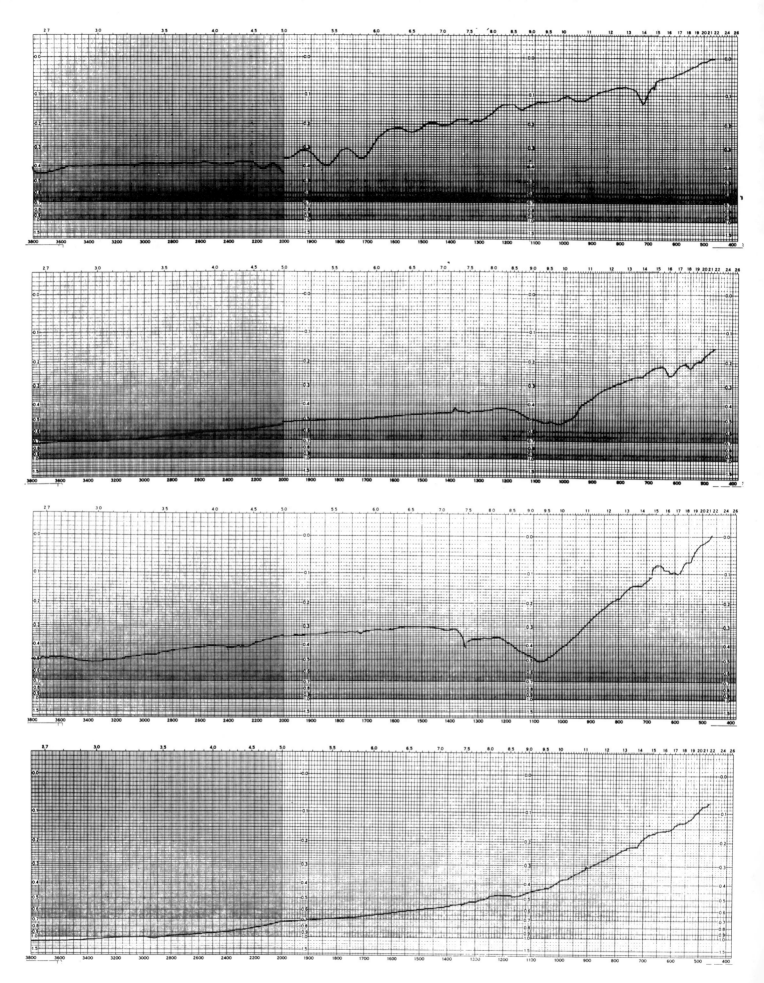

148

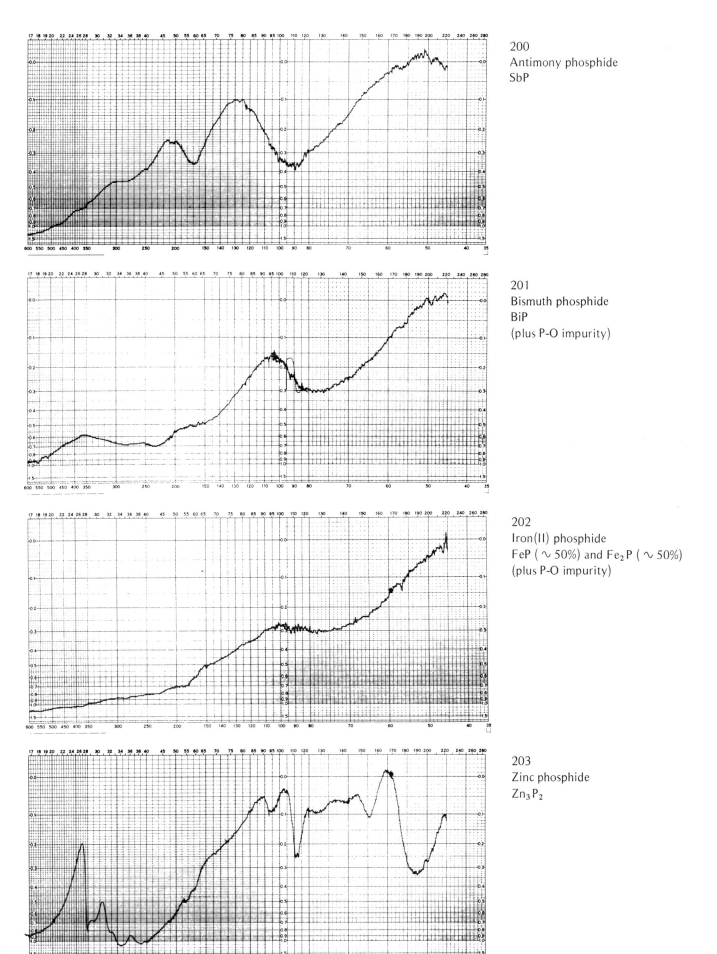

200
Antimony phosphide
SbP

201
Bismuth phosphide
BiP
(plus P-O impurity)

202
Iron(II) phosphide
FeP ($\sim$ 50%) and Fe$_2$P ($\sim$ 50%)
(plus P-O impurity)

203
Zinc phosphide
Zn$_3$P$_2$

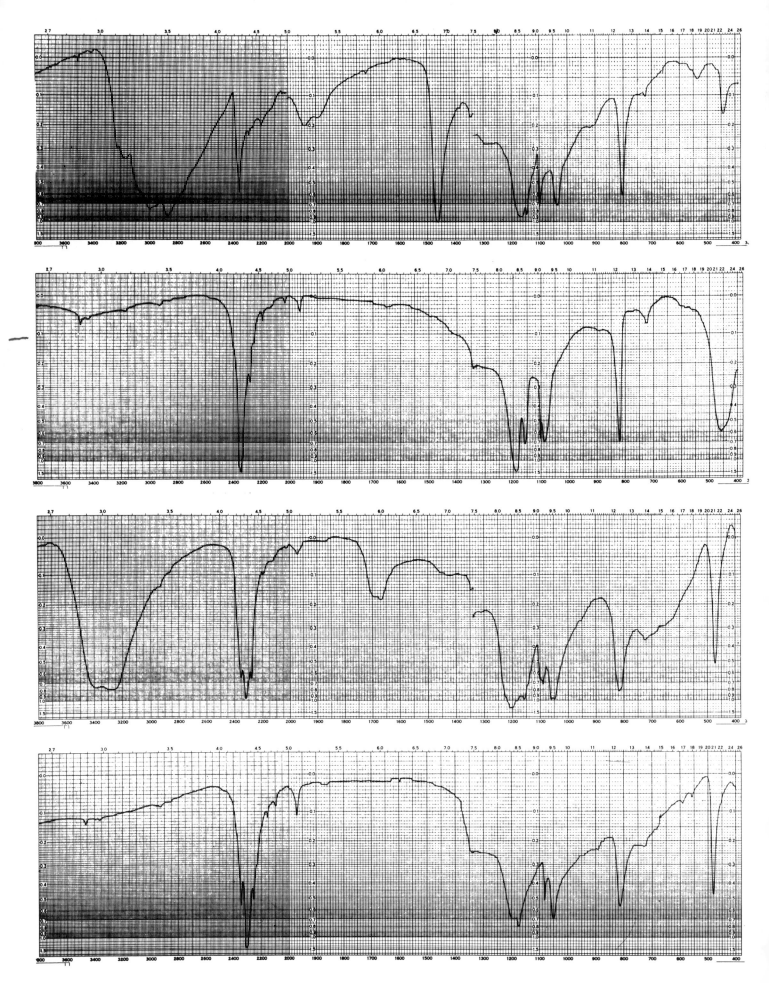

150

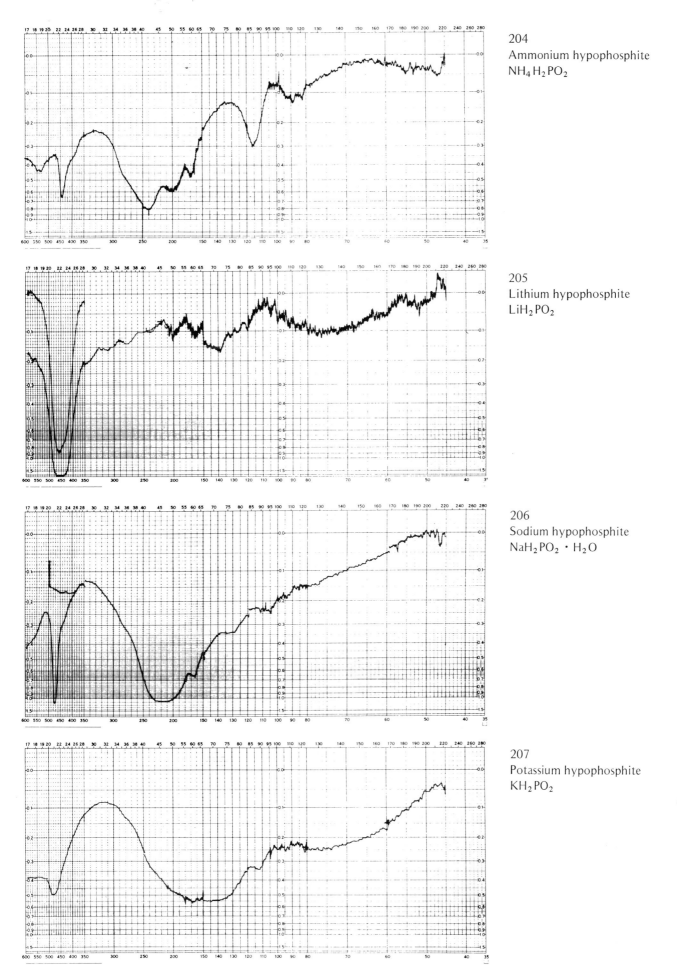

204
Ammonium hypophosphite
$NH_4H_2PO_2$

205
Lithium hypophosphite
LiH_2PO_2

206
Sodium hypophosphite
$NaH_2PO_2 \cdot H_2O$

207
Potassium hypophosphite
KH_2PO_2

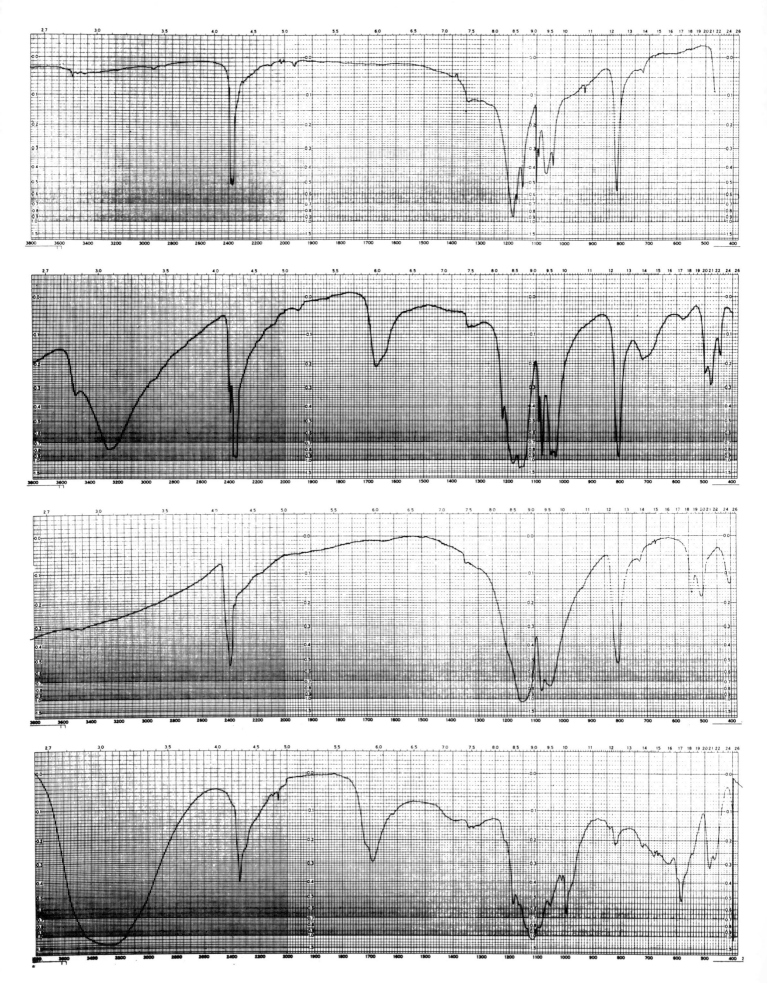

152

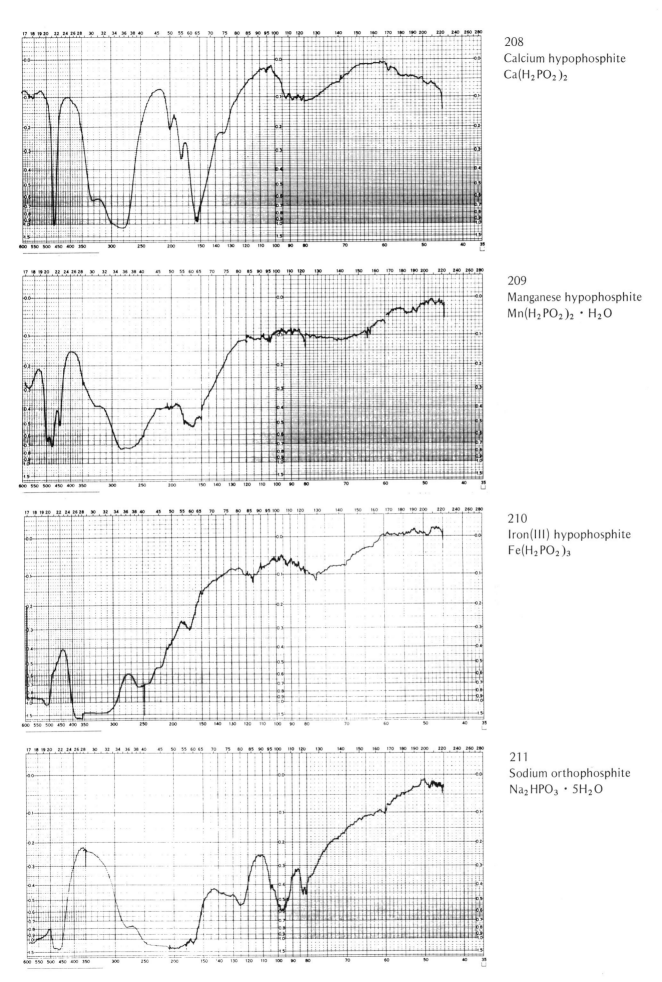

208
Calcium hypophosphite
$Ca(H_2PO_2)_2$

209
Manganese hypophosphite
$Mn(H_2PO_2)_2 \cdot H_2O$

210
Iron(III) hypophosphite
$Fe(H_2PO_2)_3$

211
Sodium orthophosphite
$Na_2HPO_3 \cdot 5H_2O$

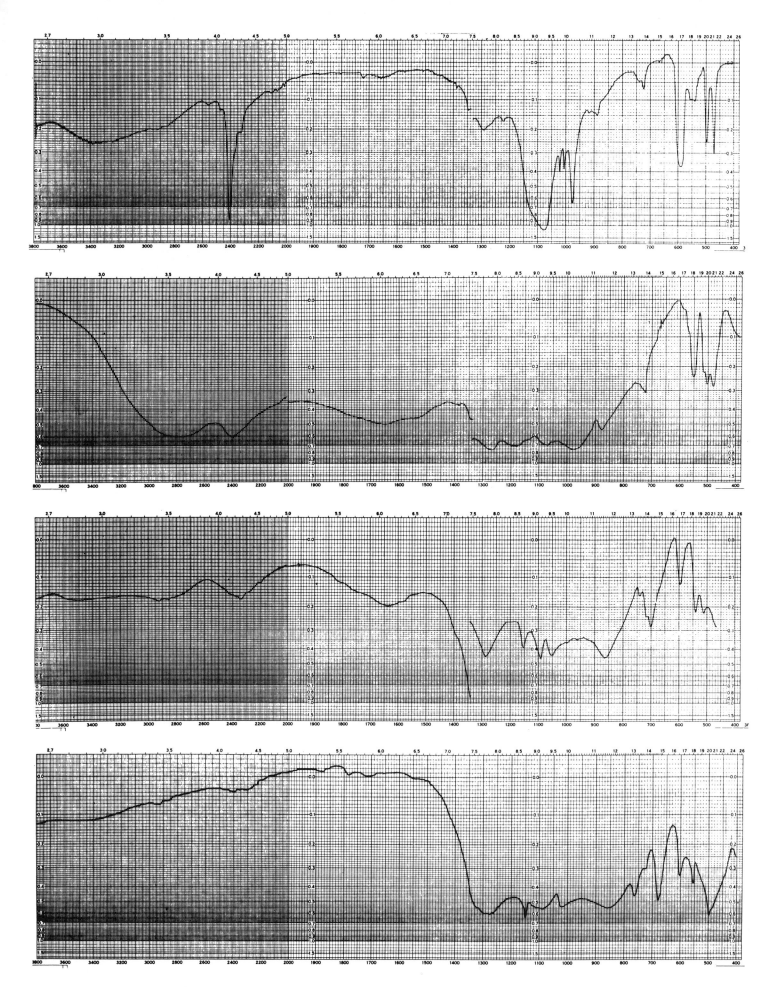

154

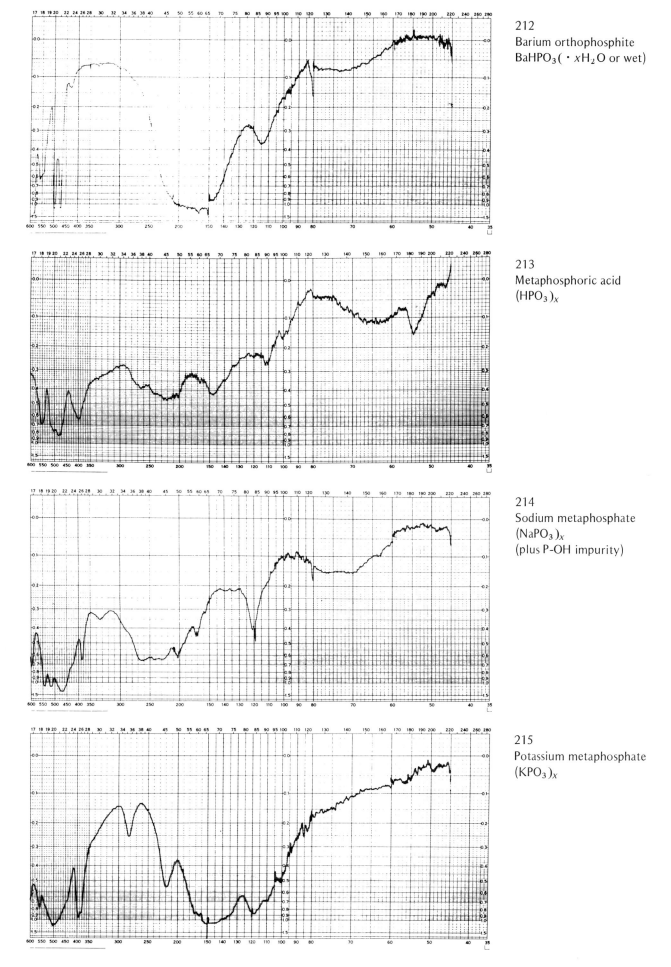

212
Barium orthophosphite
BaHPO$_3$(· xH$_2$O or wet)

213
Metaphosphoric acid
(HPO$_3$)$_x$

214
Sodium metaphosphate
(NaPO$_3$)$_x$
(plus P-OH impurity)

215
Potassium metaphosphate
(KPO$_3$)$_x$

155

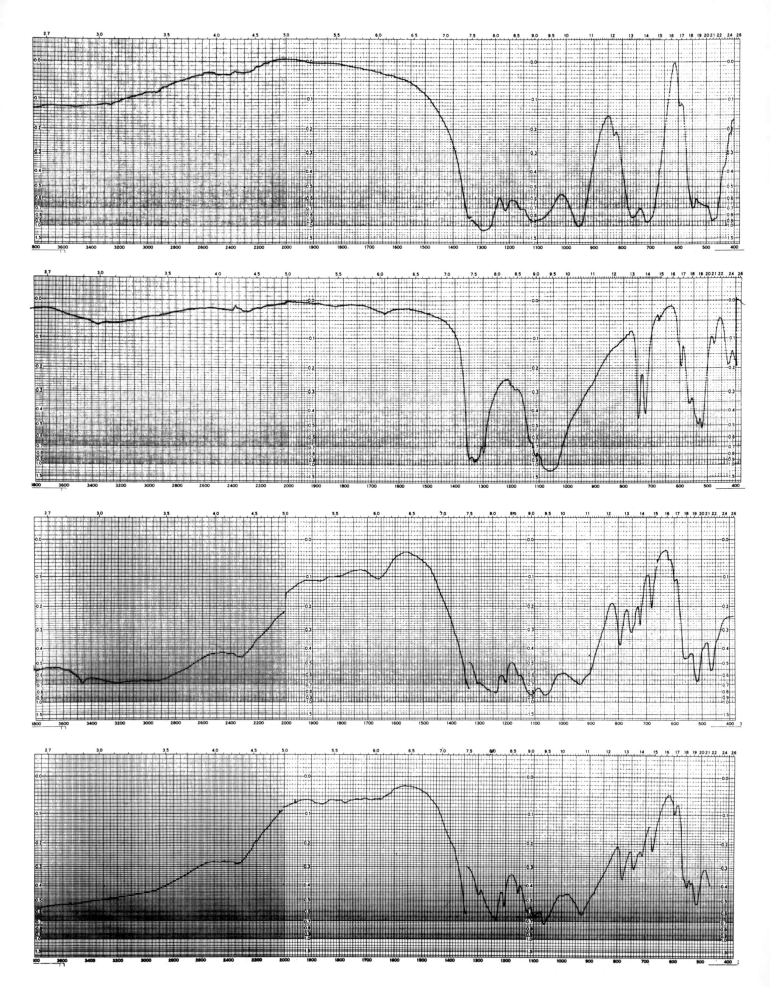

156

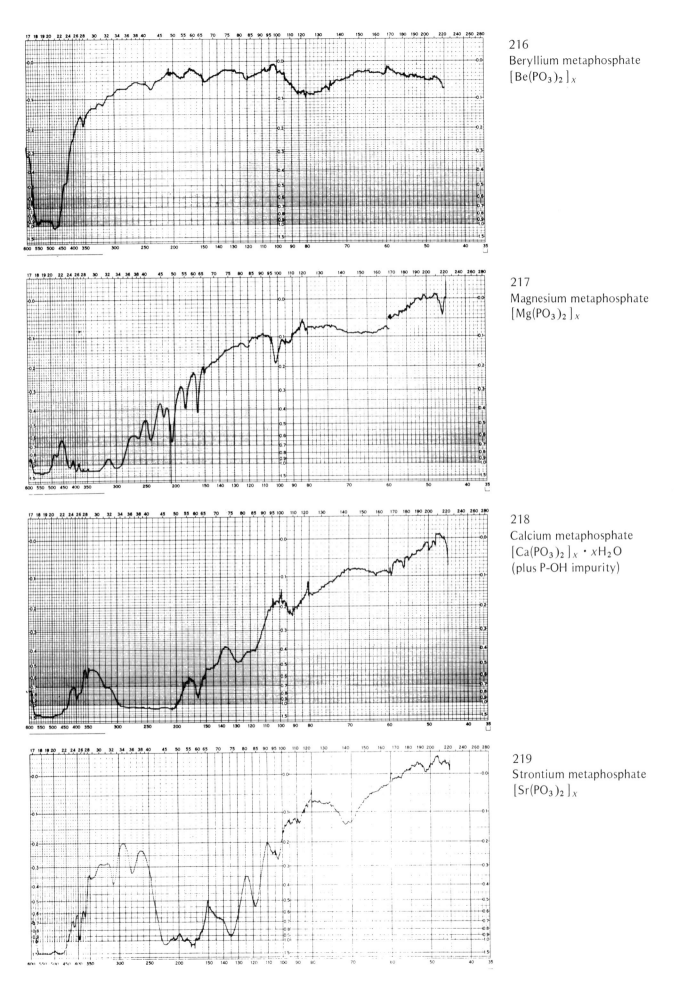

216
Beryllium metaphosphate
$[Be(PO_3)_2]_x$

217
Magnesium metaphosphate
$[Mg(PO_3)_2]_x$

218
Calcium metaphosphate
$[Ca(PO_3)_2]_x \cdot xH_2O$
(plus P-OH impurity)

219
Strontium metaphosphate
$[Sr(PO_3)_2]_x$

157

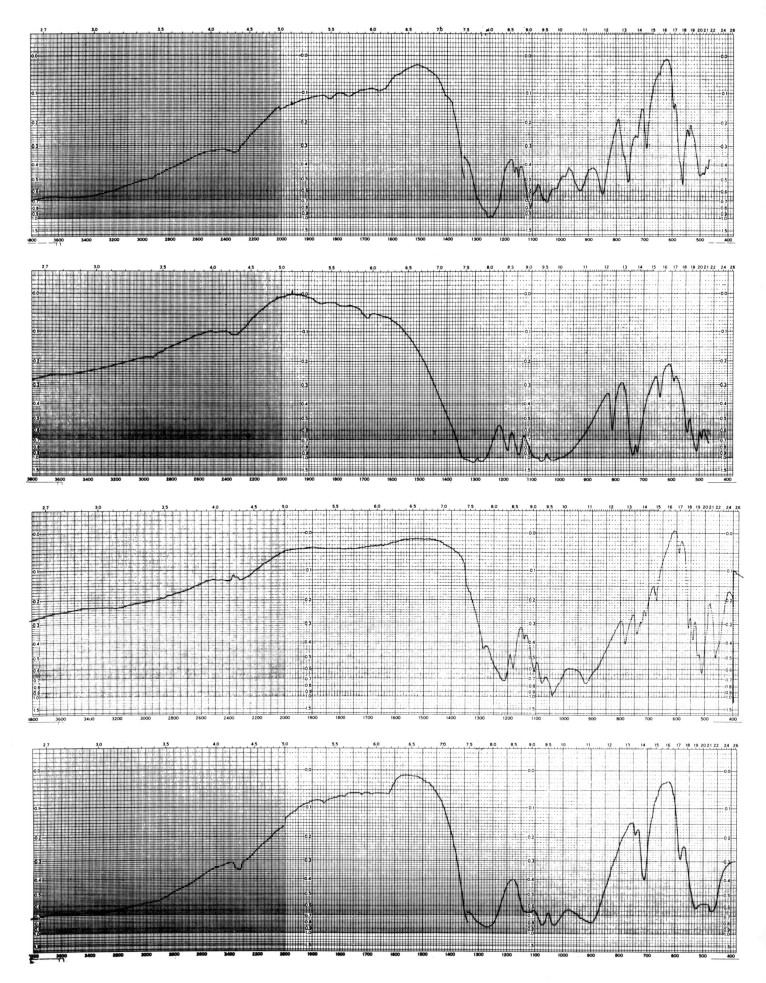

158

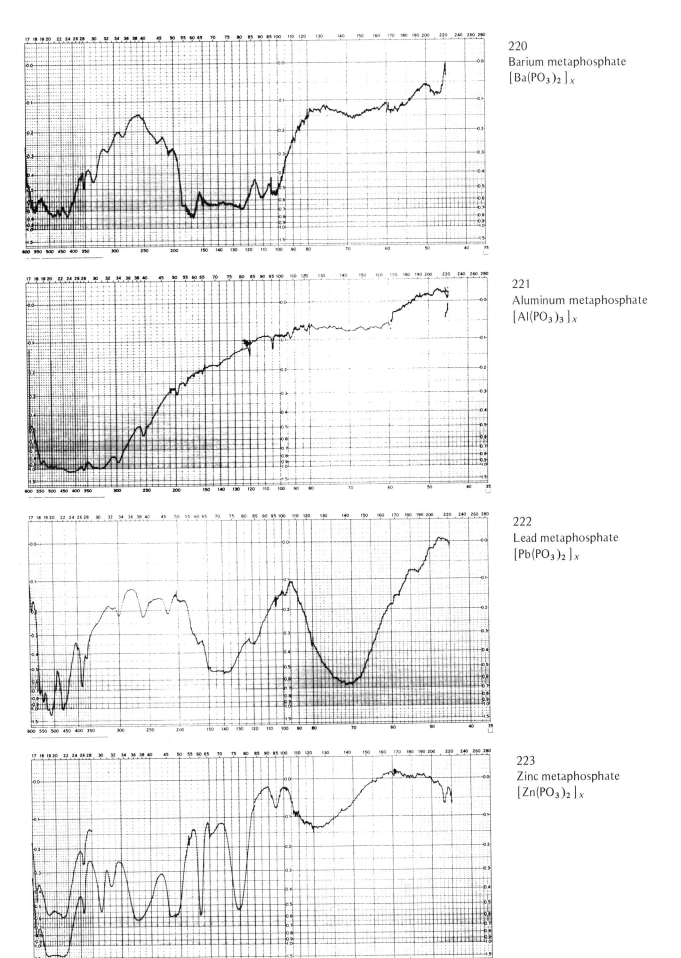

220
Barium metaphosphate
$[Ba(PO_3)_2]_x$

221
Aluminum metaphosphate
$[Al(PO_3)_3]_x$

222
Lead metaphosphate
$[Pb(PO_3)_2]_x$

223
Zinc metaphosphate
$[Zn(PO_3)_2]_x$

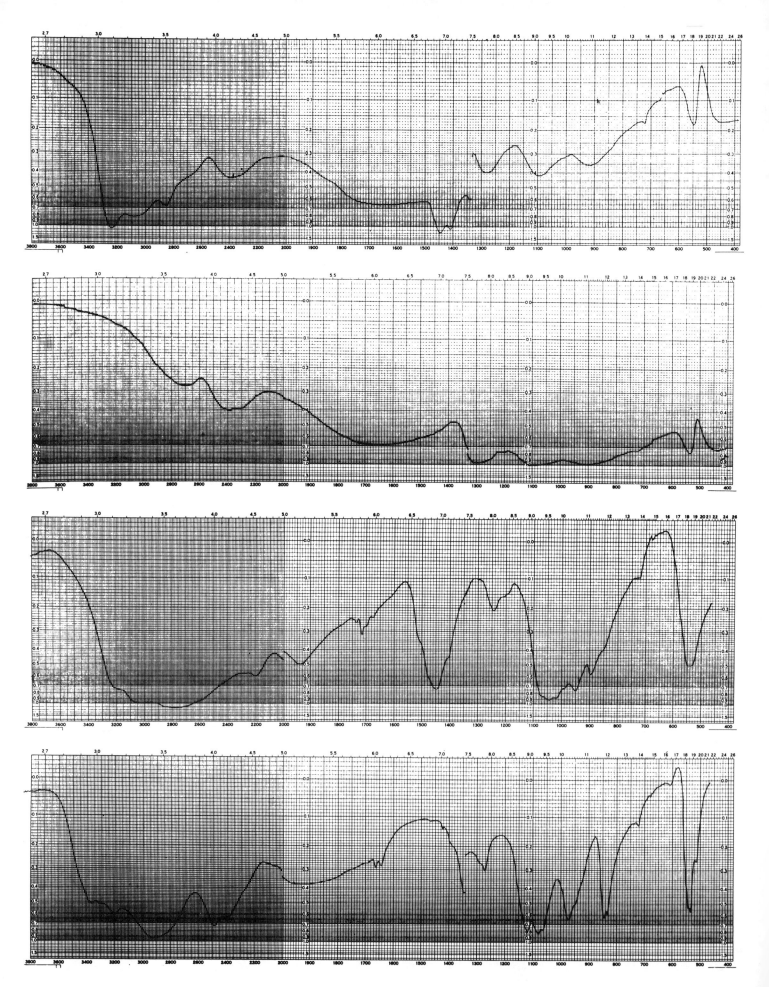

160

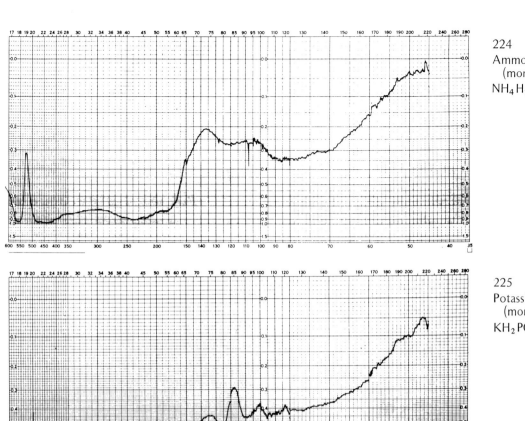

224
Ammonium orthophosphate
  (monobasic)
$NH_4H_2PO_4$

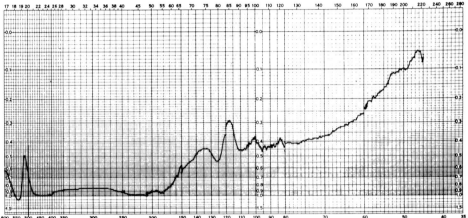

225
Potassium orthophosphate
  (monobasic)
KH_2PO_4

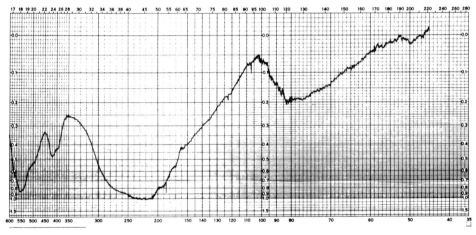

226
Ammonium orthophosphate
  (dibasic)
$(NH_4)_2HPO_4$

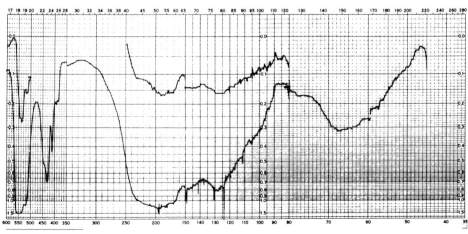

227
Potassium orthophosphate (dibasic)
$K_2HPO_4(\cdot \, xH_2O$ or wet)

161

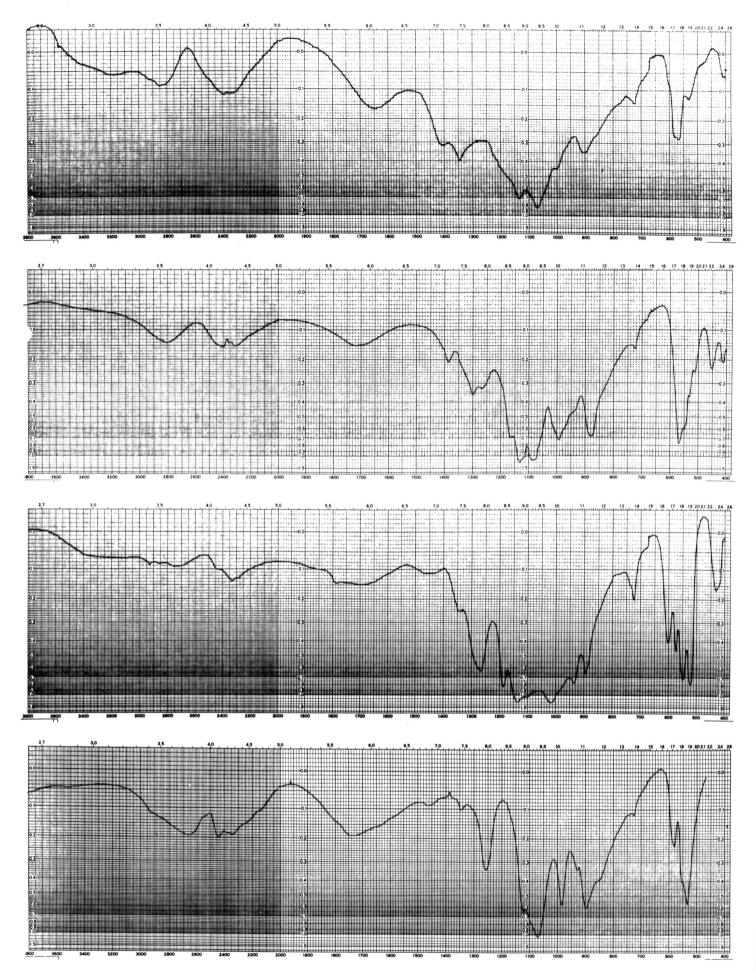

162

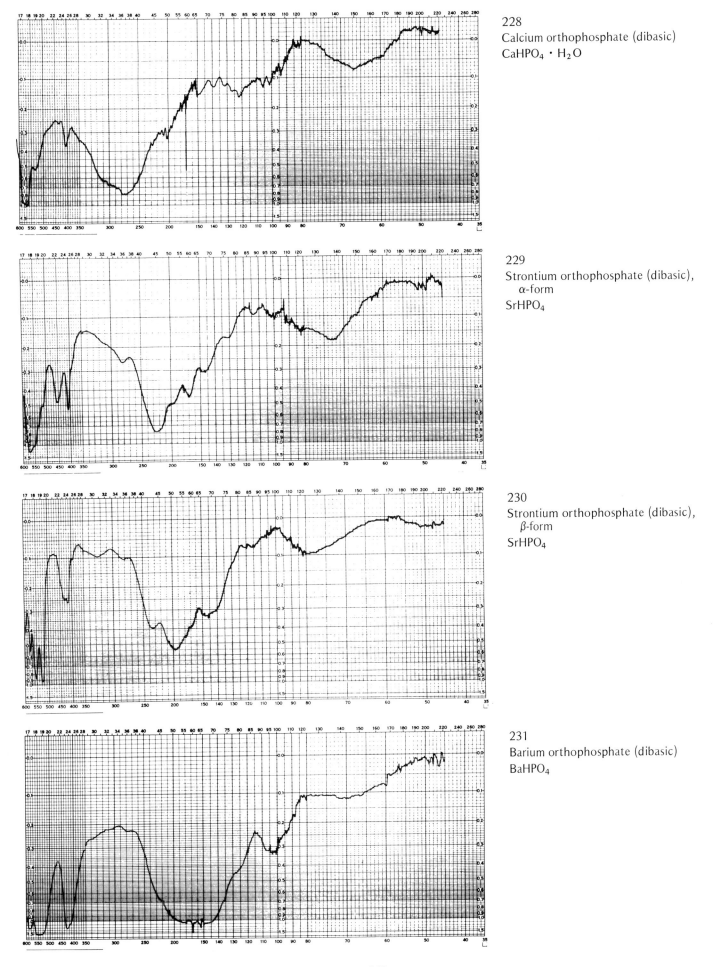

228
Calcium orthophosphate (dibasic)
CaHPO$_4$ · H$_2$O

229
Strontium orthophosphate (dibasic),
α-form
SrHPO$_4$

230
Strontium orthophosphate (dibasic),
β-form
SrHPO$_4$

231
Barium orthophosphate (dibasic)
BaHPO$_4$

163

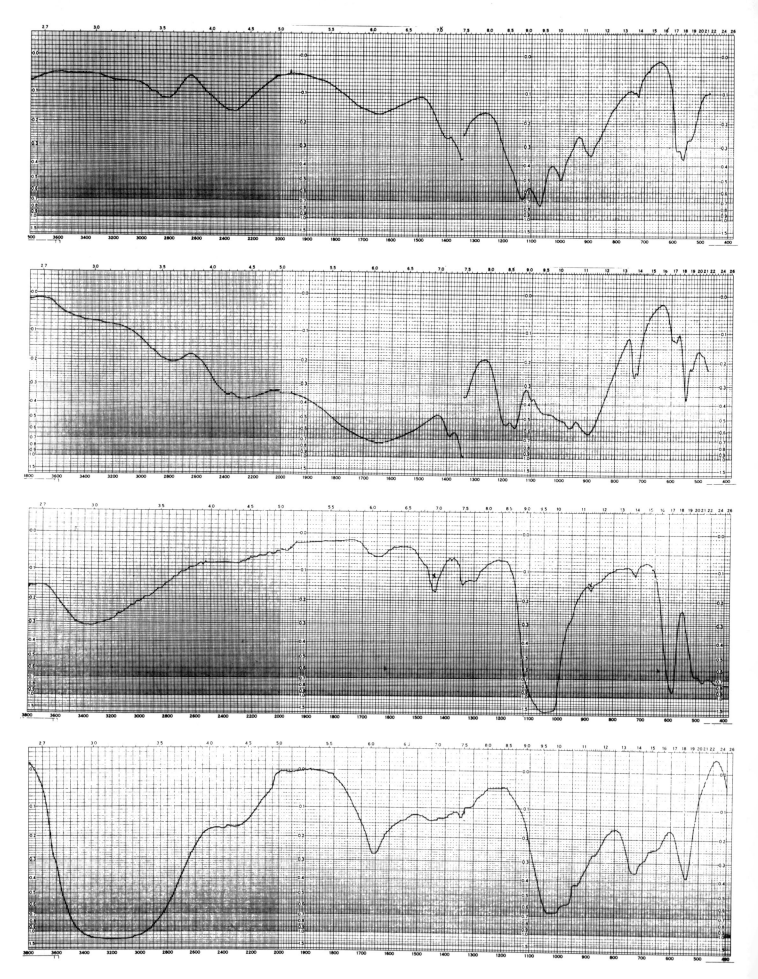

164

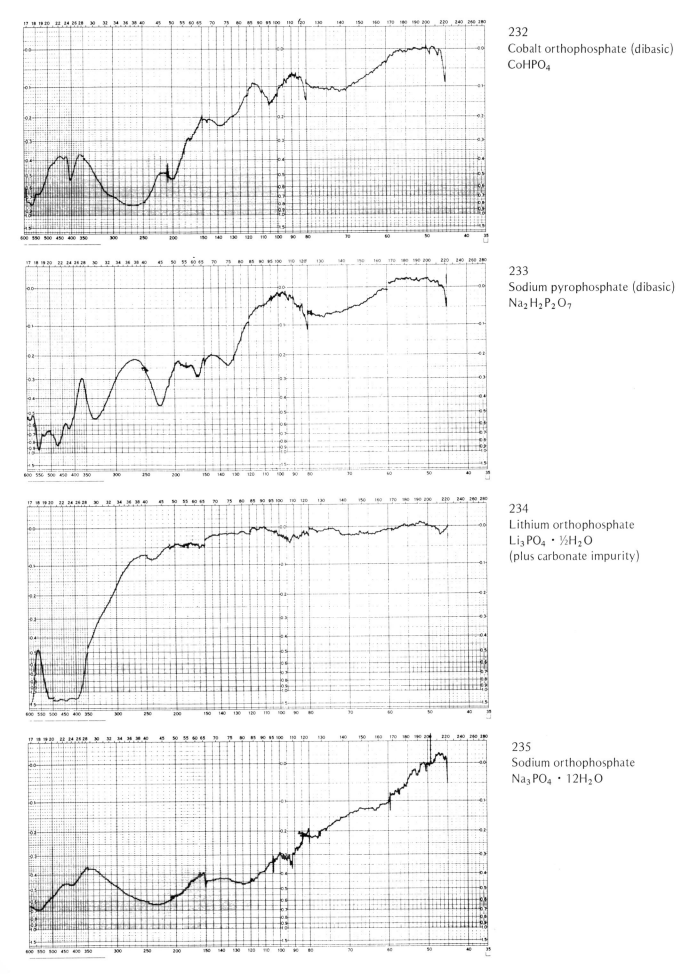

232
Cobalt orthophosphate (dibasic)
CoHPO₄

233
Sodium pyrophosphate (dibasic)
Na₂H₂P₂O₇

234
Lithium orthophosphate
Li₃PO₄ · ½H₂O
(plus carbonate impurity)

235
Sodium orthophosphate
Na₃PO₄ · 12H₂O

165

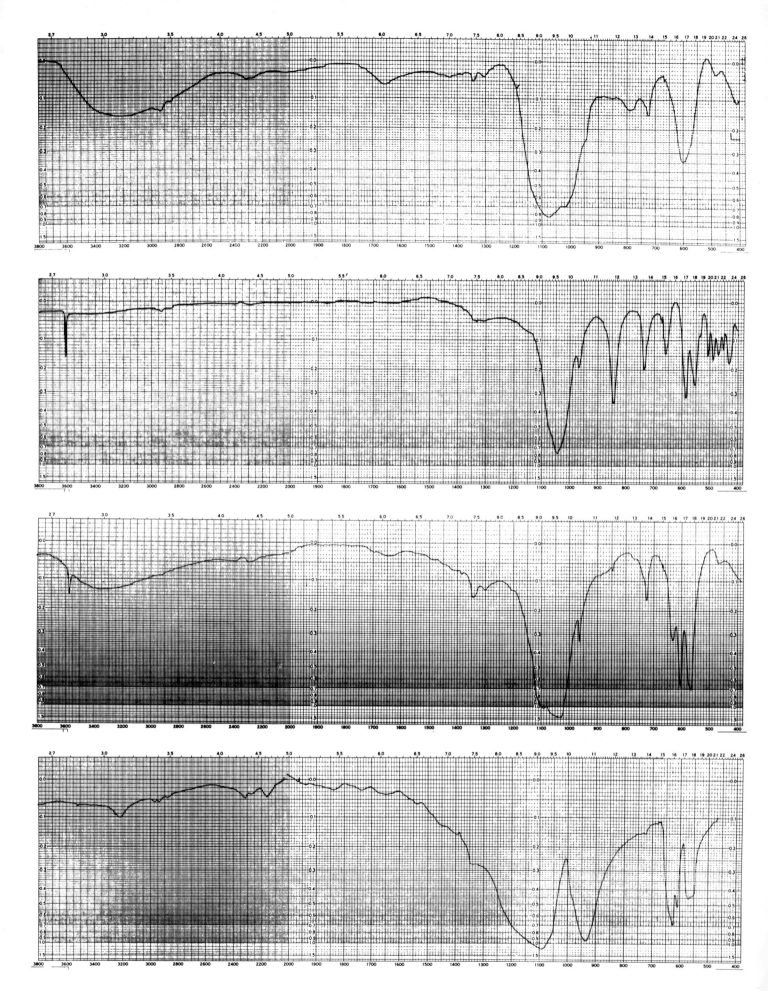

166

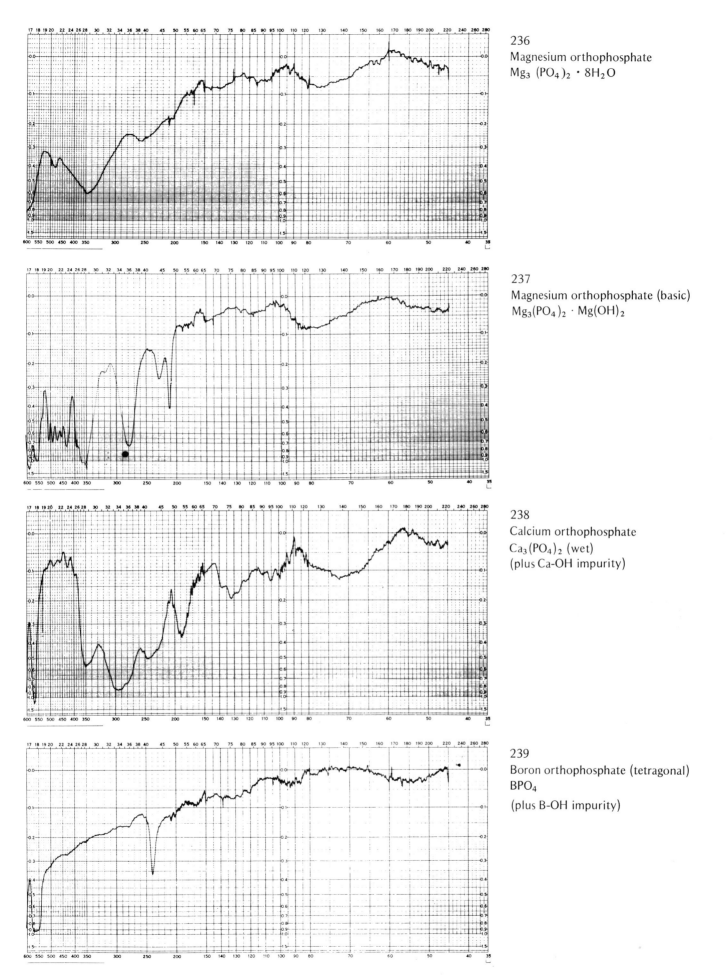

236
Magnesium orthophosphate
$Mg_3(PO_4)_2 \cdot 8H_2O$

237
Magnesium orthophosphate (basic)
$Mg_3(PO_4)_2 \cdot Mg(OH)_2$

238
Calcium orthophosphate
$Ca_3(PO_4)_2$ (wet)
(plus Ca-OH impurity)

239
Boron orthophosphate (tetragonal)
BPO_4
(plus B-OH impurity)

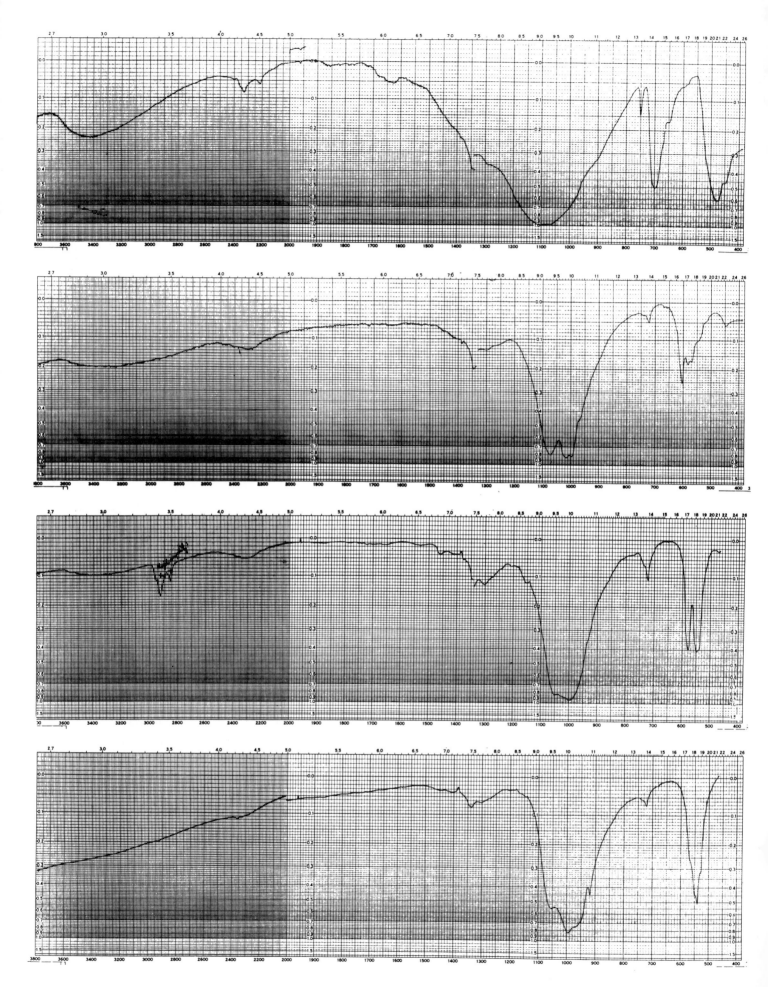

168

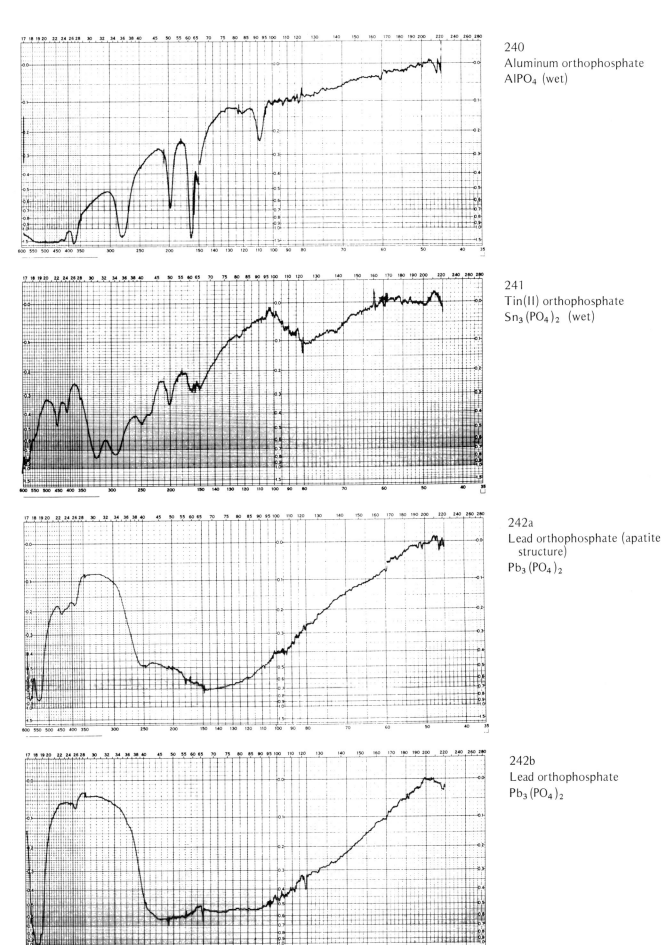

240
Aluminum orthophosphate
$AlPO_4$ (wet)

241
Tin(II) orthophosphate
$Sn_3(PO_4)_2$ (wet)

242a
Lead orthophosphate (apatite
  structure)
$Pb_3(PO_4)_2$

242b
Lead orthophosphate
$Pb_3(PO_4)_2$

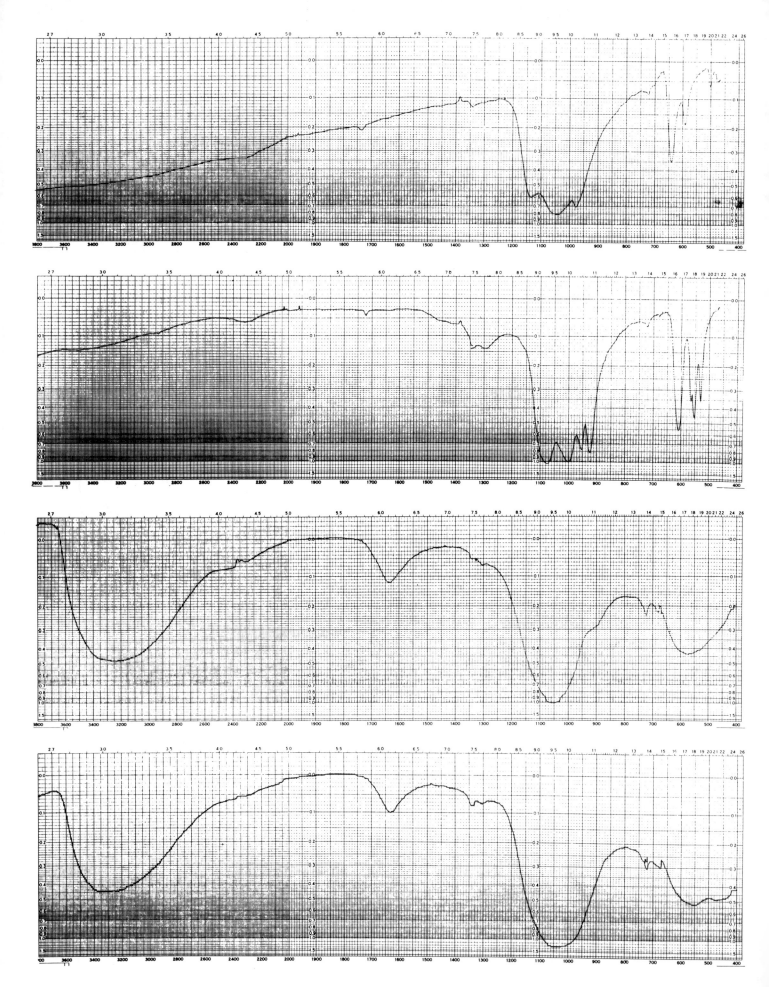

170

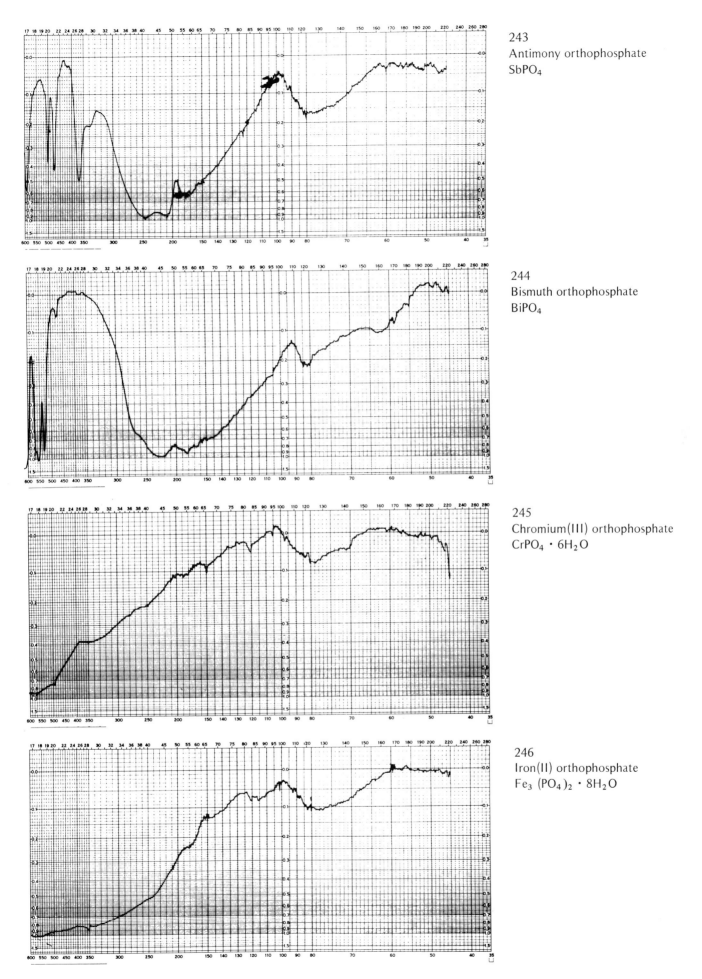

243
Antimony orthophosphate
SbPO$_4$

244
Bismuth orthophosphate
BiPO$_4$

245
Chromium(III) orthophosphate
CrPO$_4$ · 6H$_2$O

246
Iron(II) orthophosphate
Fe$_3$(PO$_4$)$_2$ · 8H$_2$O

171

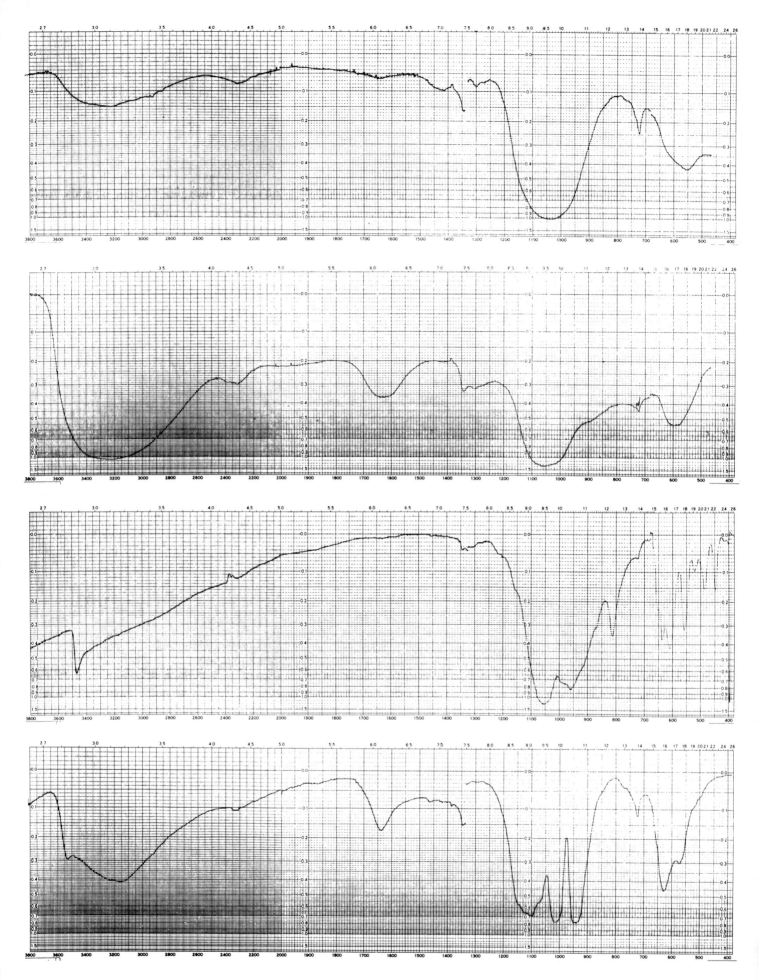

172

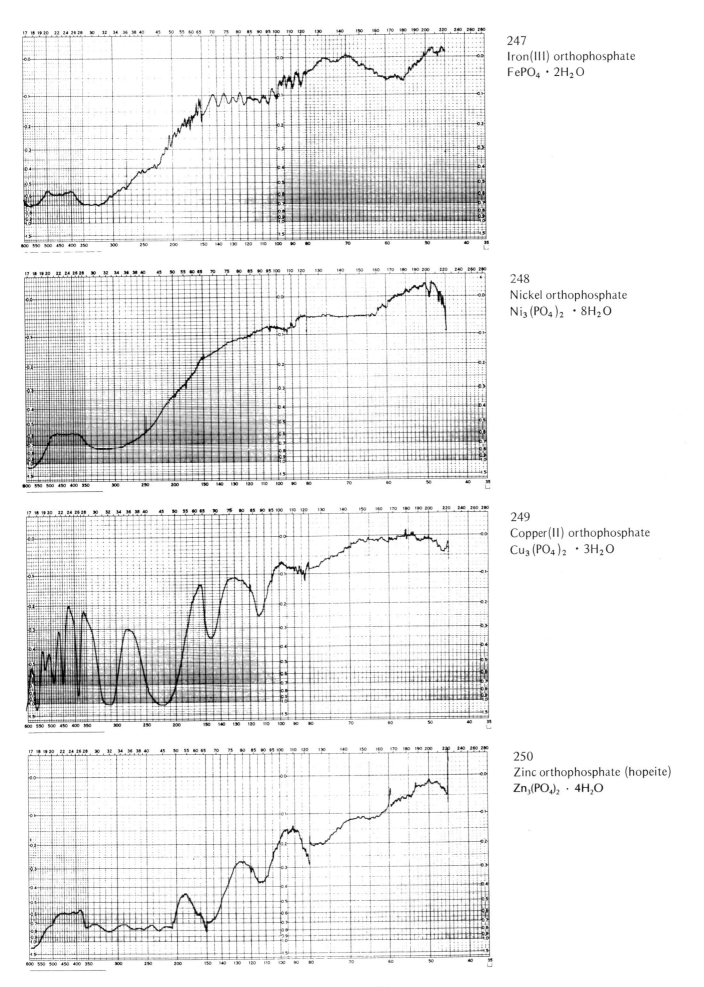

247
Iron(III) orthophosphate
$FePO_4 \cdot 2H_2O$

248
Nickel orthophosphate
$Ni_3(PO_4)_2 \cdot 8H_2O$

249
Copper(II) orthophosphate
$Cu_3(PO_4)_2 \cdot 3H_2O$

250
Zinc orthophosphate (hopeite)
$Zn_3(PO_4)_2 \cdot 4H_2O$

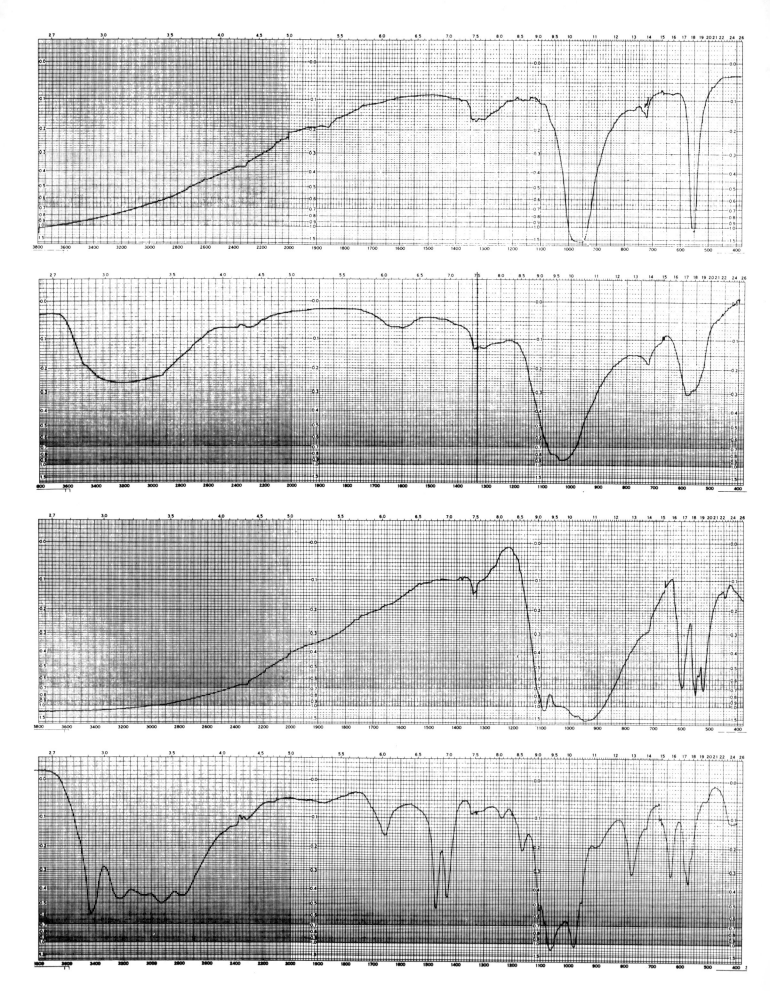

174

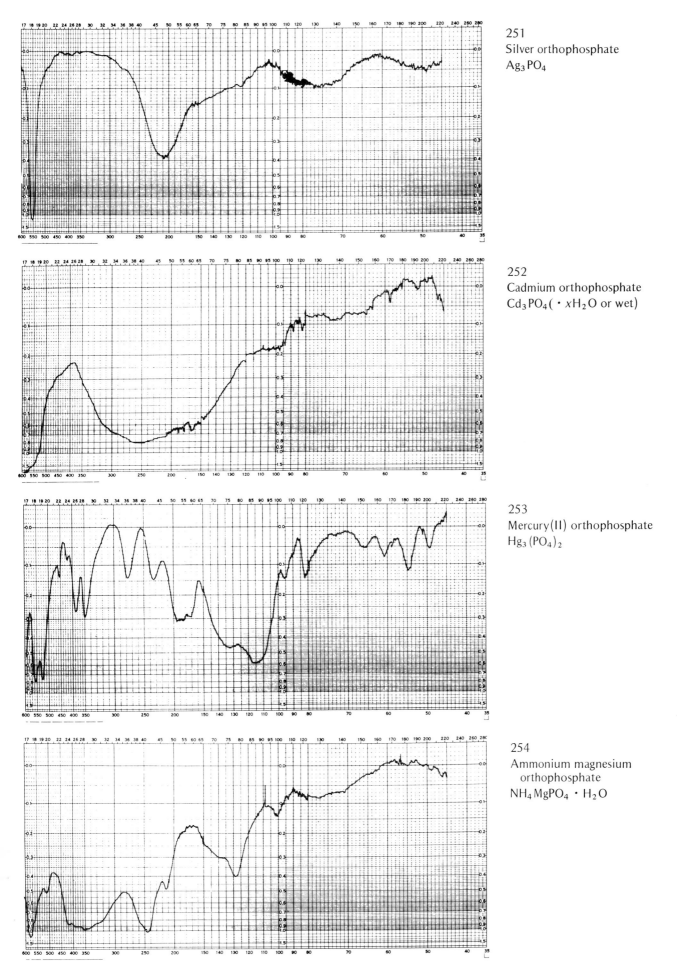

251
Silver orthophosphate
Ag_3PO_4

252
Cadmium orthophosphate
$Cd_3PO_4(\cdot\, xH_2O\ \text{or wet})$

253
Mercury(II) orthophosphate
$Hg_3(PO_4)_2$

254
Ammonium magnesium
  orthophosphate
$NH_4MgPO_4 \cdot H_2O$

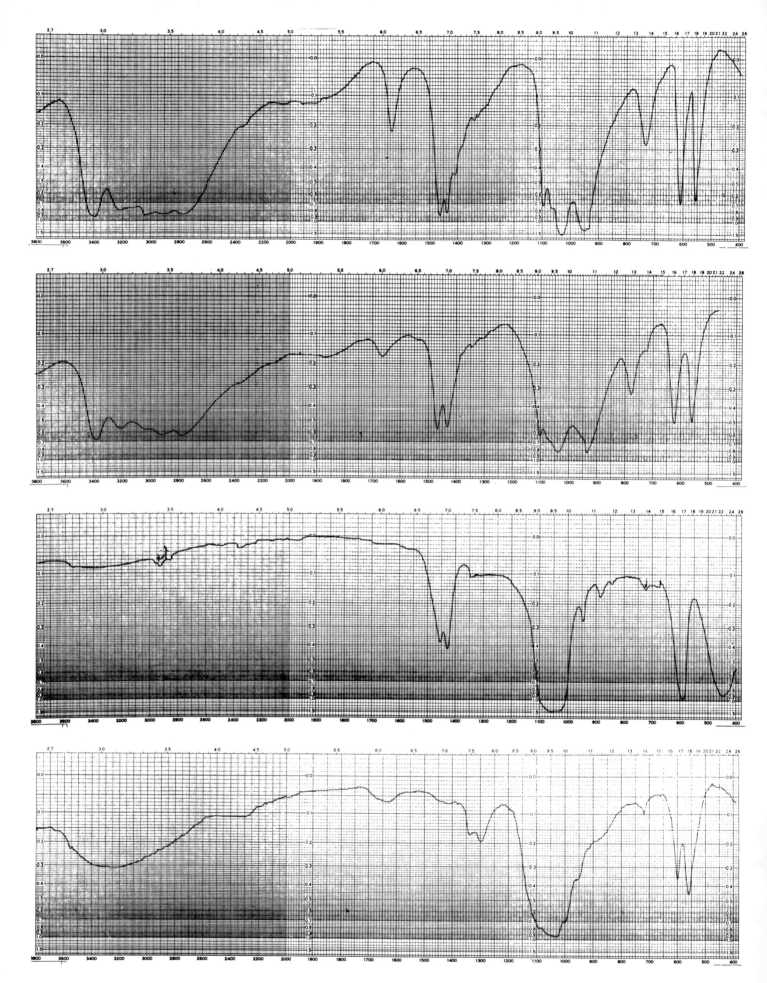

176

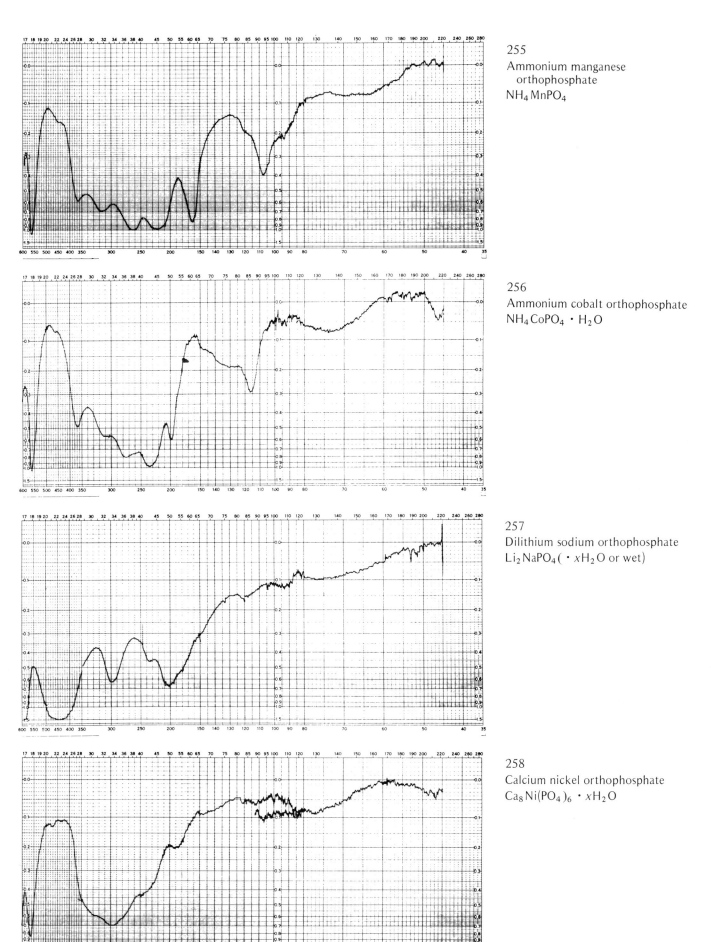

255
Ammonium manganese
  orthophosphate
$NH_4 MnPO_4$

256
Ammonium cobalt orthophosphate
$NH_4 CoPO_4 \cdot H_2O$

257
Dilithium sodium orthophosphate
$Li_2 NaPO_4 (\cdot xH_2O \text{ or wet})$

258
Calcium nickel orthophosphate
$Ca_8 Ni(PO_4)_6 \cdot xH_2O$

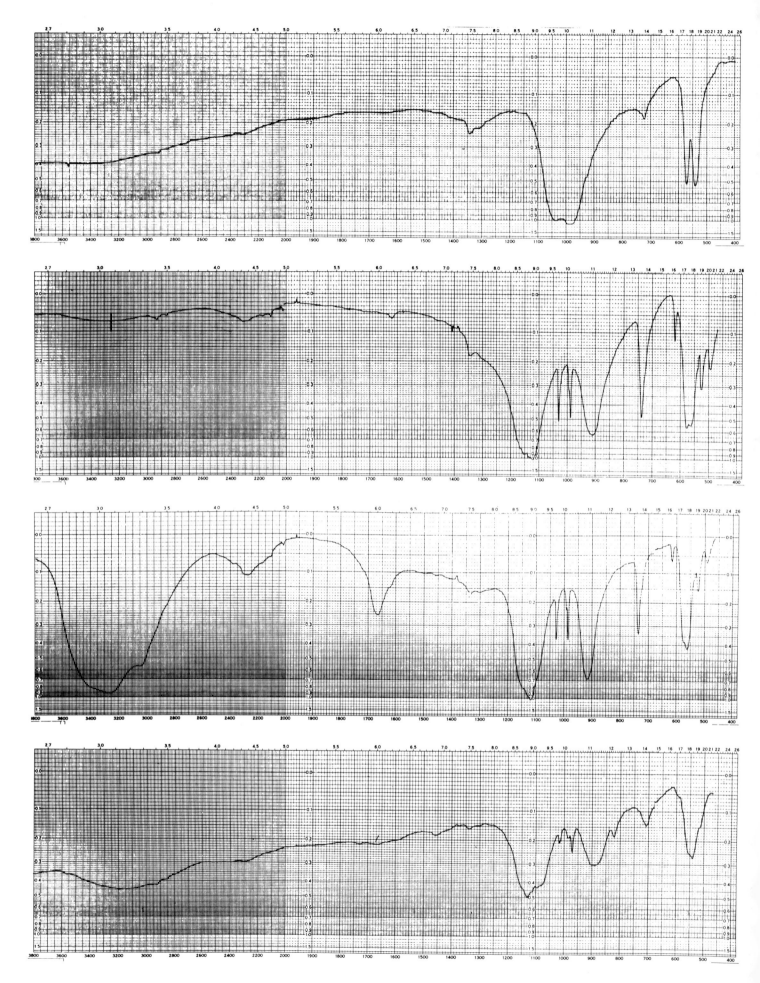

178

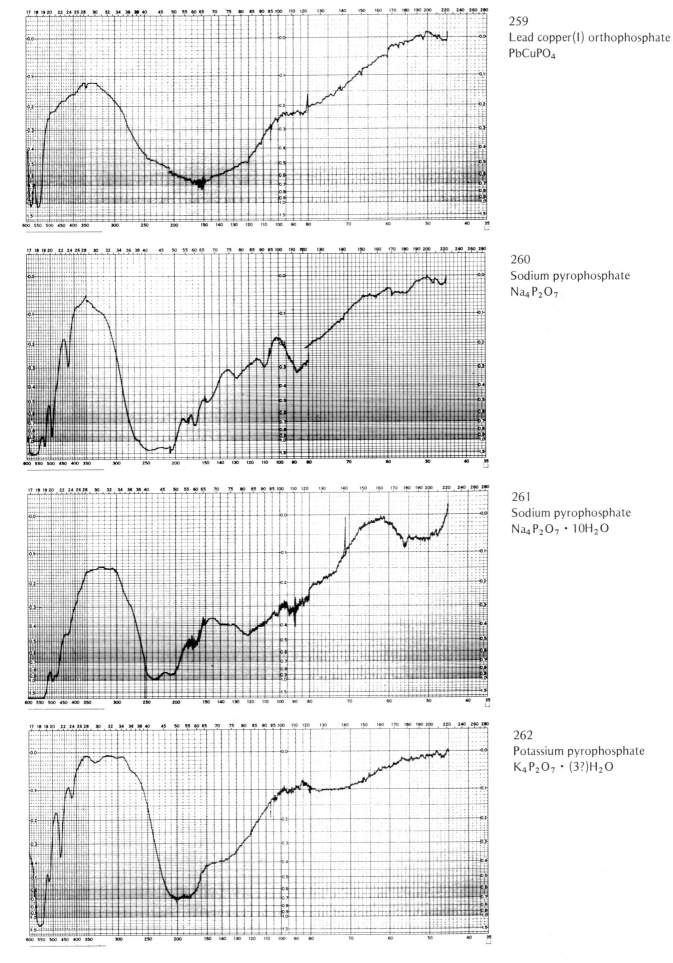

259
Lead copper(I) orthophosphate
PbCuPO$_4$

260
Sodium pyrophosphate
Na$_4$P$_2$O$_7$

261
Sodium pyrophosphate
Na$_4$P$_2$O$_7$ · 10H$_2$O

262
Potassium pyrophosphate
K$_4$P$_2$O$_7$ · (3?)H$_2$O

179

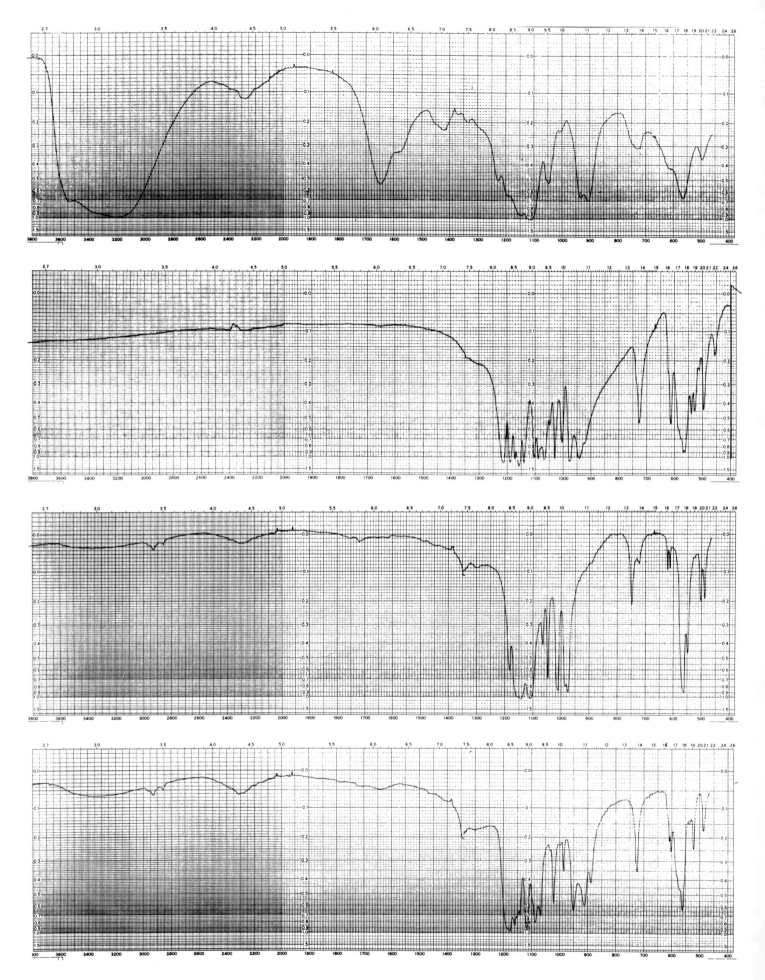

180

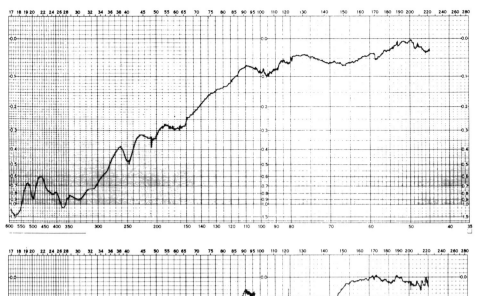

263
Magnesium pyrophosphate
$Mg_2P_2O_7 \cdot 3H_2O$

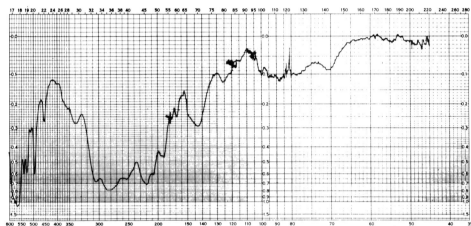

264
Calcium pyrophosphate (β-form)
$Ca_2P_2O_7$

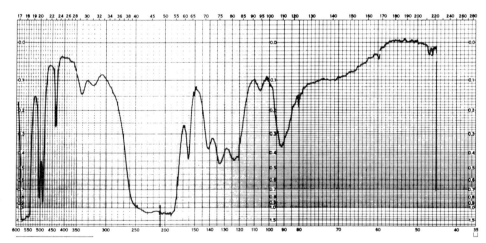

265
Strontium pyrophosphate (α-form, orthorhombic)
$Sr_2P_2O_7$

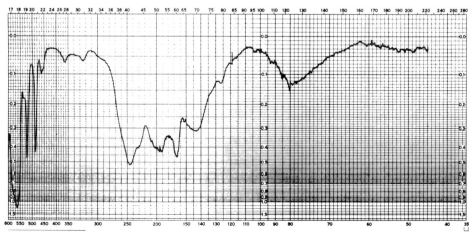

266
Strontium pyrophosphate (β-form, tetragonal)
$Sr_2P_2O_7$

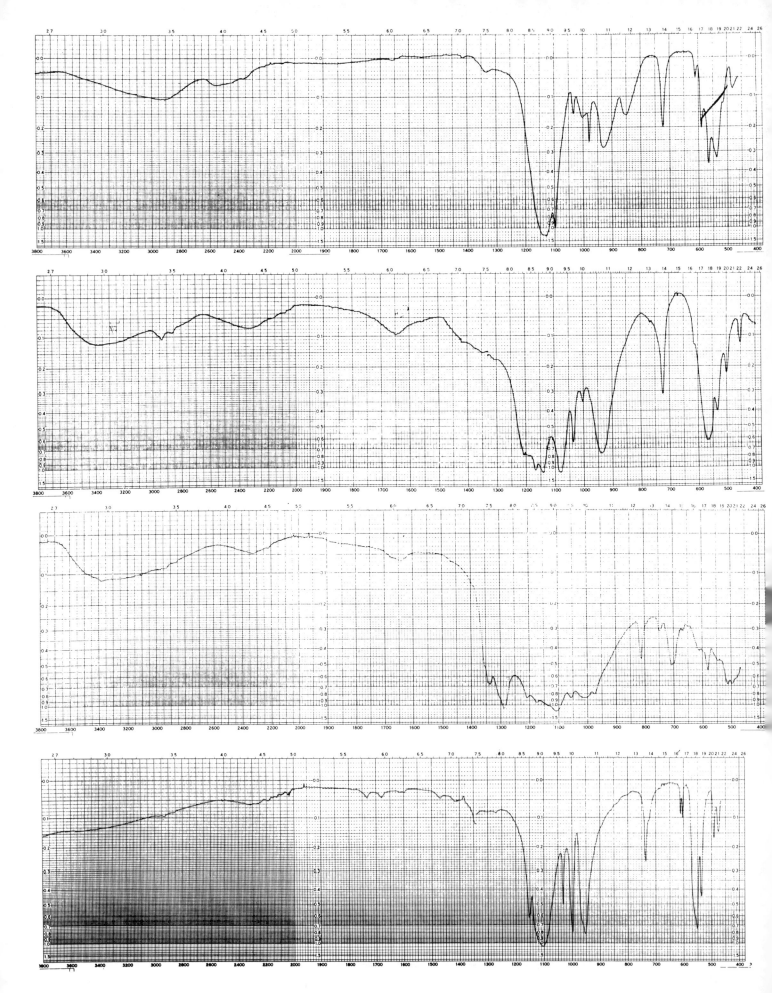

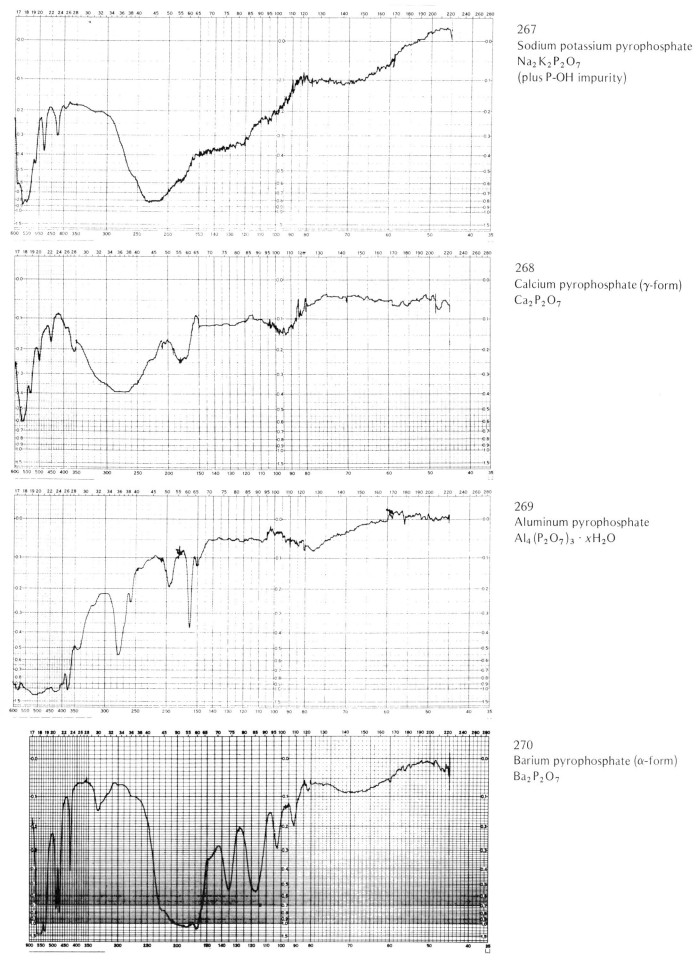

267
Sodium potassium pyrophosphate
$Na_2K_2P_2O_7$
(plus P-OH impurity)

268
Calcium pyrophosphate (γ-form)
$Ca_2P_2O_7$

269
Aluminum pyrophosphate
$Al_4(P_2O_7)_3 \cdot xH_2O$

270
Barium pyrophosphate (α-form)
$Ba_2P_2O_7$

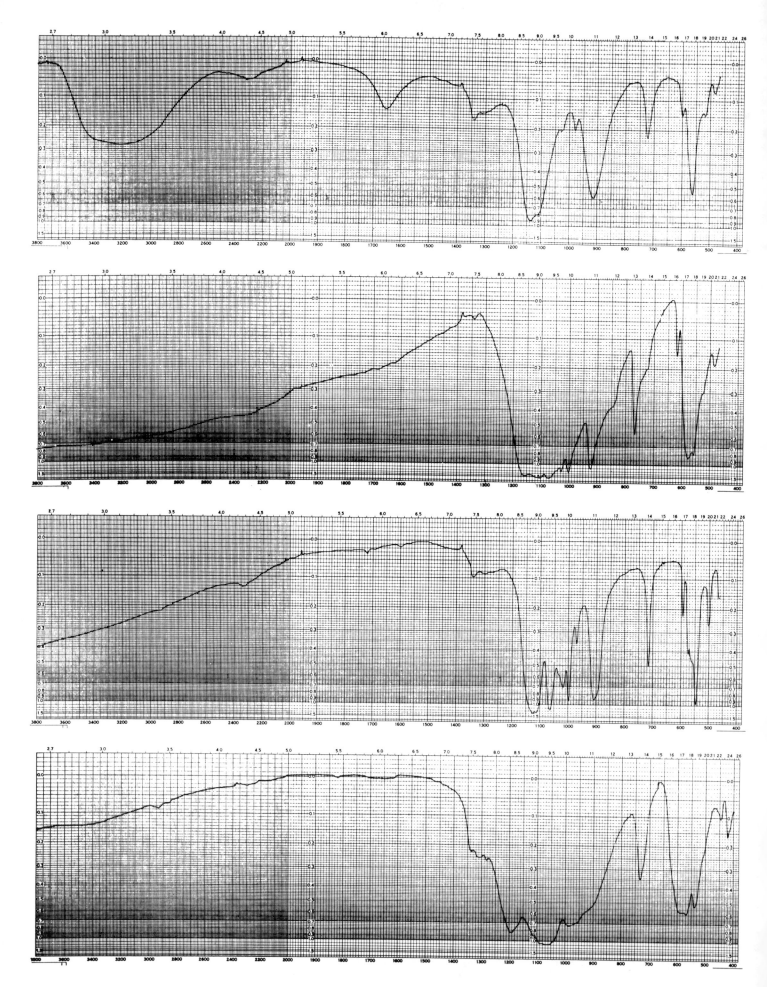

184

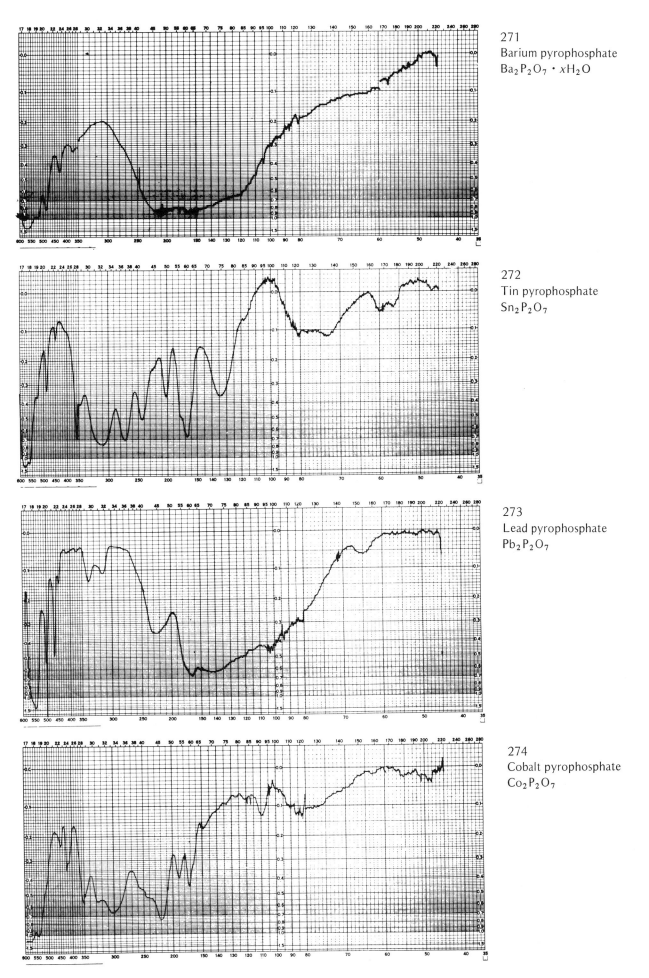

271
Barium pyrophosphate
$Ba_2P_2O_7 \cdot xH_2O$

272
Tin pyrophosphate
$Sn_2P_2O_7$

273
Lead pyrophosphate
$Pb_2P_2O_7$

274
Cobalt pyrophosphate
$Co_2P_2O_7$

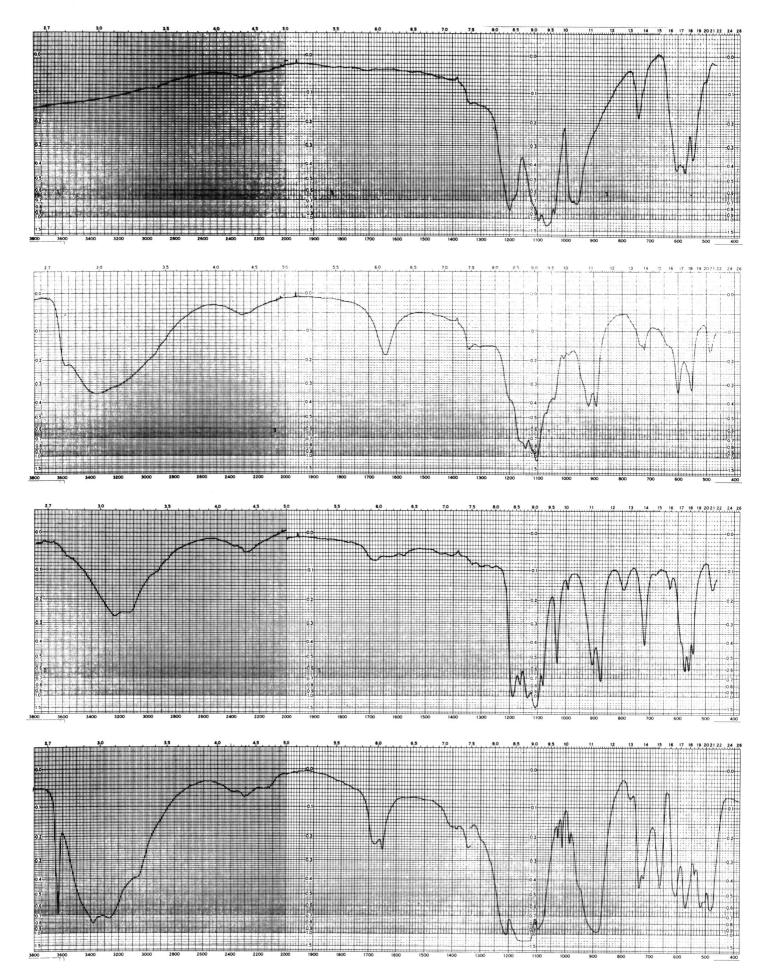

186

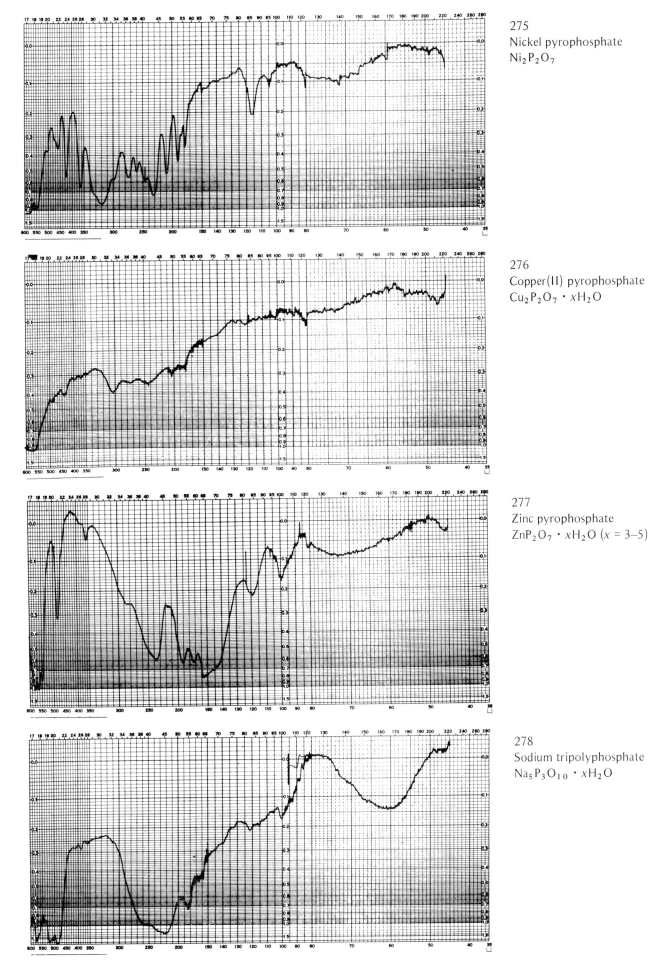

275
Nickel pyrophosphate
$Ni_2P_2O_7$

276
Copper(II) pyrophosphate
$Cu_2P_2O_7 \cdot xH_2O$

277
Zinc pyrophosphate
$ZnP_2O_7 \cdot xH_2O$ ($x = 3–5$)

278
Sodium tripolyphosphate
$Na_5P_3O_{10} \cdot xH_2O$

187

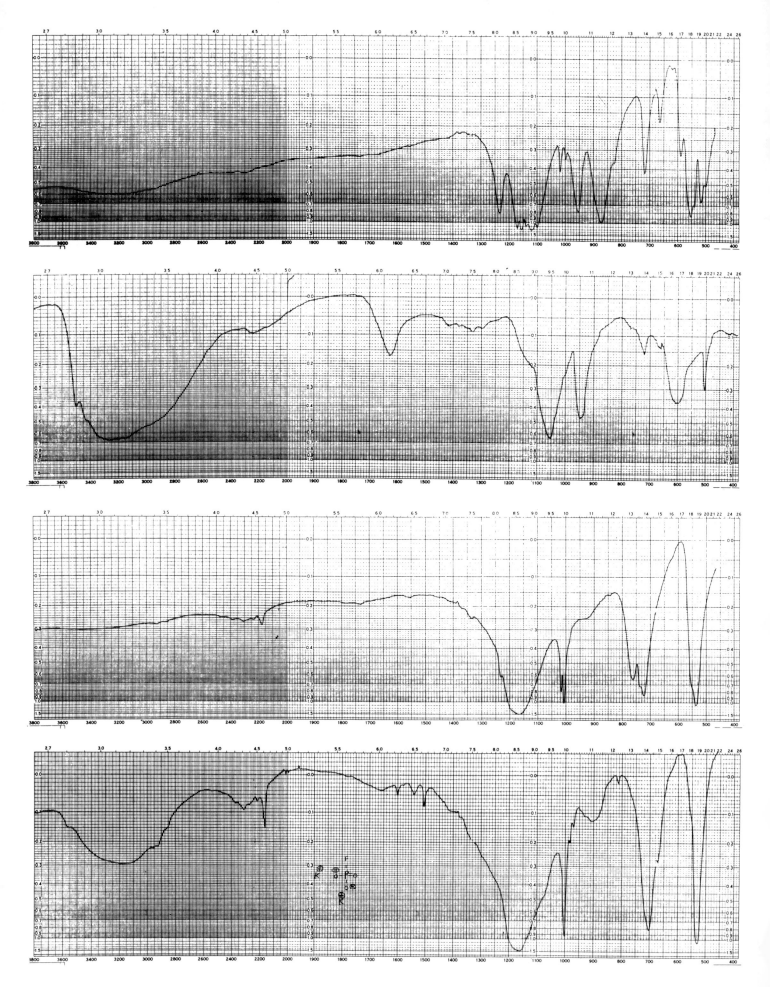

188

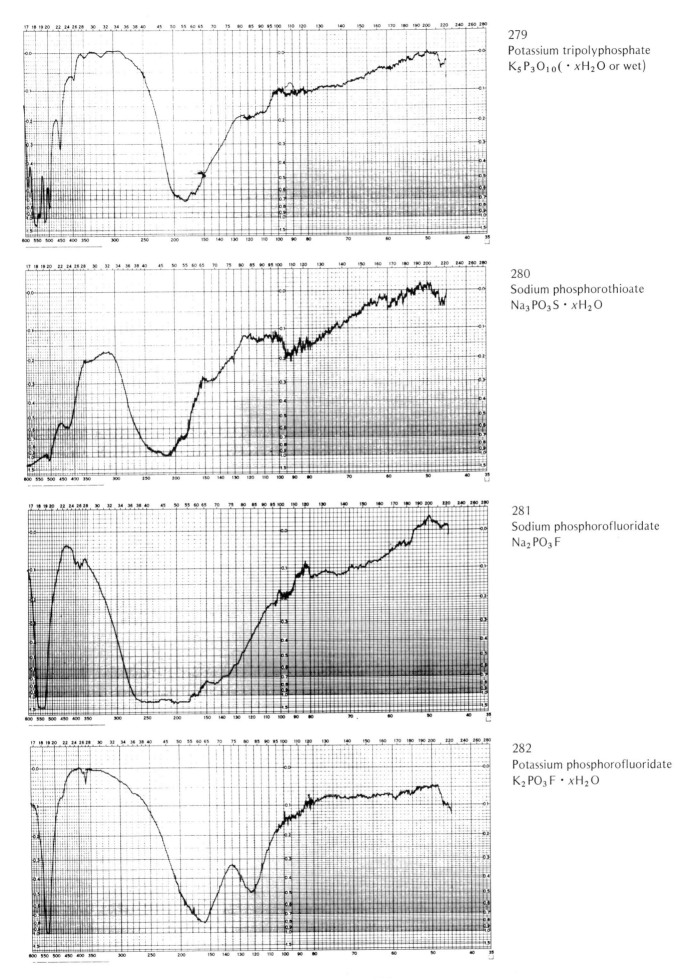

279
Potassium tripolyphosphate
$K_5P_3O_{10}(\cdot\, xH_2O$ or wet)

280
Sodium phosphorothioate
$Na_3PO_3S \cdot xH_2O$

281
Sodium phosphorofluoridate
Na_2PO_3F

282
Potassium phosphorofluoridate
$K_2PO_3F \cdot xH_2O$

189

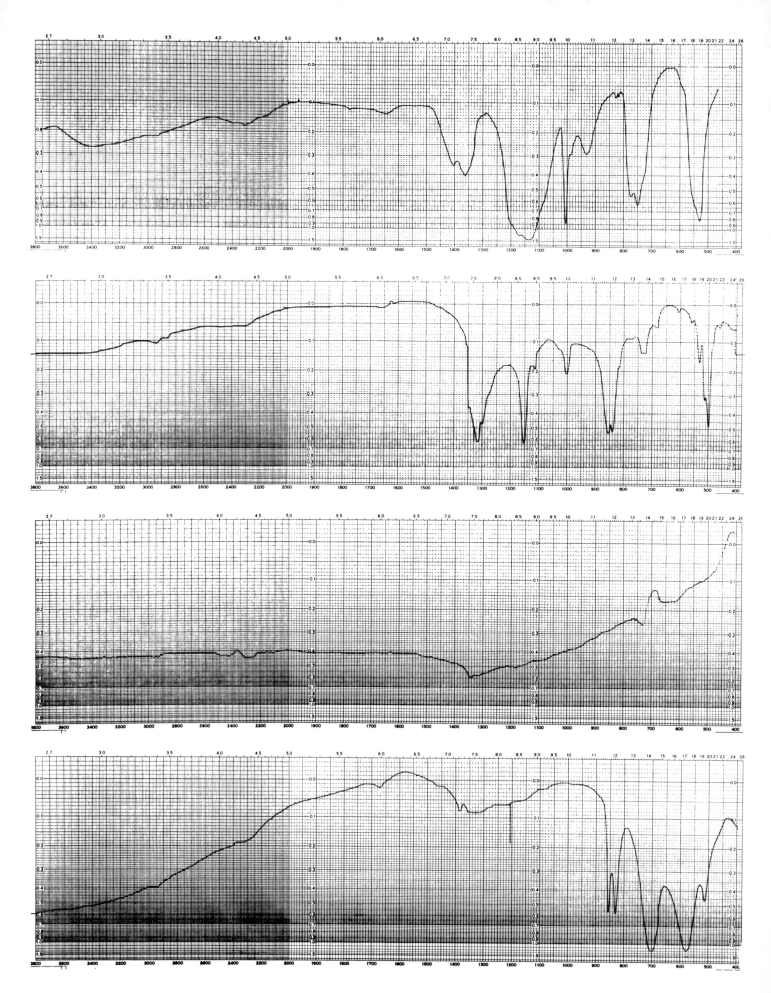

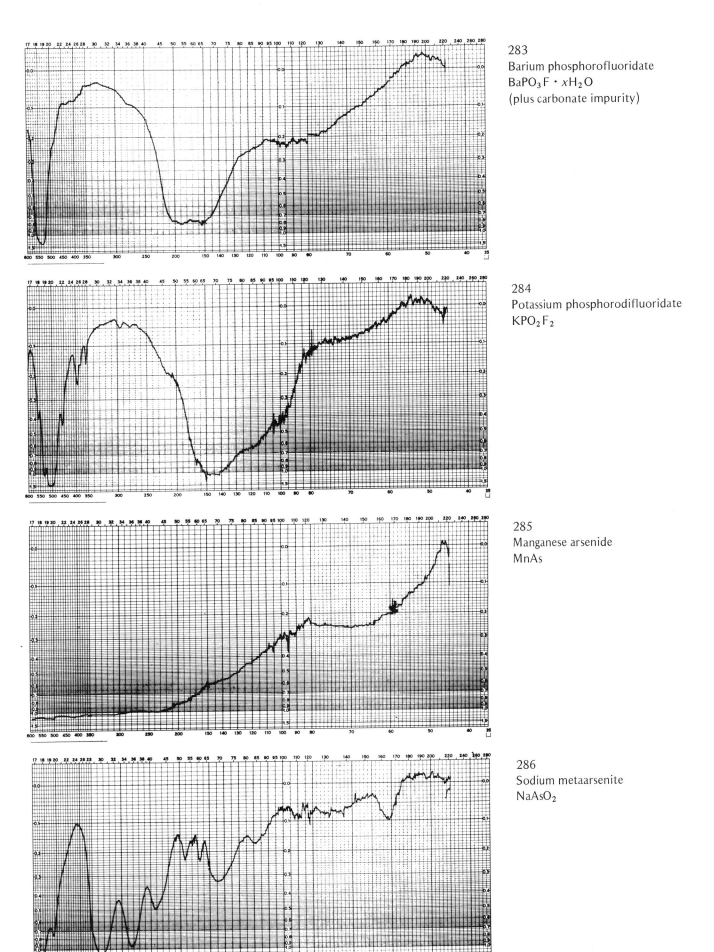

283
Barium phosphorofluoridate
$BaPO_3F \cdot xH_2O$
(plus carbonate impurity)

284
Potassium phosphorodifluoridate
KPO_2F_2

285
Manganese arsenide
MnAs

286
Sodium metaarsenite
$NaAsO_2$

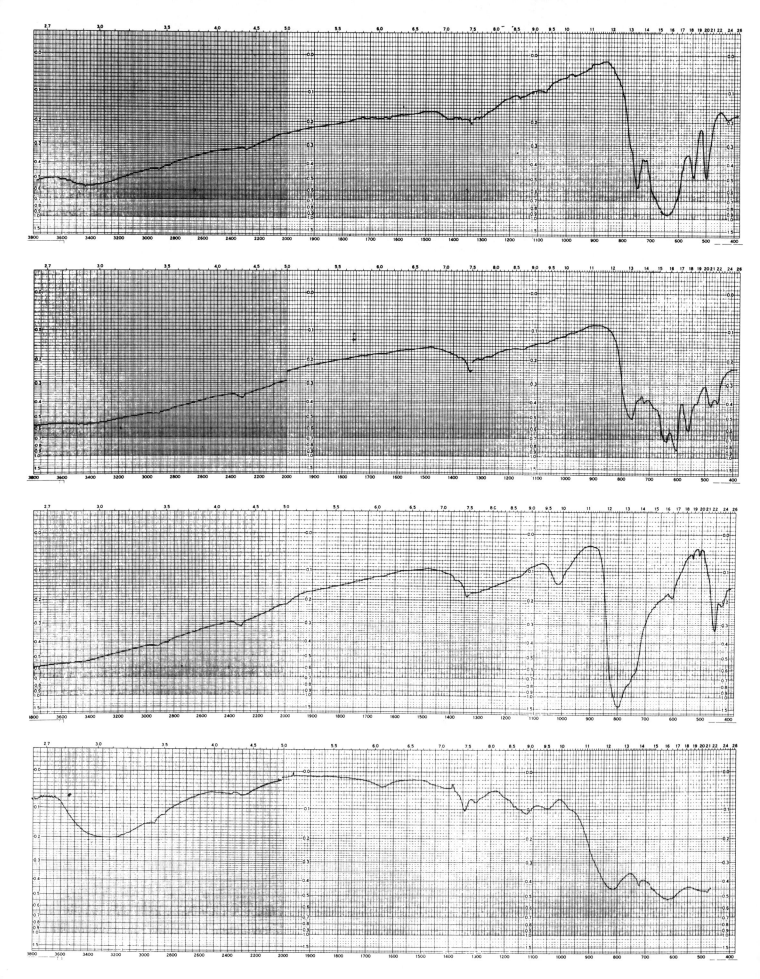

192

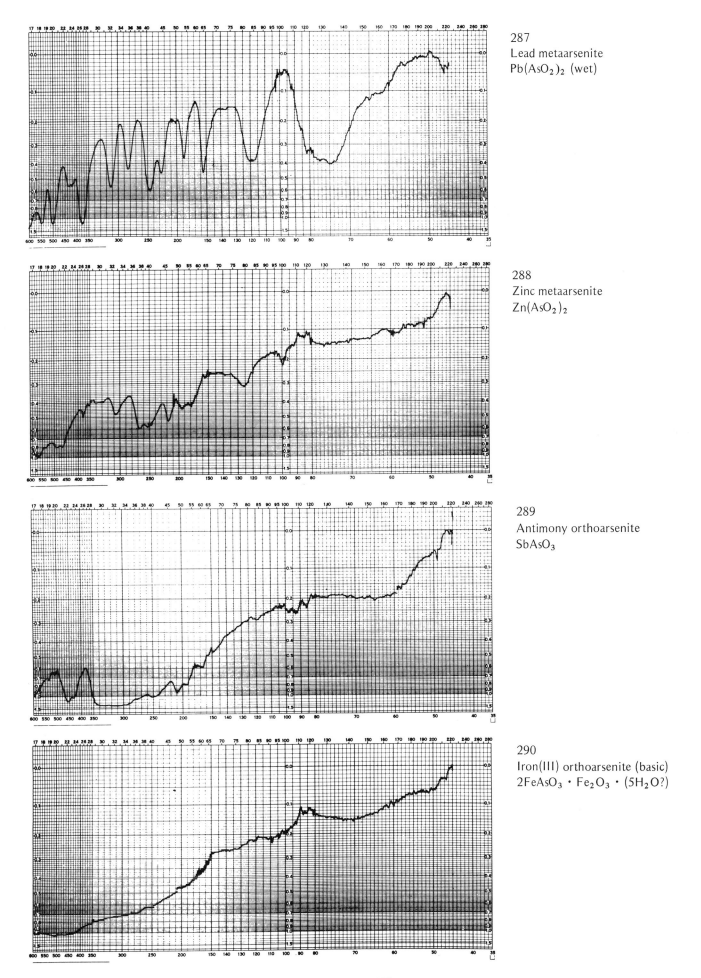

287
Lead metaarsenite
Pb(AsO₂)₂ (wet)

288
Zinc metaarsenite
Zn(AsO₂)₂

289
Antimony orthoarsenite
SbAsO₃

290
Iron(III) orthoarsenite (basic)
2FeAsO₃ · Fe₂O₃ · (5H₂O?)

193

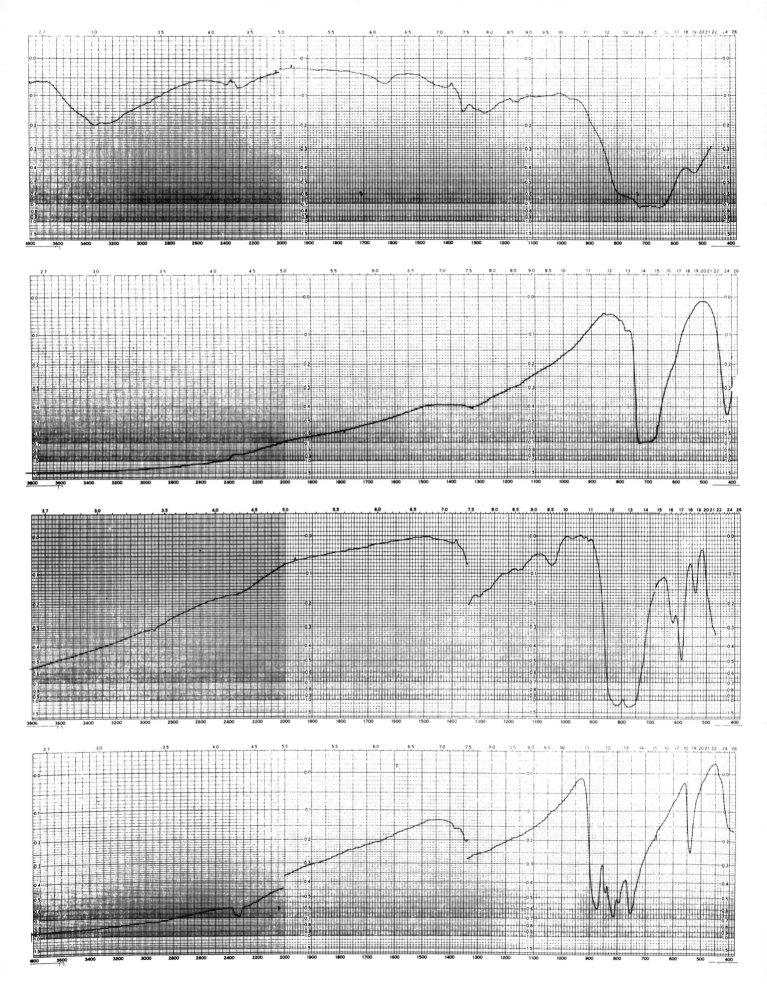

194

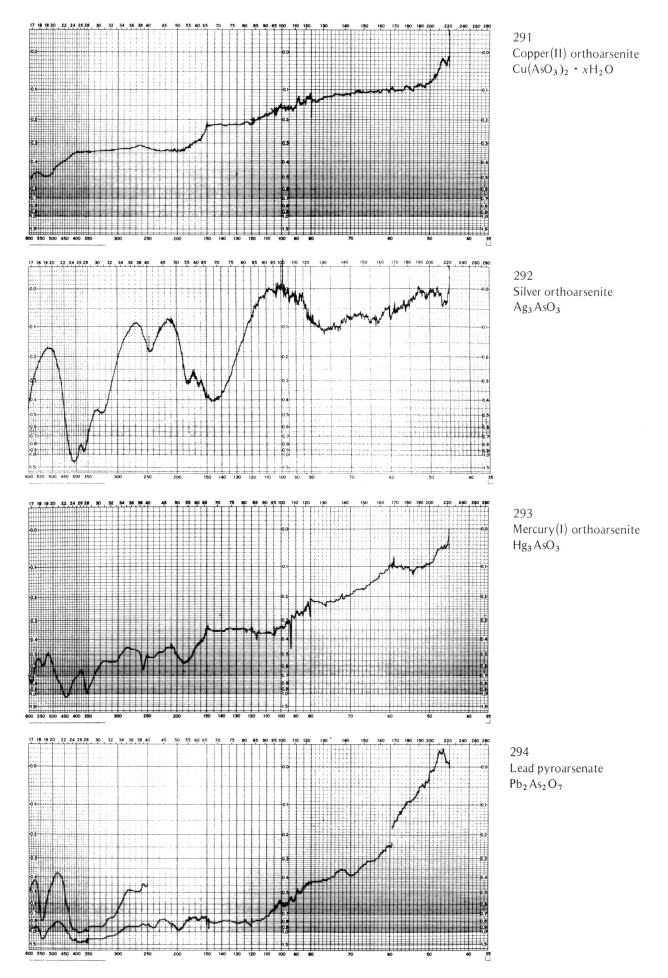

291
Copper(II) orthoarsenite
$Cu(AsO_3)_2 \cdot xH_2O$

292
Silver orthoarsenite
Ag_3AsO_3

293
Mercury(I) orthoarsenite
Hg_3AsO_3

294
Lead pyroarsenate
$Pb_2As_2O_7$

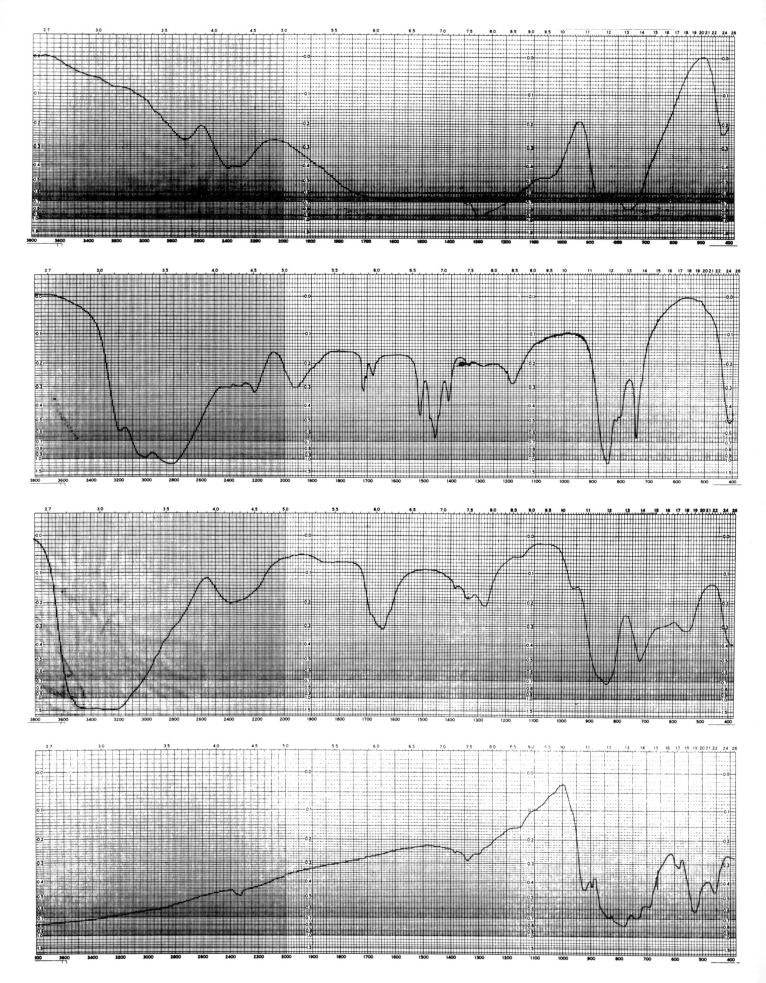

196

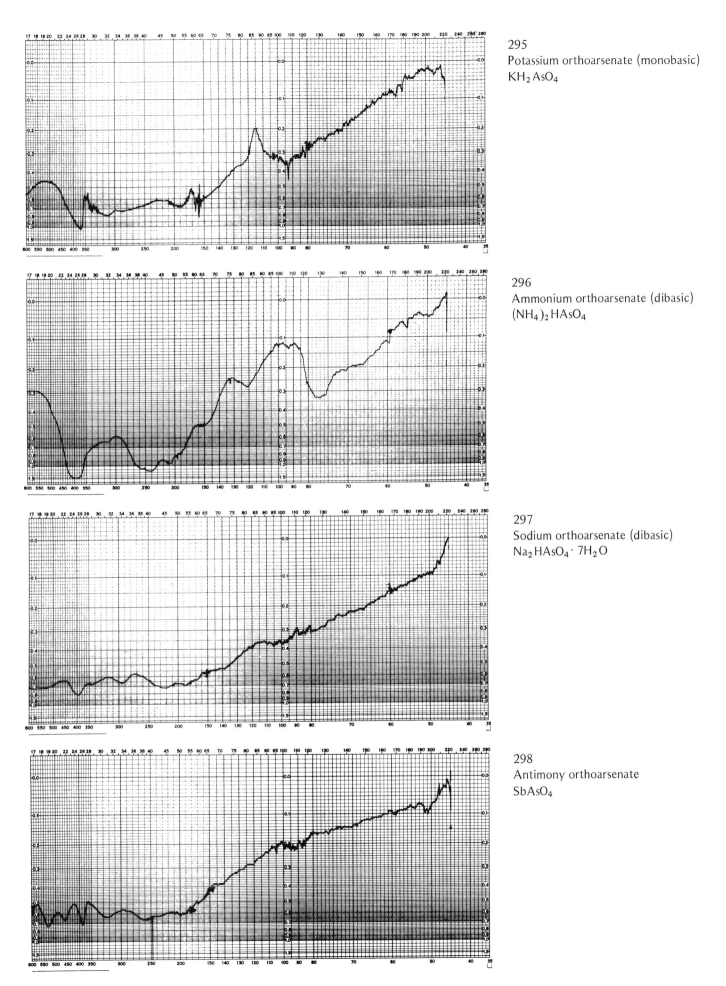

295
Potassium orthoarsenate (monobasic)
KH_2AsO_4

296
Ammonium orthoarsenate (dibasic)
$(NH_4)_2HAsO_4$

297
Sodium orthoarsenate (dibasic)
$Na_2HAsO_4 \cdot 7H_2O$

298
Antimony orthoarsenate
$SbAsO_4$

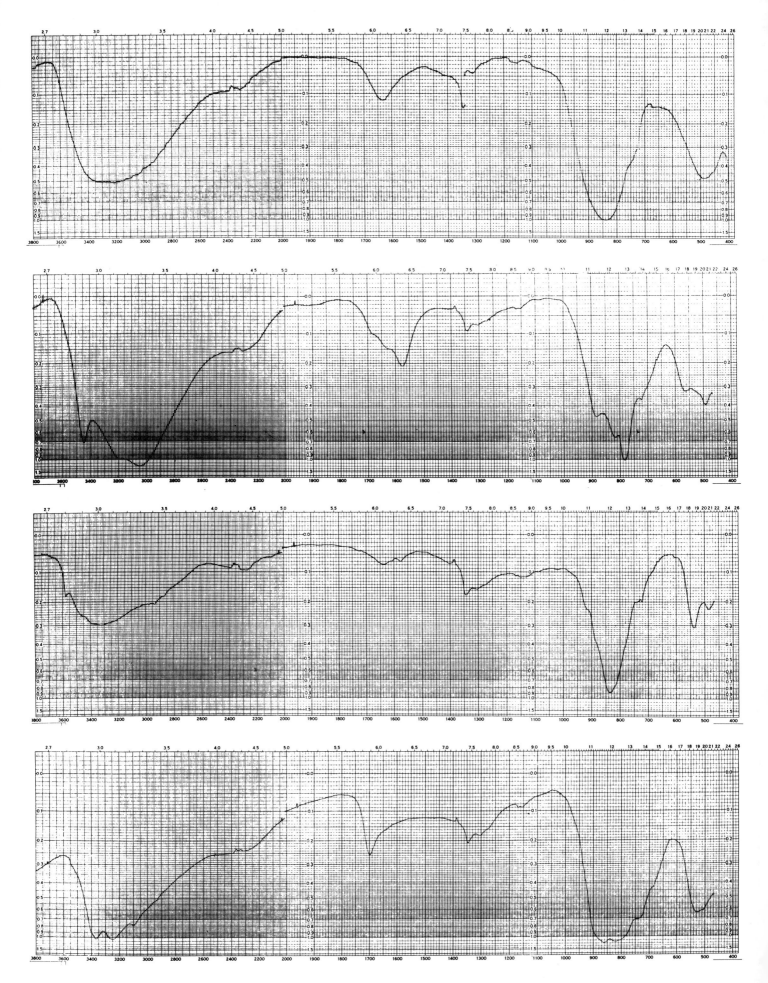

198

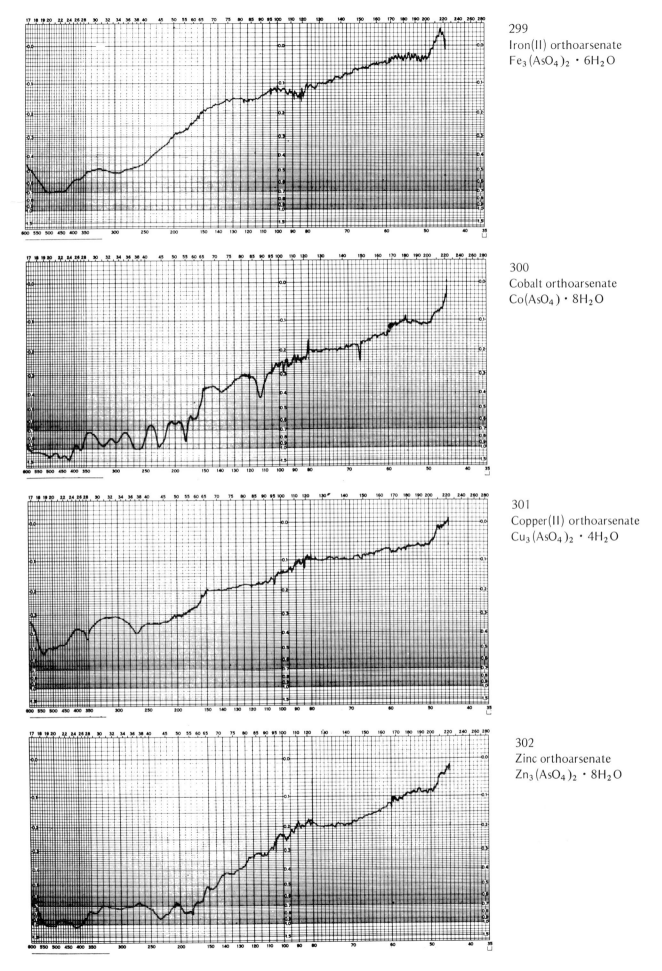

299
Iron(II) orthoarsenate
$Fe_3(AsO_4)_2 \cdot 6H_2O$

300
Cobalt orthoarsenate
$Co(AsO_4) \cdot 8H_2O$

301
Copper(II) orthoarsenate
$Cu_3(AsO_4)_2 \cdot 4H_2O$

302
Zinc orthoarsenate
$Zn_3(AsO_4)_2 \cdot 8H_2O$

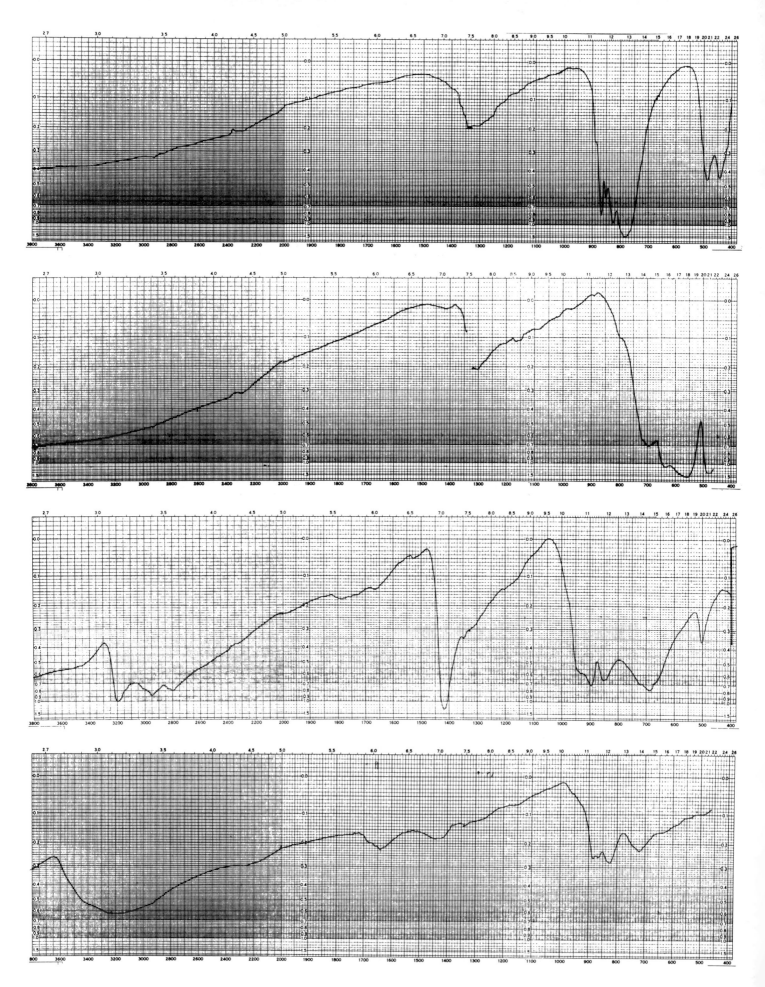

200

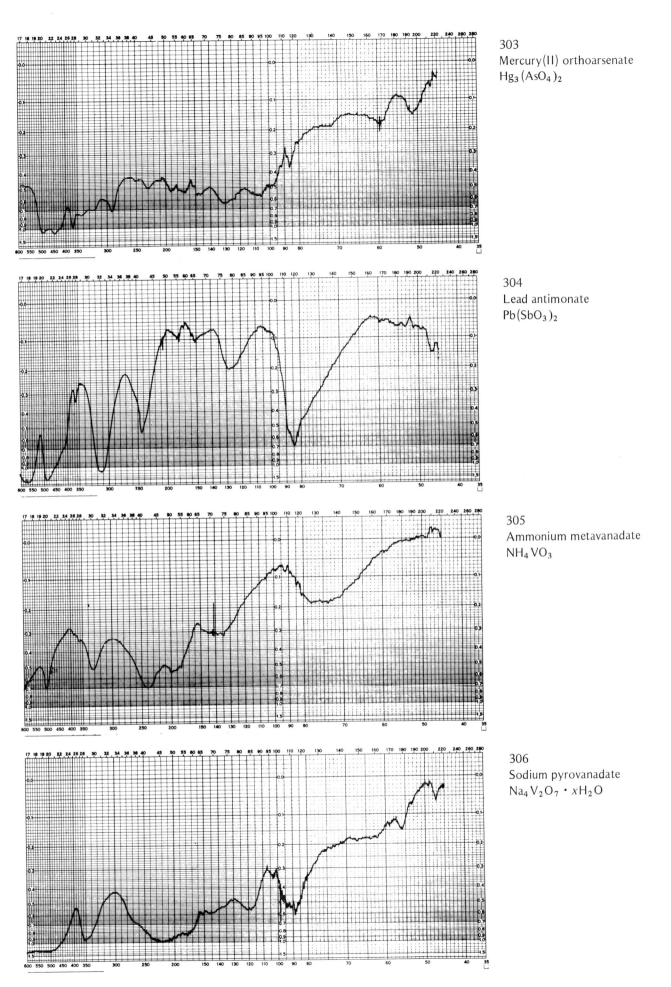

303
Mercury(II) orthoarsenate
$Hg_3(AsO_4)_2$

304
Lead antimonate
$Pb(SbO_3)_2$

305
Ammonium metavanadate
NH_4VO_3

306
Sodium pyrovanadate
$Na_4V_2O_7 \cdot xH_2O$

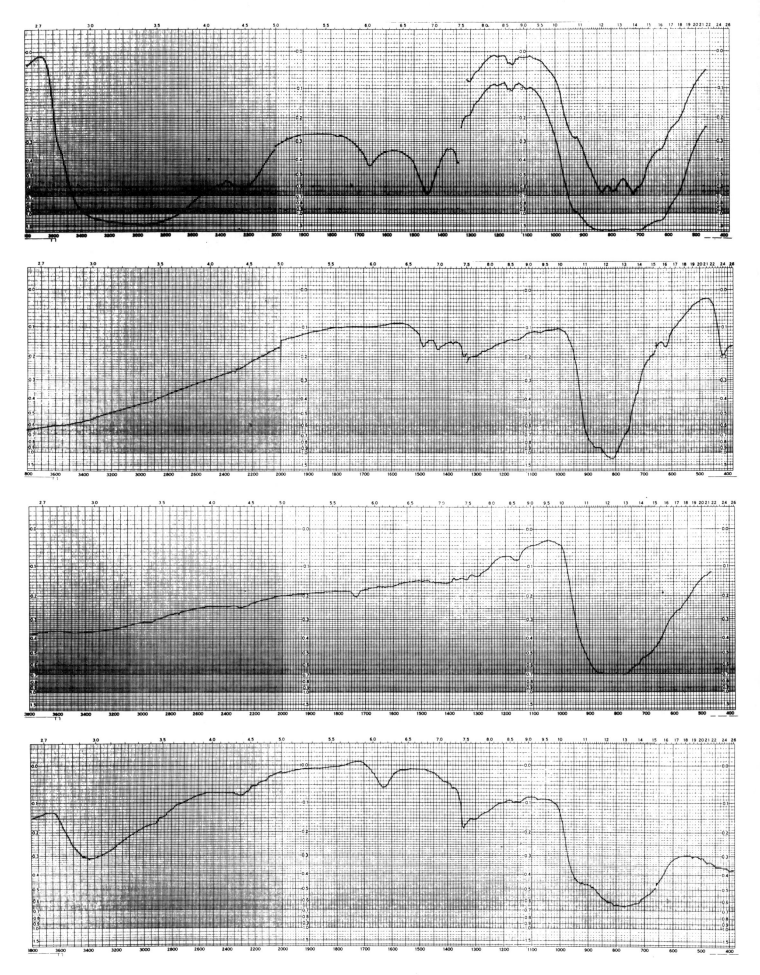

202

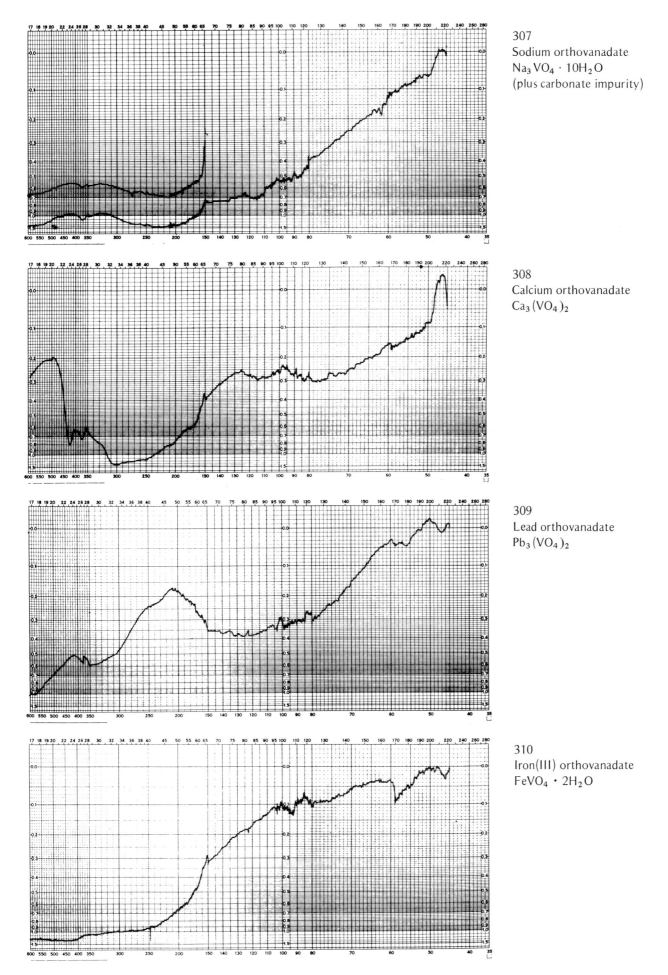

307
Sodium orthovanadate
Na$_3$VO$_4$ · 10H$_2$O
(plus carbonate impurity)

308
Calcium orthovanadate
Ca$_3$(VO$_4$)$_2$

309
Lead orthovanadate
Pb$_3$(VO$_4$)$_2$

310
Iron(III) orthovanadate
FeVO$_4$ · 2H$_2$O

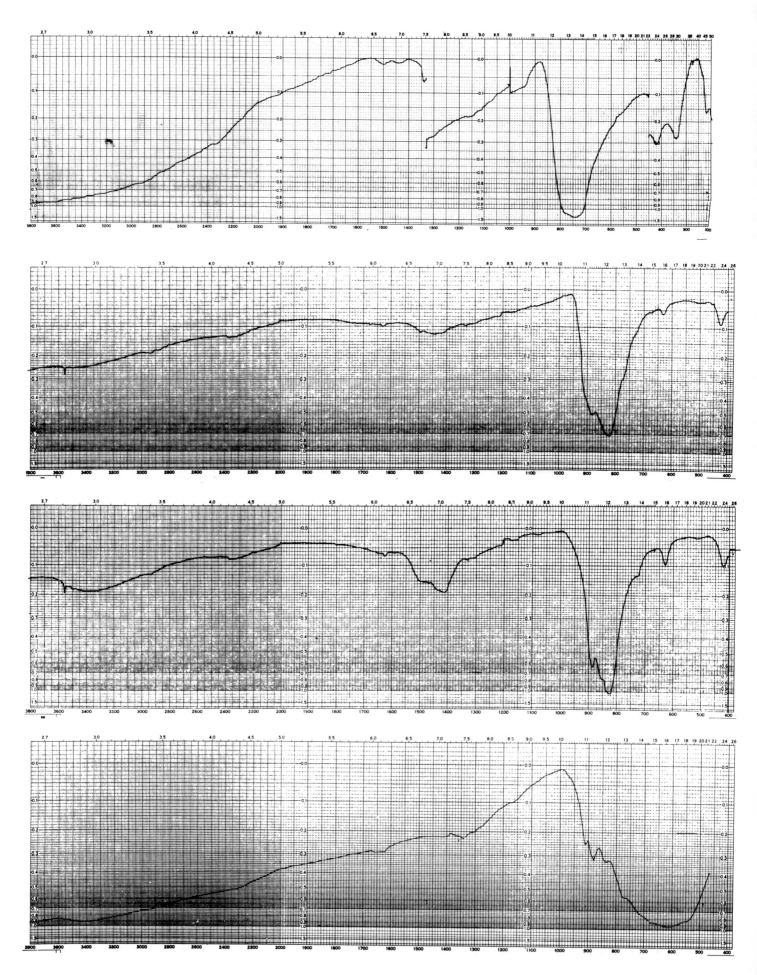

204

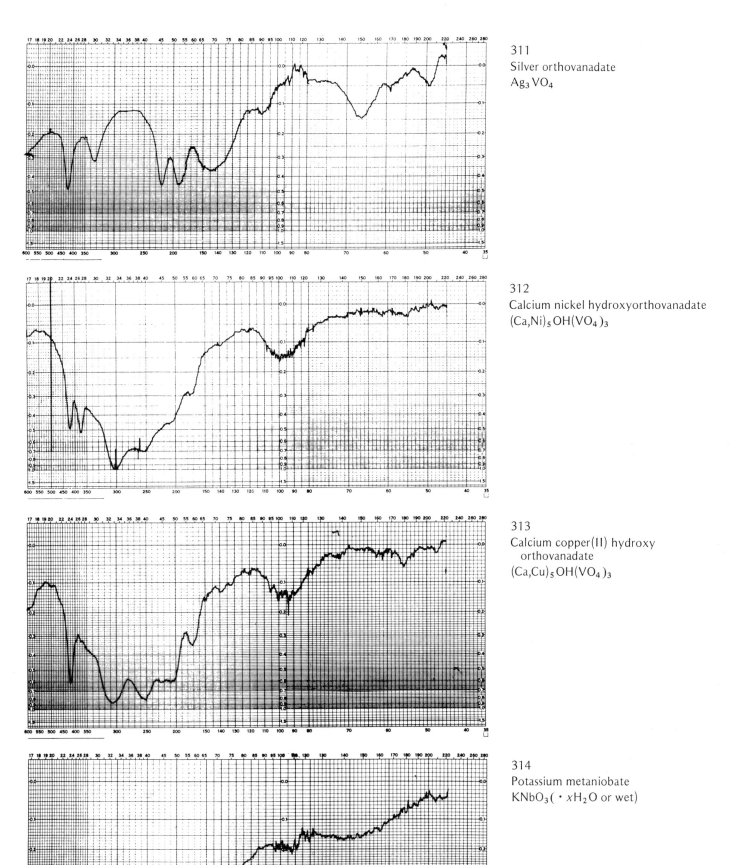

311
Silver orthovanadate
Ag_3VO_4

312
Calcium nickel hydroxyorthovanadate
$(Ca,Ni)_5OH(VO_4)_3$

313
Calcium copper(II) hydroxy
  orthovanadate
$(Ca,Cu)_5OH(VO_4)_3$

314
Potassium metaniobate
$KNbO_3(\cdot xH_2O$ or wet$)$

205

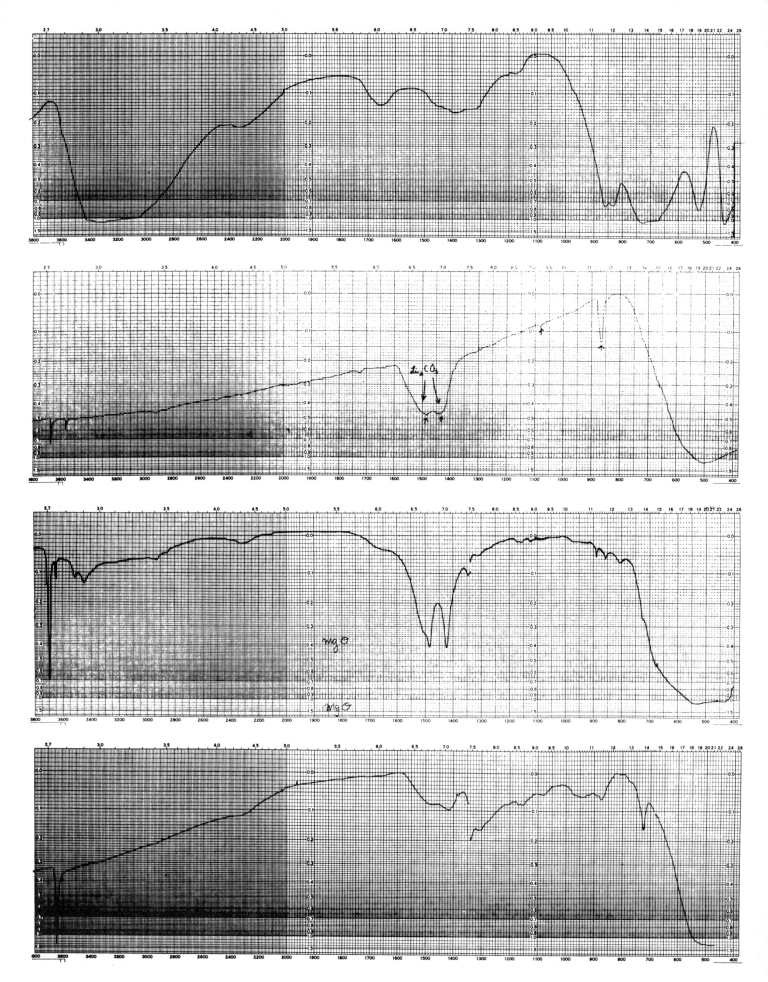

206

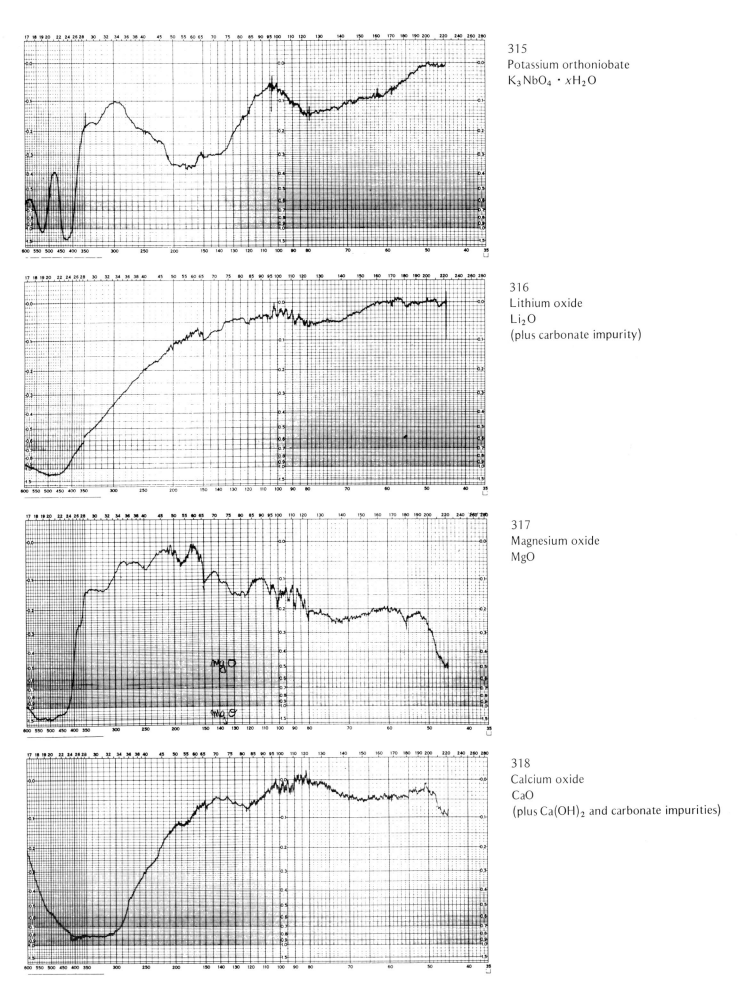

315
Potassium orthoniobate
$K_3NbO_4 \cdot xH_2O$

316
Lithium oxide
Li_2O
(plus carbonate impurity)

317
Magnesium oxide
MgO

318
Calcium oxide
CaO
(plus $Ca(OH)_2$ and carbonate impurities)

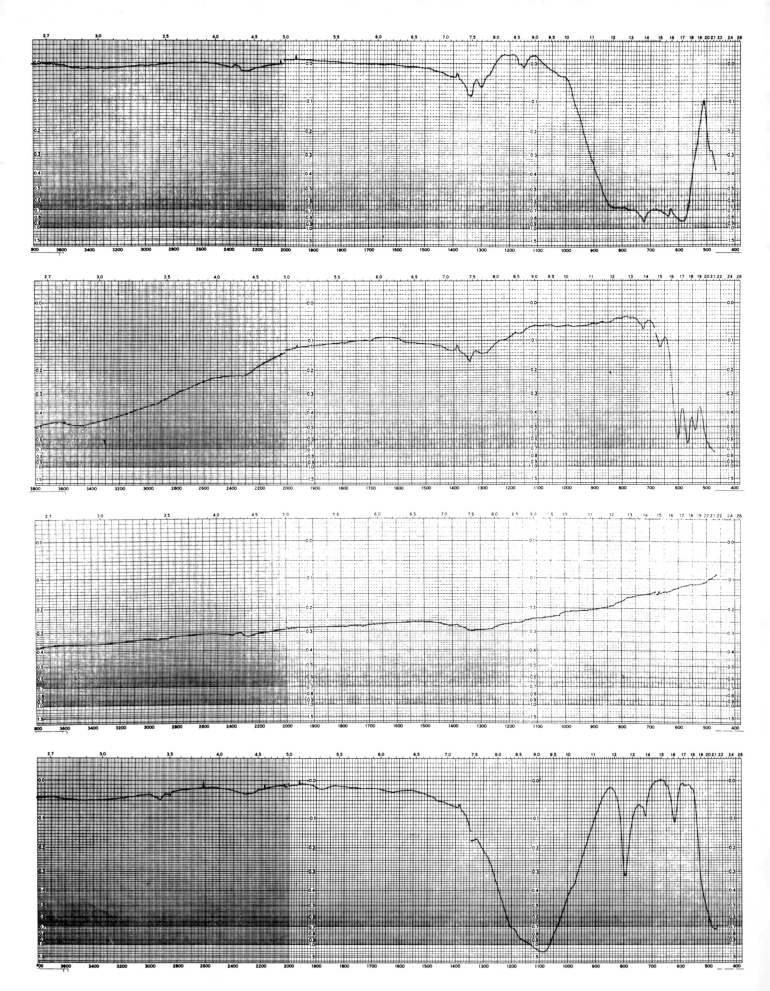

208

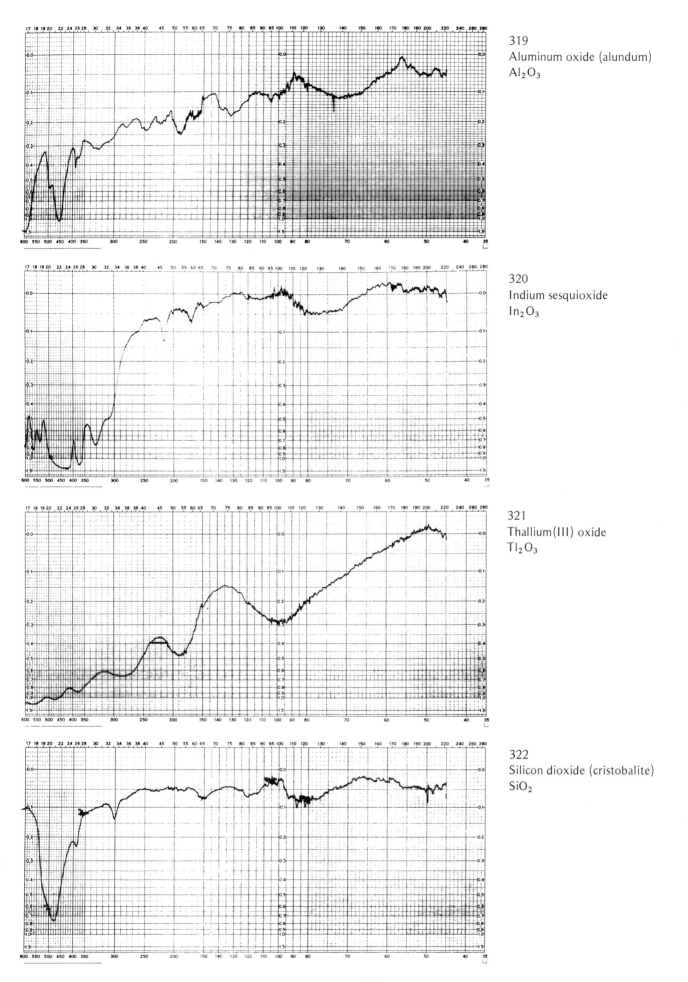

319
Aluminum oxide (alundum)
Al_2O_3

320
Indium sesquioxide
In_2O_3

321
Thallium(III) oxide
Tl_2O_3

322
Silicon dioxide (cristobalite)
SiO_2

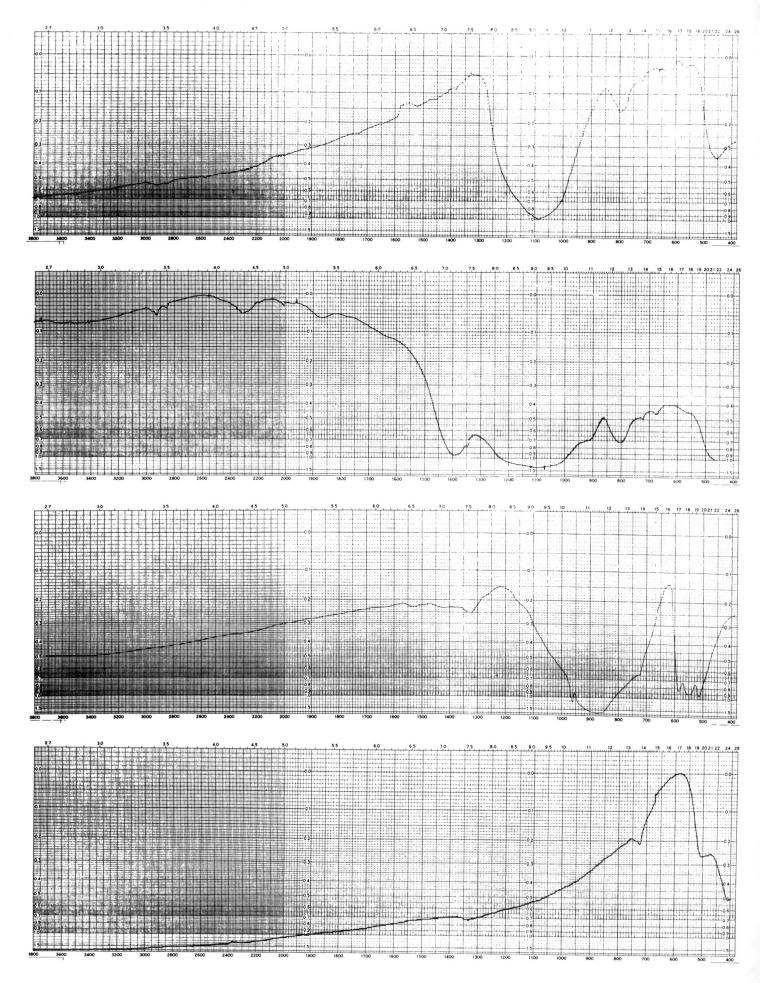

210

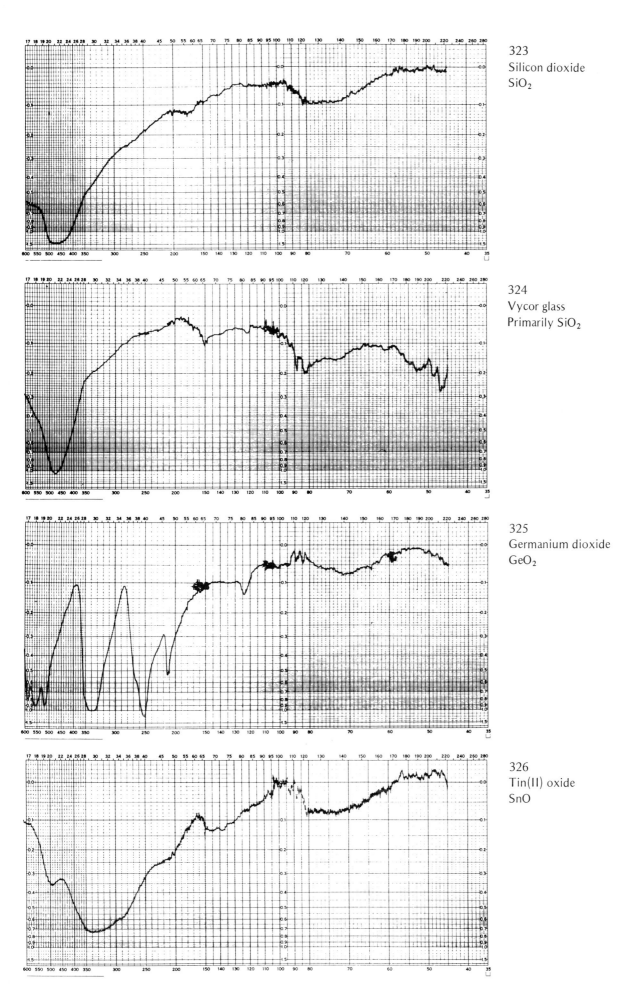

323
Silicon dioxide
SiO$_2$

324
Vycor glass
Primarily SiO$_2$

325
Germanium dioxide
GeO$_2$

326
Tin(II) oxide
SnO

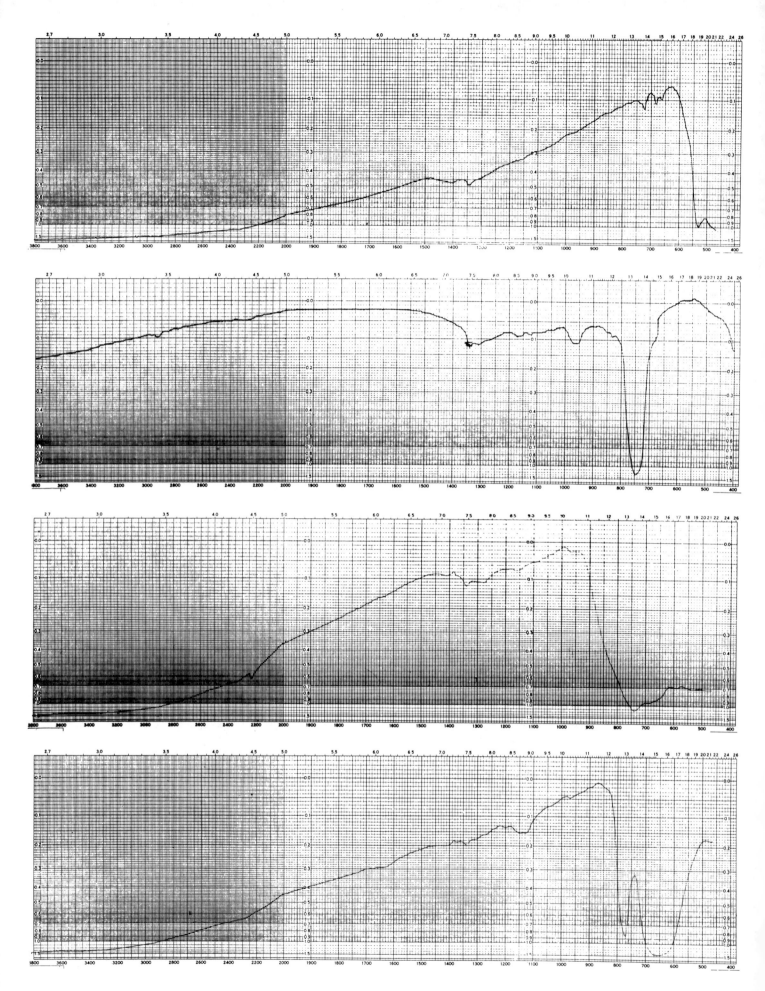

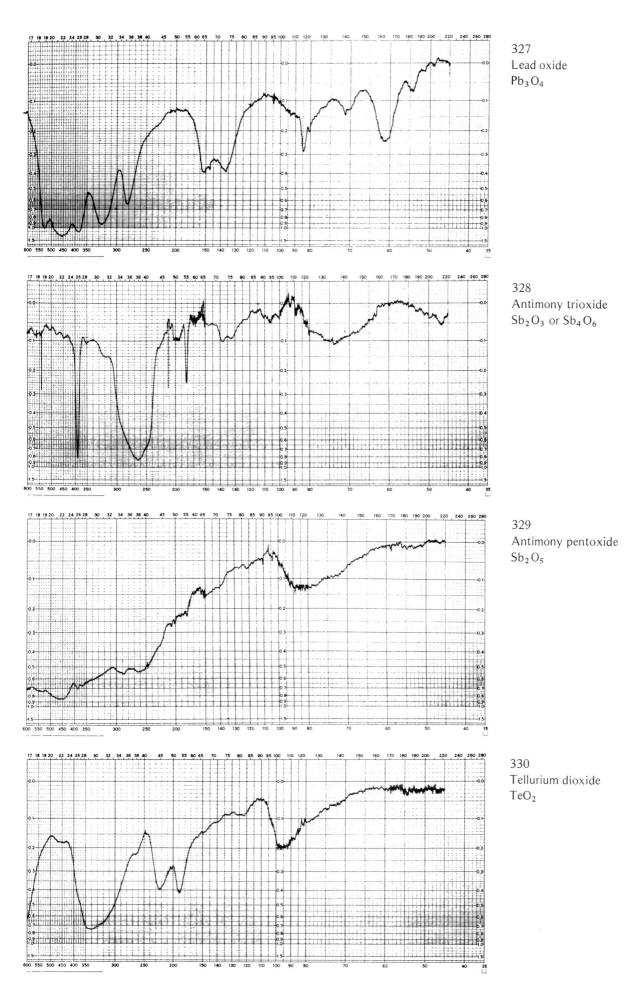

327
Lead oxide
Pb_3O_4

328
Antimony trioxide
Sb_2O_3 or Sb_4O_6

329
Antimony pentoxide
Sb_2O_5

330
Tellurium dioxide
TeO_2

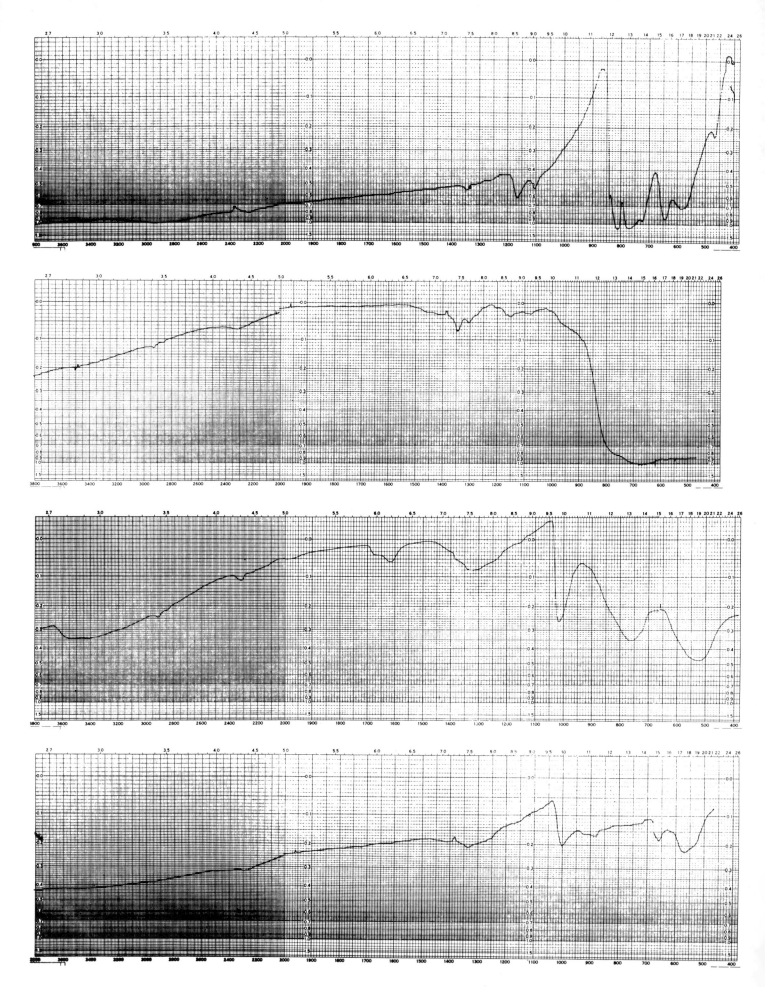

214

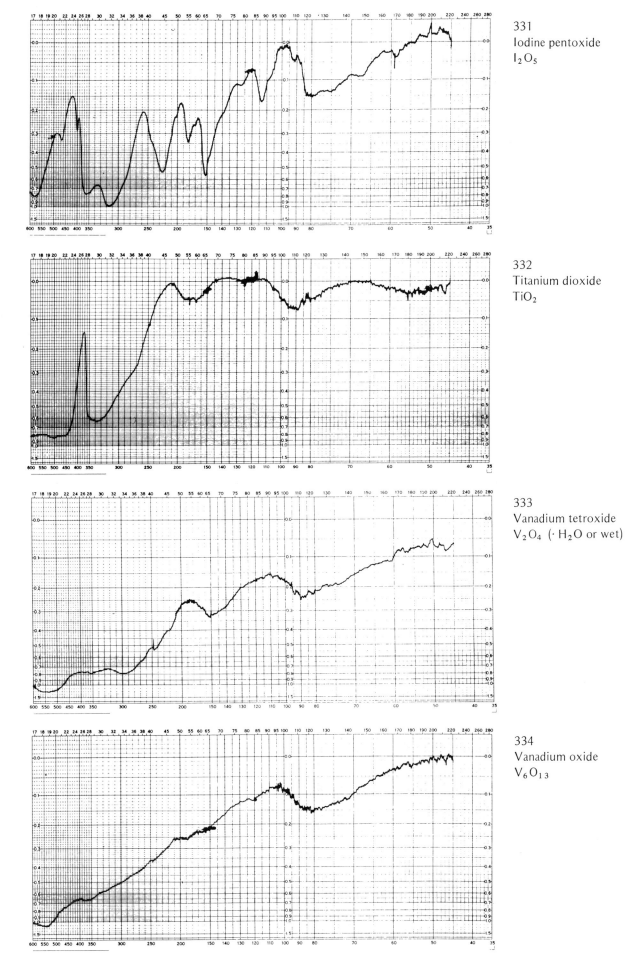

331
Iodine pentoxide
I_2O_5

332
Titanium dioxide
TiO_2

333
Vanadium tetroxide
V_2O_4 (· H_2O or wet)

334
Vanadium oxide
V_6O_{13}

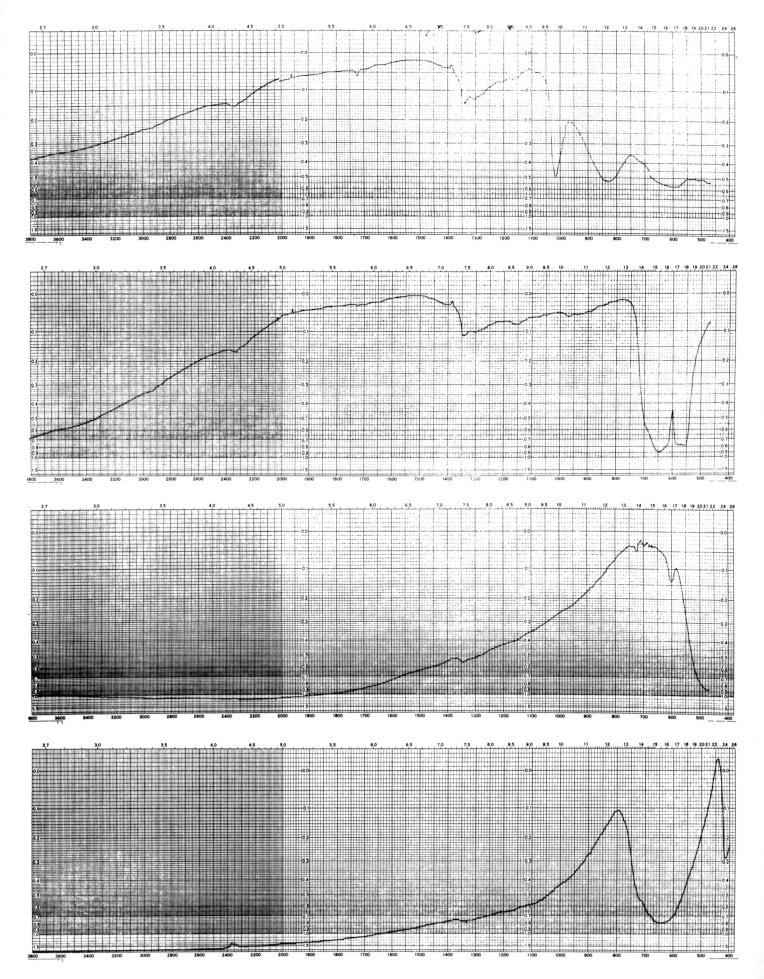

216

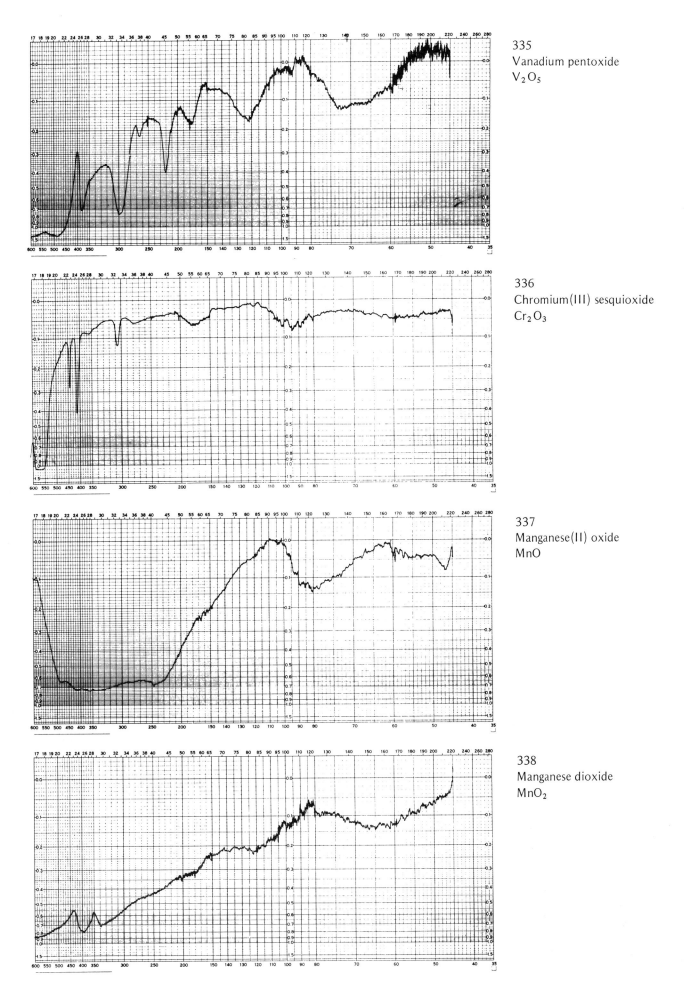

335
Vanadium pentoxide
V_2O_5

336
Chromium(III) sesquioxide
Cr_2O_3

337
Manganese(II) oxide
MnO

338
Manganese dioxide
MnO_2

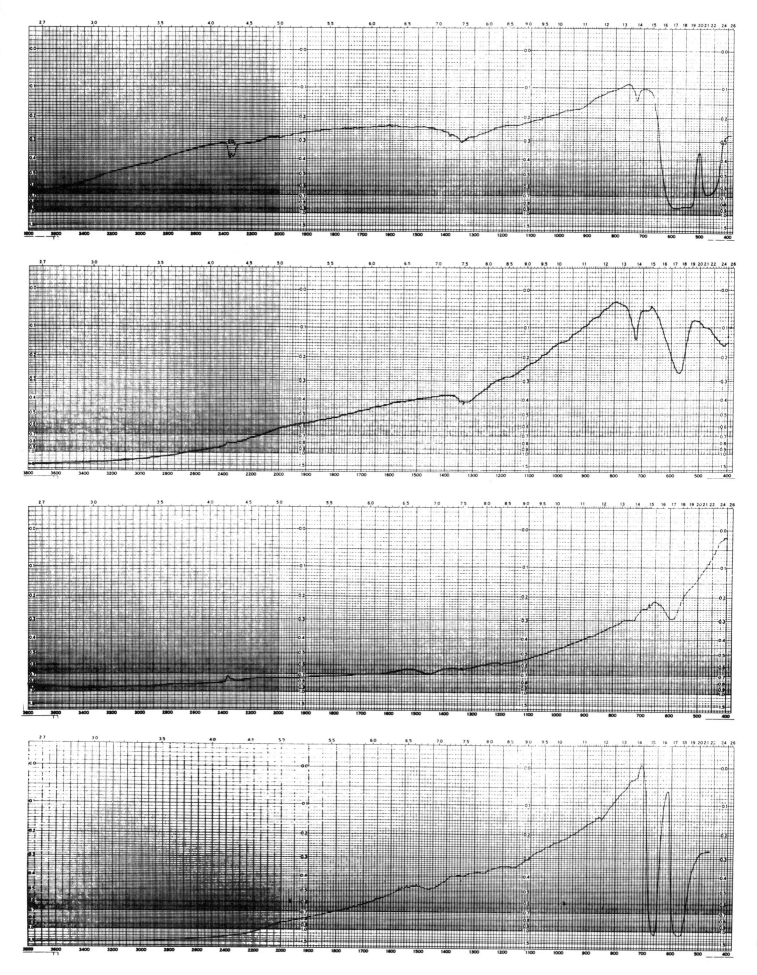

218

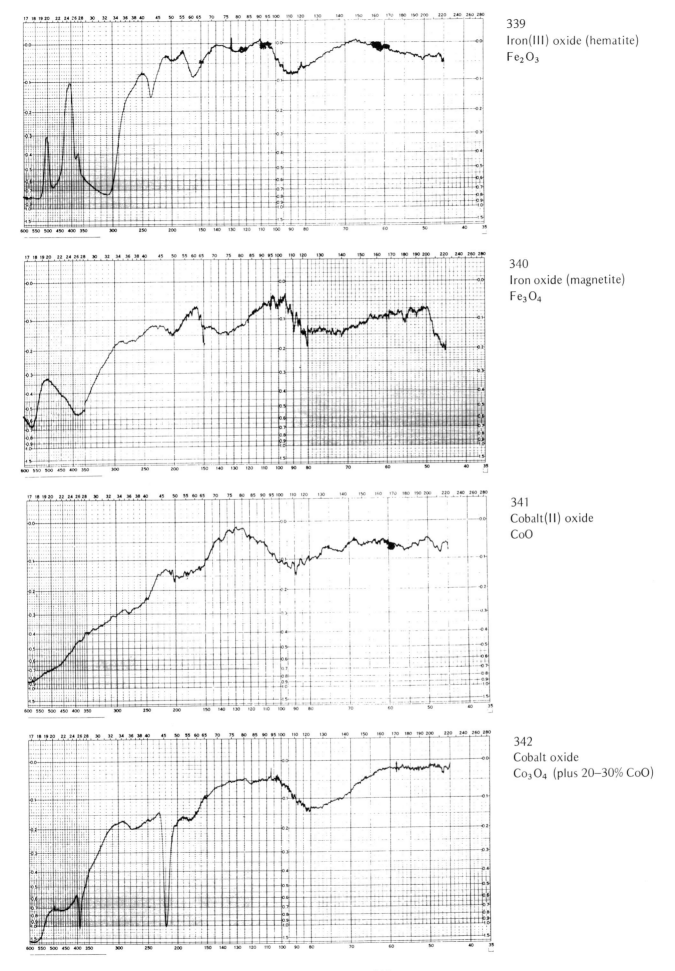

339
Iron(III) oxide (hematite)
Fe_2O_3

340
Iron oxide (magnetite)
Fe_3O_4

341
Cobalt(II) oxide
CoO

342
Cobalt oxide
Co_3O_4 (plus 20–30% CoO)

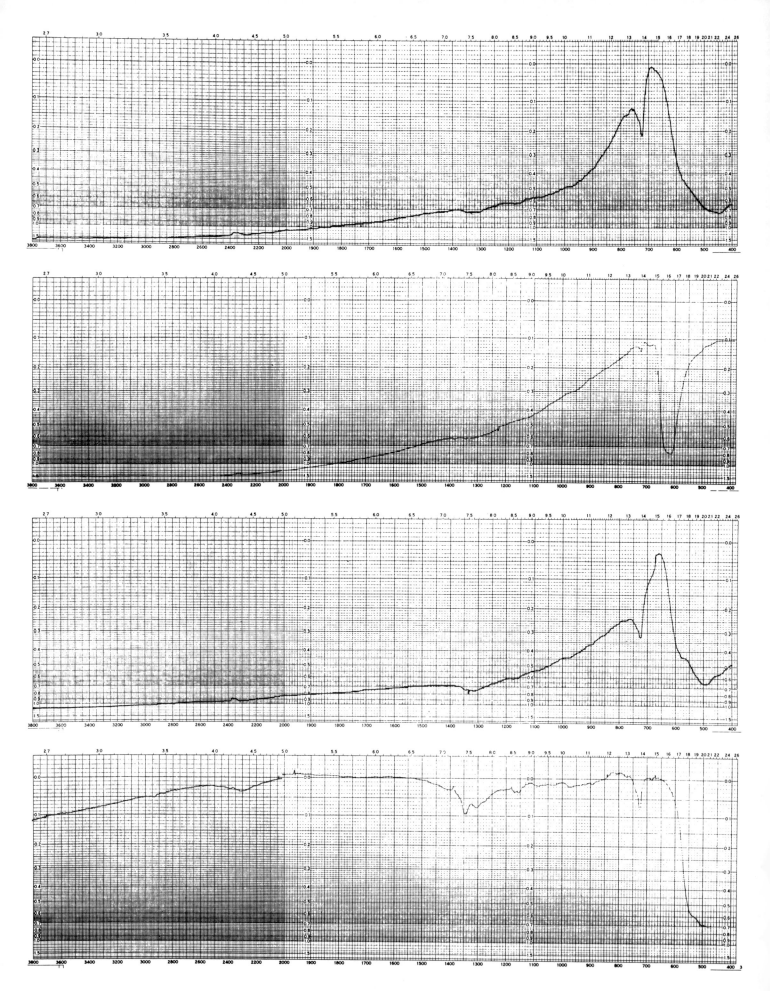

220

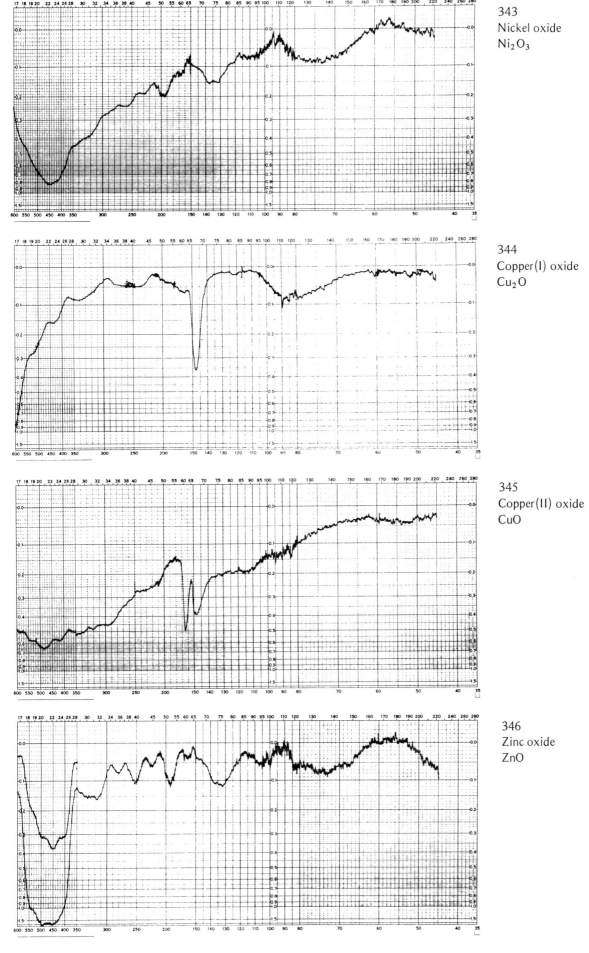

343
Nickel oxide
Ni$_2$O$_3$

344
Copper(I) oxide
Cu$_2$O

345
Copper(II) oxide
CuO

346
Zinc oxide
ZnO

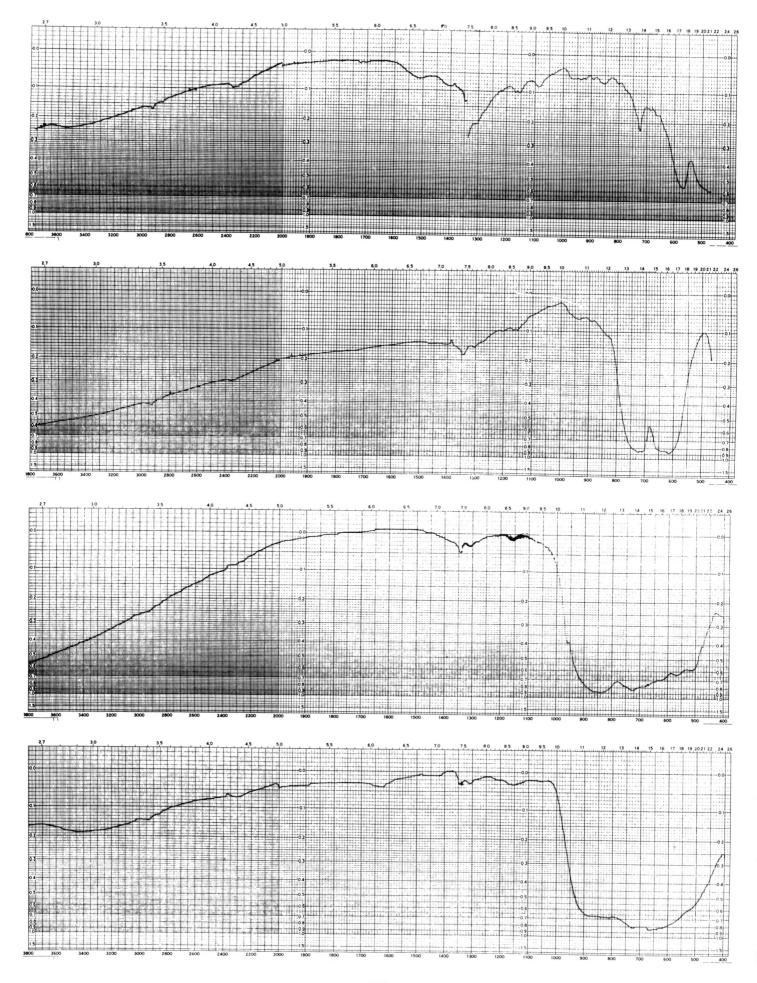

222

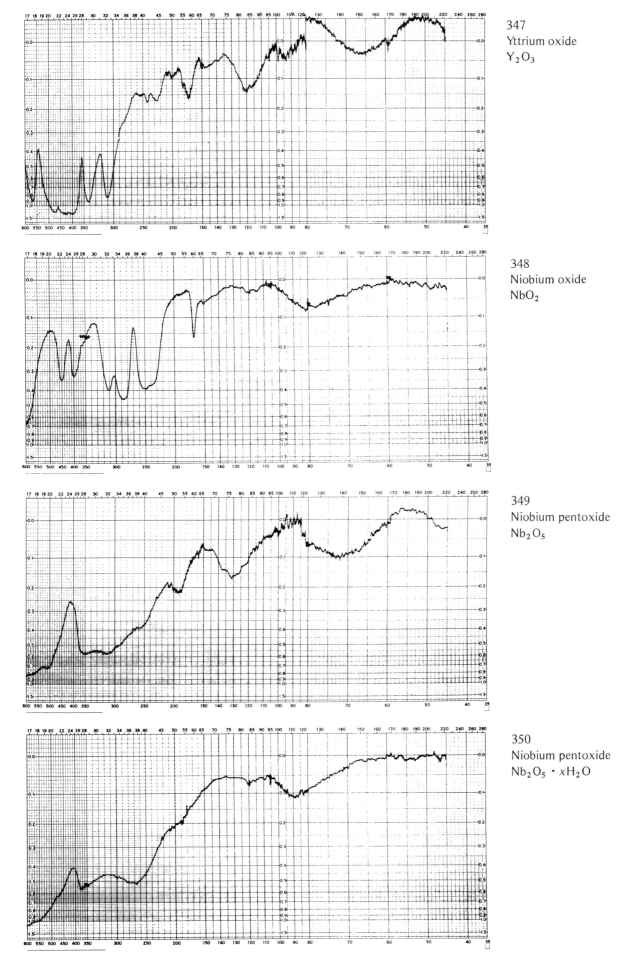

347
Yttrium oxide
Y$_2$O$_3$

348
Niobium oxide
NbO$_2$

349
Niobium pentoxide
Nb$_2$O$_5$

350
Niobium pentoxide
Nb$_2$O$_5$ · xH$_2$O

223

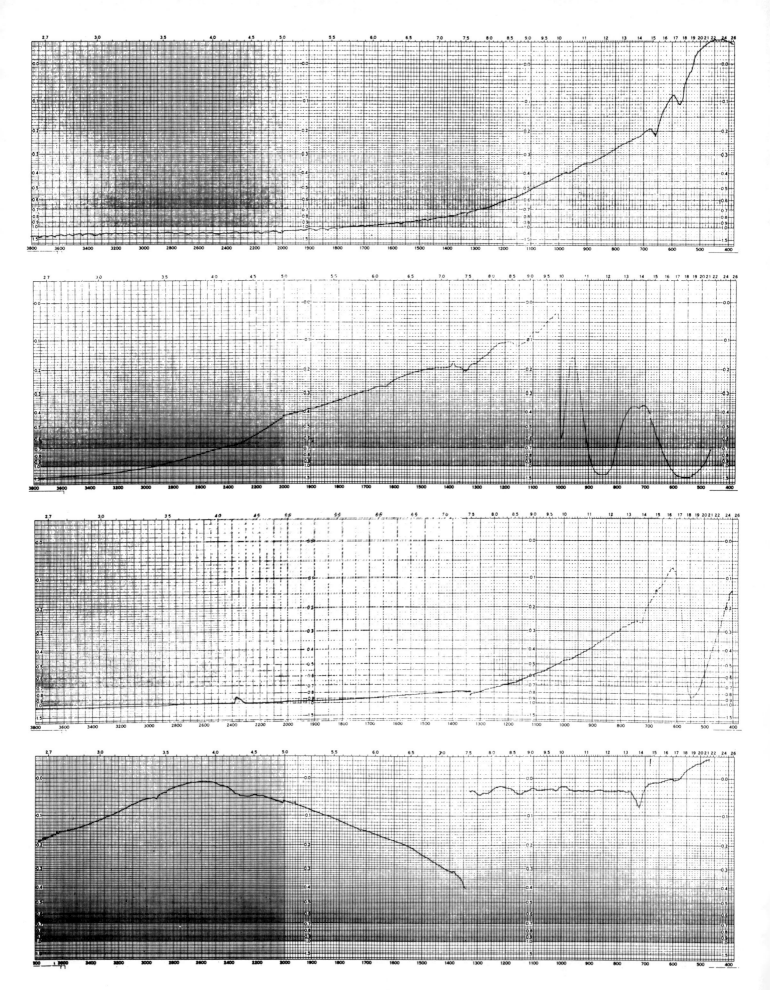

224

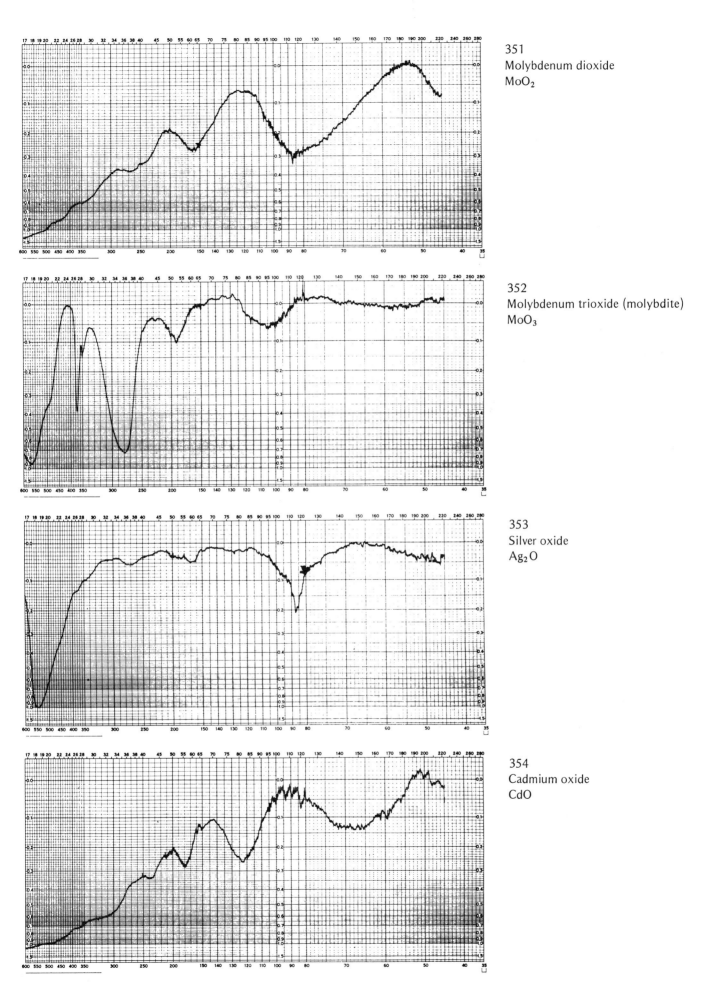

351
Molybdenum dioxide
MoO$_2$

352
Molybdenum trioxide (molybdite)
MoO$_3$

353
Silver oxide
Ag$_2$O

354
Cadmium oxide
CdO

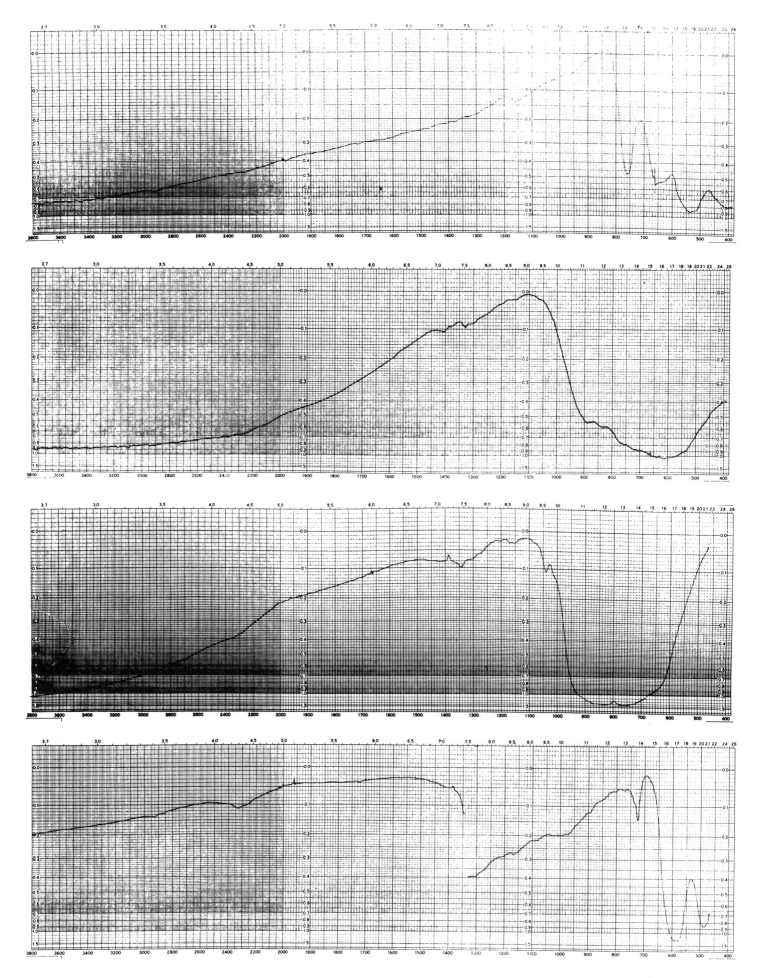

226

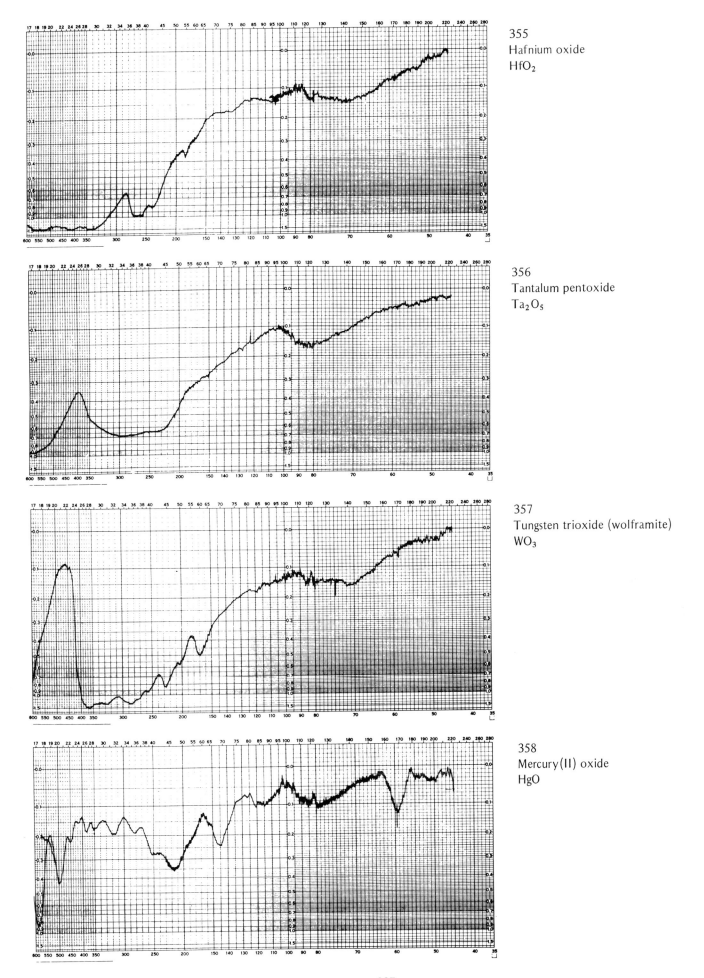

355
Hafnium oxide
HfO$_2$

356
Tantalum pentoxide
Ta$_2$O$_5$

357
Tungsten trioxide (wolframite)
WO$_3$

358
Mercury(II) oxide
HgO

227

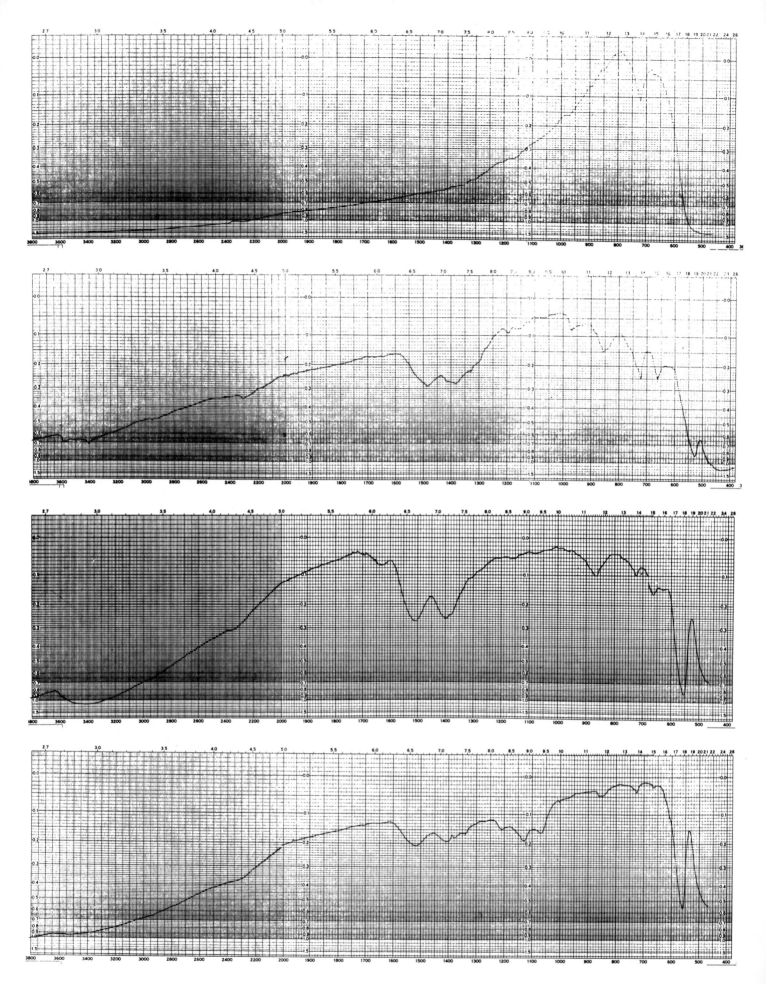

228

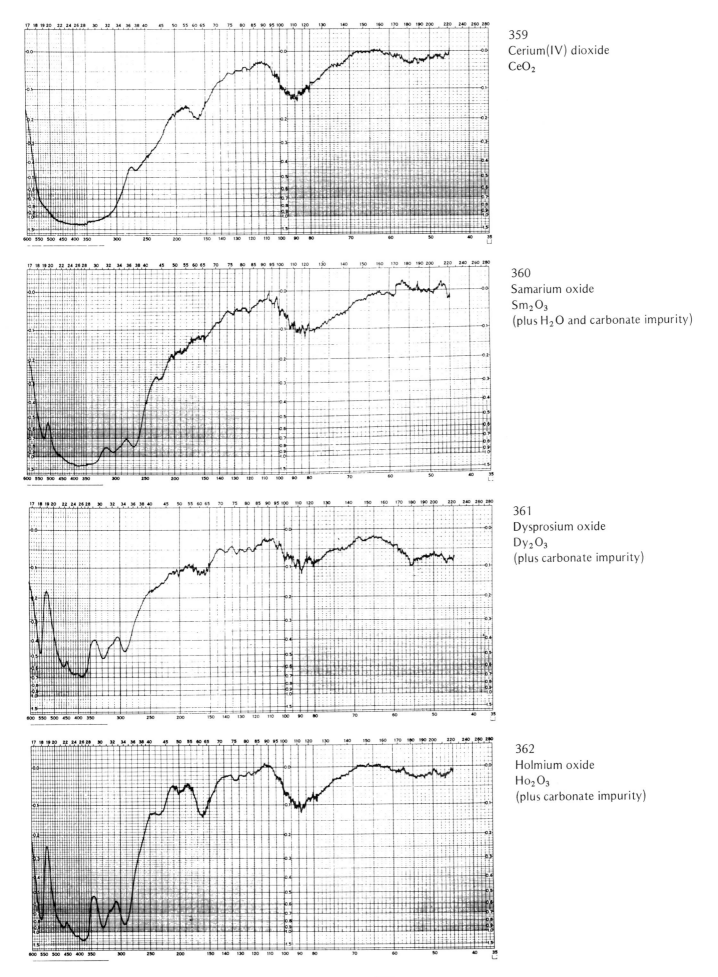

359
Cerium(IV) dioxide
CeO_2

360
Samarium oxide
Sm_2O_3
(plus H_2O and carbonate impurity)

361
Dysprosium oxide
Dy_2O_3
(plus carbonate impurity)

362
Holmium oxide
Ho_2O_3
(plus carbonate impurity)

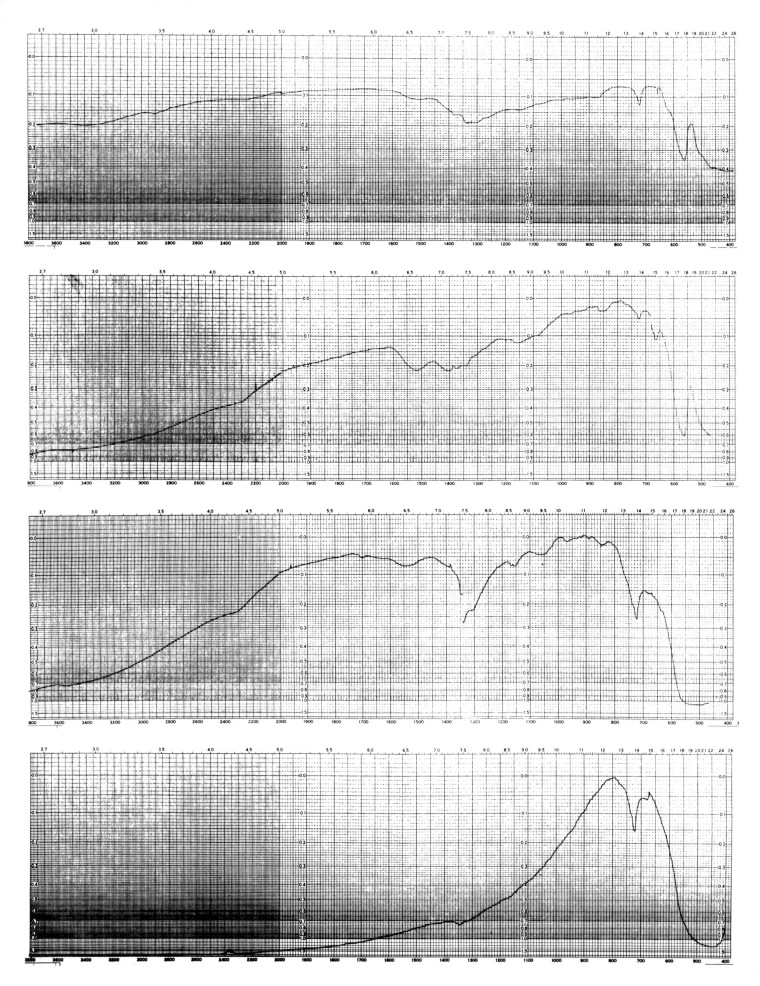

230

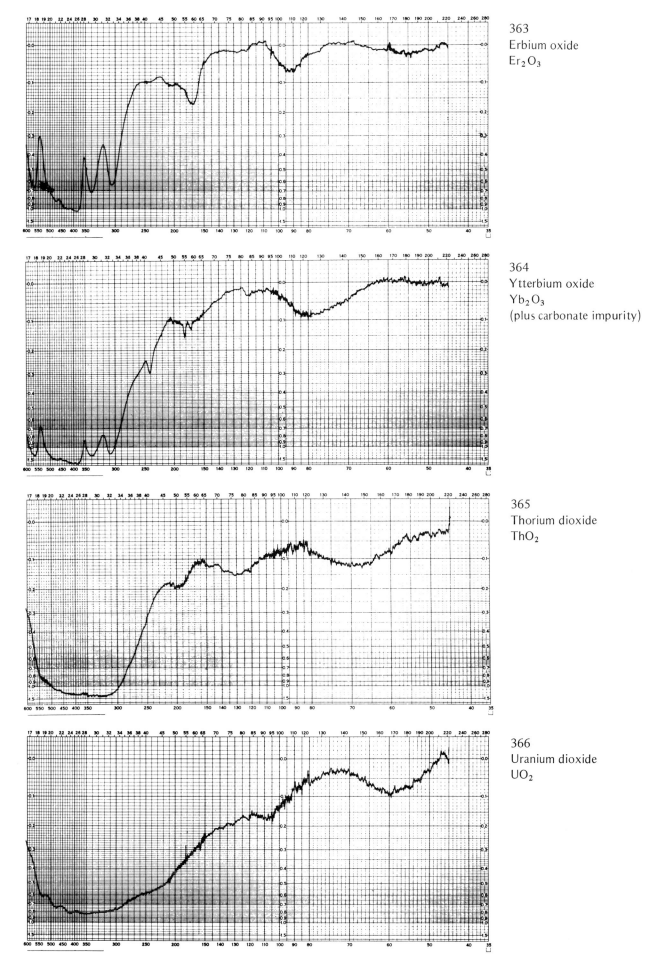

363
Erbium oxide
Er_2O_3

364
Ytterbium oxide
Yb_2O_3
(plus carbonate impurity)

365
Thorium dioxide
ThO_2

366
Uranium dioxide
UO_2

231

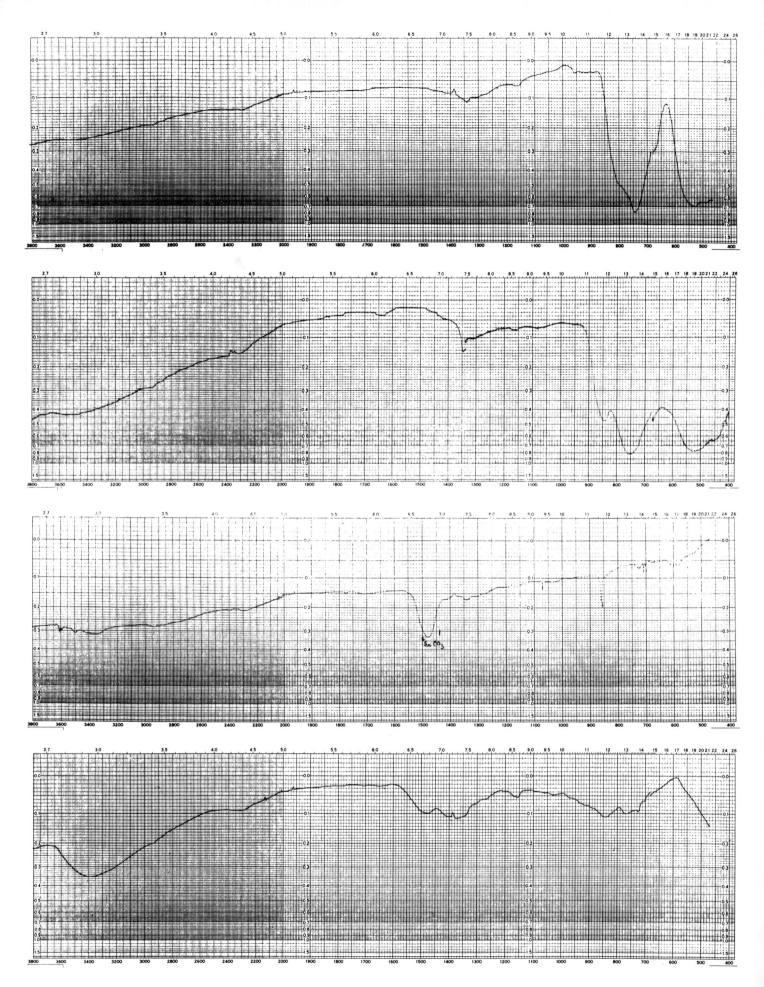

232

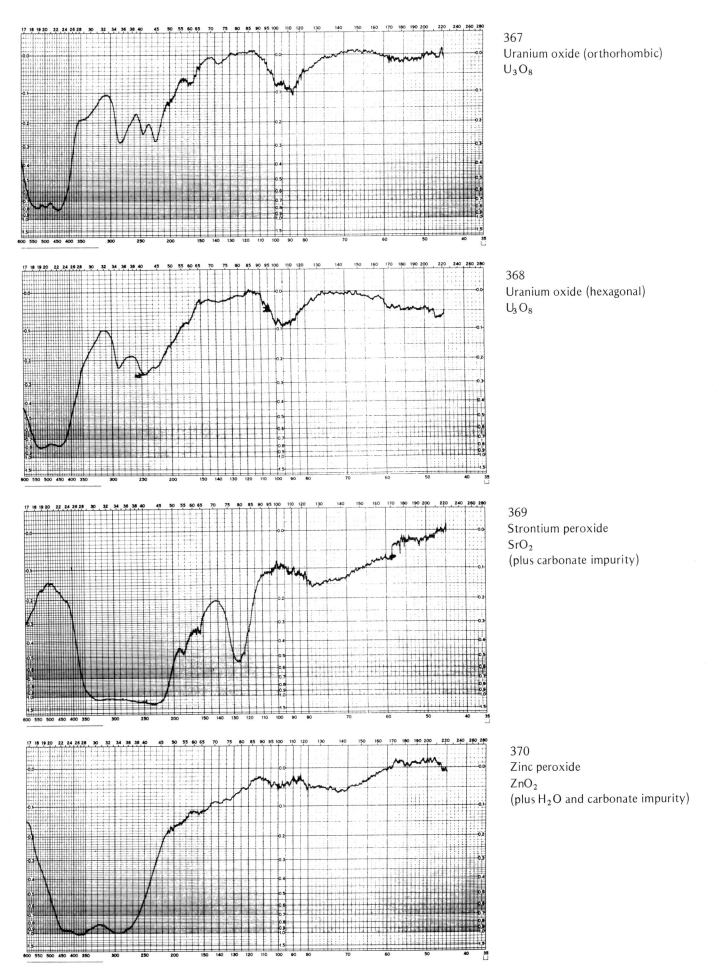

367
Uranium oxide (orthorhombic)
U_3O_8

368
Uranium oxide (hexagonal)
U_3O_8

369
Strontium peroxide
SrO_2
(plus carbonate impurity)

370
Zinc peroxide
ZnO_2
(plus H_2O and carbonate impurity)

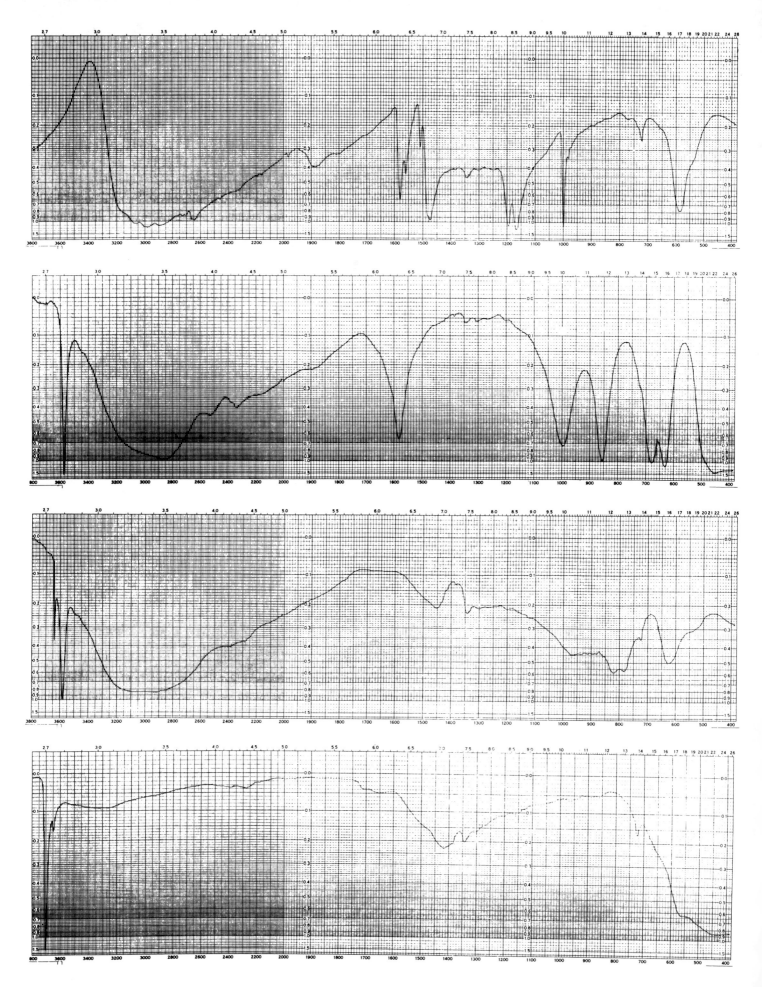

234

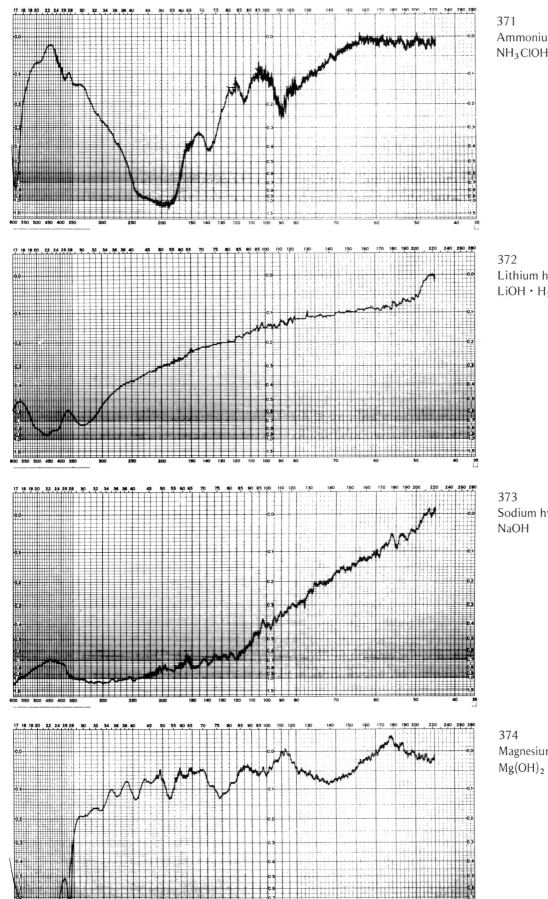

371
Ammonium hydroxide hydrochloride
NH₃ClOH

372
Lithium hydroxide
LiOH · H₂O

373
Sodium hydroxide
NaOH

374
Magnesium hydroxide
Mg(OH)₂

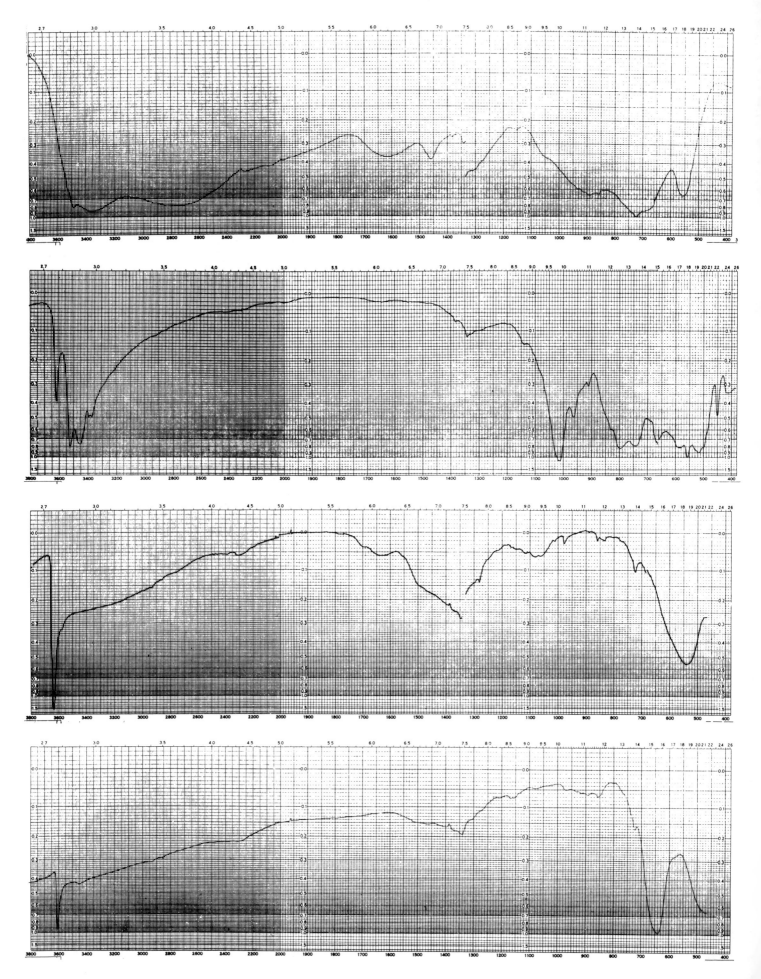

236

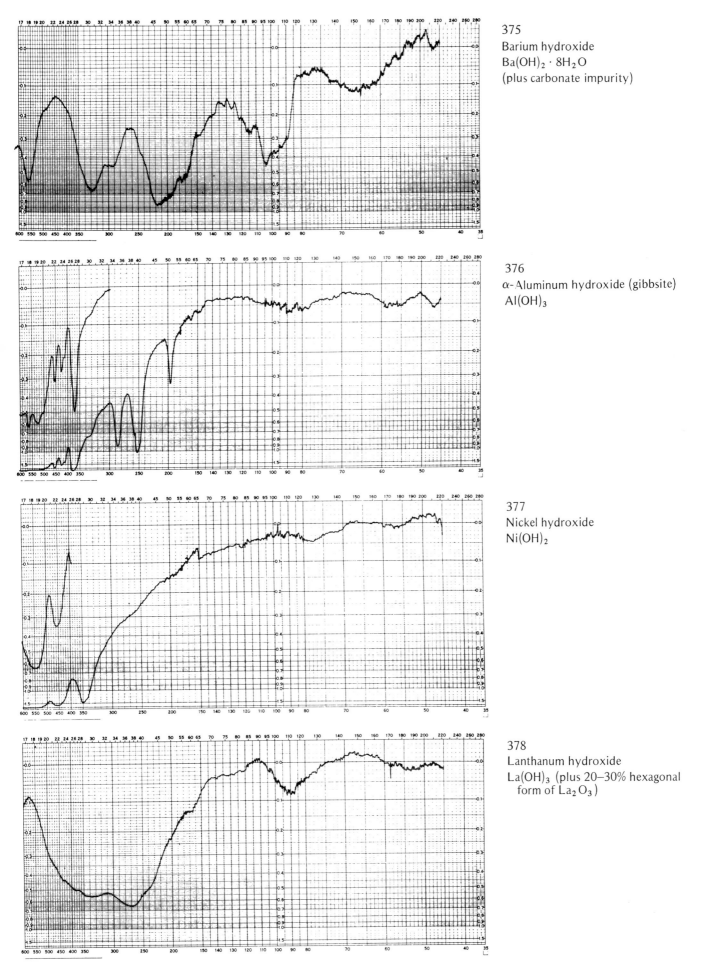

375
Barium hydroxide
Ba(OH)$_2$ · 8H$_2$O
(plus carbonate impurity)

376
α-Aluminum hydroxide (gibbsite)
Al(OH)$_3$

377
Nickel hydroxide
Ni(OH)$_2$

378
Lanthanum hydroxide
La(OH)$_3$ (plus 20–30% hexagonal
   form of La$_2$O$_3$)

237

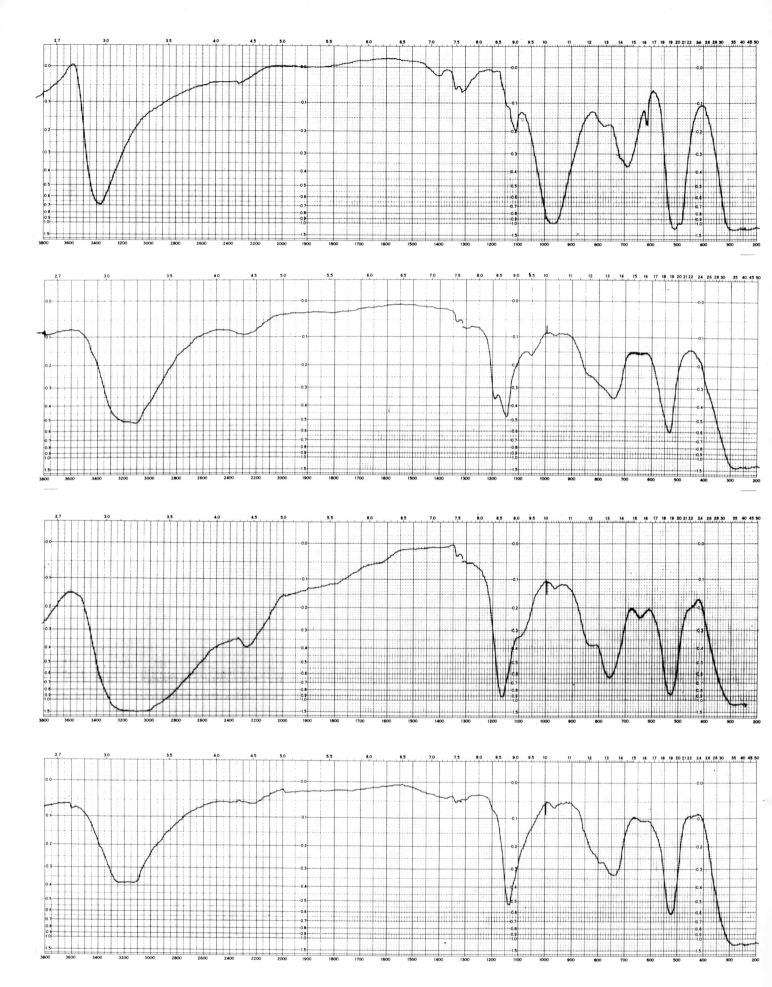

238

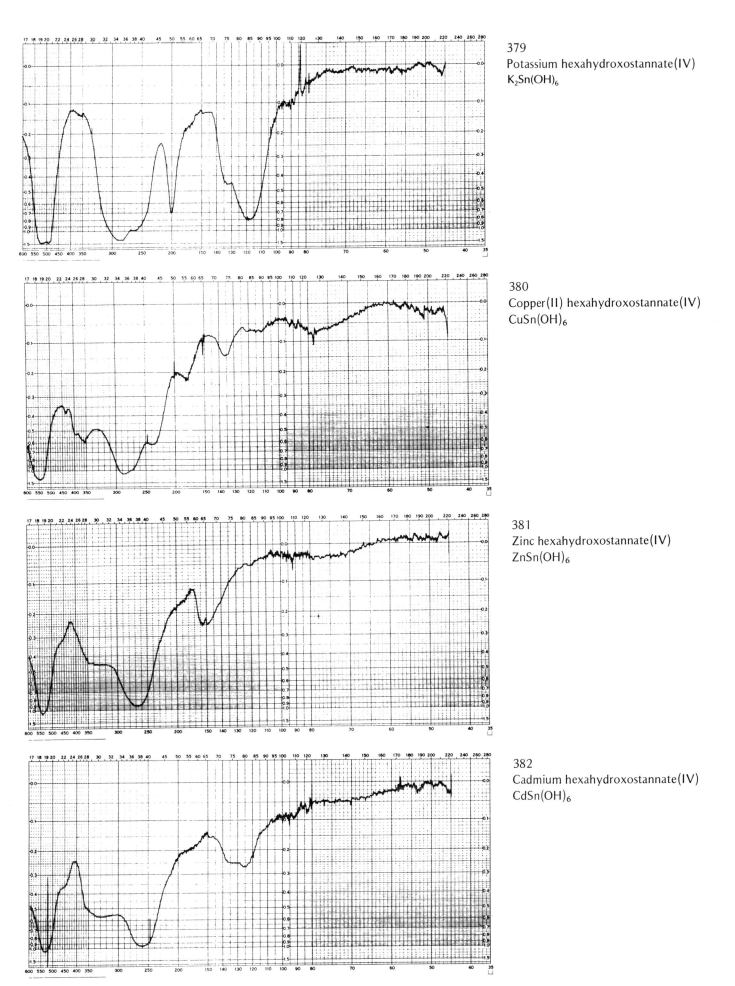

379
Potassium hexahydroxostannate(IV)
$K_2Sn(OH)_6$

380
Copper(II) hexahydroxostannate(IV)
$CuSn(OH)_6$

381
Zinc hexahydroxostannate(IV)
$ZnSn(OH)_6$

382
Cadmium hexahydroxostannate(IV)
$CdSn(OH)_6$

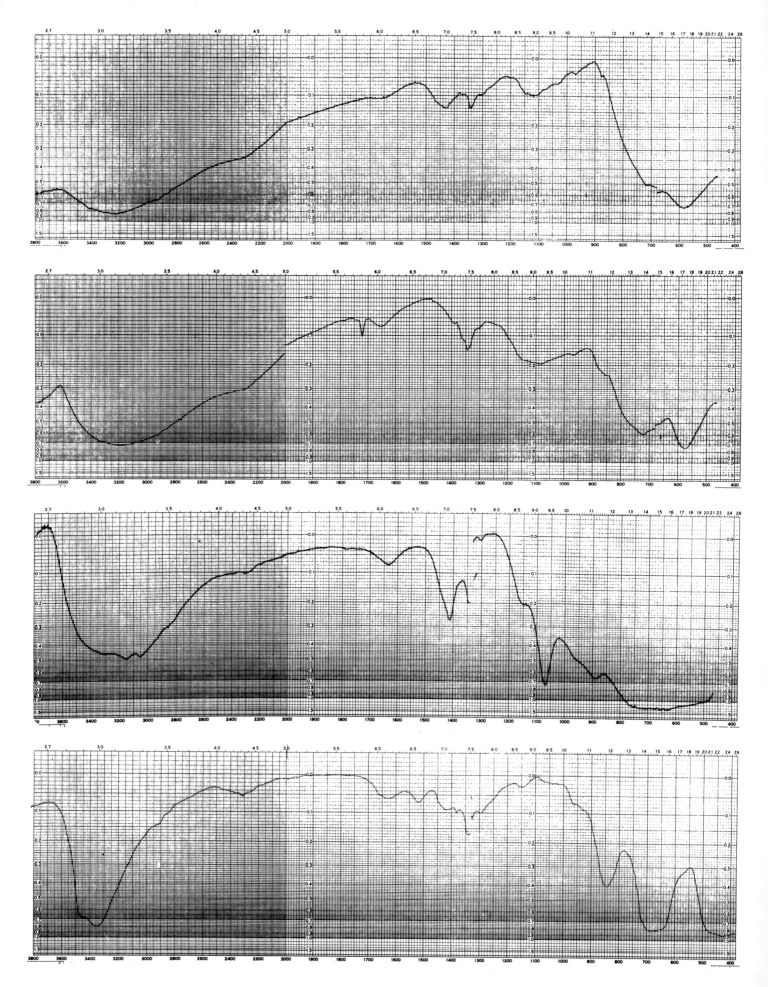

240

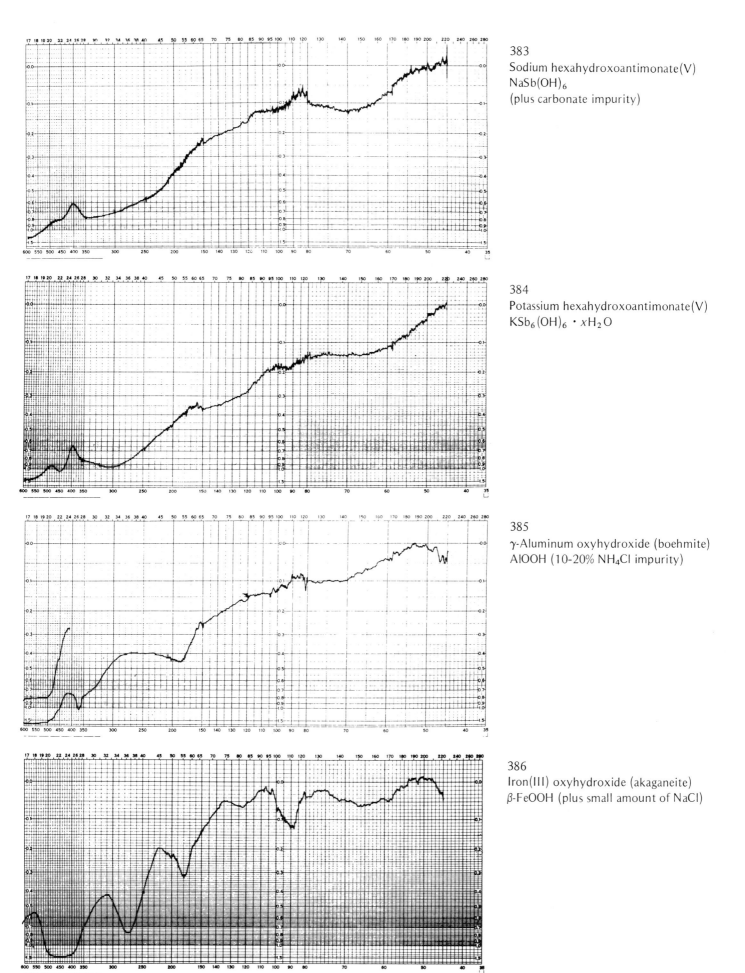

383
Sodium hexahydroxoantimonate(V)
$NaSb(OH)_6$
(plus carbonate impurity)

384
Potassium hexahydroxoantimonate(V)
$KSb_6(OH)_6 \cdot xH_2O$

385
γ-Aluminum oxyhydroxide (boehmite)
$AlOOH$ (10-20% NH_4Cl impurity)

386
Iron(III) oxyhydroxide (akaganeite)
β-FeOOH (plus small amount of NaCl)

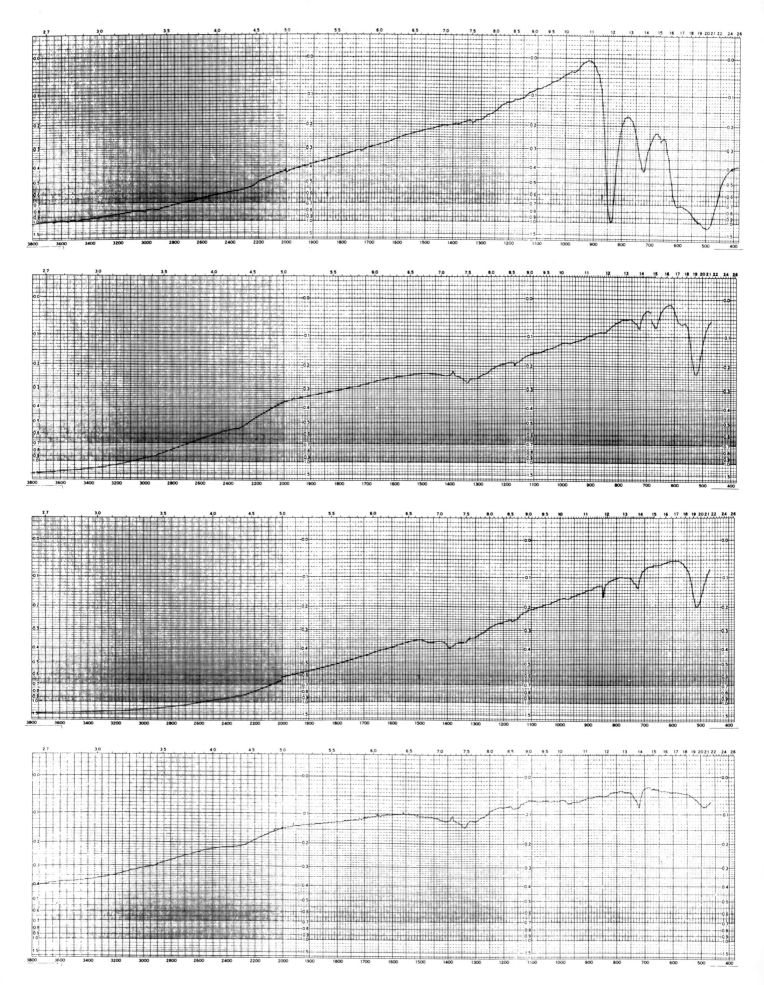

242

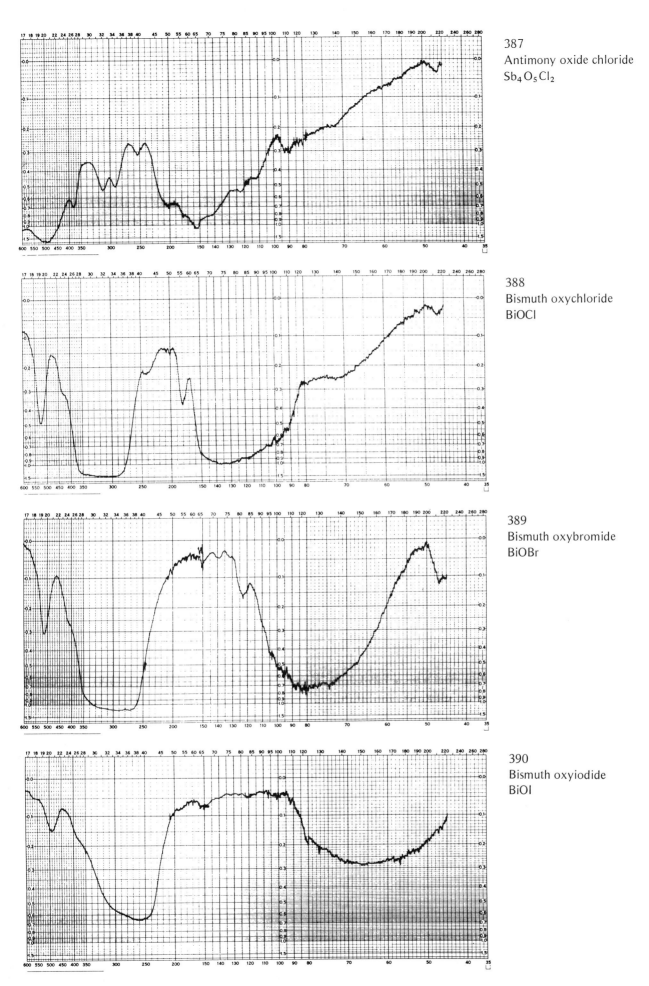

387
Antimony oxide chloride
$Sb_4O_5Cl_2$

388
Bismuth oxychloride
BiOCl

389
Bismuth oxybromide
BiOBr

390
Bismuth oxyiodide
BiOI

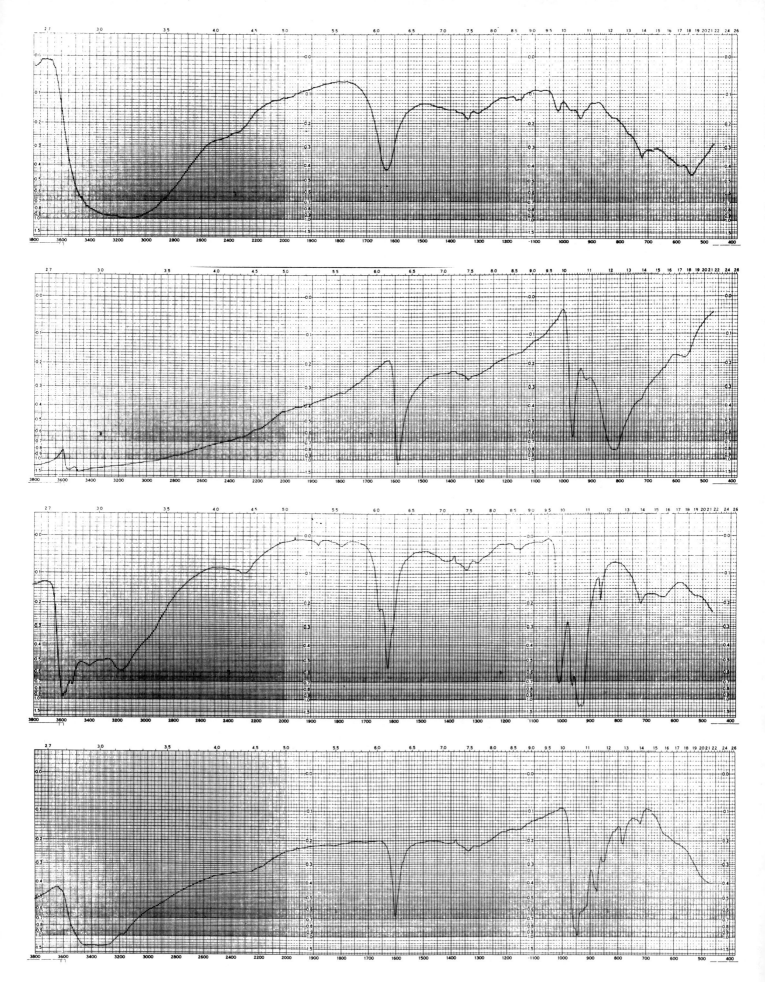

244

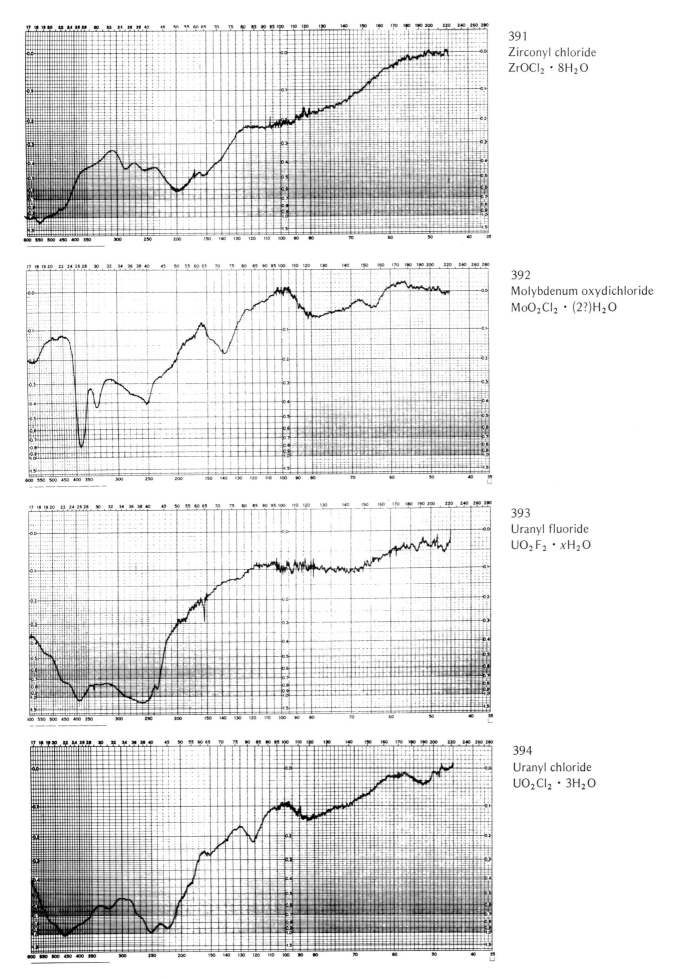

391
Zirconyl chloride
$ZrOCl_2 \cdot 8H_2O$

392
Molybdenum oxydichloride
$MoO_2Cl_2 \cdot (2?)H_2O$

393
Uranyl fluoride
$UO_2F_2 \cdot xH_2O$

394
Uranyl chloride
$UO_2Cl_2 \cdot 3H_2O$

245

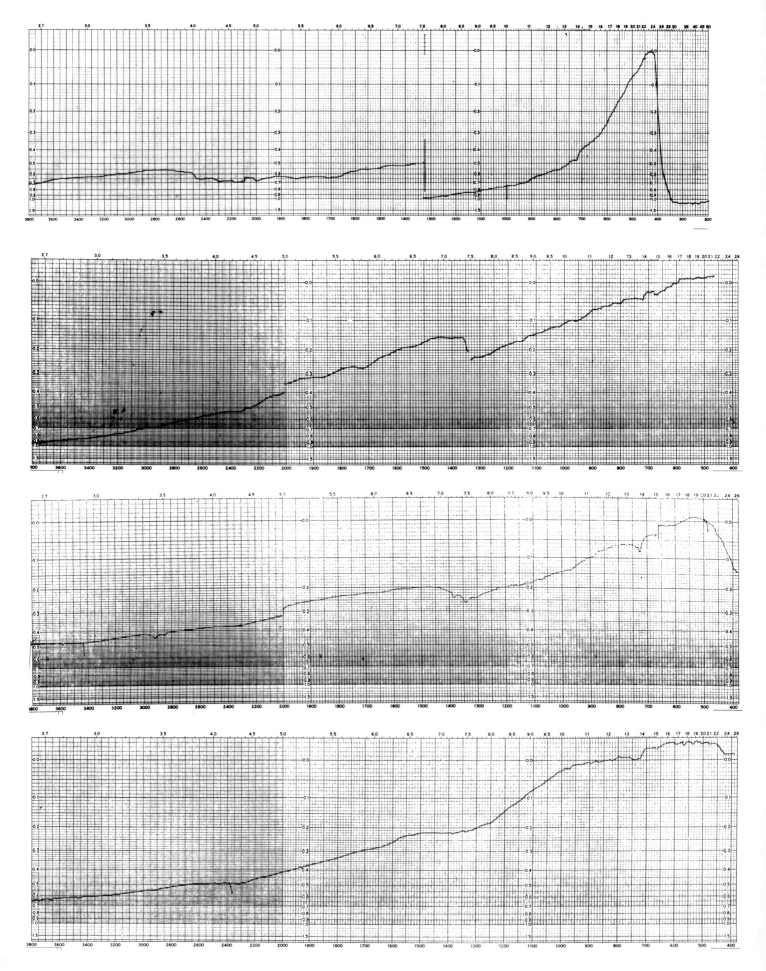

246

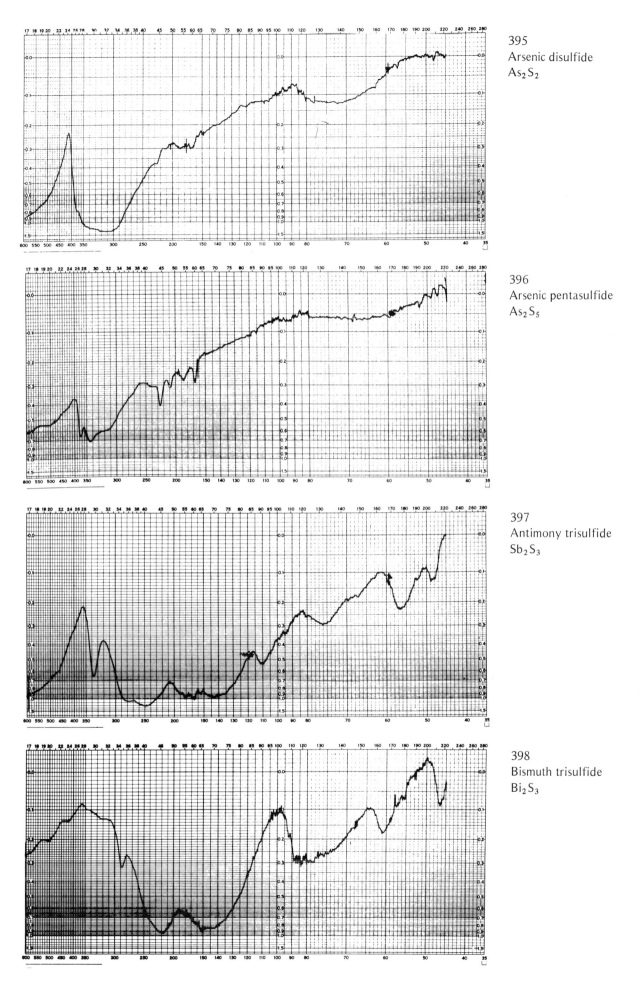

395
Arsenic disulfide
As$_2$S$_2$

396
Arsenic pentasulfide
As$_2$S$_5$

397
Antimony trisulfide
Sb$_2$S$_3$

398
Bismuth trisulfide
Bi$_2$S$_3$

247

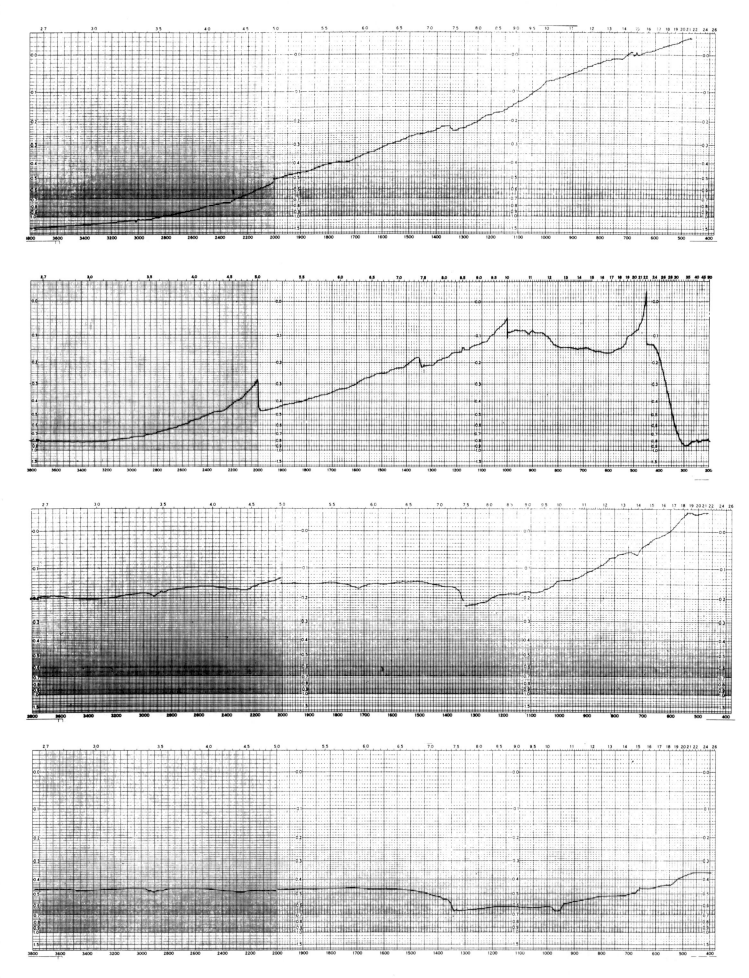

248

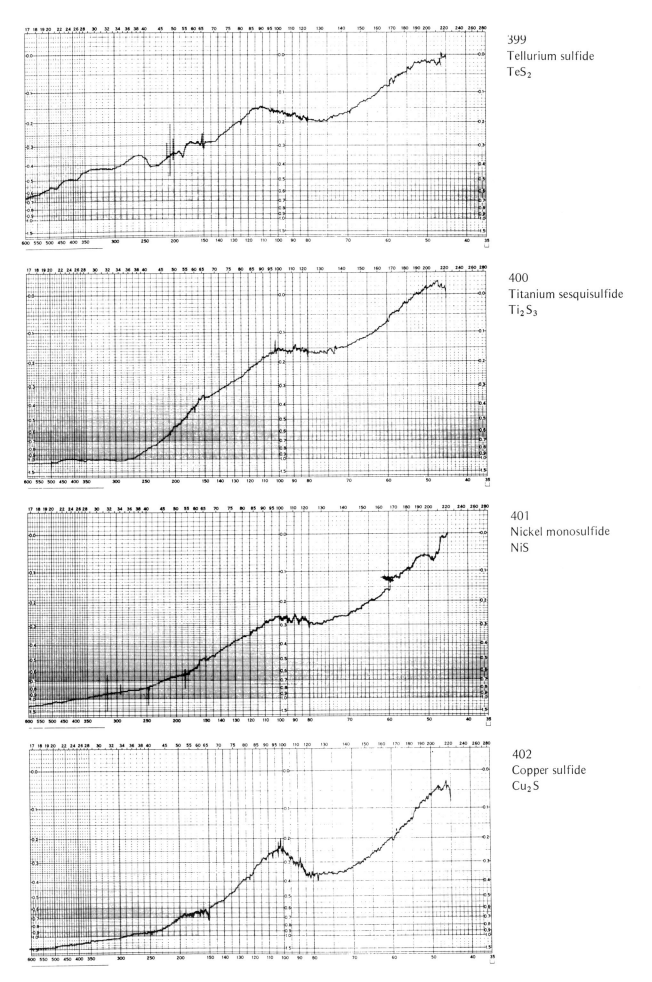

399
Tellurium sulfide
TeS_2

400
Titanium sesquisulfide
Ti_2S_3

401
Nickel monosulfide
NiS

402
Copper sulfide
Cu_2S

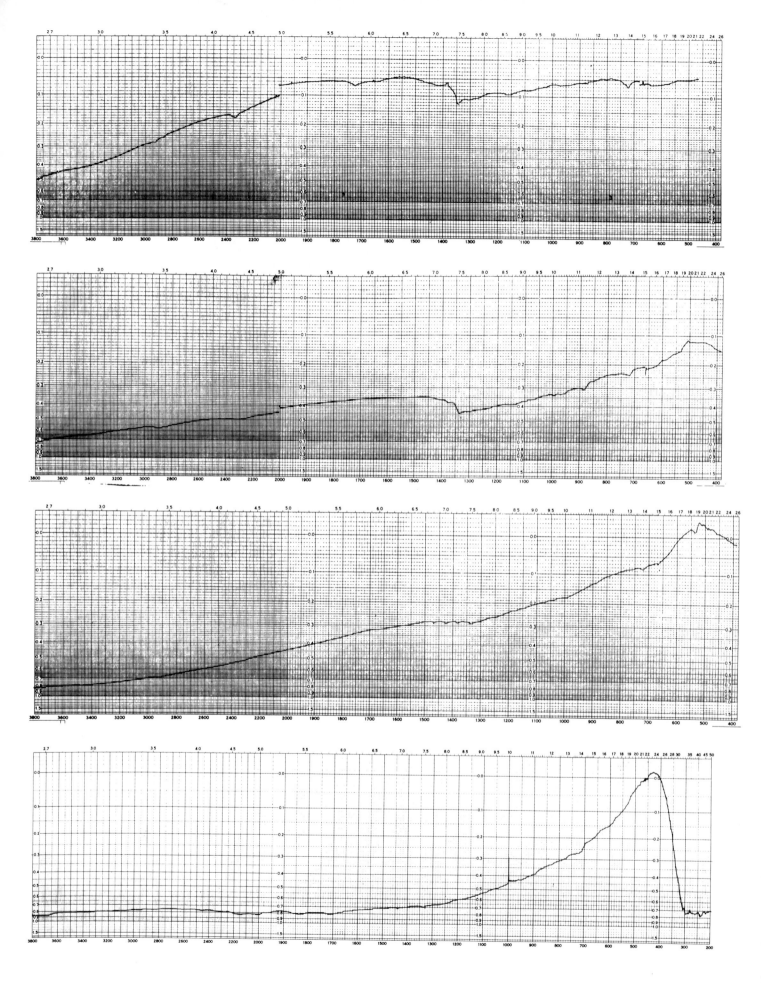

250

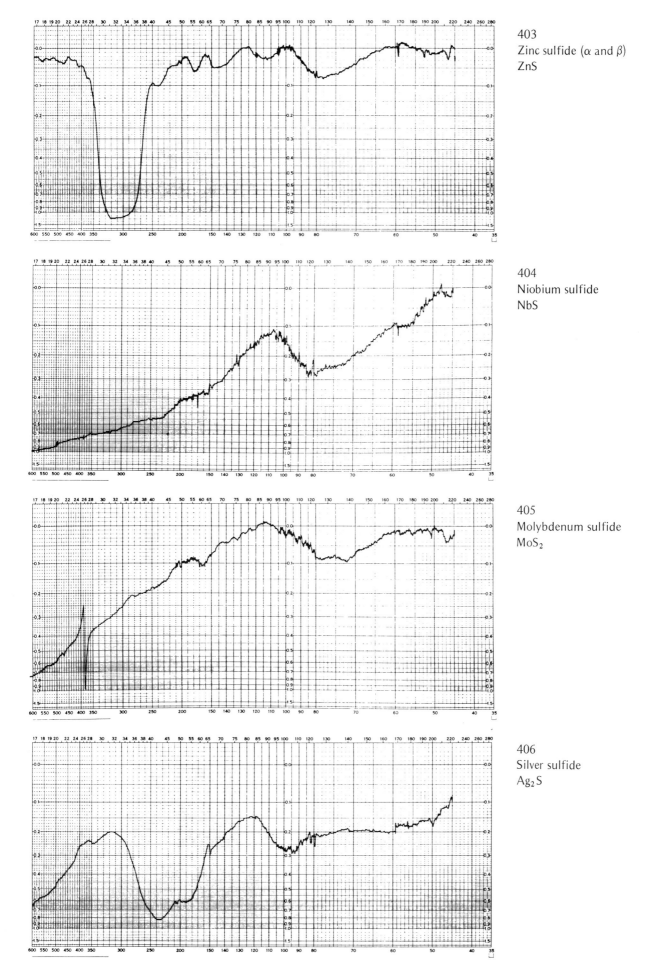

403
Zinc sulfide (α and β)
ZnS

404
Niobium sulfide
NbS

405
Molybdenum sulfide
MoS_2

406
Silver sulfide
Ag_2S

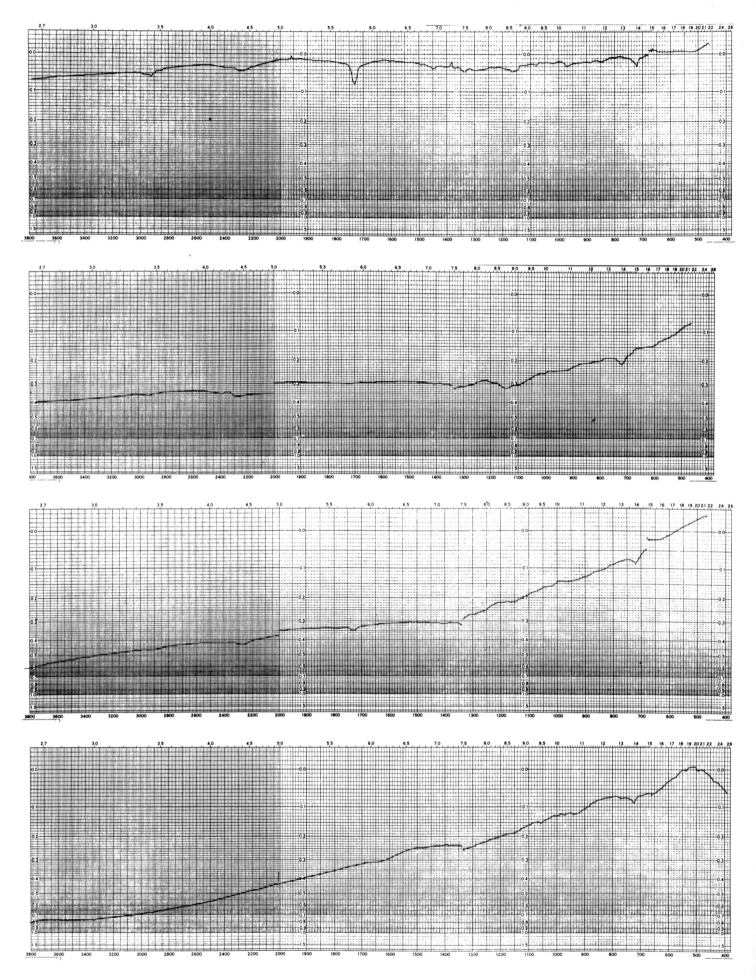

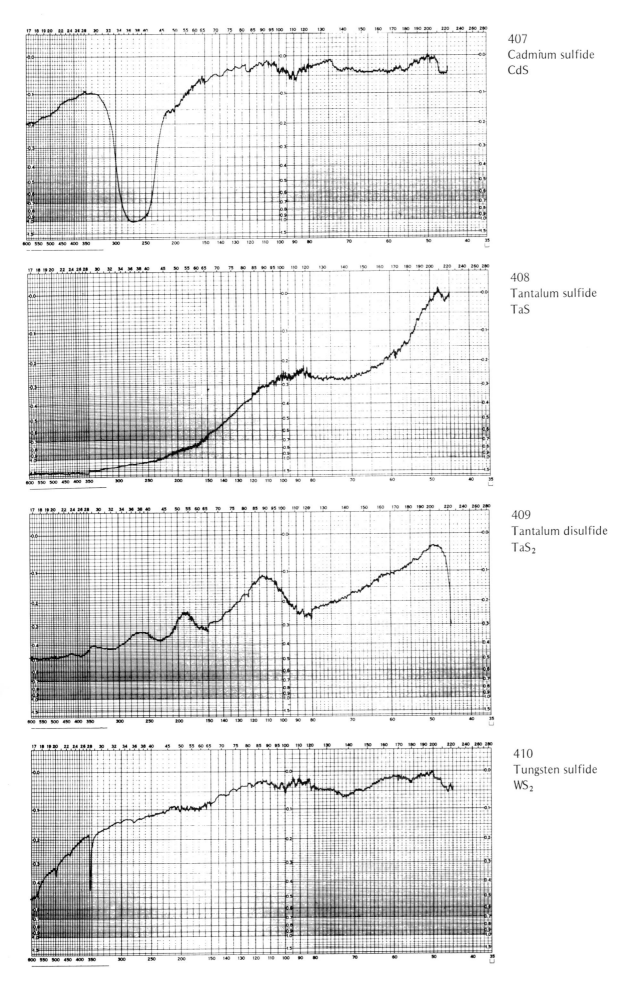

407
Cadmium sulfide
CdS

408
Tantalum sulfide
TaS

409
Tantalum disulfide
TaS$_2$

410
Tungsten sulfide
WS$_2$

253

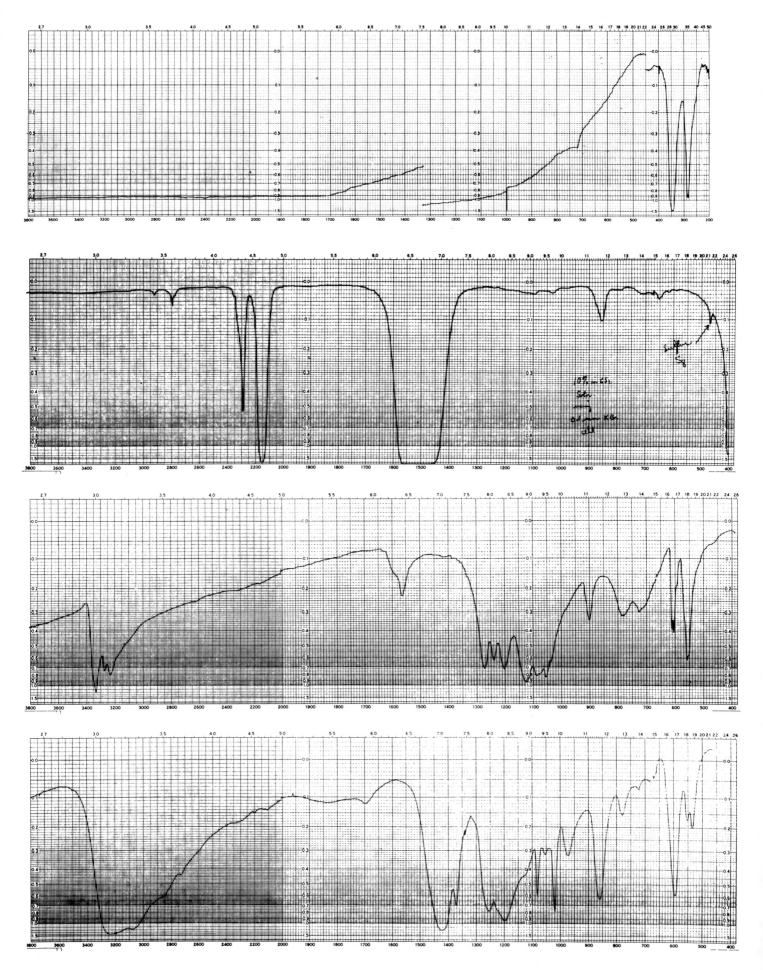

254

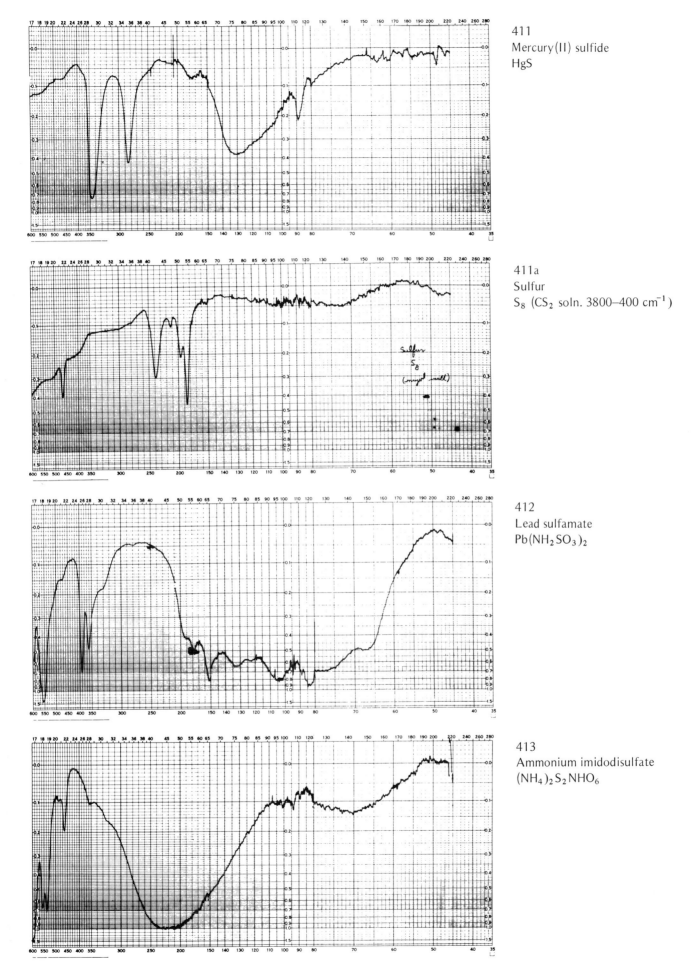

411
Mercury(II) sulfide
HgS

411a
Sulfur
S_8 (CS$_2$ soln. 3800–400 cm^{-1})

412
Lead sulfamate
$Pb(NH_2SO_3)_2$

413
Ammonium imidodisulfate
$(NH_4)_2S_2NHO_6$

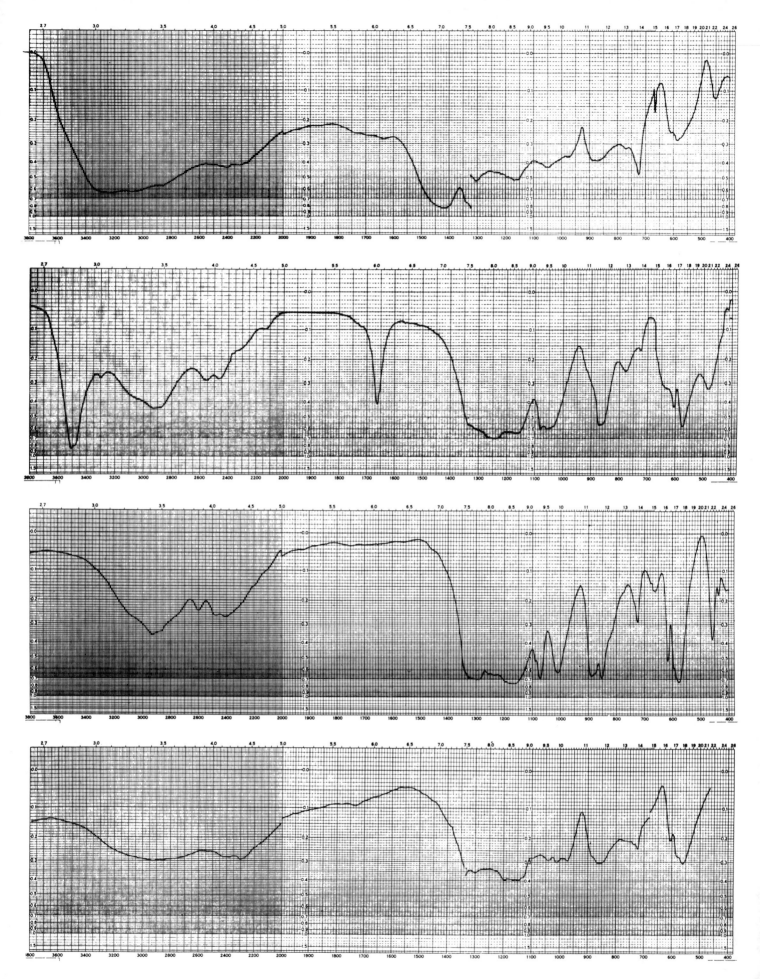

256

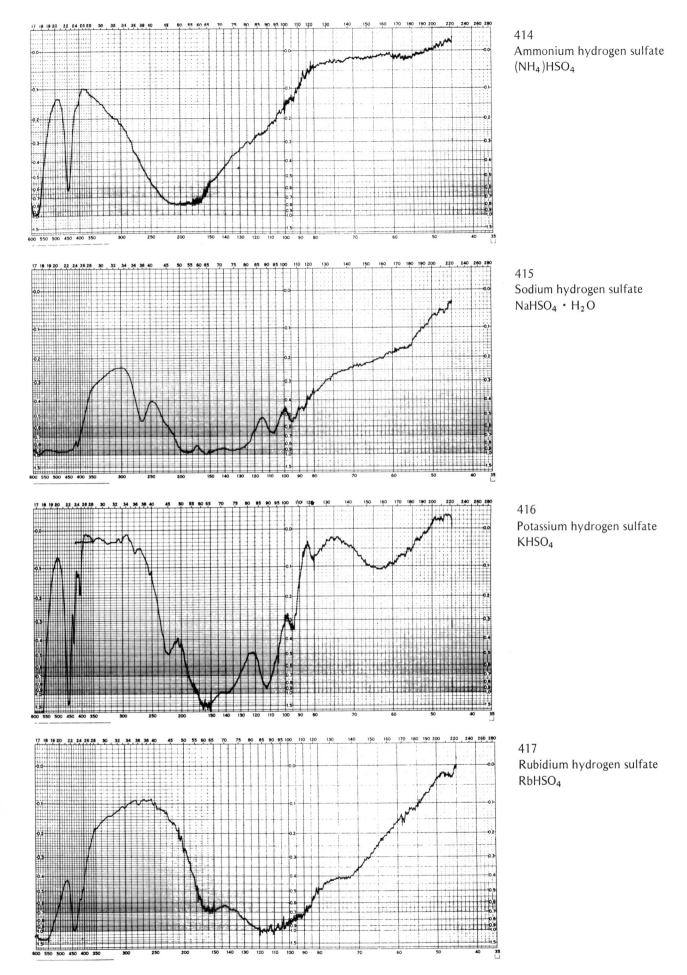

414
Ammonium hydrogen sulfate
$(NH_4)HSO_4$

415
Sodium hydrogen sulfate
$NaHSO_4 \cdot H_2O$

416
Potassium hydrogen sulfate
$KHSO_4$

417
Rubidium hydrogen sulfate
$RbHSO_4$

257

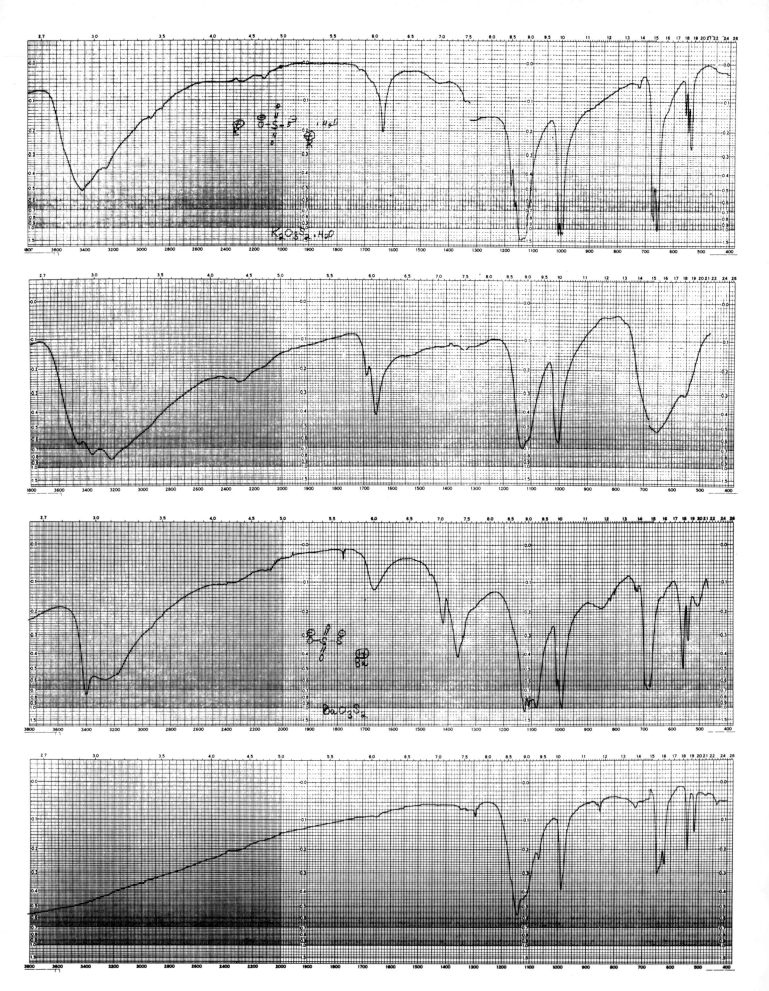

K₂O₃S₂·H₂O

BaO₃S₂

258

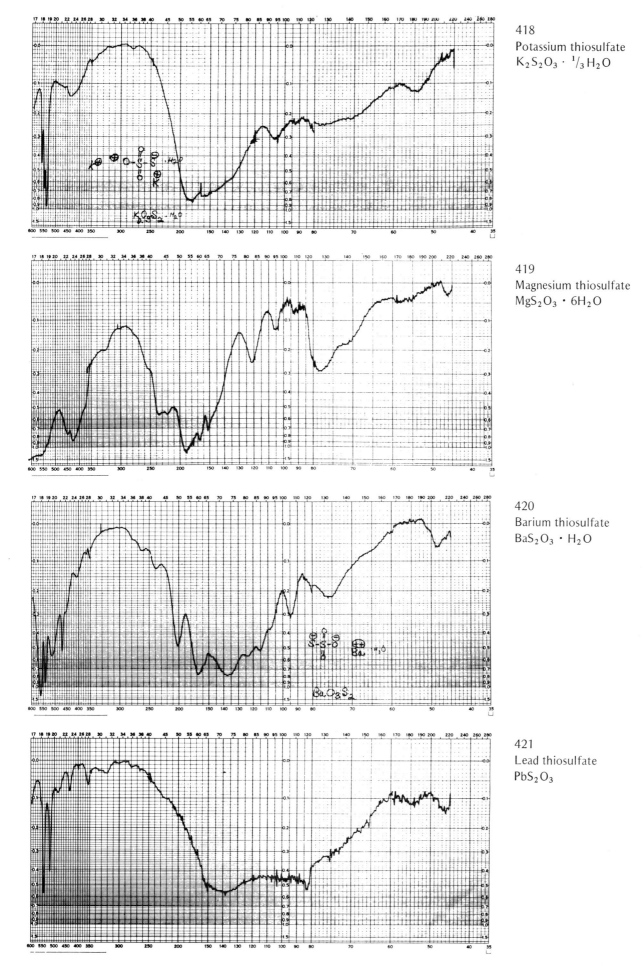

418
Potassium thiosulfate
$K_2S_2O_3 \cdot {}^1/_3 H_2O$

419
Magnesium thiosulfate
$MgS_2O_3 \cdot 6H_2O$

420
Barium thiosulfate
$BaS_2O_3 \cdot H_2O$

421
Lead thiosulfate
PbS_2O_3

259

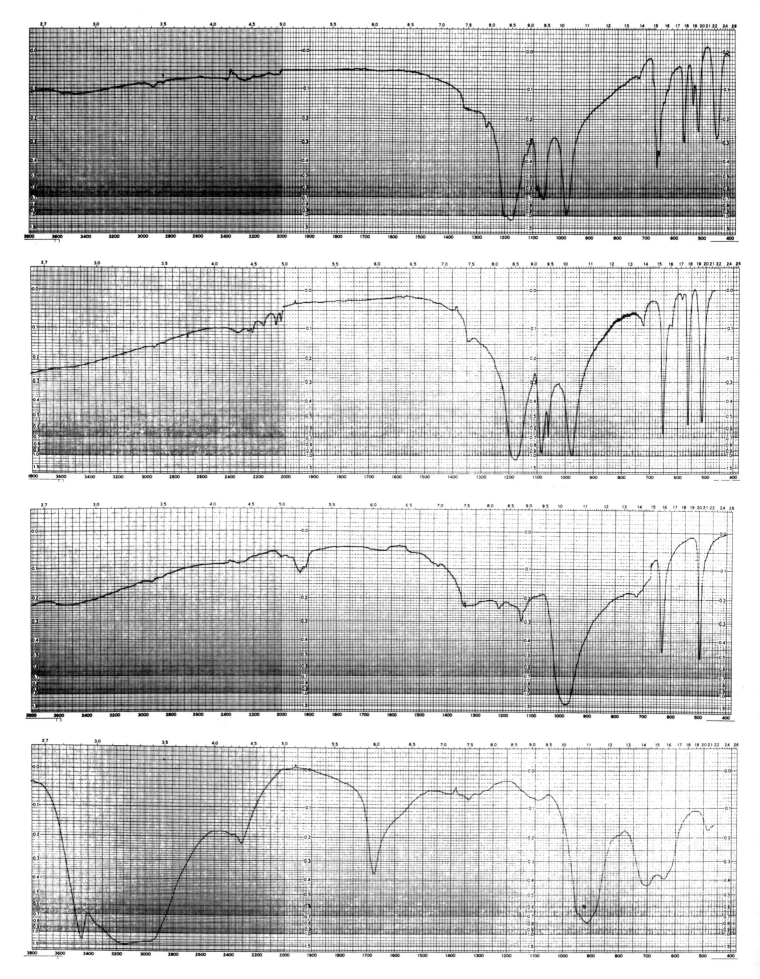

260

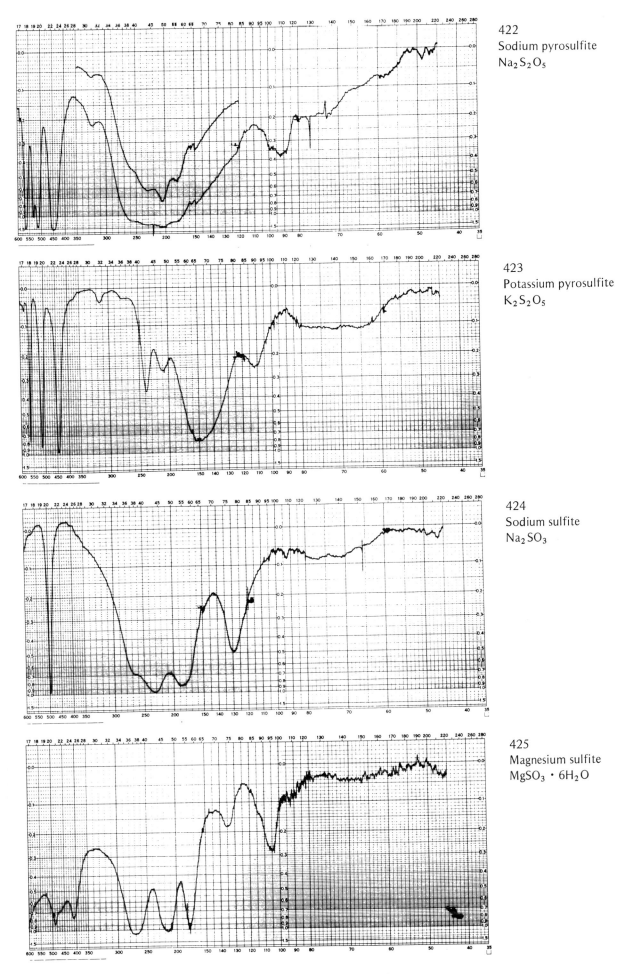

422
Sodium pyrosulfite
$Na_2S_2O_5$

423
Potassium pyrosulfite
$K_2S_2O_5$

424
Sodium sulfite
Na_2SO_3

425
Magnesium sulfite
$MgSO_3 \cdot 6H_2O$

261

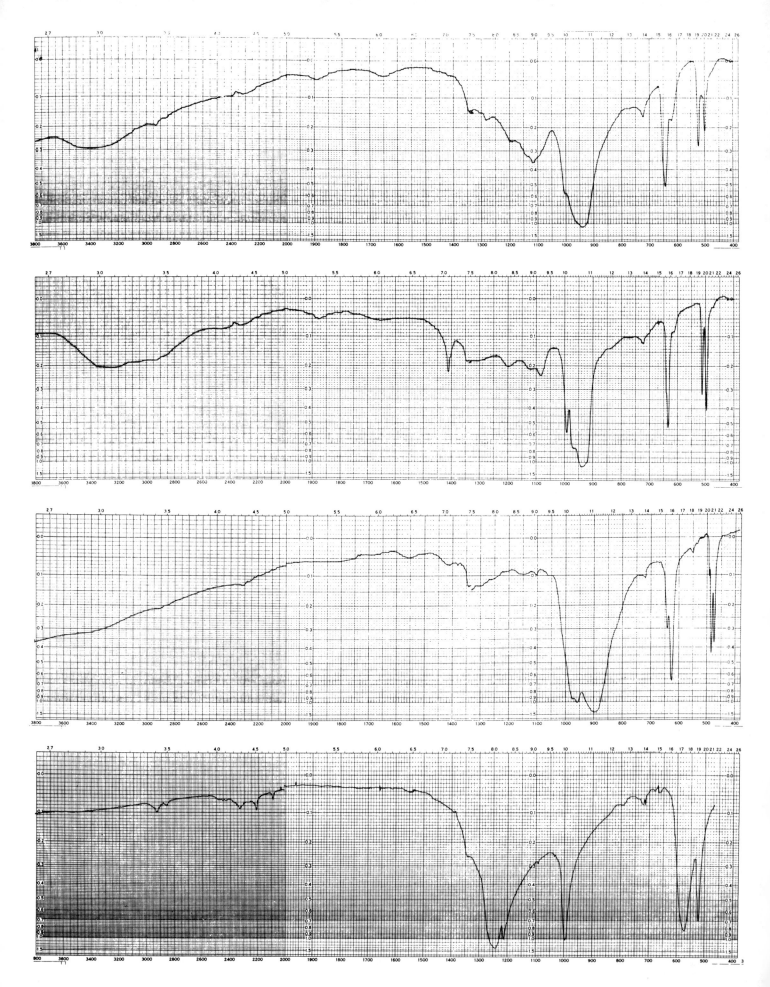

262

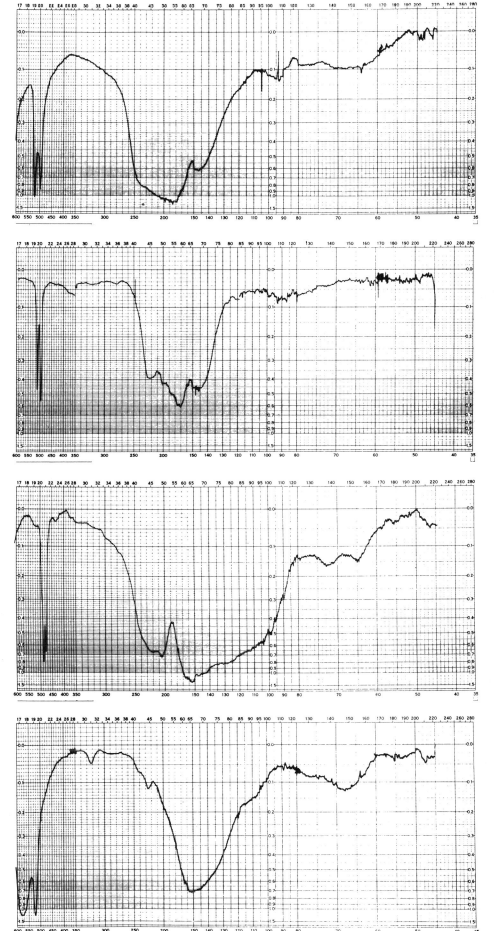

426
Strontium sulfite
SrSO$_3$ (wet)
(plus sulfate impurity)

427
Barium sulfite
BaSO$_3$ (wet)
(plus carbonate and sulfate impurities)

428
Lead sulfite
PbSO$_3$

429
Potassium dithionate
K$_2$S$_2$O$_6$

263

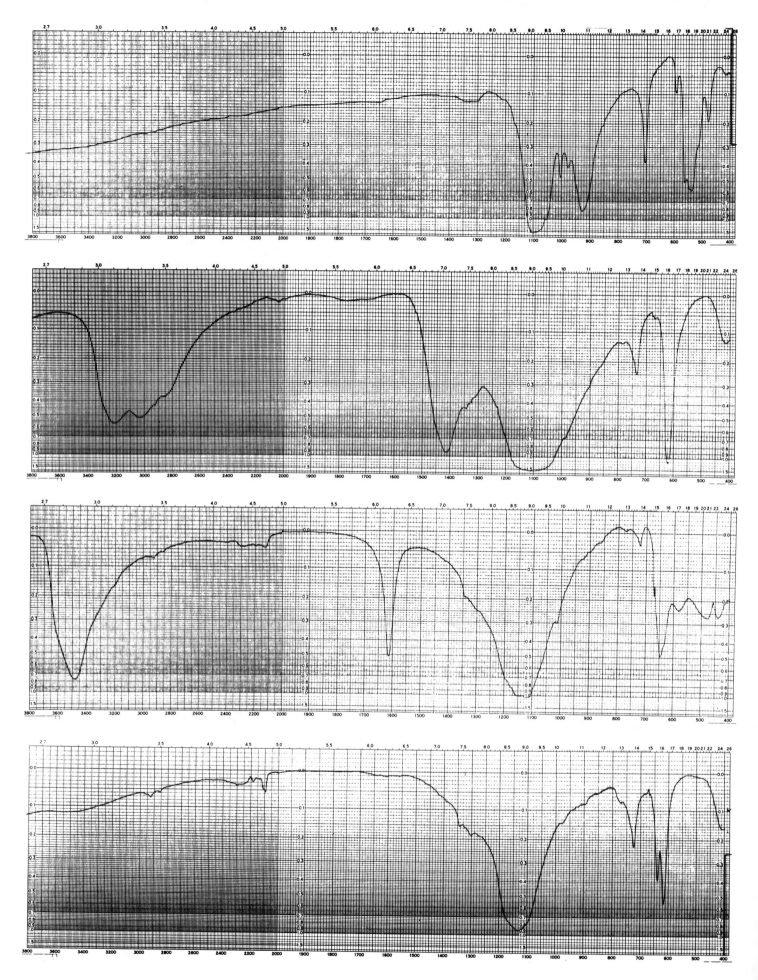

264

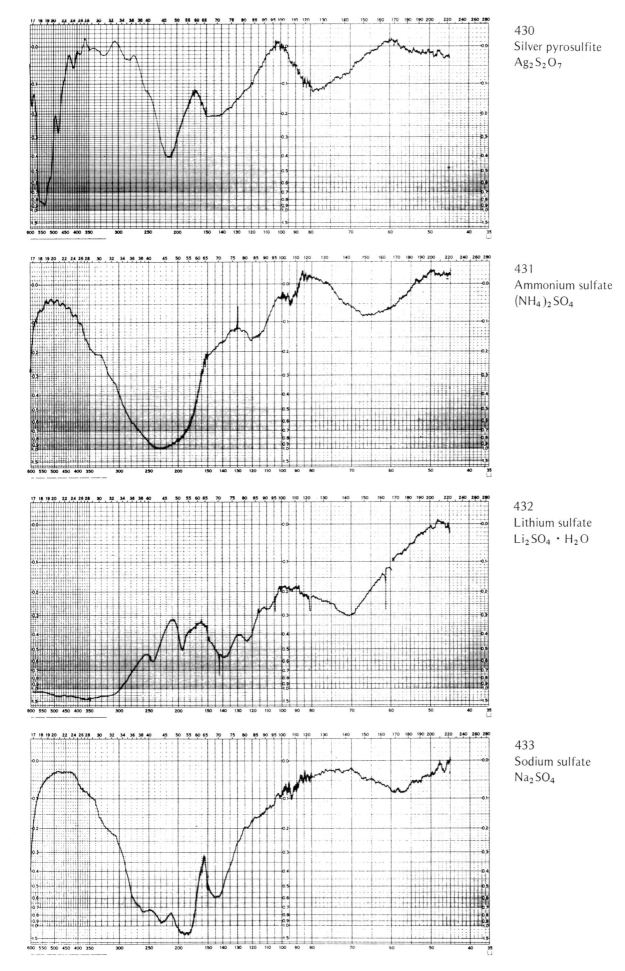

430
Silver pyrosulfite
$Ag_2S_2O_7$

431
Ammonium sulfate
$(NH_4)_2SO_4$

432
Lithium sulfate
$Li_2SO_4 \cdot H_2O$

433
Sodium sulfate
Na_2SO_4

265

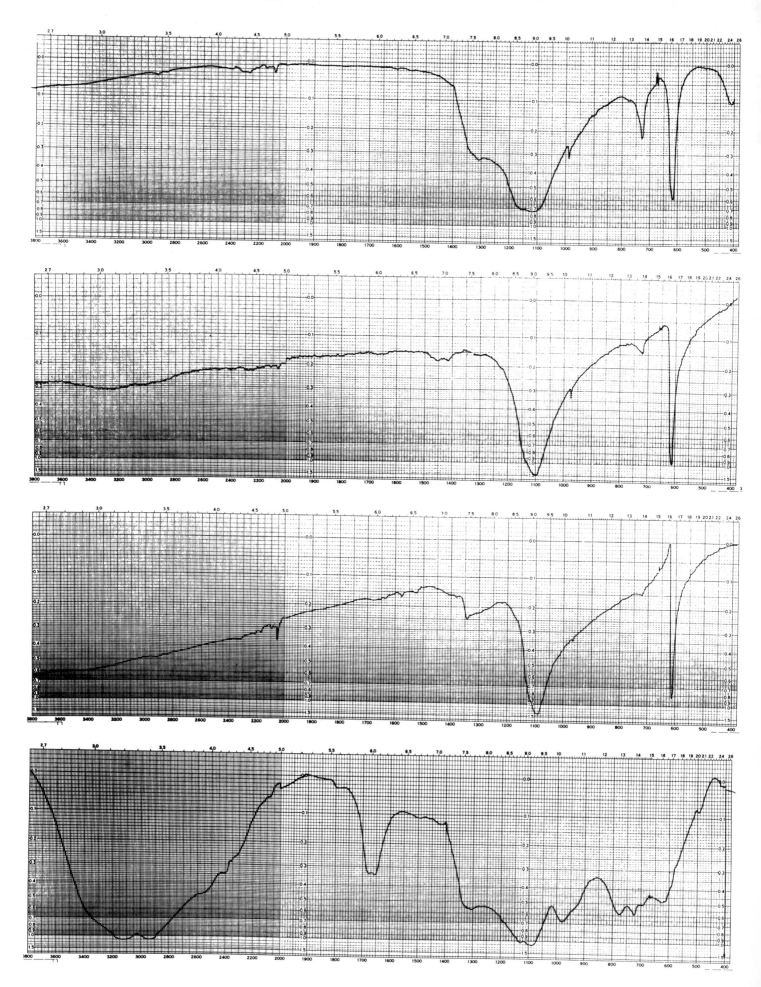

266

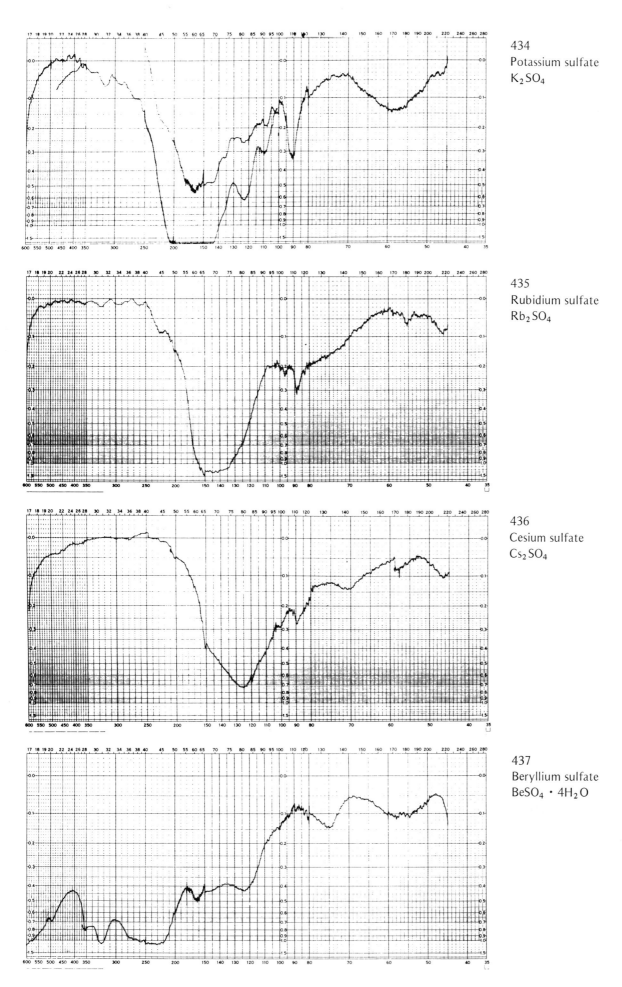

434
Potassium sulfate
K_2SO_4

435
Rubidium sulfate
Rb_2SO_4

436
Cesium sulfate
Cs_2SO_4

437
Beryllium sulfate
$BeSO_4 \cdot 4H_2O$

267

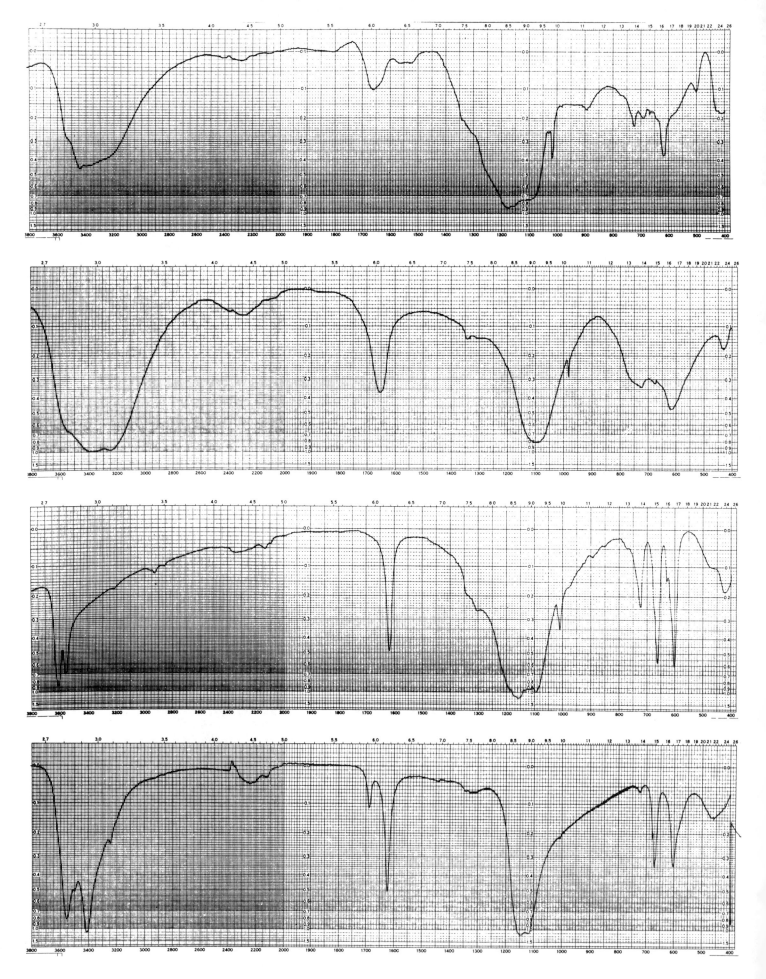

268

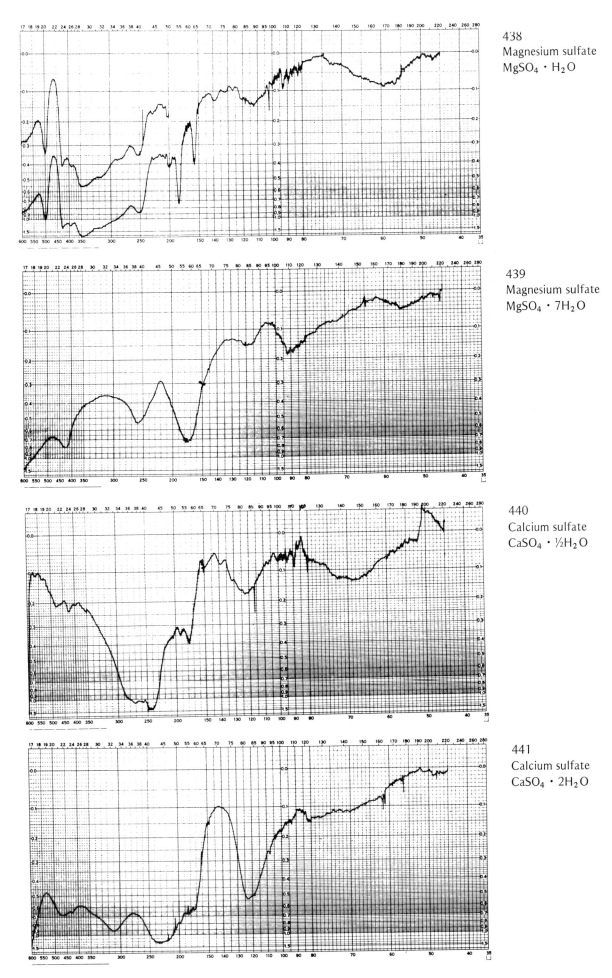

438
Magnesium sulfate
$MgSO_4 \cdot H_2O$

439
Magnesium sulfate
$MgSO_4 \cdot 7H_2O$

440
Calcium sulfate
$CaSO_4 \cdot \frac{1}{2}H_2O$

441
Calcium sulfate
$CaSO_4 \cdot 2H_2O$

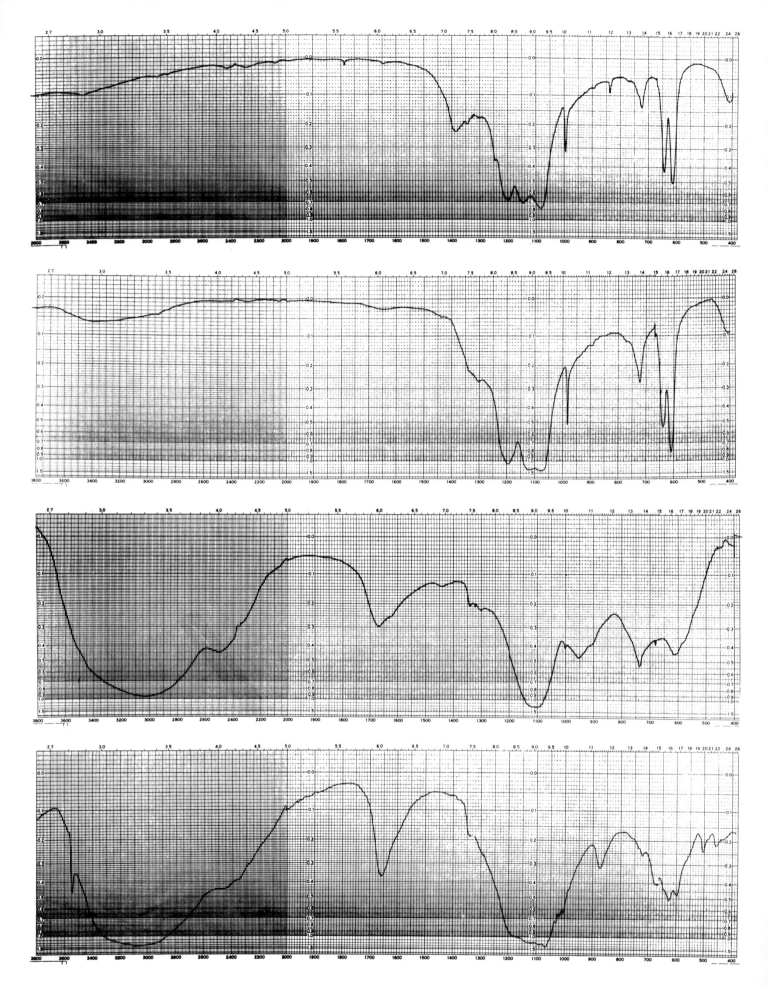

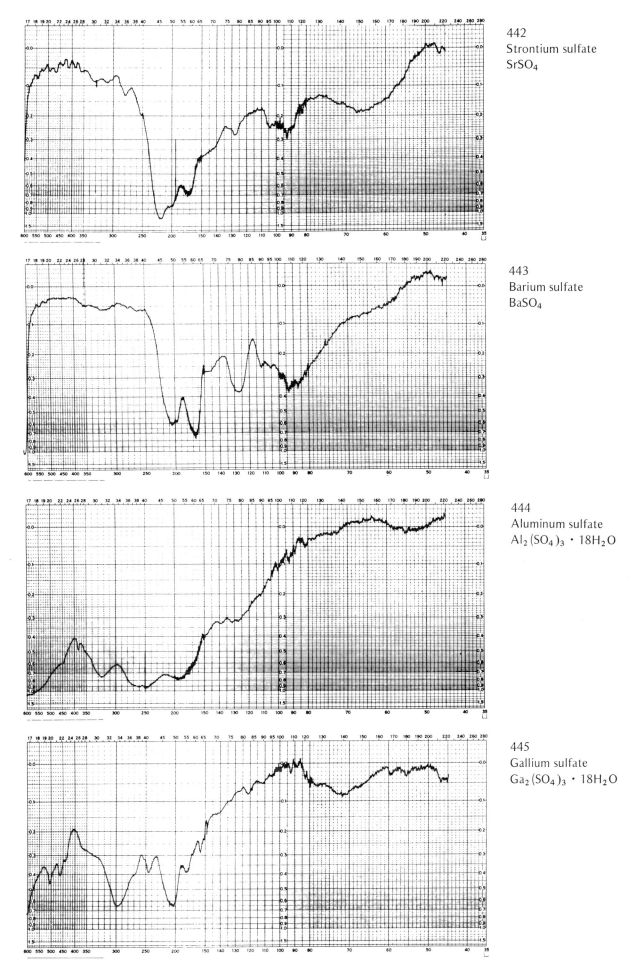

442
Strontium sulfate
SrSO$_4$

443
Barium sulfate
BaSO$_4$

444
Aluminum sulfate
Al$_2$(SO$_4$)$_3$ · 18H$_2$O

445
Gallium sulfate
Ga$_2$(SO$_4$)$_3$ · 18H$_2$O

271

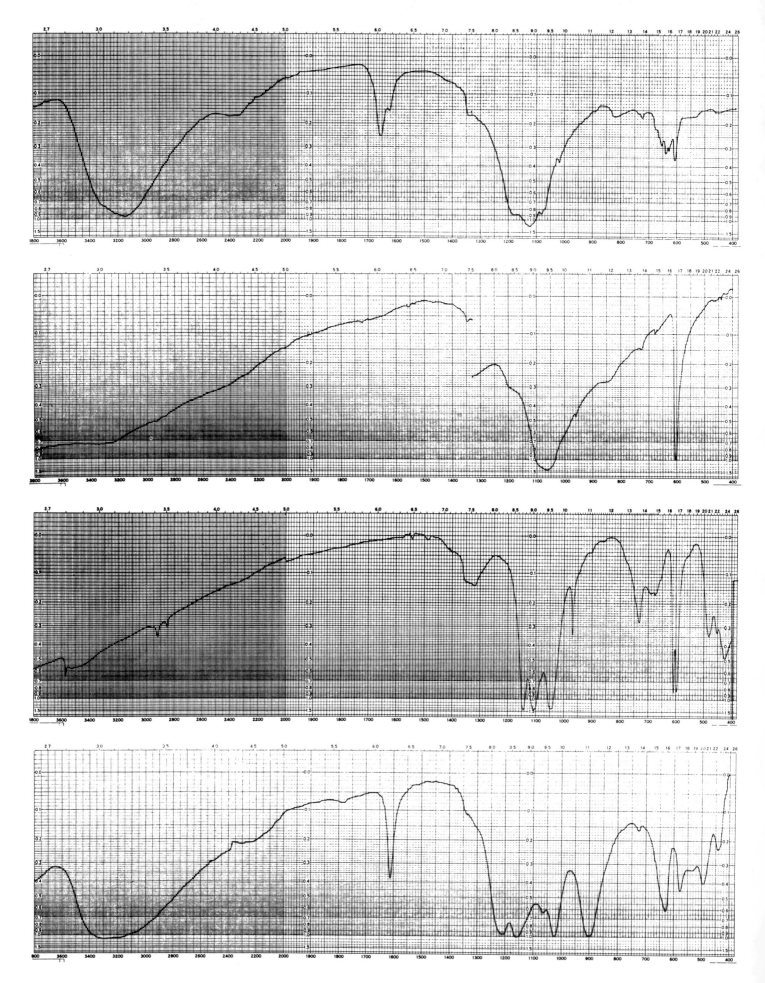

272

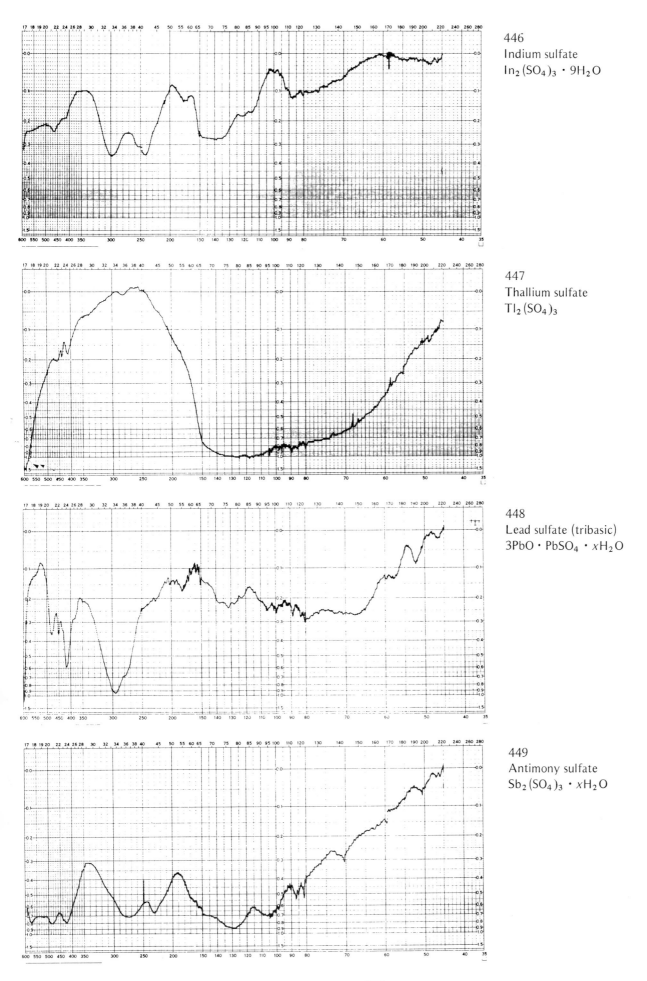

446
Indium sulfate
$In_2(SO_4)_3 \cdot 9H_2O$

447
Thallium sulfate
$Tl_2(SO_4)_3$

448
Lead sulfate (tribasic)
$3PbO \cdot PbSO_4 \cdot xH_2O$

449
Antimony sulfate
$Sb_2(SO_4)_3 \cdot xH_2O$

273

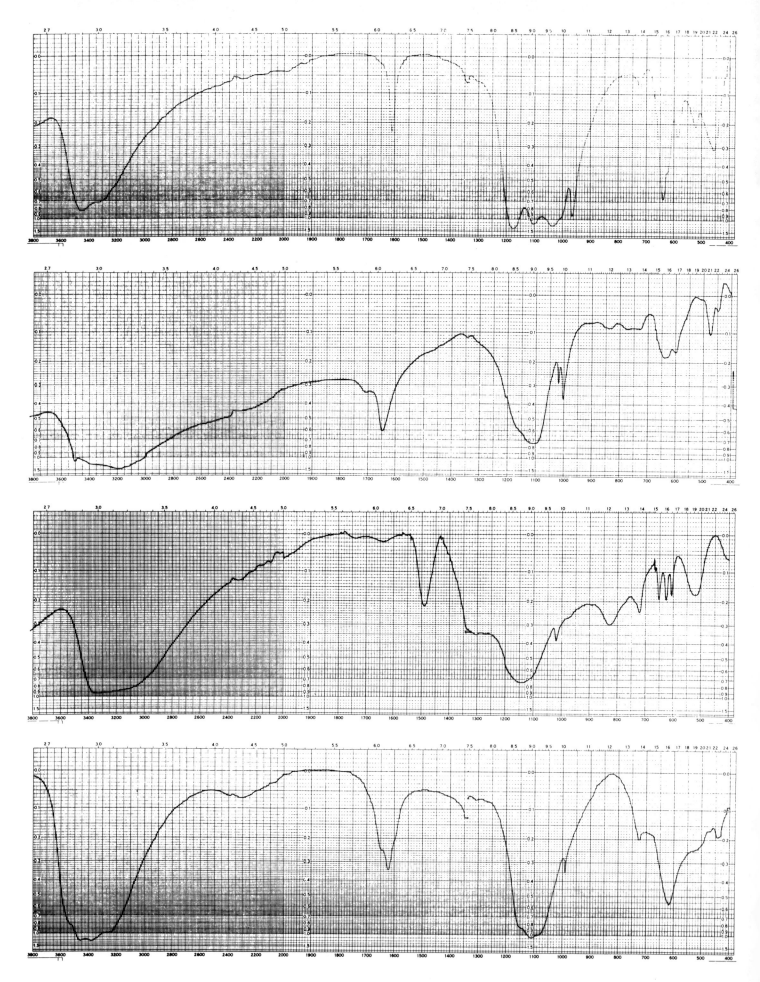

274

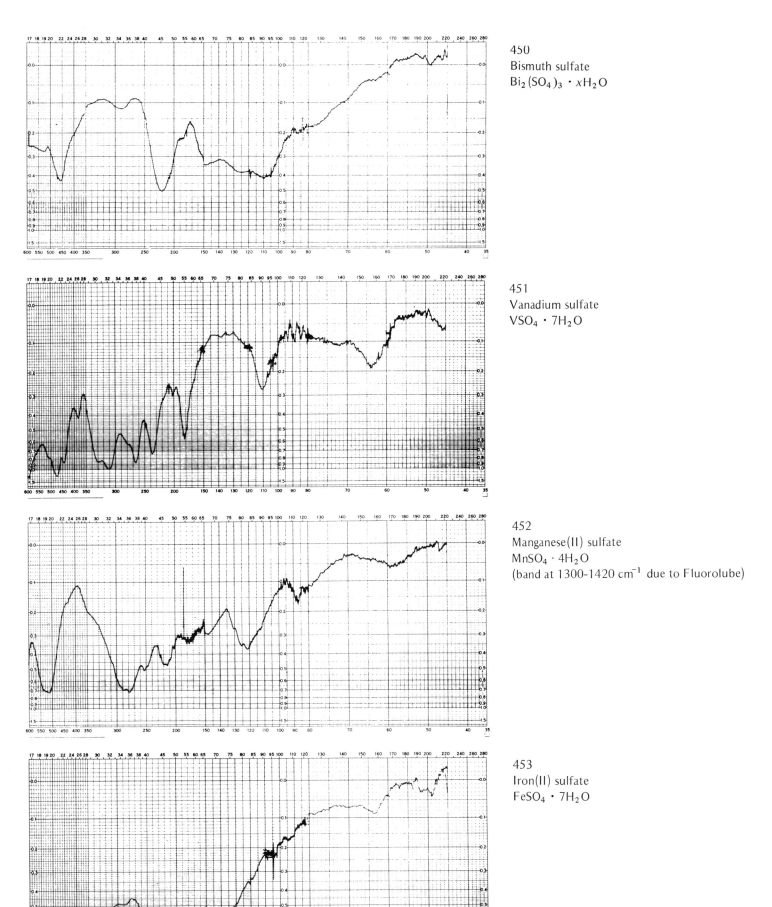

450
Bismuth sulfate
$Bi_2(SO_4)_3 \cdot xH_2O$

451
Vanadium sulfate
$VSO_4 \cdot 7H_2O$

452
Manganese(II) sulfate
$MnSO_4 \cdot 4H_2O$
(band at 1300-1420 cm^{-1} due to Fluorolube)

453
Iron(II) sulfate
$FeSO_4 \cdot 7H_2O$

275

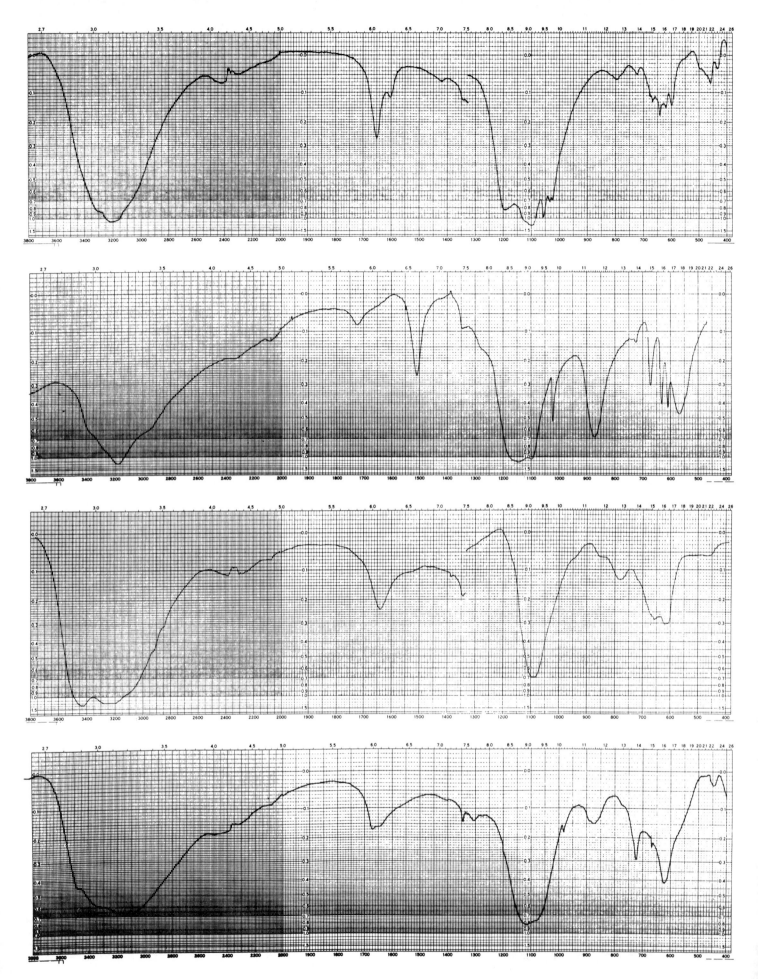

276

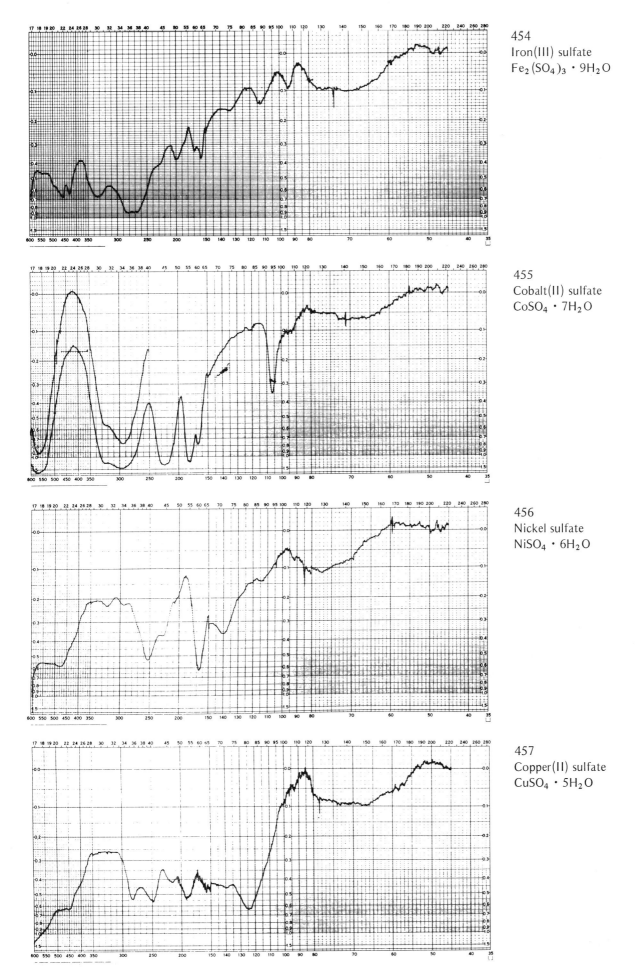

454
Iron(III) sulfate
$Fe_2(SO_4)_3 \cdot 9H_2O$

455
Cobalt(II) sulfate
$CoSO_4 \cdot 7H_2O$

456
Nickel sulfate
$NiSO_4 \cdot 6H_2O$

457
Copper(II) sulfate
$CuSO_4 \cdot 5H_2O$

277

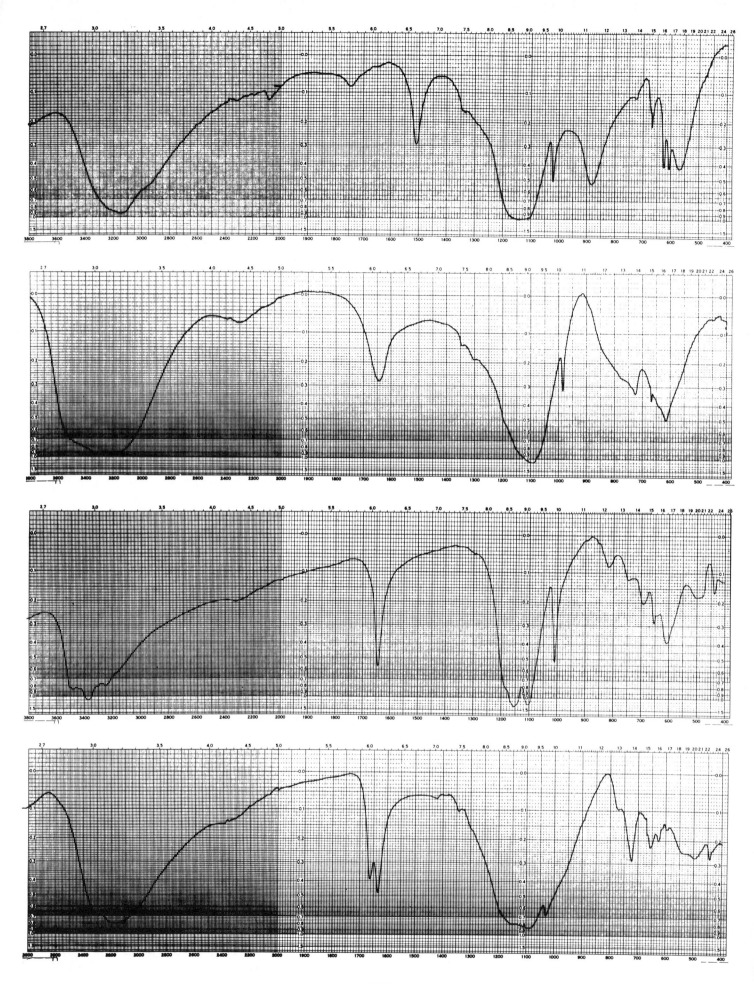

278

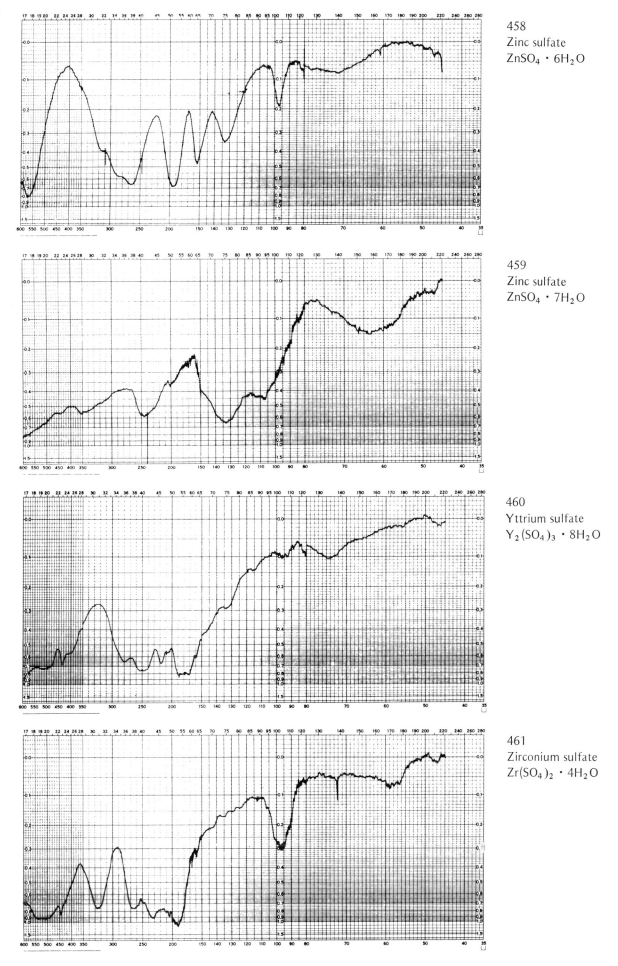

458
Zinc sulfate
$ZnSO_4 \cdot 6H_2O$

459
Zinc sulfate
$ZnSO_4 \cdot 7H_2O$

460
Yttrium sulfate
$Y_2(SO_4)_3 \cdot 8H_2O$

461
Zirconium sulfate
$Zr(SO_4)_2 \cdot 4H_2O$

279

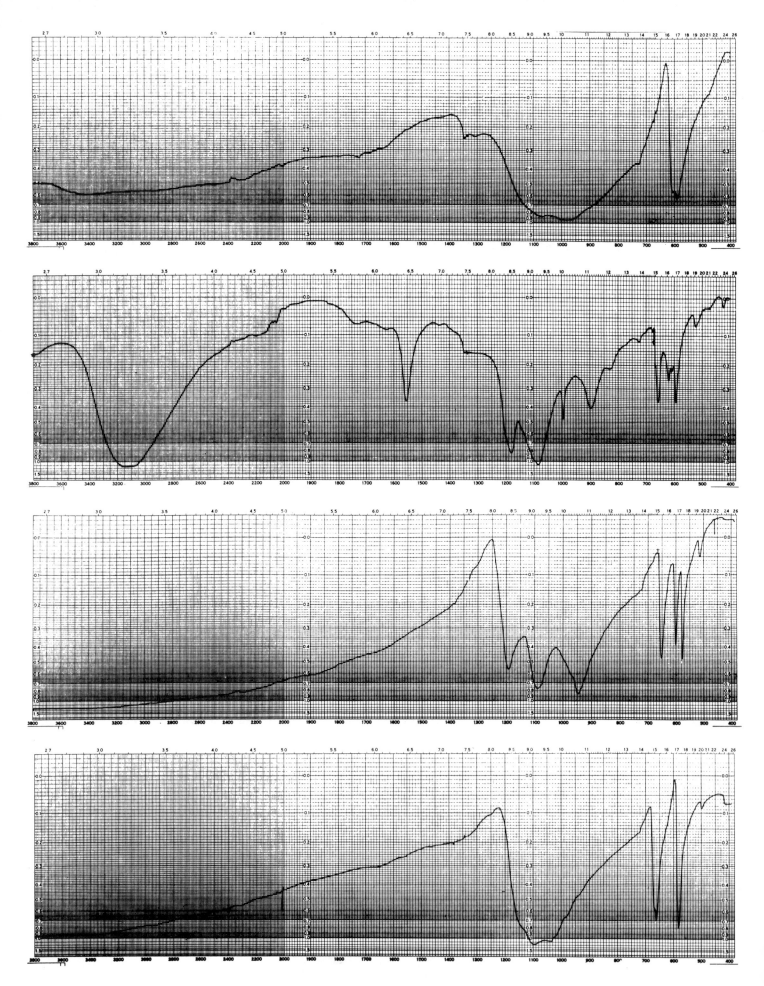

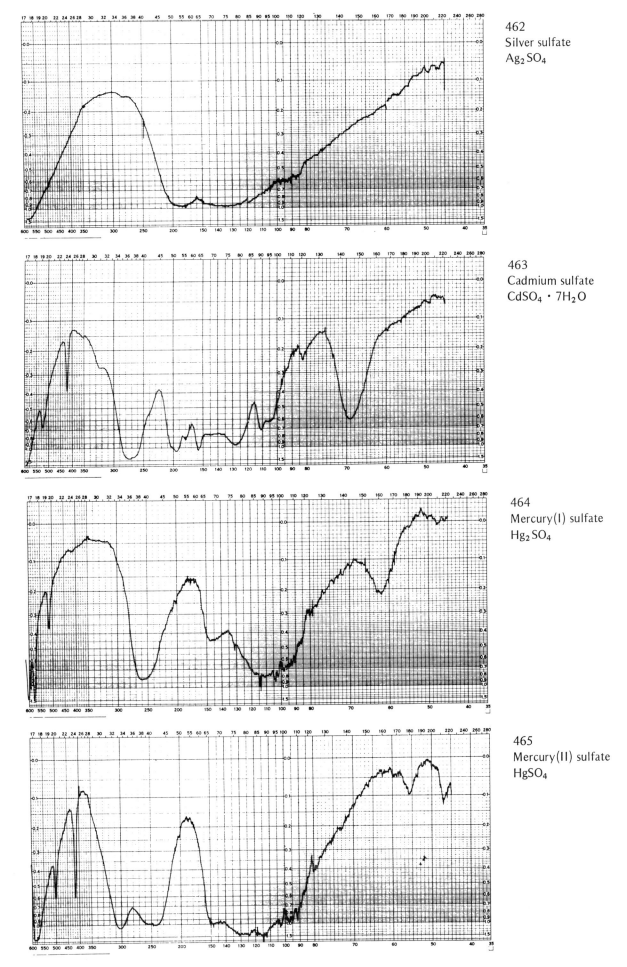

462
Silver sulfate
Ag$_2$SO$_4$

463
Cadmium sulfate
CdSO$_4$ · 7H$_2$O

464
Mercury(I) sulfate
Hg$_2$SO$_4$

465
Mercury(II) sulfate
HgSO$_4$

281

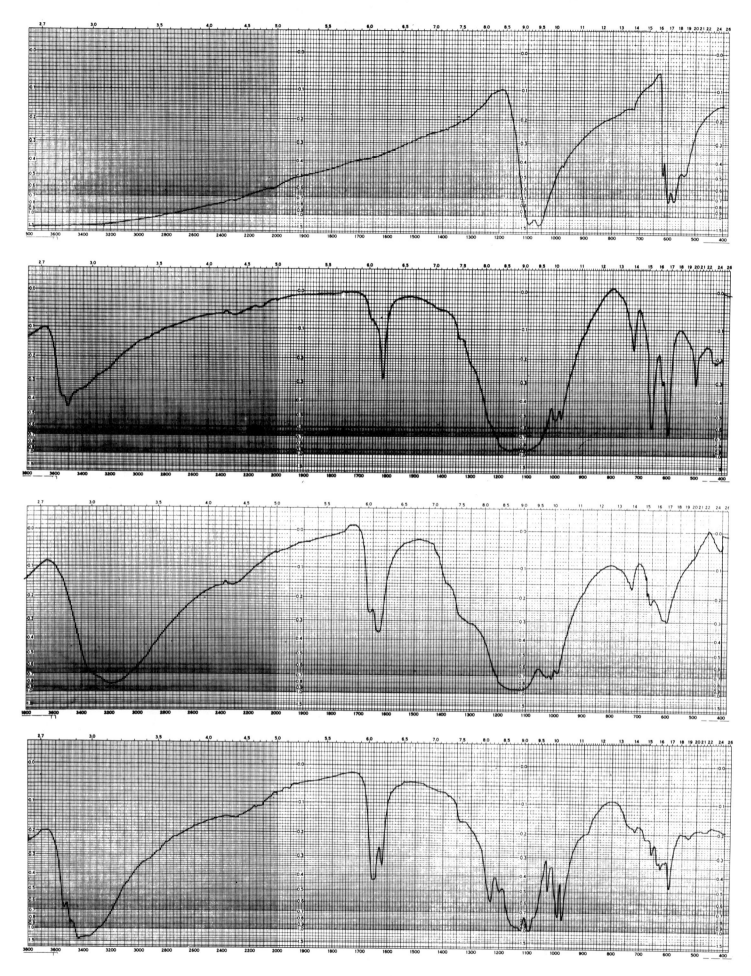

282

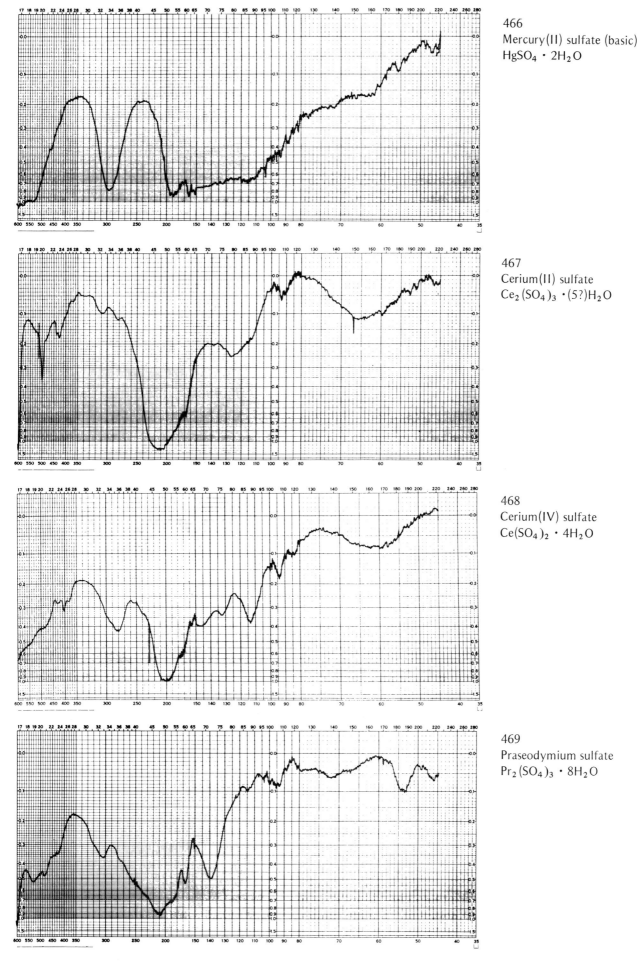

466
Mercury(II) sulfate (basic)
$HgSO_4 \cdot 2H_2O$

467
Cerium(II) sulfate
$Ce_2(SO_4)_3 \cdot (5?)H_2O$

468
Cerium(IV) sulfate
$Ce(SO_4)_2 \cdot 4H_2O$

469
Praseodymium sulfate
$Pr_2(SO_4)_3 \cdot 8H_2O$

283

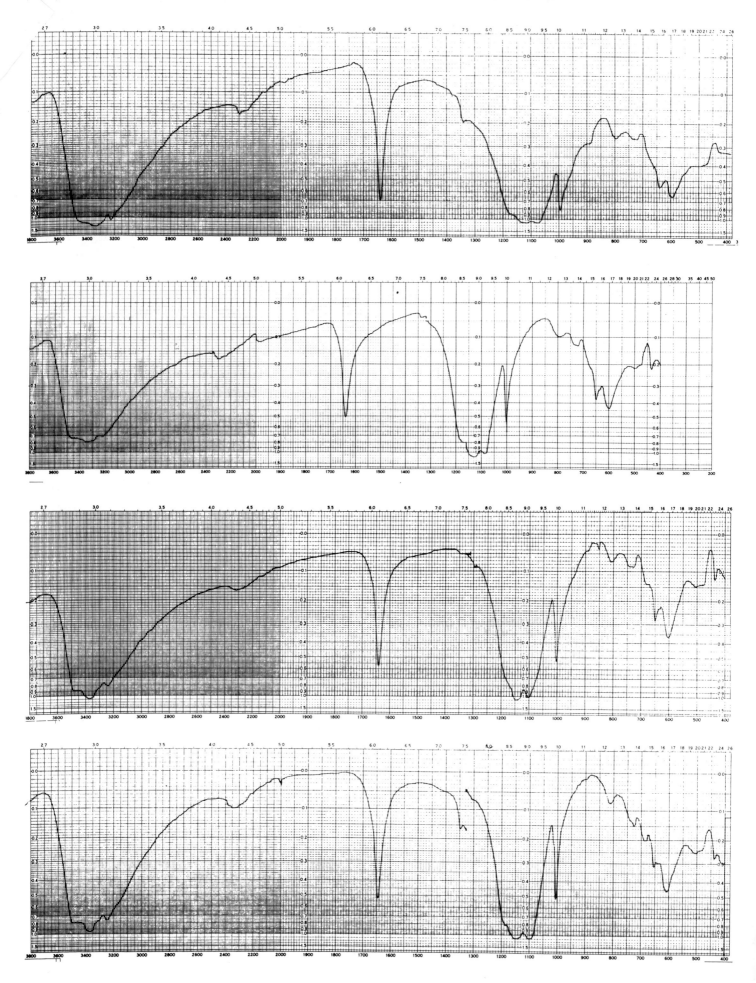

284

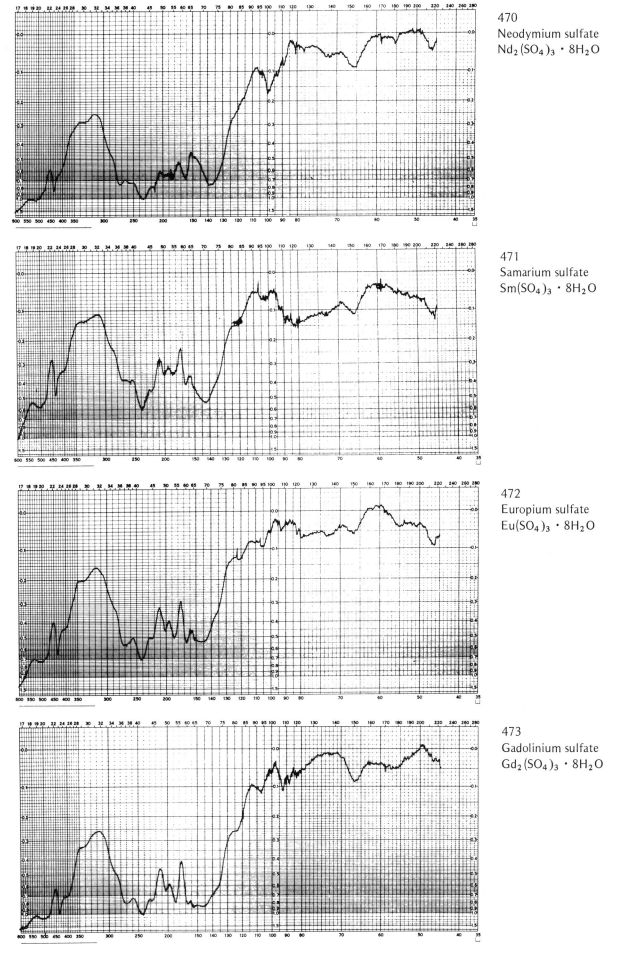

470
Neodymium sulfate
$Nd_2(SO_4)_3 \cdot 8H_2O$

471
Samarium sulfate
$Sm(SO_4)_3 \cdot 8H_2O$

472
Europium sulfate
$Eu(SO_4)_3 \cdot 8H_2O$

473
Gadolinium sulfate
$Gd_2(SO_4)_3 \cdot 8H_2O$

285

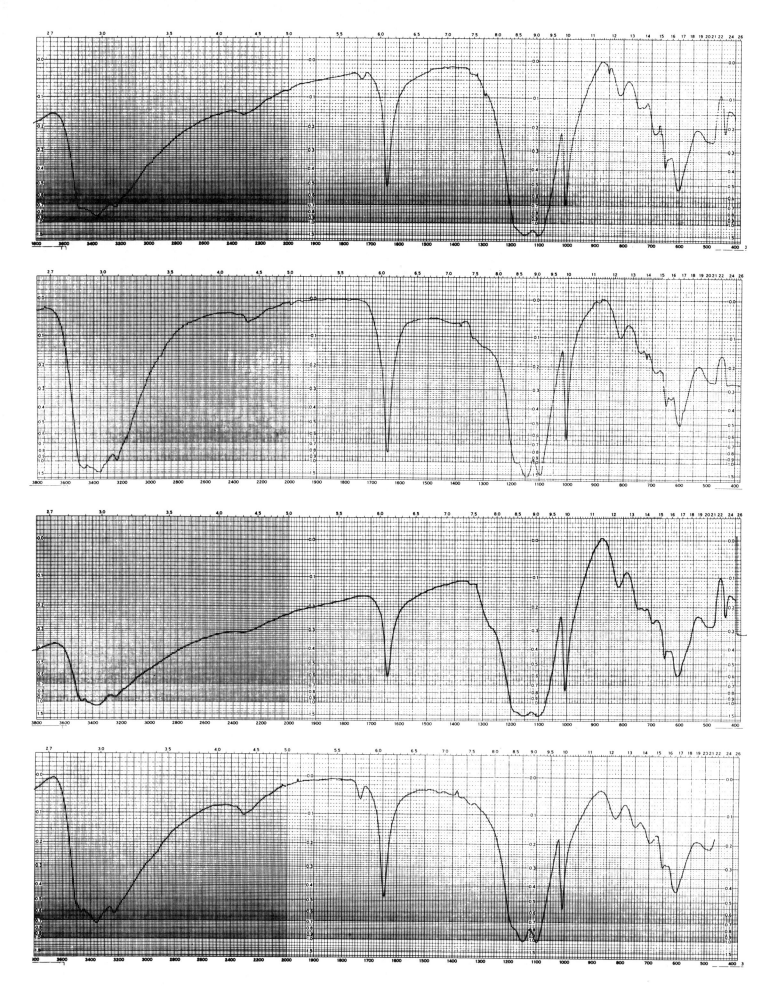

286

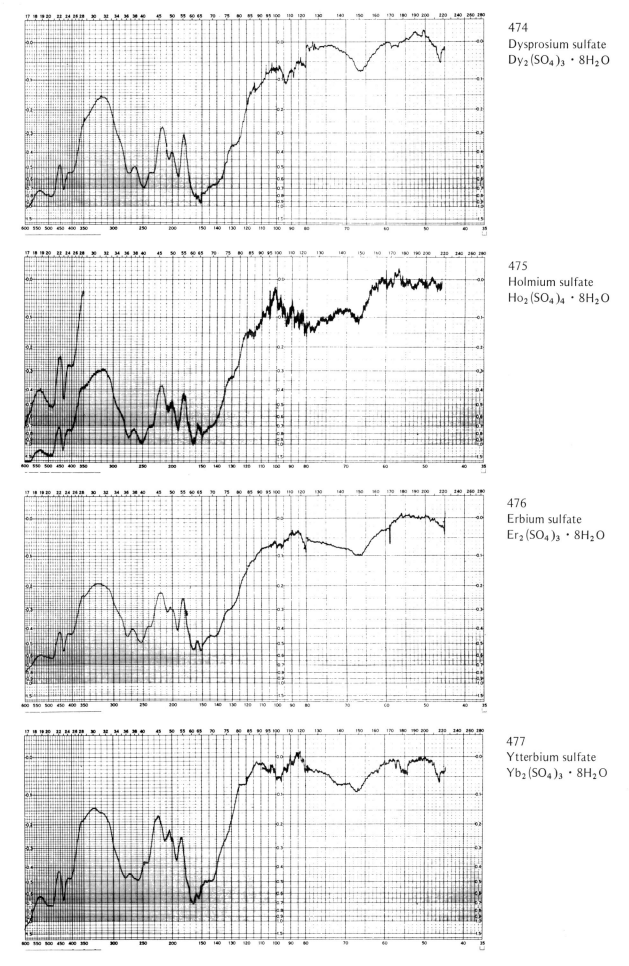

474
Dysprosium sulfate
$Dy_2(SO_4)_3 \cdot 8H_2O$

475
Holmium sulfate
$Ho_2(SO_4)_4 \cdot 8H_2O$

476
Erbium sulfate
$Er_2(SO_4)_3 \cdot 8H_2O$

477
Ytterbium sulfate
$Yb_2(SO_4)_3 \cdot 8H_2O$

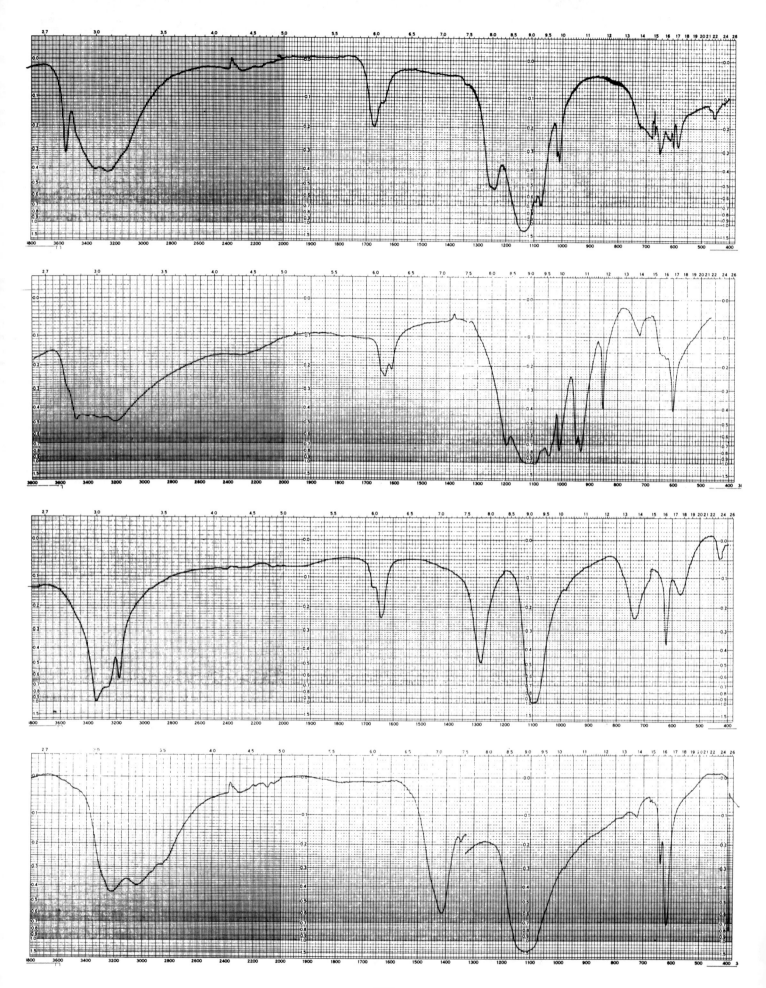

288

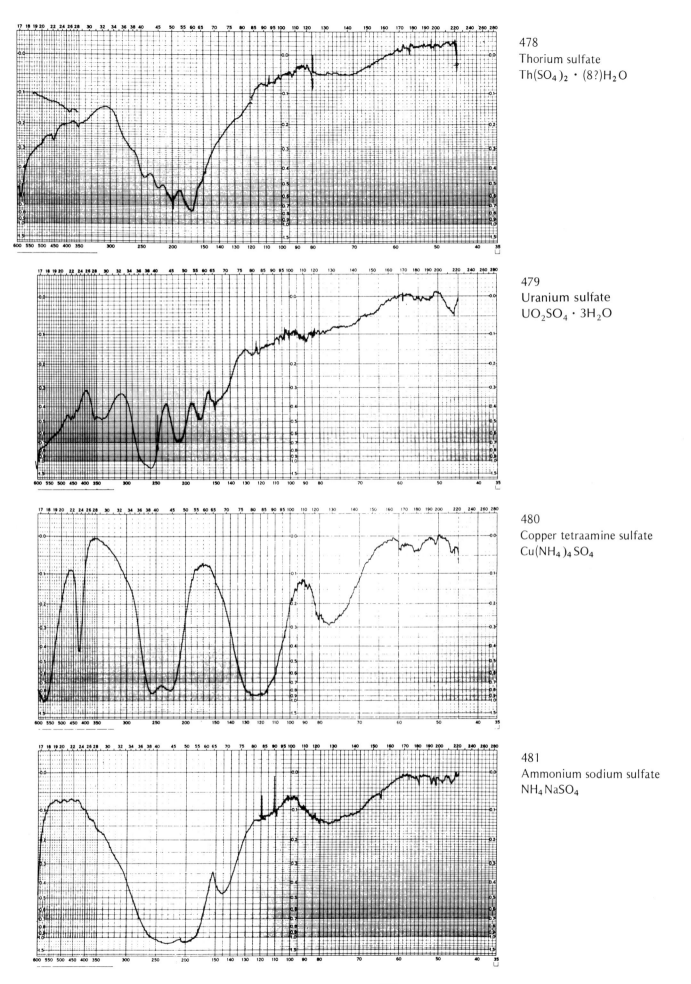

478
Thorium sulfate
$Th(SO_4)_2 \cdot (8?)H_2O$

479
Uranium sulfate
$UO_2SO_4 \cdot 3H_2O$

480
Copper tetraamine sulfate
$Cu(NH_4)_4SO_4$

481
Ammonium sodium sulfate
NH_4NaSO_4

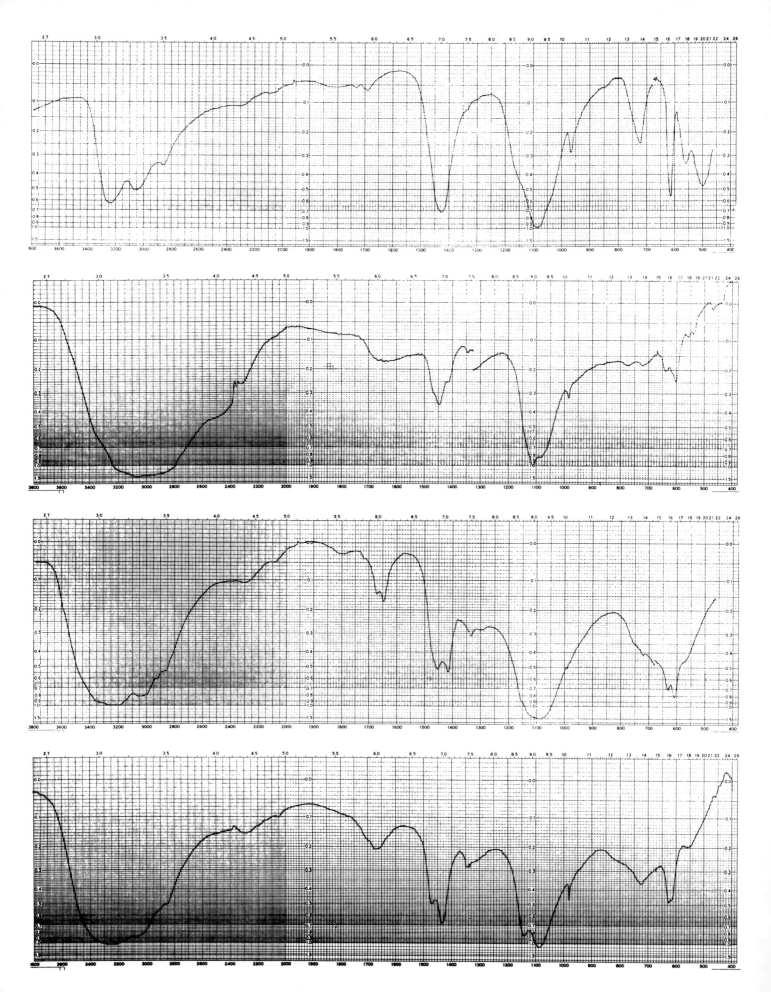

290

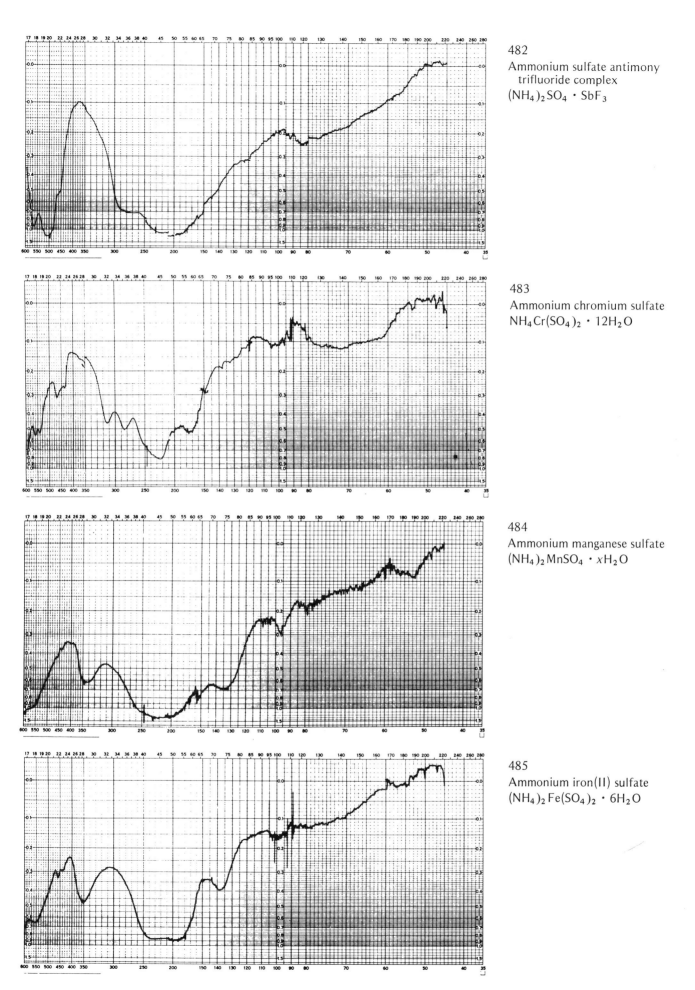

482
Ammonium sulfate antimony
  trifluoride complex
$(NH_4)_2SO_4 \cdot SbF_3$

483
Ammonium chromium sulfate
$NH_4Cr(SO_4)_2 \cdot 12H_2O$

484
Ammonium manganese sulfate
$(NH_4)_2MnSO_4 \cdot xH_2O$

485
Ammonium iron(II) sulfate
$(NH_4)_2Fe(SO_4)_2 \cdot 6H_2O$

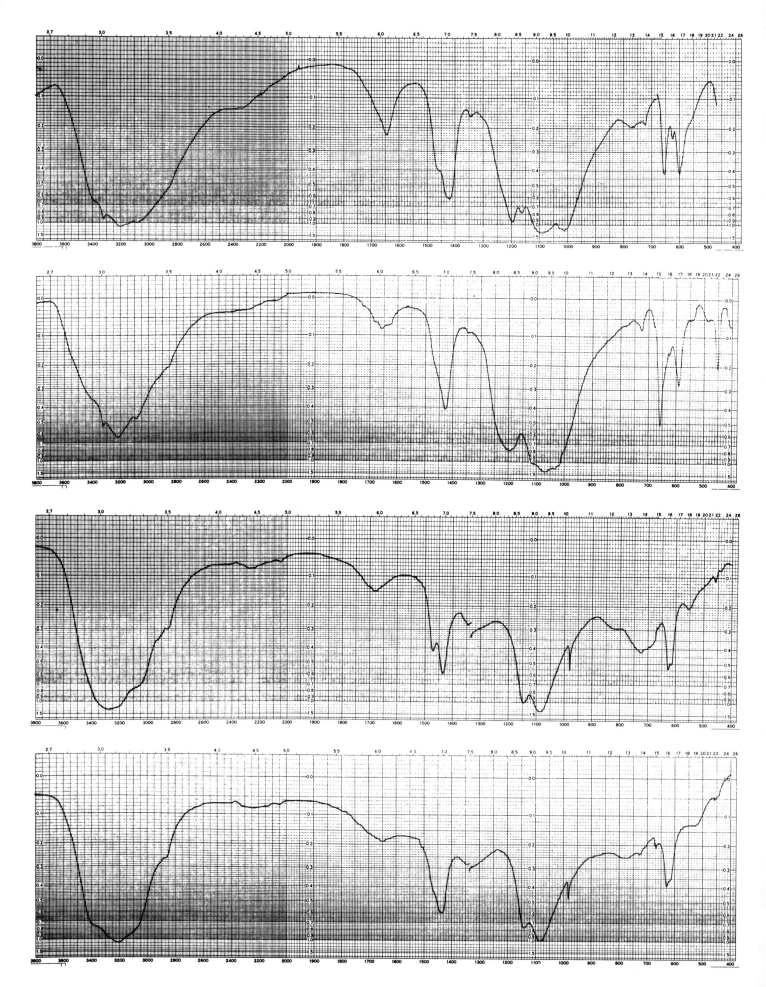

292

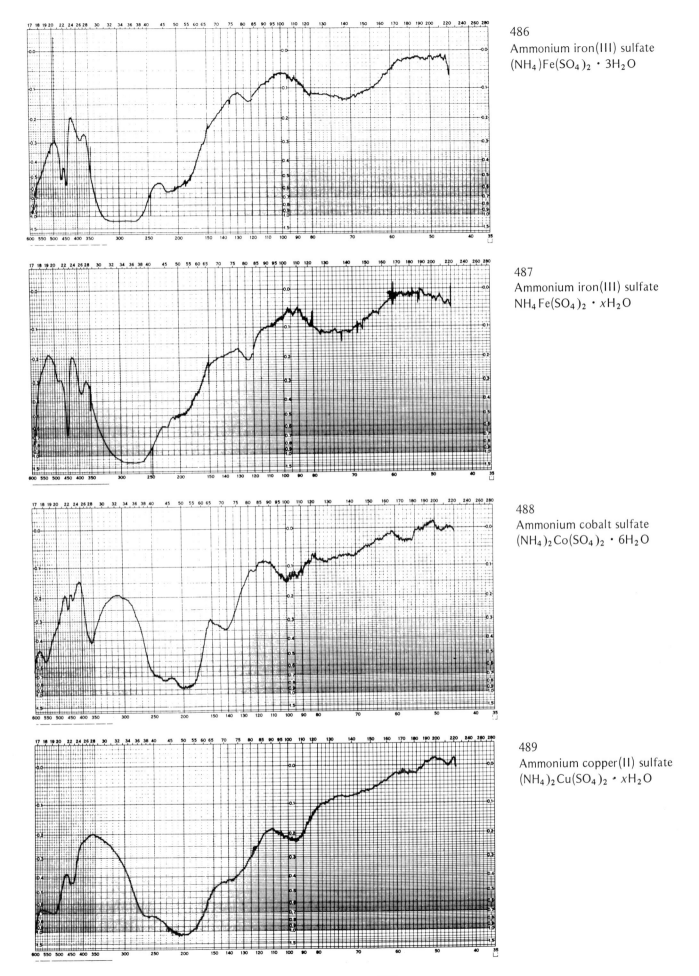

486
Ammonium iron(III) sulfate
(NH₄)Fe(SO₄)₂ · 3H₂O

487
Ammonium iron(III) sulfate
NH₄Fe(SO₄)₂ · xH₂O

488
Ammonium cobalt sulfate
(NH₄)₂Co(SO₄)₂ · 6H₂O

489
Ammonium copper(II) sulfate
(NH₄)₂Cu(SO₄)₂ · xH₂O

293

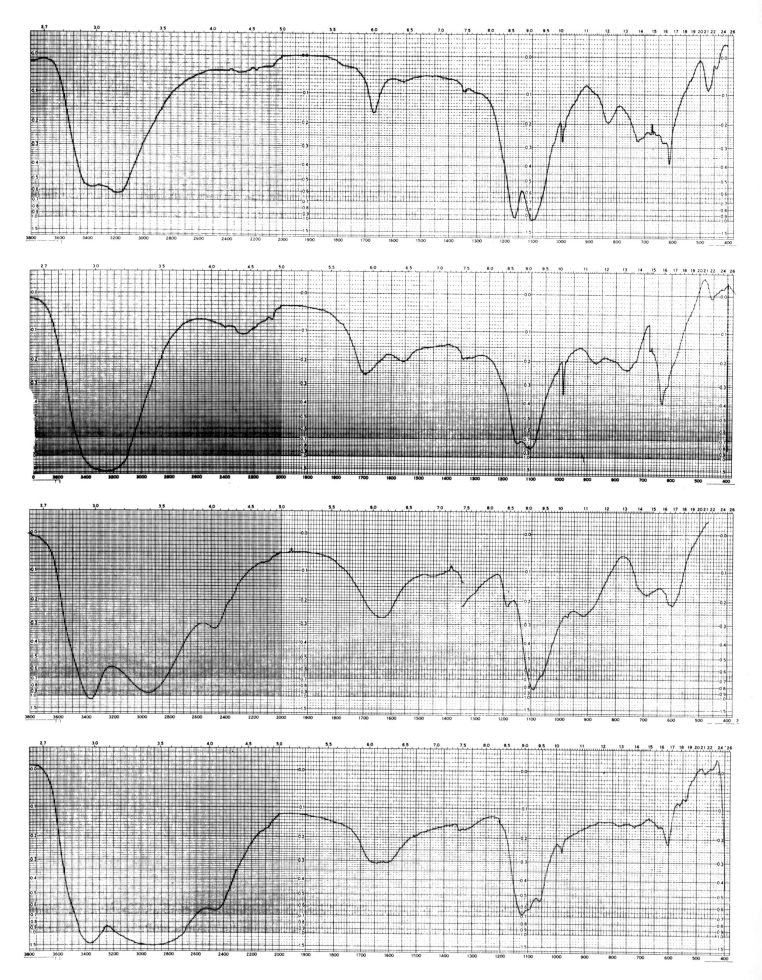

294

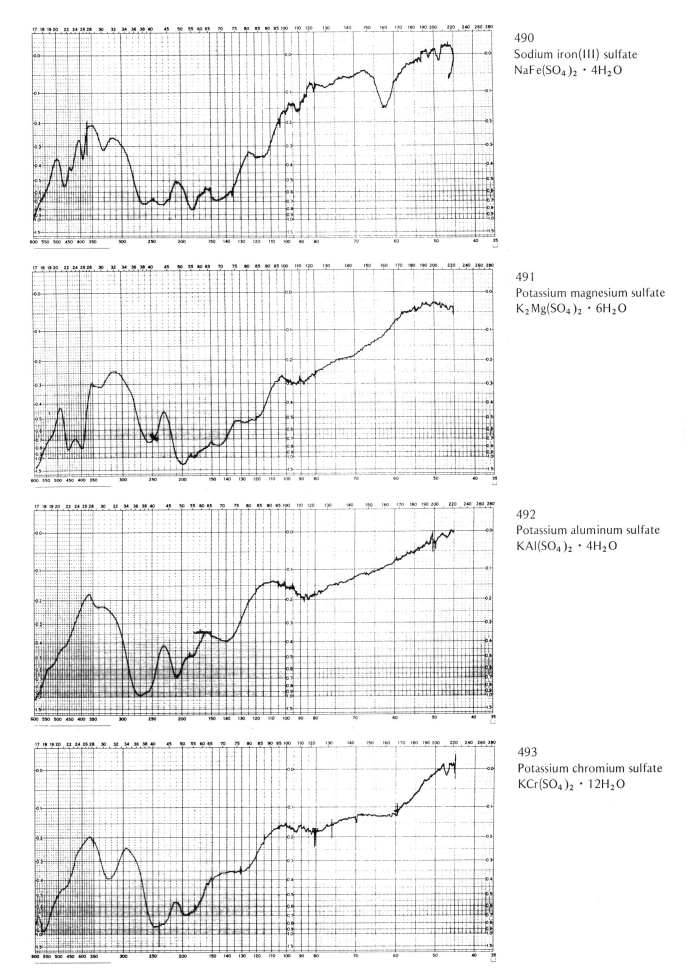

490
Sodium iron(III) sulfate
$NaFe(SO_4)_2 \cdot 4H_2O$

491
Potassium magnesium sulfate
$K_2Mg(SO_4)_2 \cdot 6H_2O$

492
Potassium aluminum sulfate
$KAl(SO_4)_2 \cdot 4H_2O$

493
Potassium chromium sulfate
$KCr(SO_4)_2 \cdot 12H_2O$

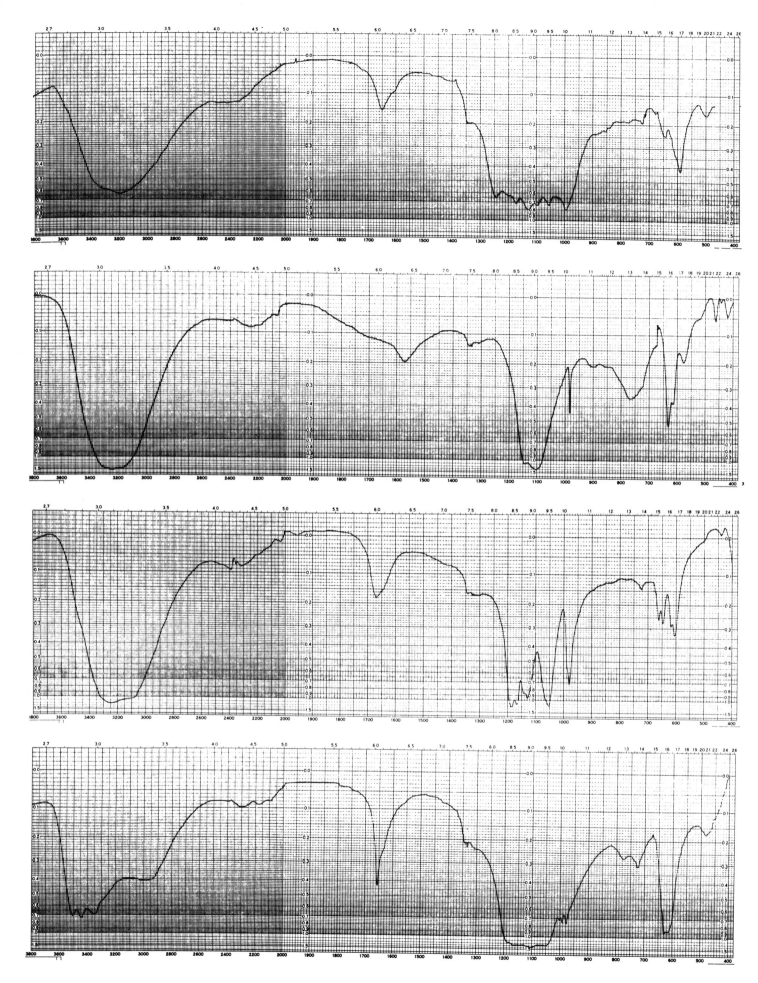

296

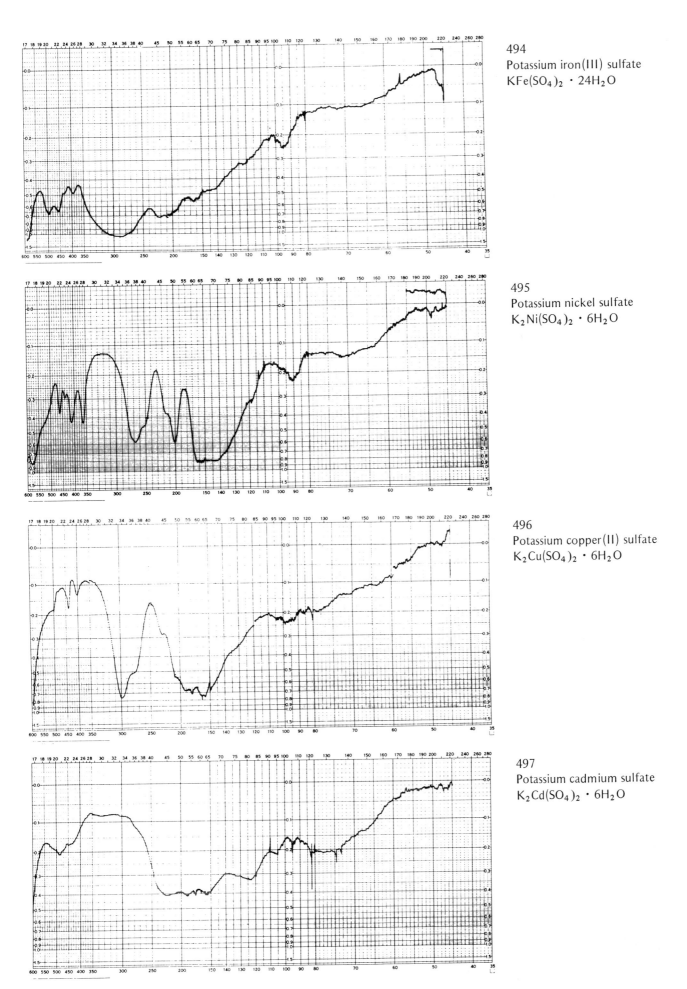

494
Potassium iron(III) sulfate
$KFe(SO_4)_2 \cdot 24H_2O$

495
Potassium nickel sulfate
$K_2Ni(SO_4)_2 \cdot 6H_2O$

496
Potassium copper(II) sulfate
$K_2Cu(SO_4)_2 \cdot 6H_2O$

497
Potassium cadmium sulfate
$K_2Cd(SO_4)_2 \cdot 6H_2O$

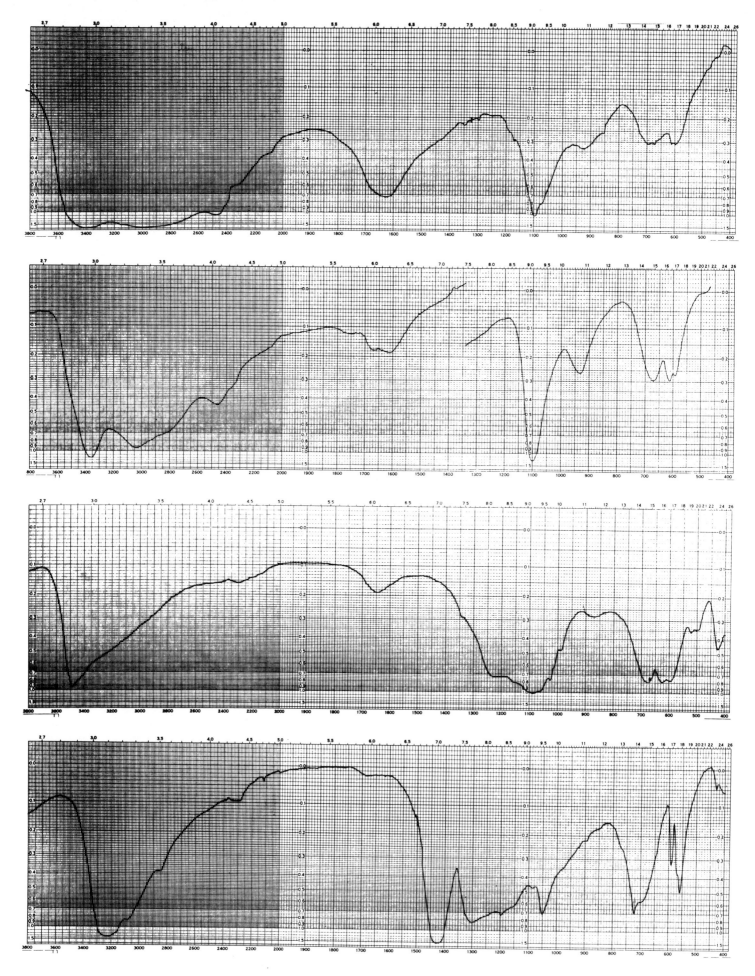

298

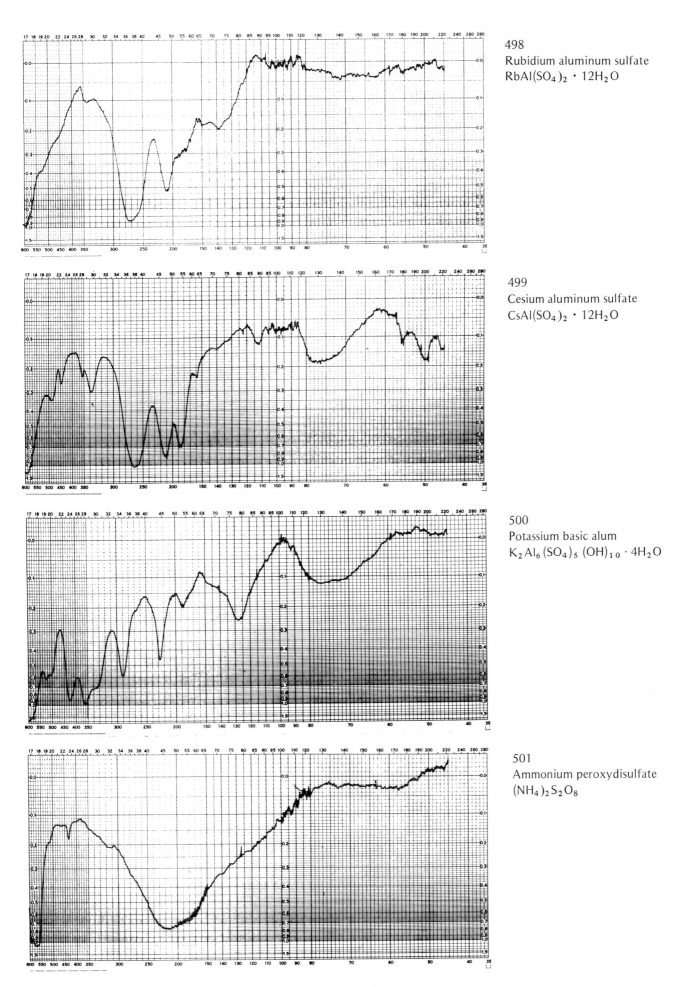

498
Rubidium aluminum sulfate
$RbAl(SO_4)_2 \cdot 12H_2O$

499
Cesium aluminum sulfate
$CsAl(SO_4)_2 \cdot 12H_2O$

500
Potassium basic alum
$K_2Al_6(SO_4)_5(OH)_{10} \cdot 4H_2O$

501
Ammonium peroxydisulfate
$(NH_4)_2S_2O_8$

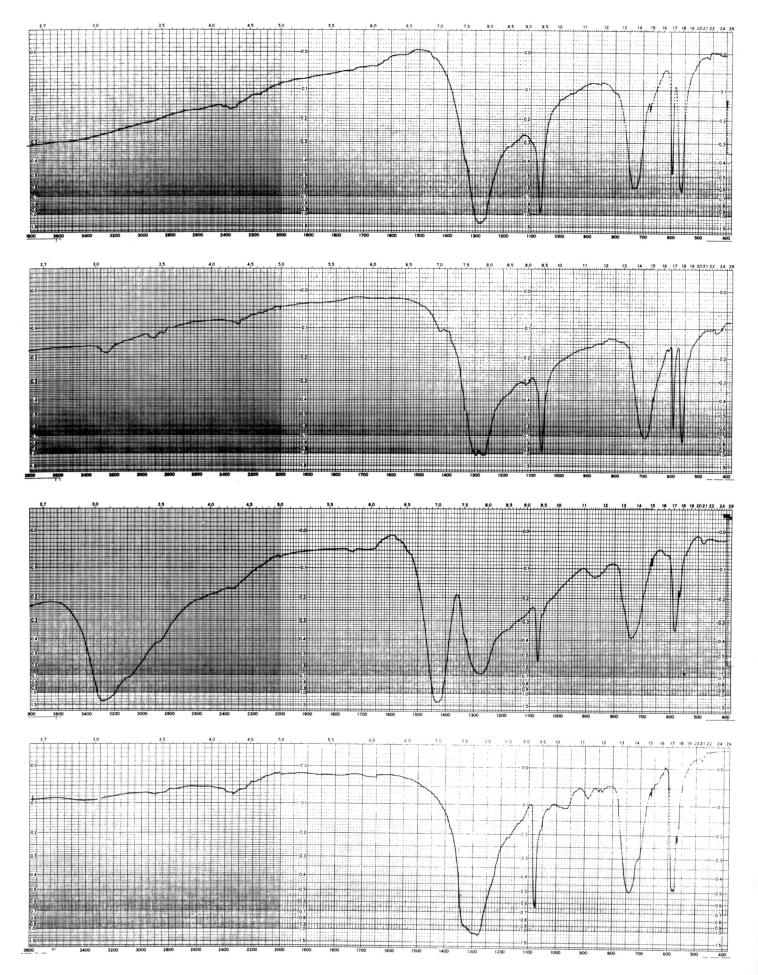

300

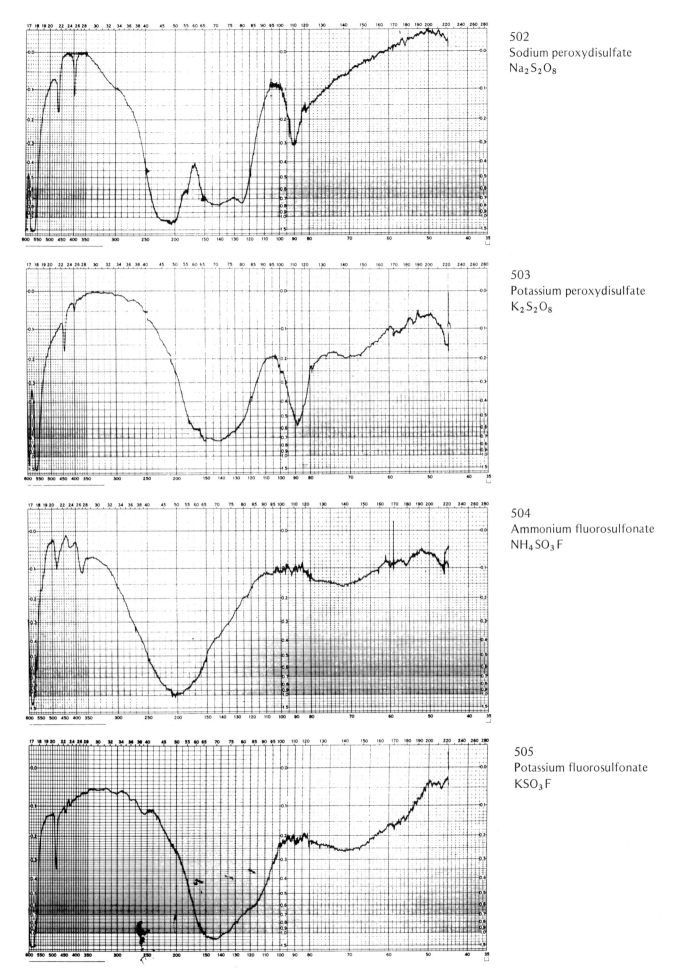

502
Sodium peroxydisulfate
Na$_2$S$_2$O$_8$

503
Potassium peroxydisulfate
K$_2$S$_2$O$_8$

504
Ammonium fluorosulfonate
NH$_4$SO$_3$F

505
Potassium fluorosulfonate
KSO$_3$F

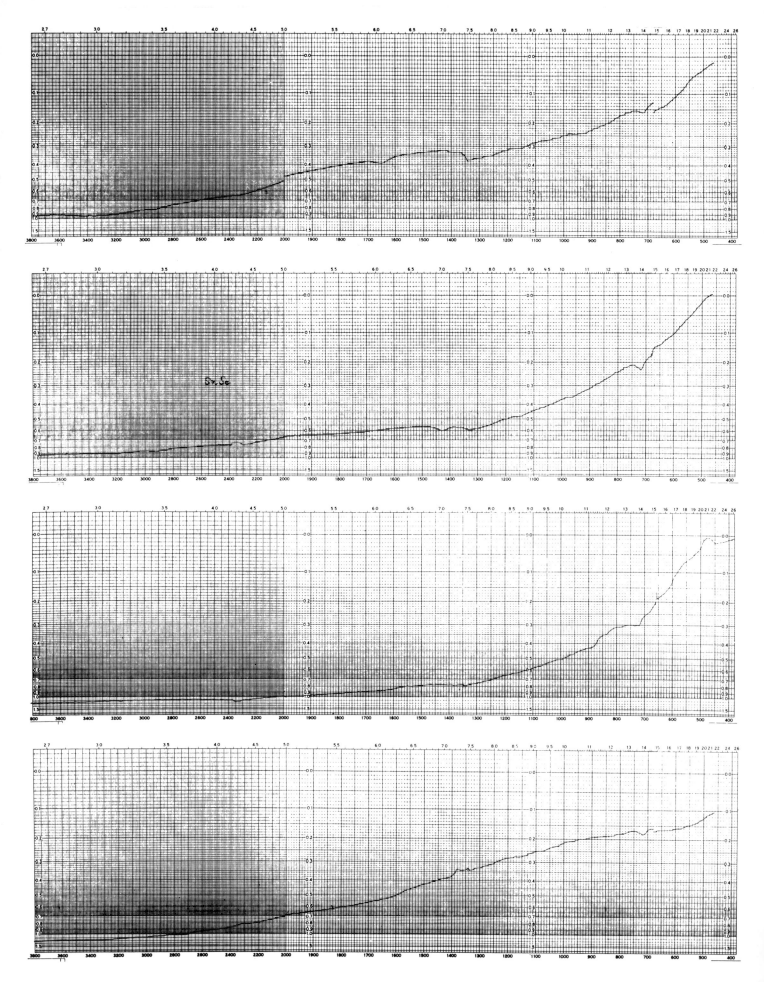

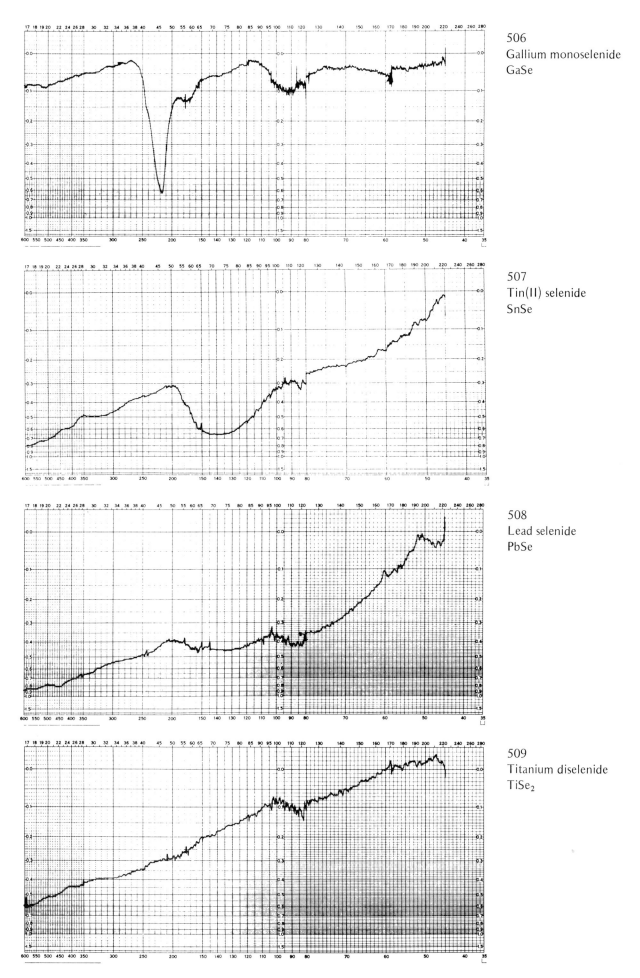

506
Gallium monoselenide
GaSe

507
Tin(II) selenide
SnSe

508
Lead selenide
PbSe

509
Titanium diselenide
TiSe$_2$

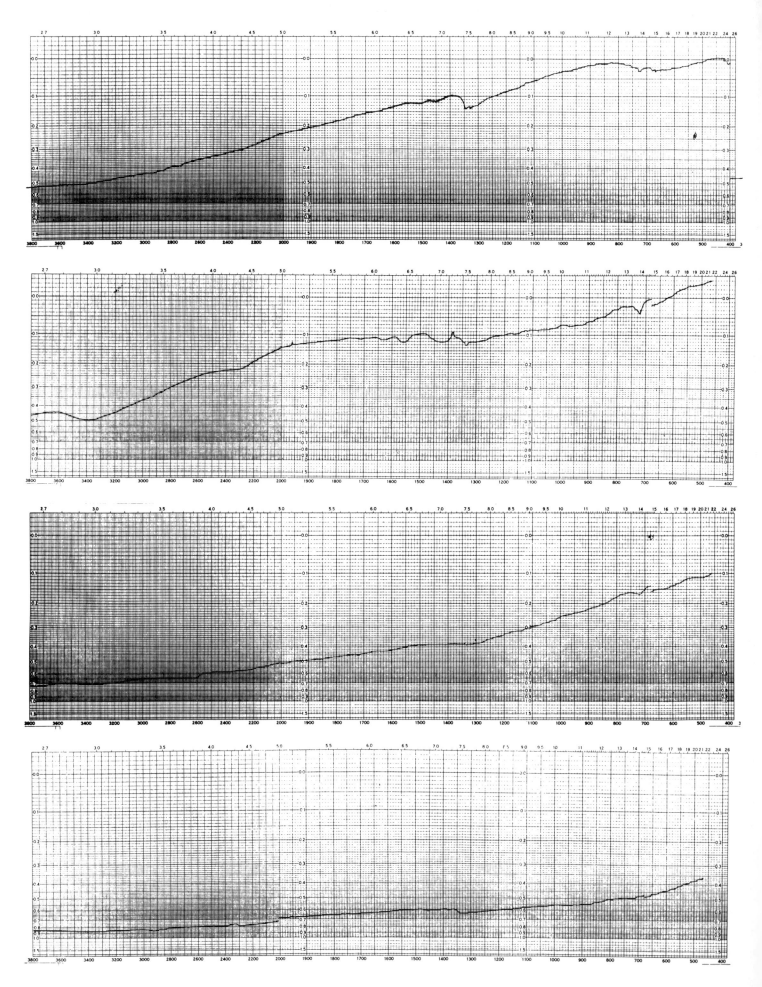

304

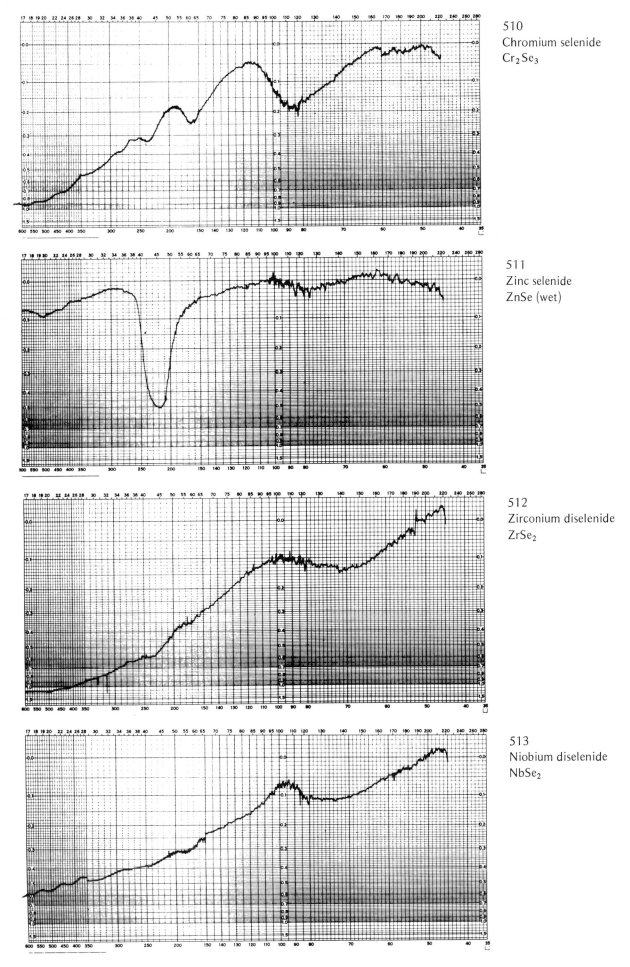

510
Chromium selenide
Cr_2Se_3

511
Zinc selenide
ZnSe (wet)

512
Zirconium diselenide
$ZrSe_2$

513
Niobium diselenide
$NbSe_2$

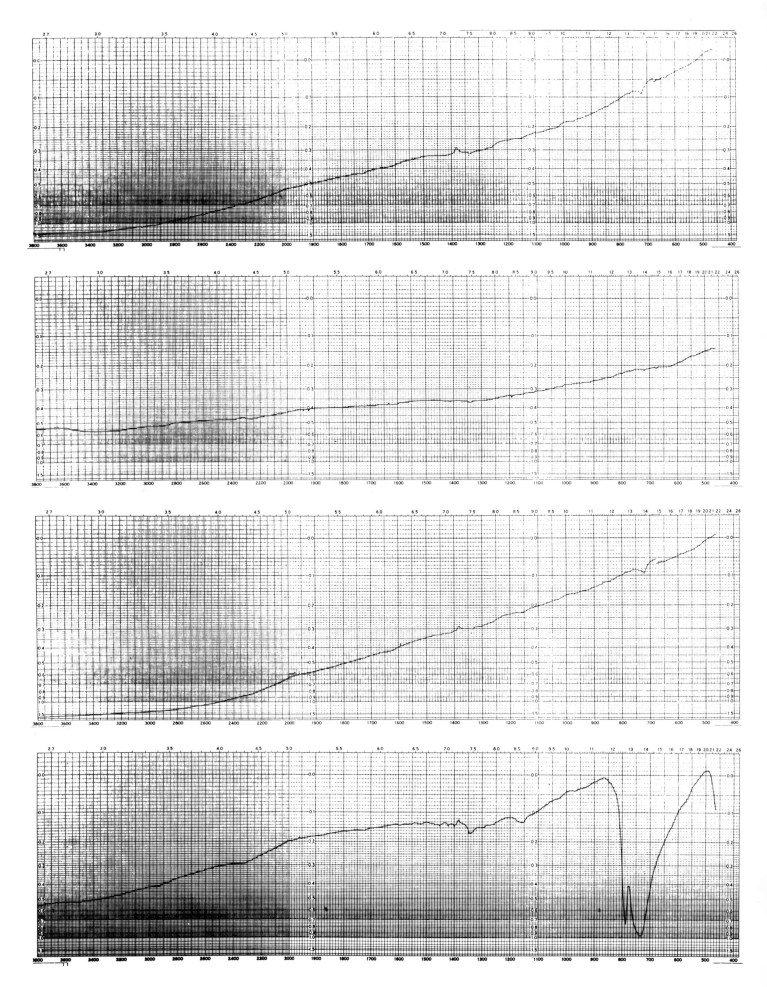

306

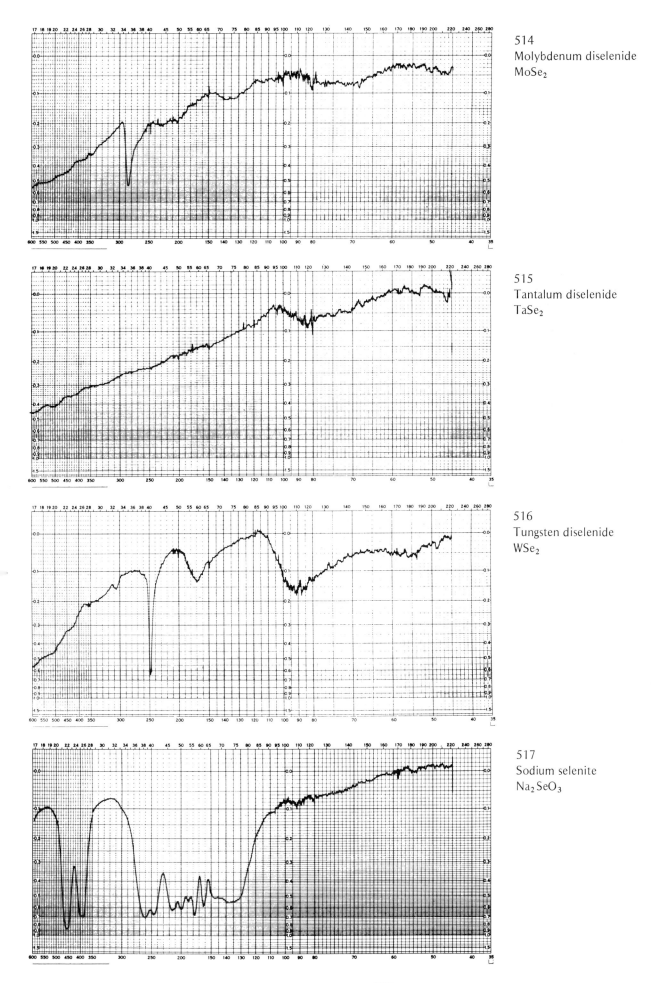

514
Molybdenum diselenide
$MoSe_2$

515
Tantalum diselenide
$TaSe_2$

516
Tungsten diselenide
WSe_2

517
Sodium selenite
Na_2SeO_3

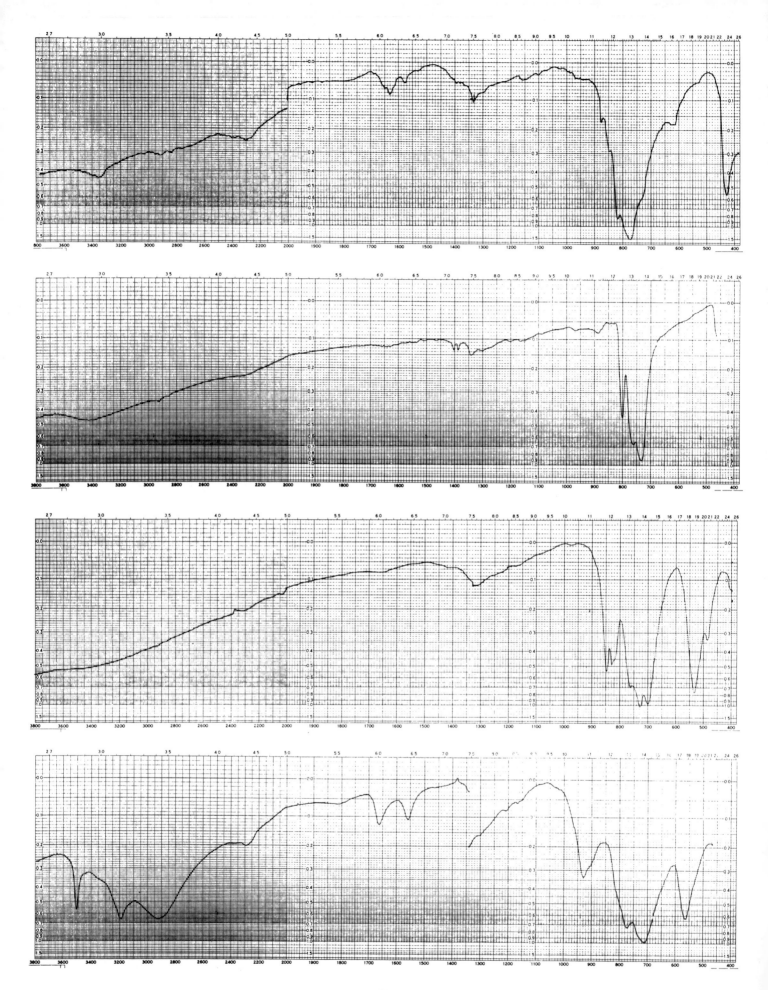

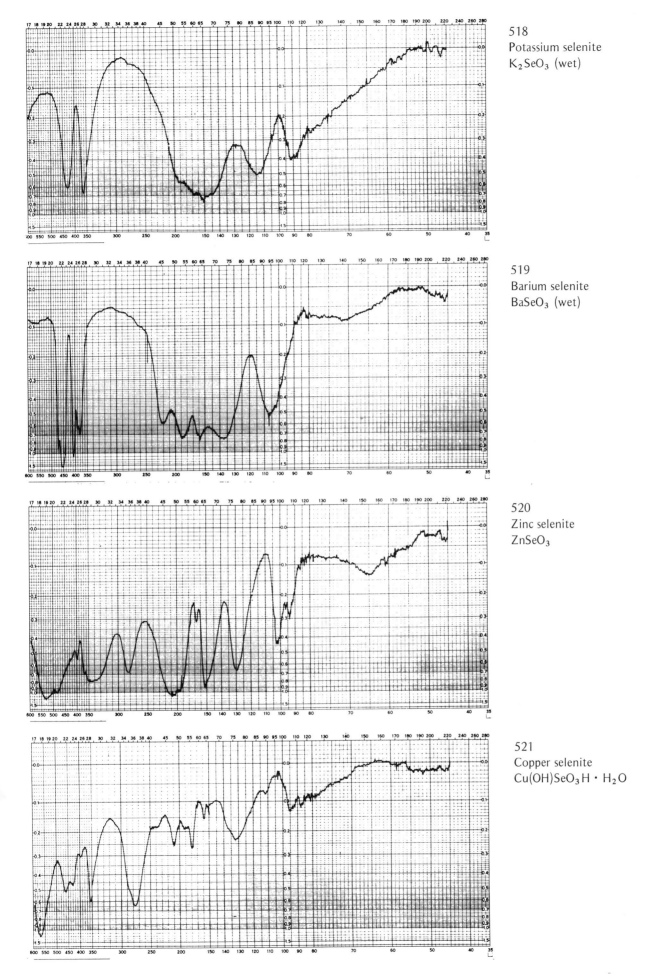

518
Potassium selenite
K_2SeO_3 (wet)

519
Barium selenite
$BaSeO_3$ (wet)

520
Zinc selenite
$ZnSeO_3$

521
Copper selenite
$Cu(OH)SeO_3H \cdot H_2O$

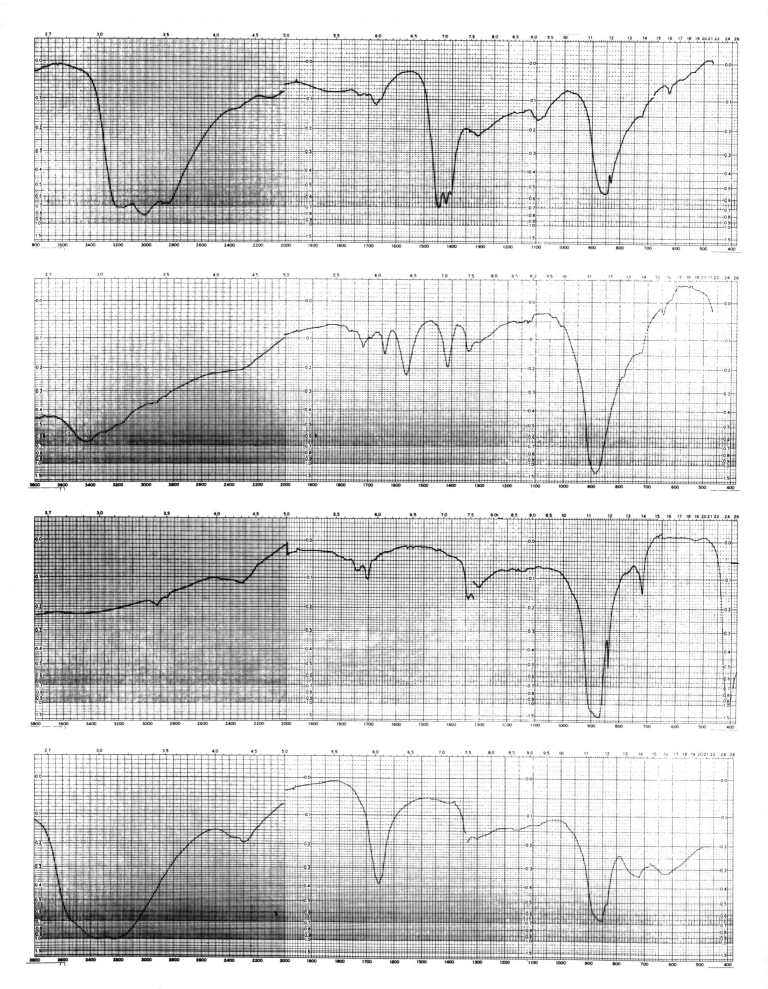

310

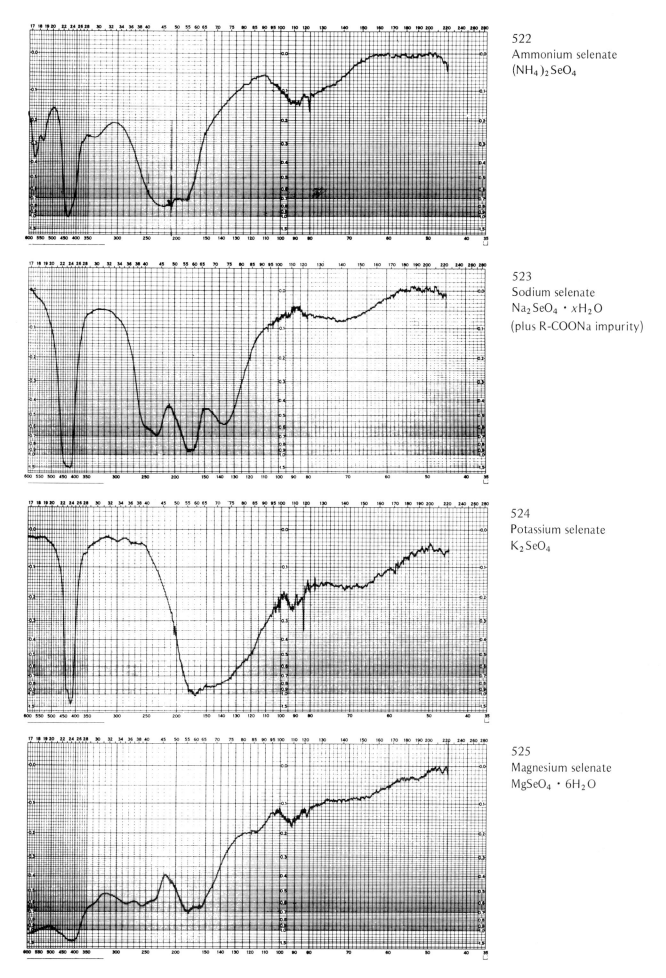

522
Ammonium selenate
$(NH_4)_2SeO_4$

523
Sodium selenate
$Na_2SeO_4 \cdot xH_2O$
(plus R-COONa impurity)

524
Potassium selenate
K_2SeO_4

525
Magnesium selenate
$MgSeO_4 \cdot 6H_2O$

311

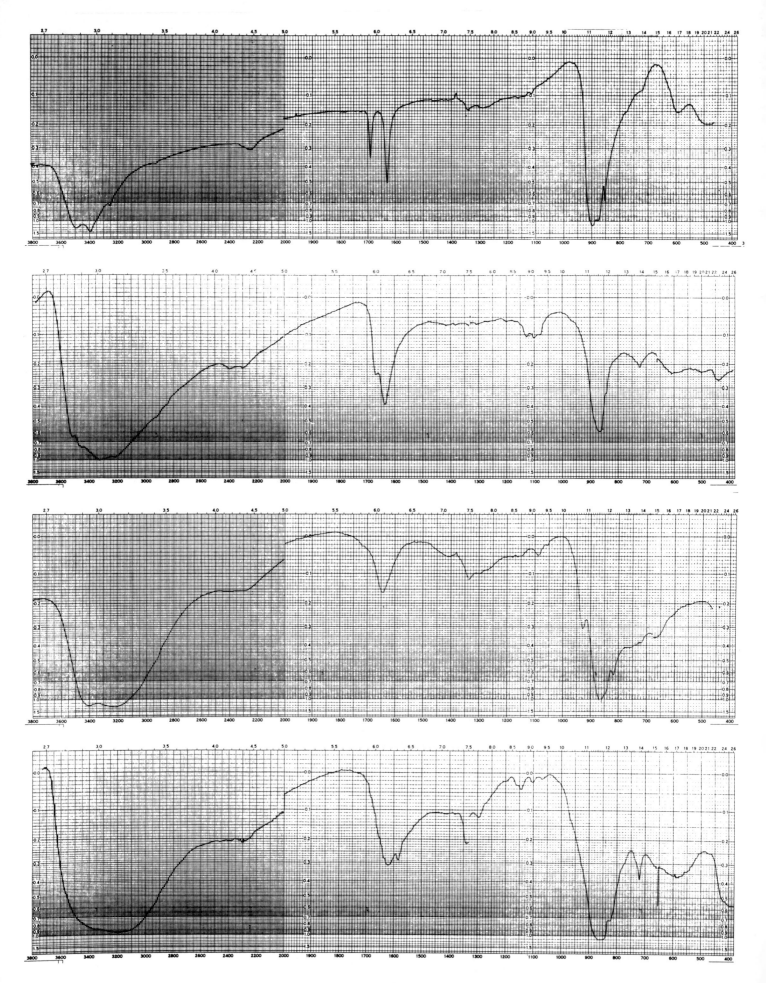

312

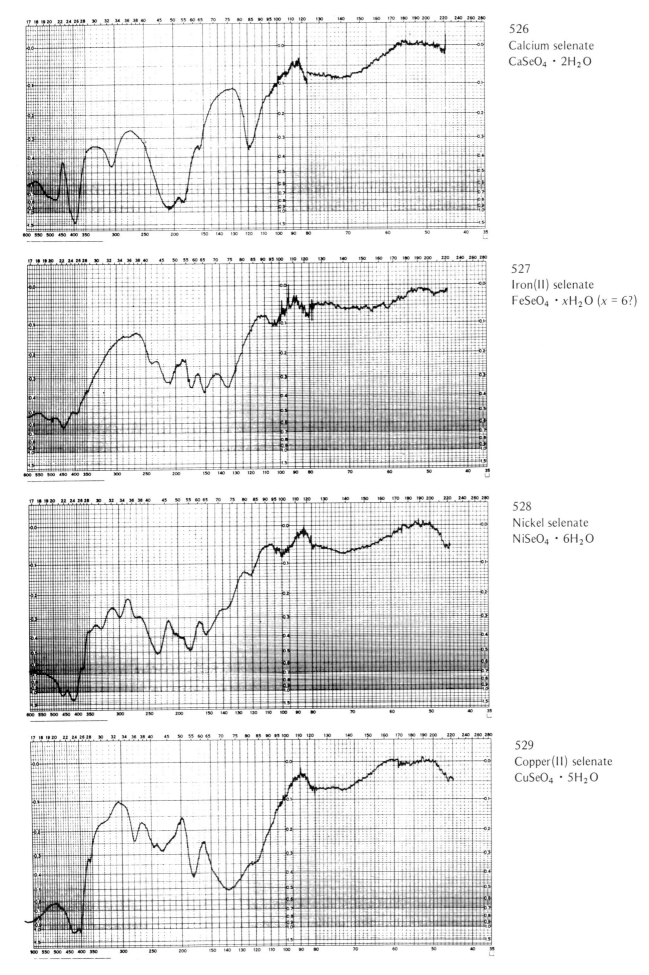

526
Calcium selenate
$CaSeO_4 \cdot 2H_2O$

527
Iron(II) selenate
$FeSeO_4 \cdot xH_2O$ ($x = 6?$)

528
Nickel selenate
$NiSeO_4 \cdot 6H_2O$

529
Copper(II) selenate
$CuSeO_4 \cdot 5H_2O$

313

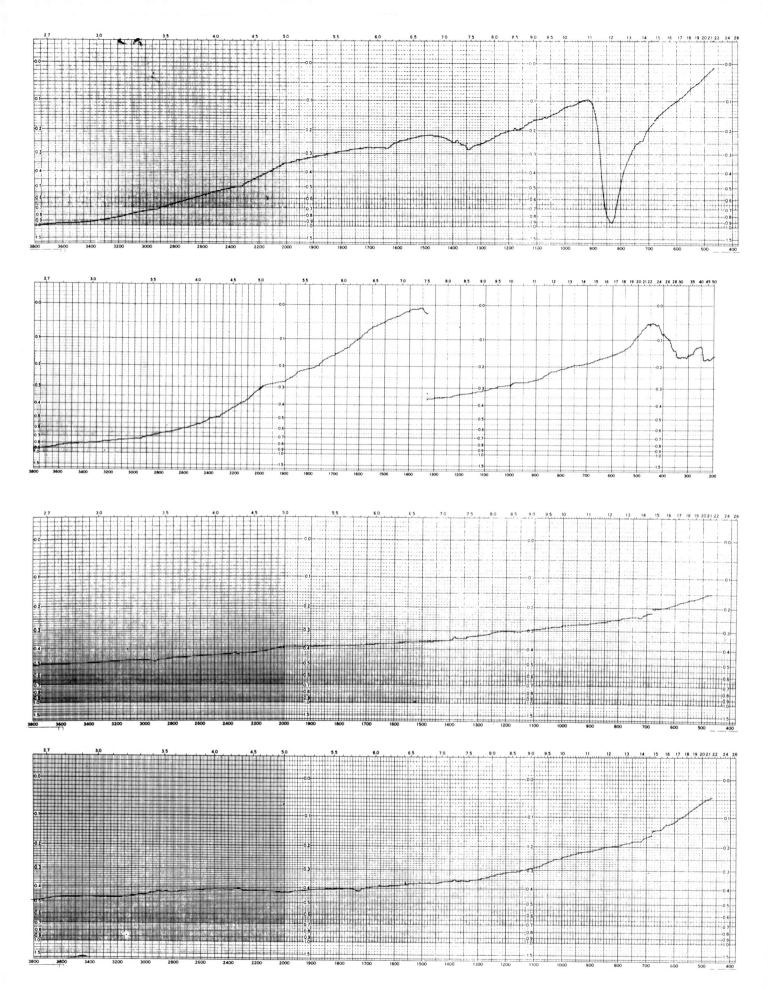

314

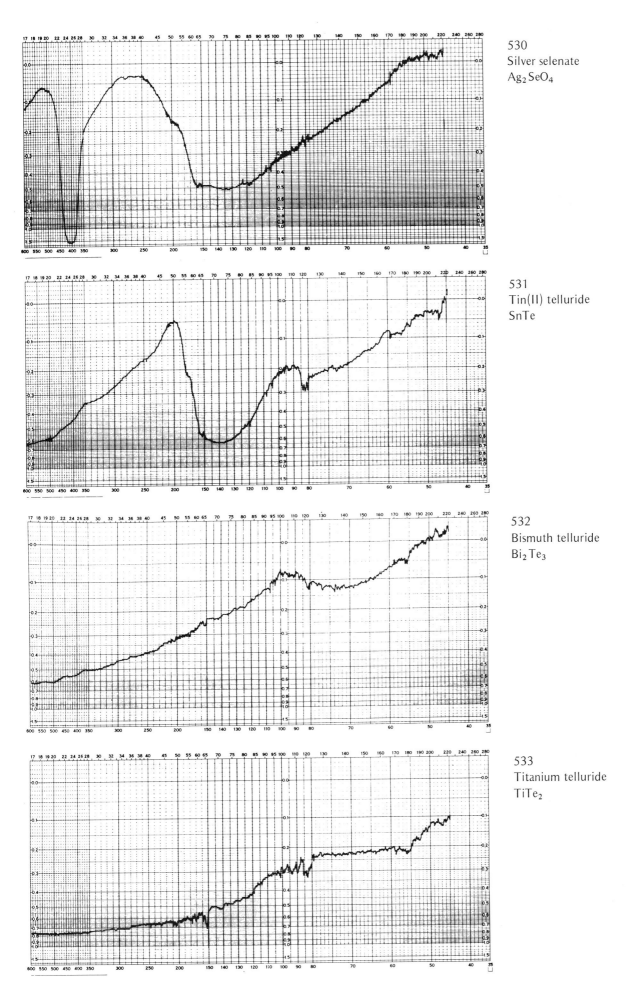

530
Silver selenate
Ag_2SeO_4

531
Tin(II) telluride
SnTe

532
Bismuth telluride
Bi_2Te_3

533
Titanium telluride
$TiTe_2$

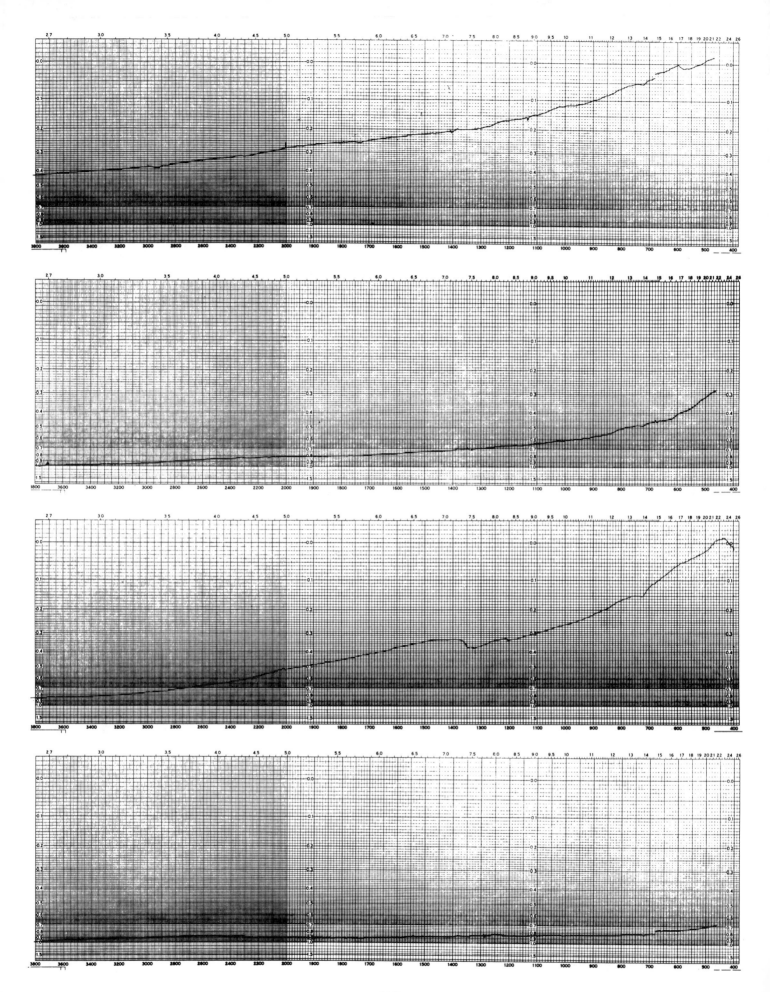

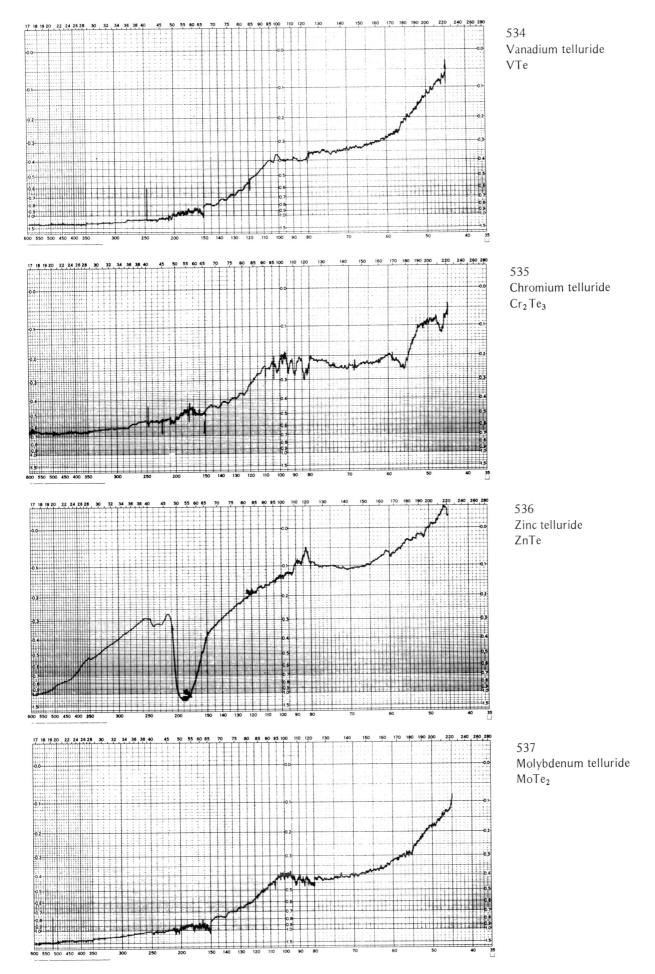

534
Vanadium telluride
VTe

535
Chromium telluride
Cr$_2$Te$_3$

536
Zinc telluride
ZnTe

537
Molybdenum telluride
MoTe$_2$

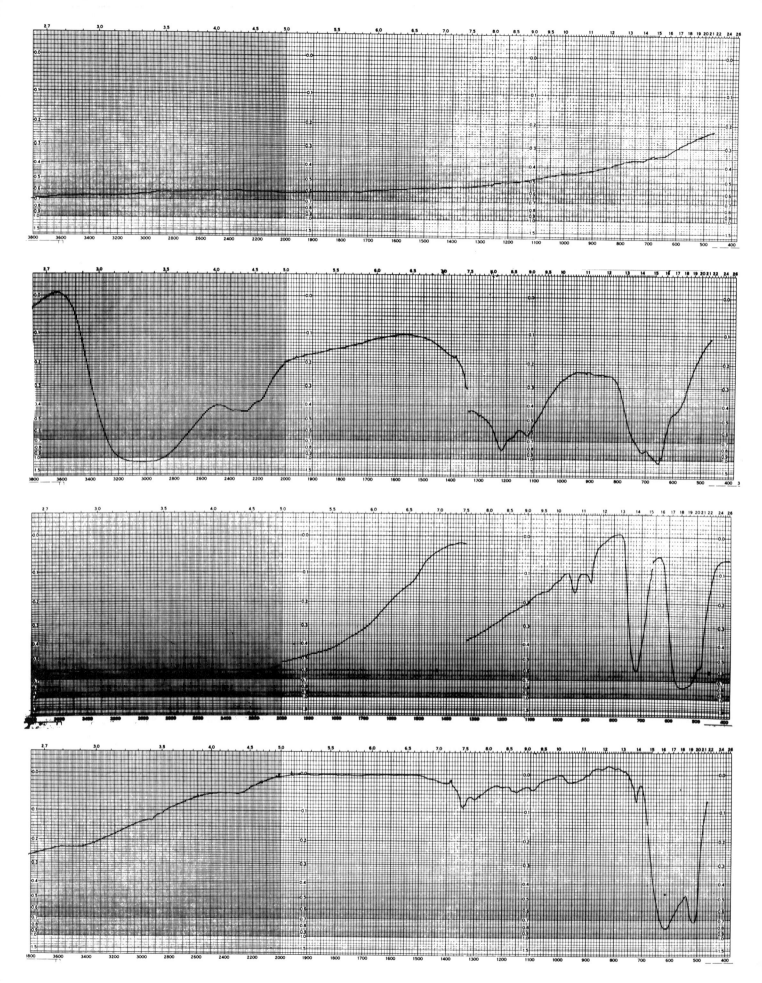

318

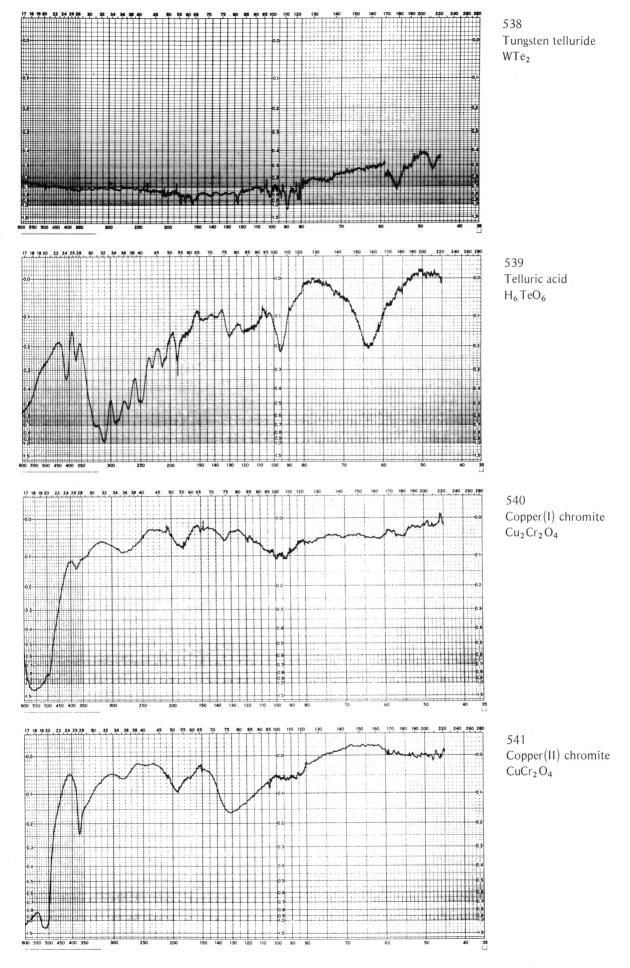

538
Tungsten telluride
WTe$_2$

539
Telluric acid
H$_6$TeO$_6$

540
Copper(I) chromite
Cu$_2$Cr$_2$O$_4$

541
Copper(II) chromite
CuCr$_2$O$_4$

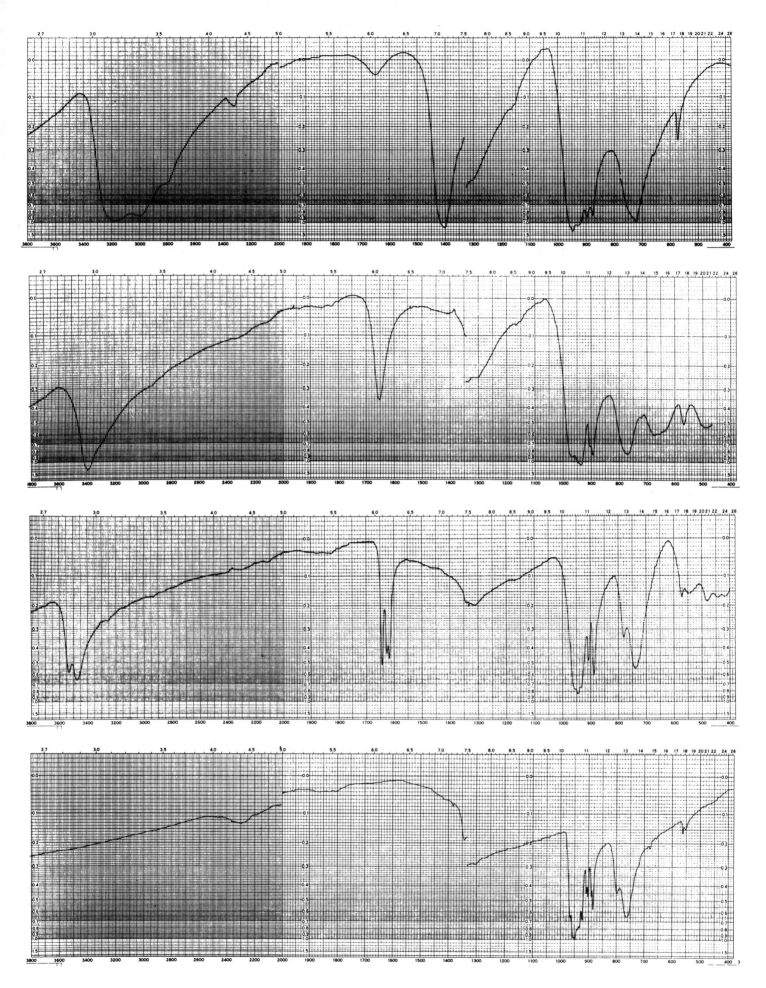

320

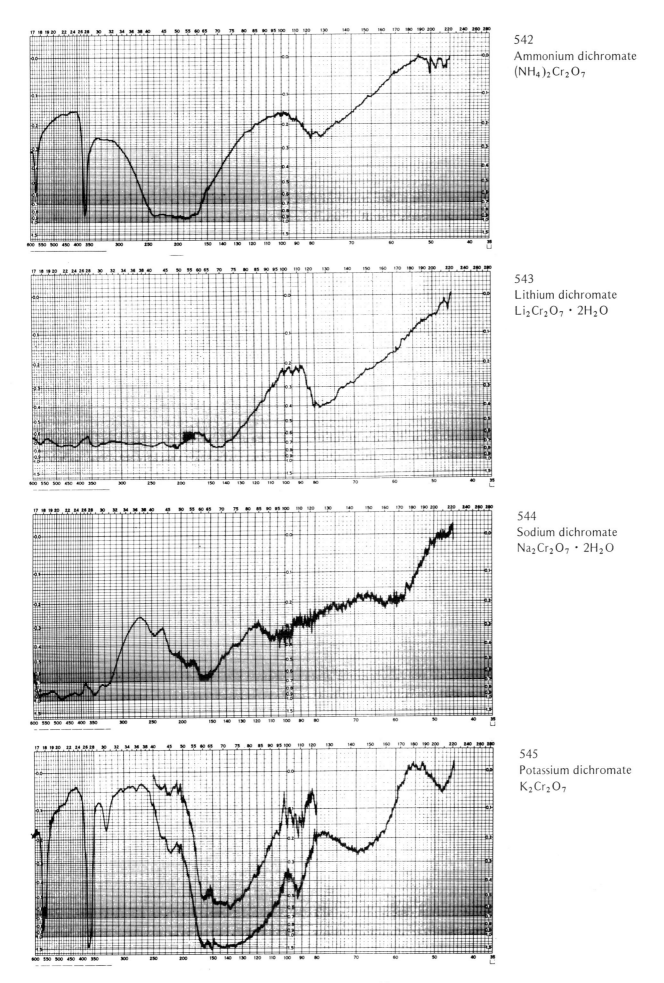

542
Ammonium dichromate
$(NH_4)_2Cr_2O_7$

543
Lithium dichromate
$Li_2Cr_2O_7 \cdot 2H_2O$

544
Sodium dichromate
$Na_2Cr_2O_7 \cdot 2H_2O$

545
Potassium dichromate
$K_2Cr_2O_7$

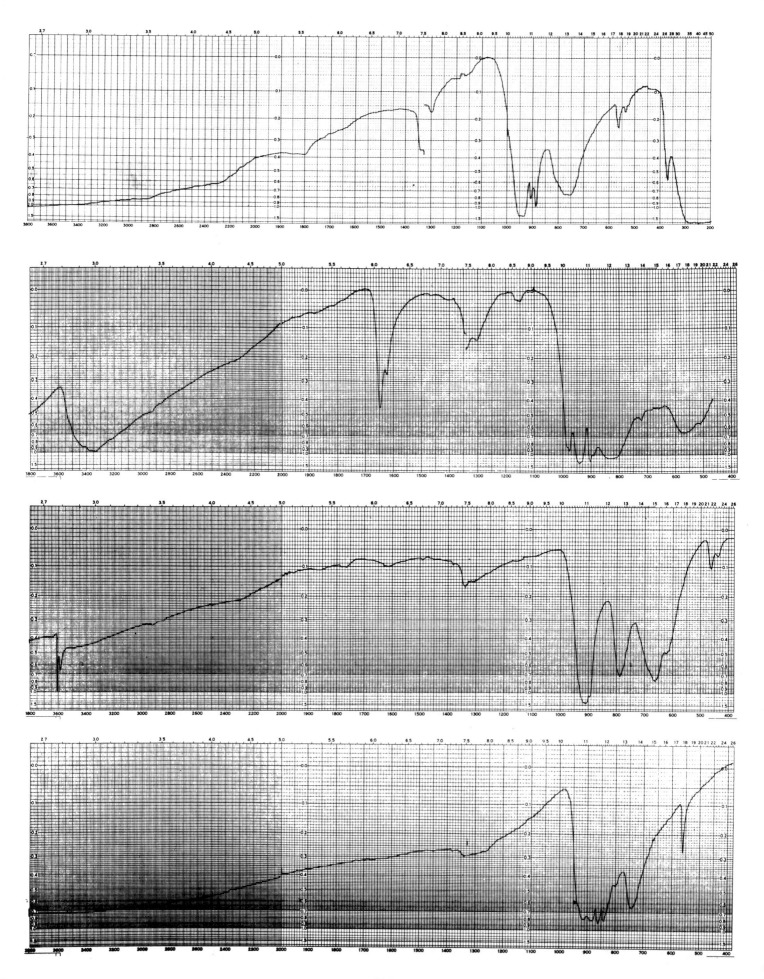

322

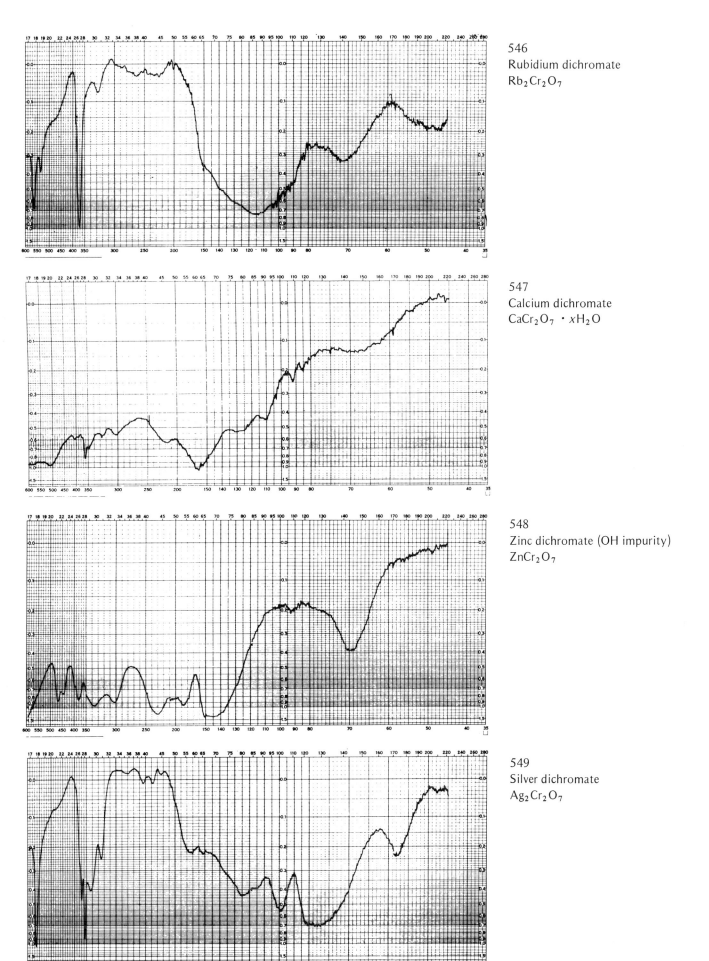

546
Rubidium dichromate
$Rb_2Cr_2O_7$

547
Calcium dichromate
$CaCr_2O_7 \cdot xH_2O$

548
Zinc dichromate (OH impurity)
$ZnCr_2O_7$

549
Silver dichromate
$Ag_2Cr_2O_7$

323

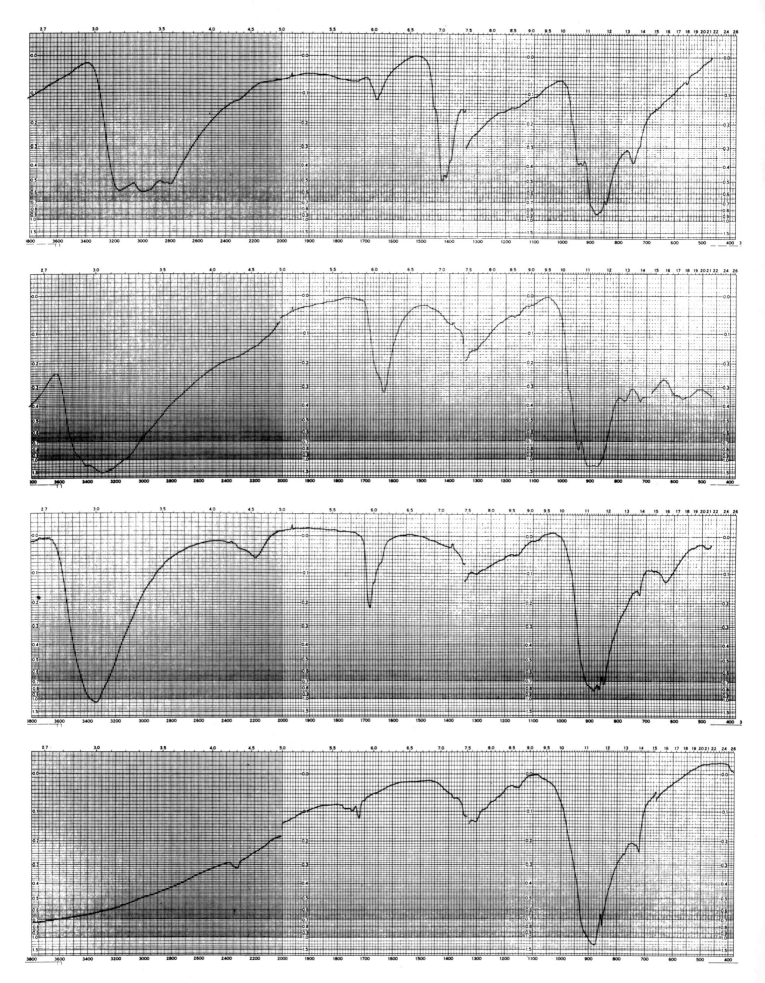

324

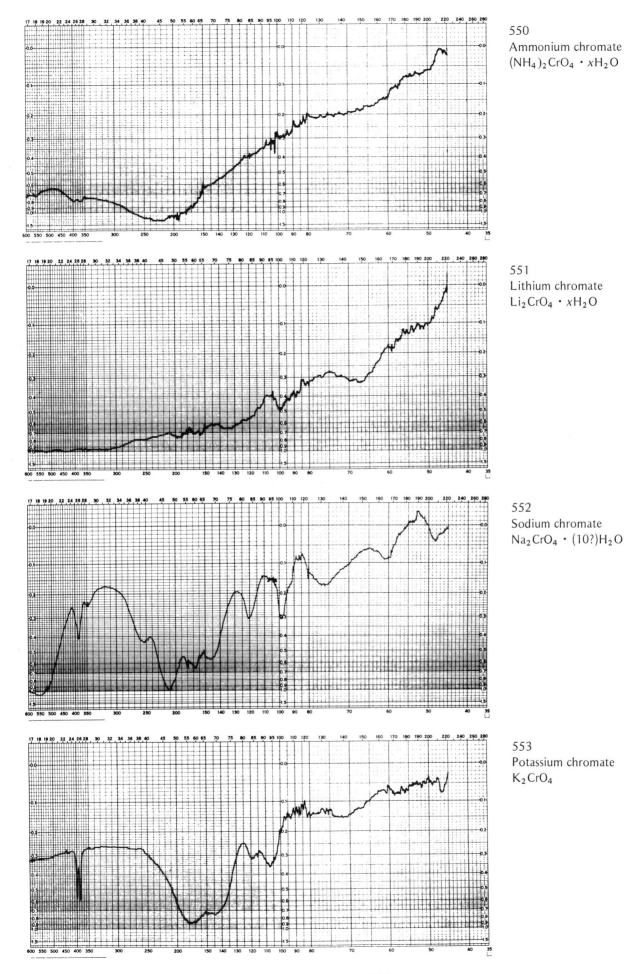

550
Ammonium chromate
$(NH_4)_2CrO_4 \cdot xH_2O$

551
Lithium chromate
$Li_2CrO_4 \cdot xH_2O$

552
Sodium chromate
$Na_2CrO_4 \cdot (10?)H_2O$

553
Potassium chromate
K_2CrO_4

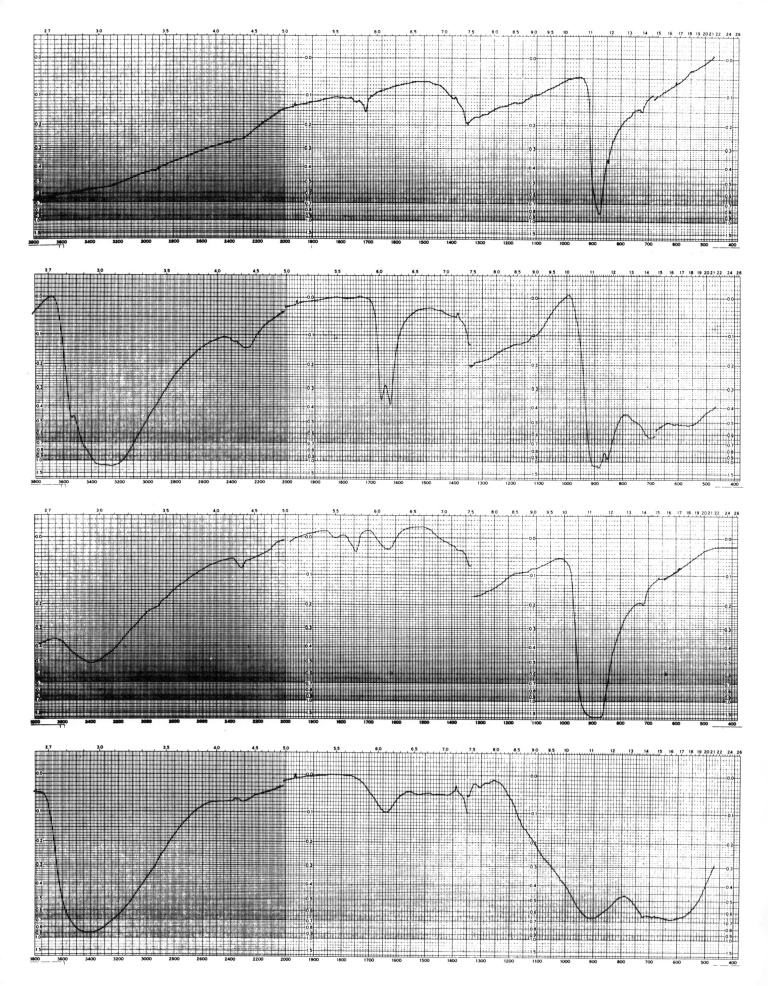

326

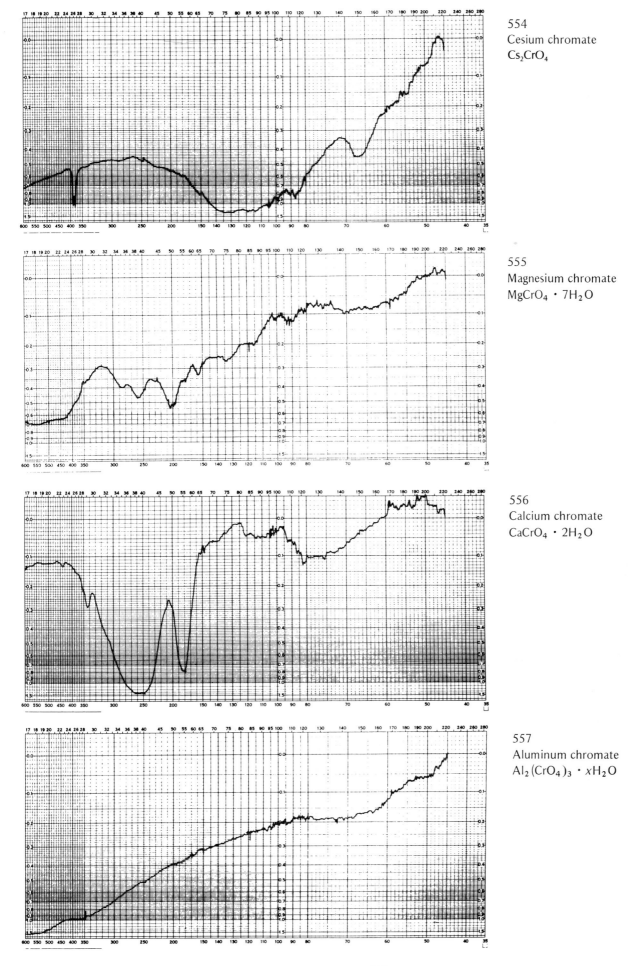

554
Cesium chromate
Cs_2CrO_4

555
Magnesium chromate
$MgCrO_4 \cdot 7H_2O$

556
Calcium chromate
$CaCrO_4 \cdot 2H_2O$

557
Aluminum chromate
$Al_2(CrO_4)_3 \cdot xH_2O$

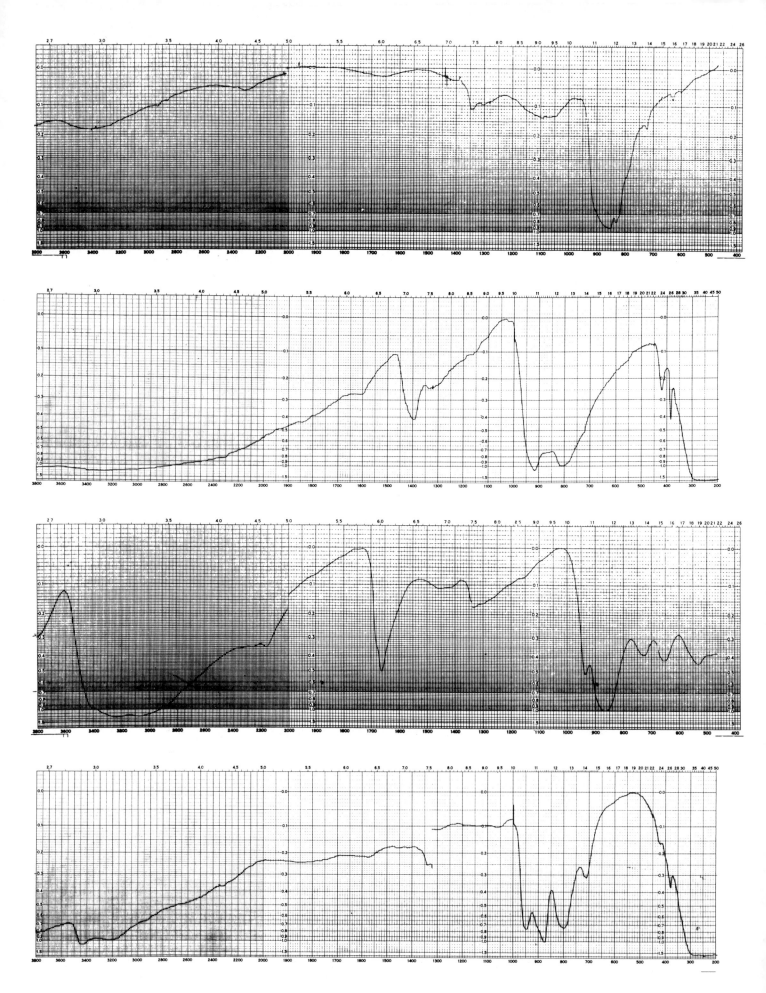

328

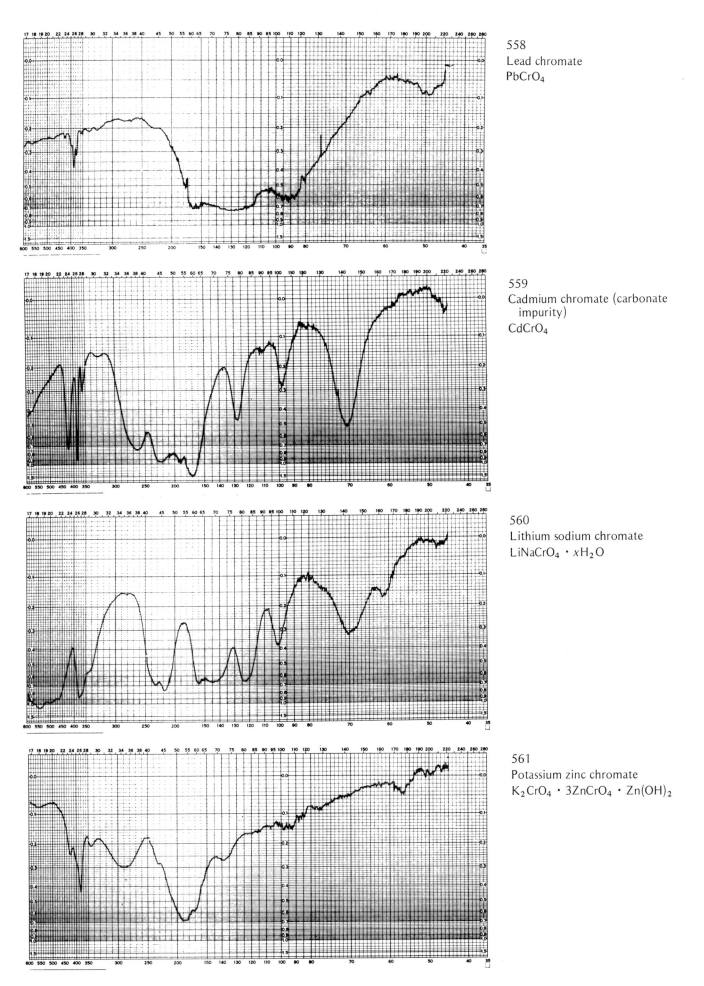

558
Lead chromate
PbCrO$_4$

559
Cadmium chromate (carbonate impurity)
CdCrO$_4$

560
Lithium sodium chromate
LiNaCrO$_4$ · xH$_2$O

561
Potassium zinc chromate
K$_2$CrO$_4$ · 3ZnCrO$_4$ · Zn(OH)$_2$

329

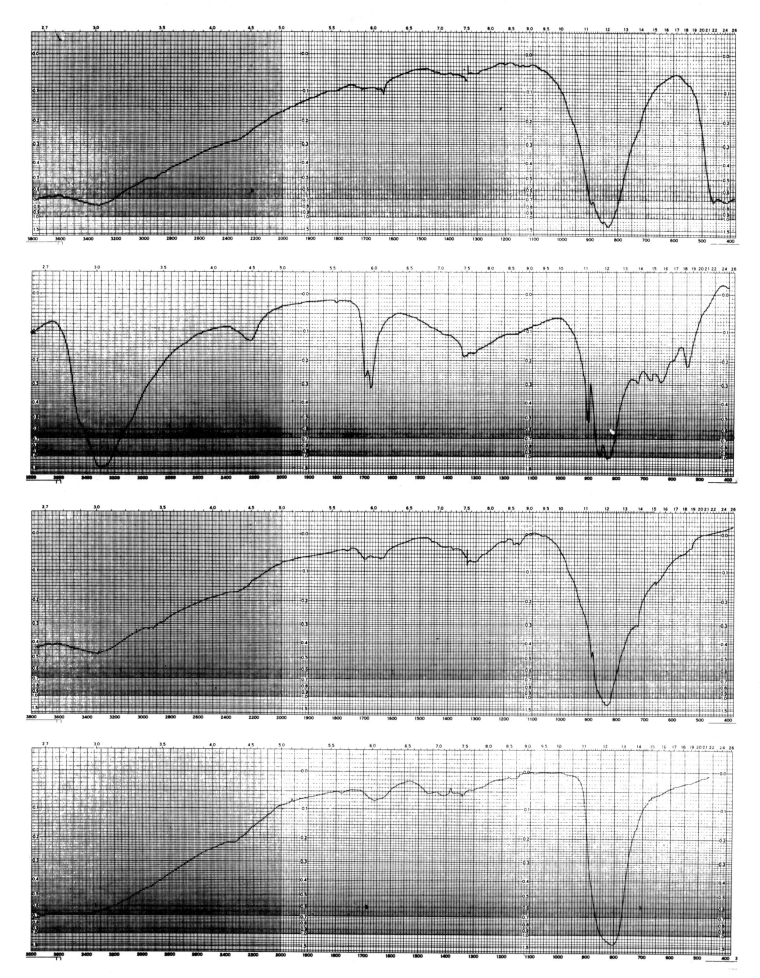

330

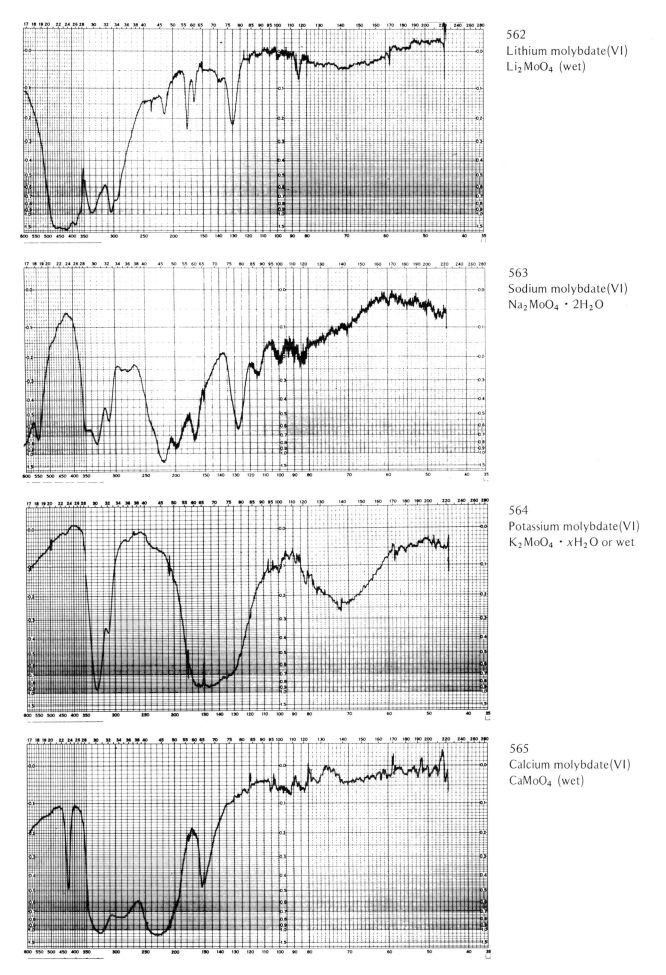

562
Lithium molybdate(VI)
Li_2MoO_4 (wet)

563
Sodium molybdate(VI)
$Na_2MoO_4 \cdot 2H_2O$

564
Potassium molybdate(VI)
$K_2MoO_4 \cdot xH_2O$ or wet

565
Calcium molybdate(VI)
$CaMoO_4$ (wet)

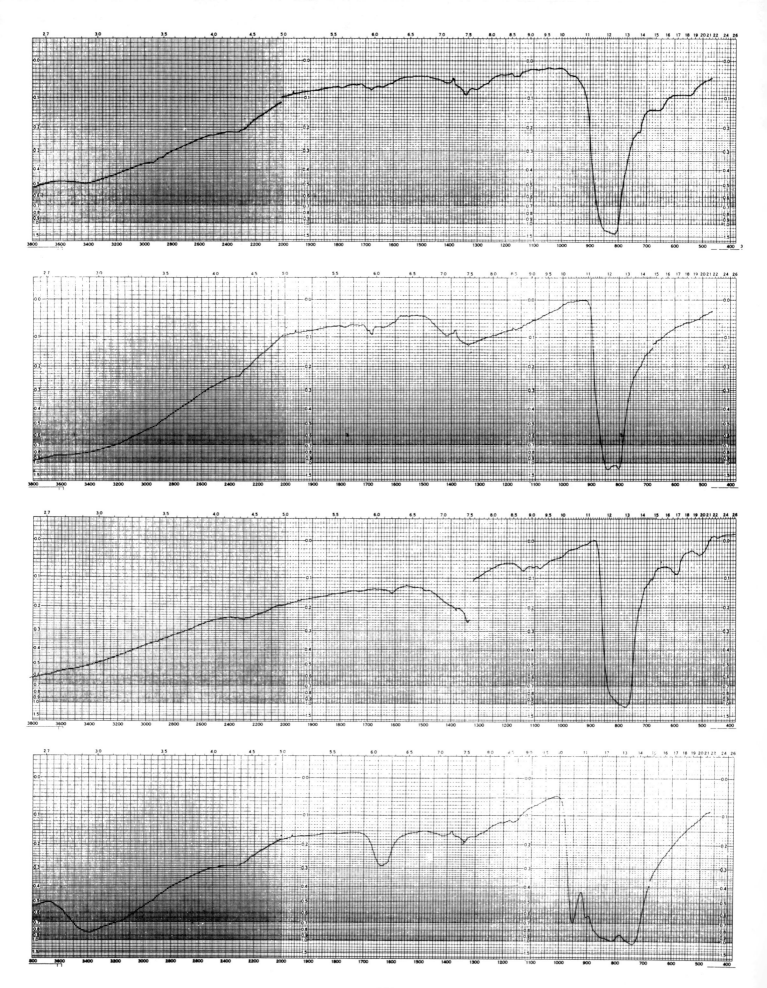

332

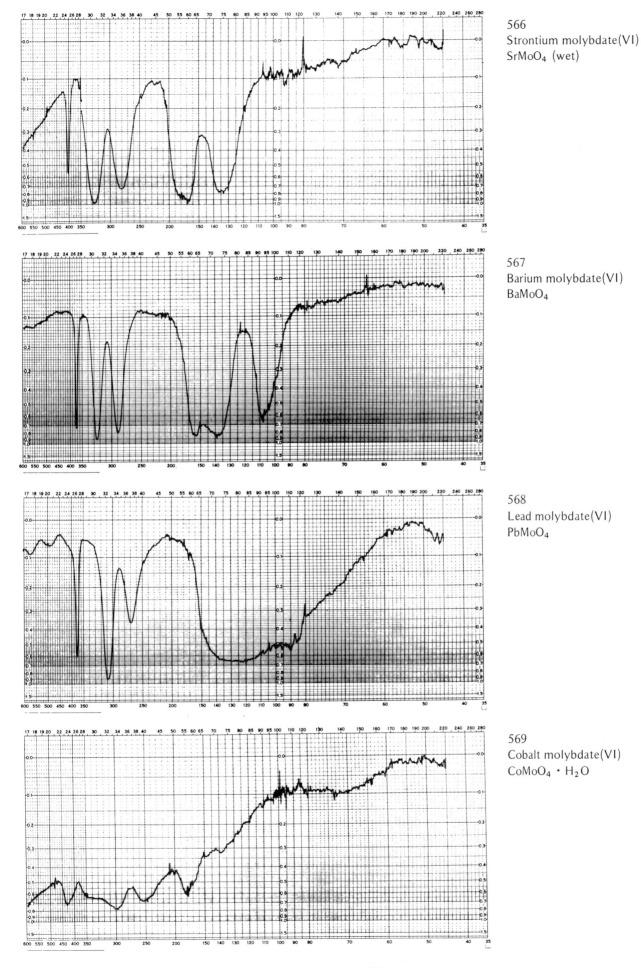

566
Strontium molybdate(VI)
$SrMoO_4$ (wet)

567
Barium molybdate(VI)
$BaMoO_4$

568
Lead molybdate(VI)
$PbMoO_4$

569
Cobalt molybdate(VI)
$CoMoO_4 \cdot H_2O$

333

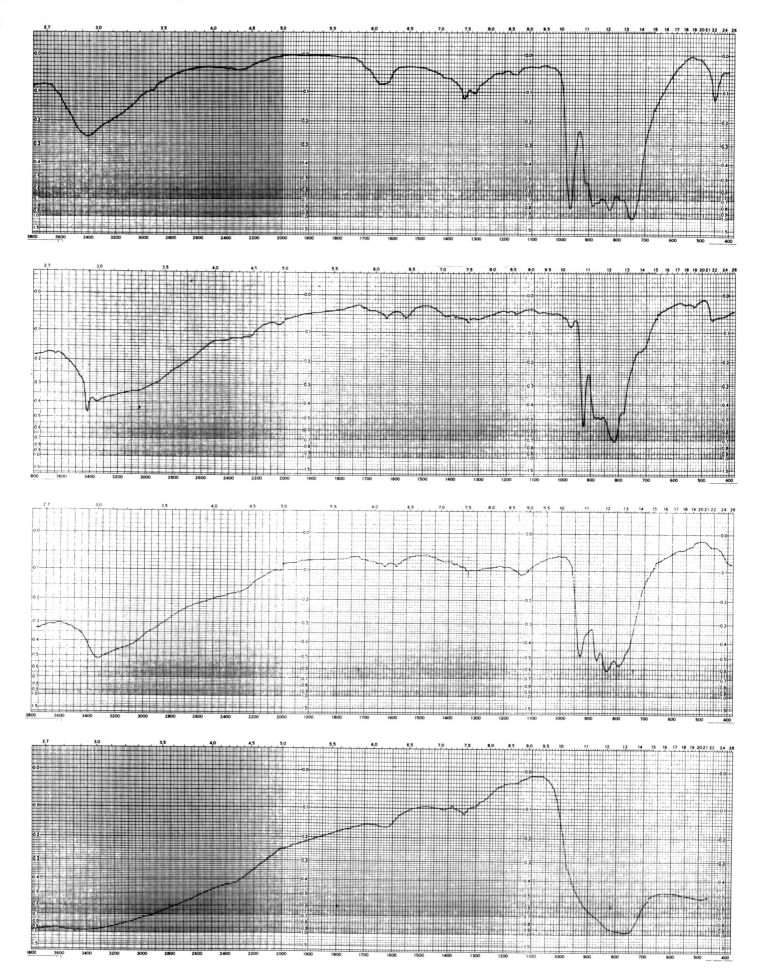

334

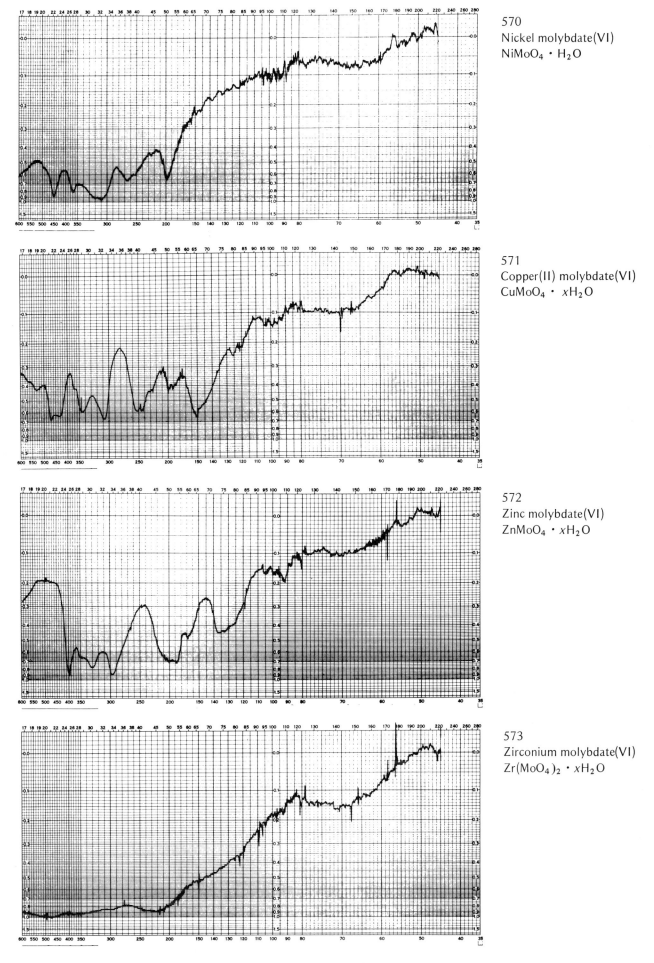

570
Nickel molybdate(VI)
$NiMoO_4 \cdot H_2O$

571
Copper(II) molybdate(VI)
$CuMoO_4 \cdot xH_2O$

572
Zinc molybdate(VI)
$ZnMoO_4 \cdot xH_2O$

573
Zirconium molybdate(VI)
$Zr(MoO_4)_2 \cdot xH_2O$

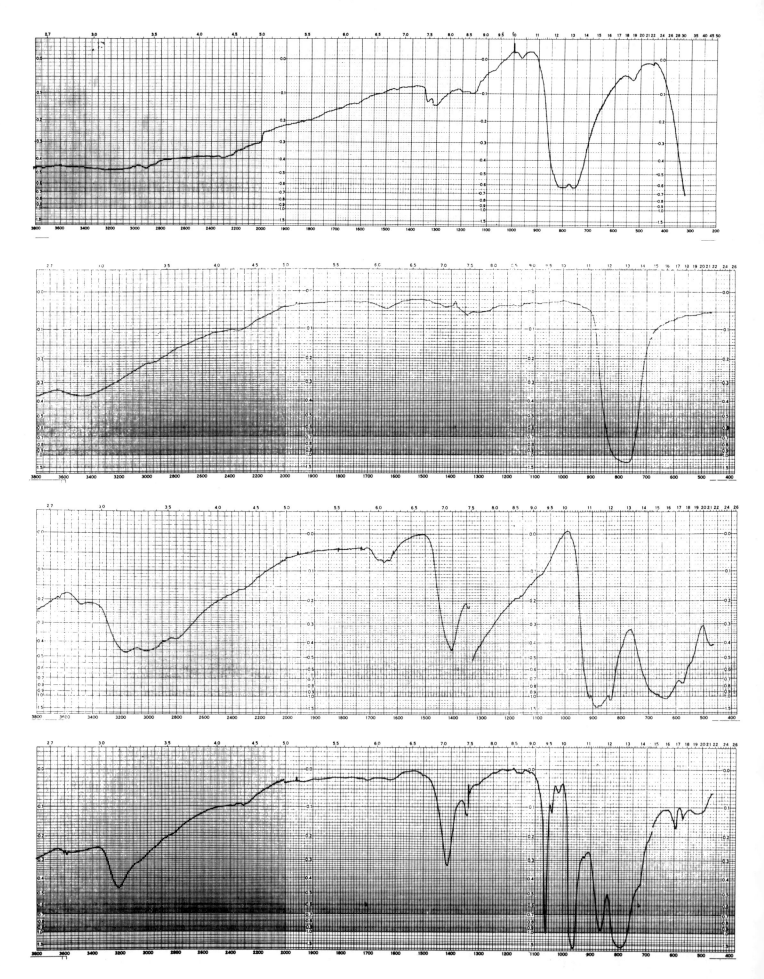

336

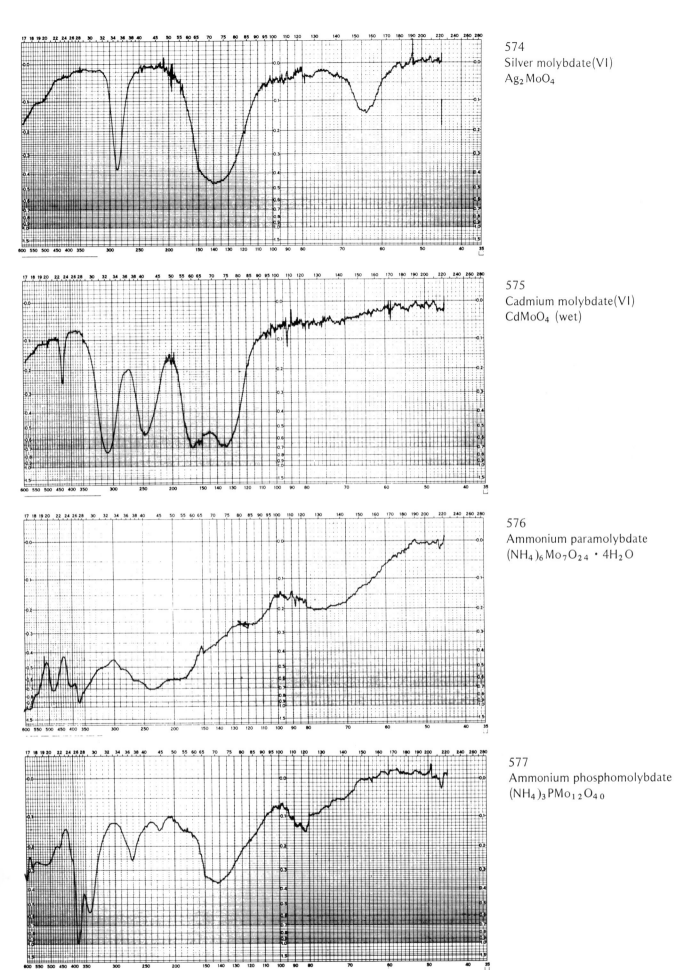

574
Silver molybdate(VI)
Ag_2MoO_4

575
Cadmium molybdate(VI)
$CdMoO_4$ (wet)

576
Ammonium paramolybdate
$(NH_4)_6Mo_7O_{24} \cdot 4H_2O$

577
Ammonium phosphomolybdate
$(NH_4)_3PMo_{12}O_{40}$

337

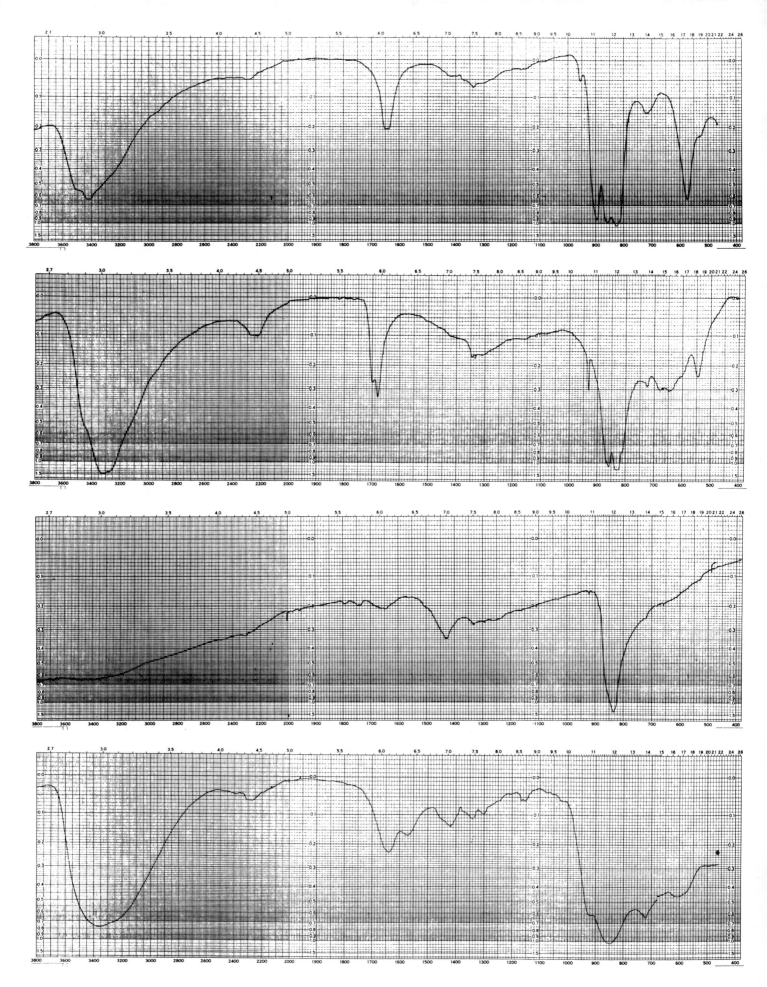

338

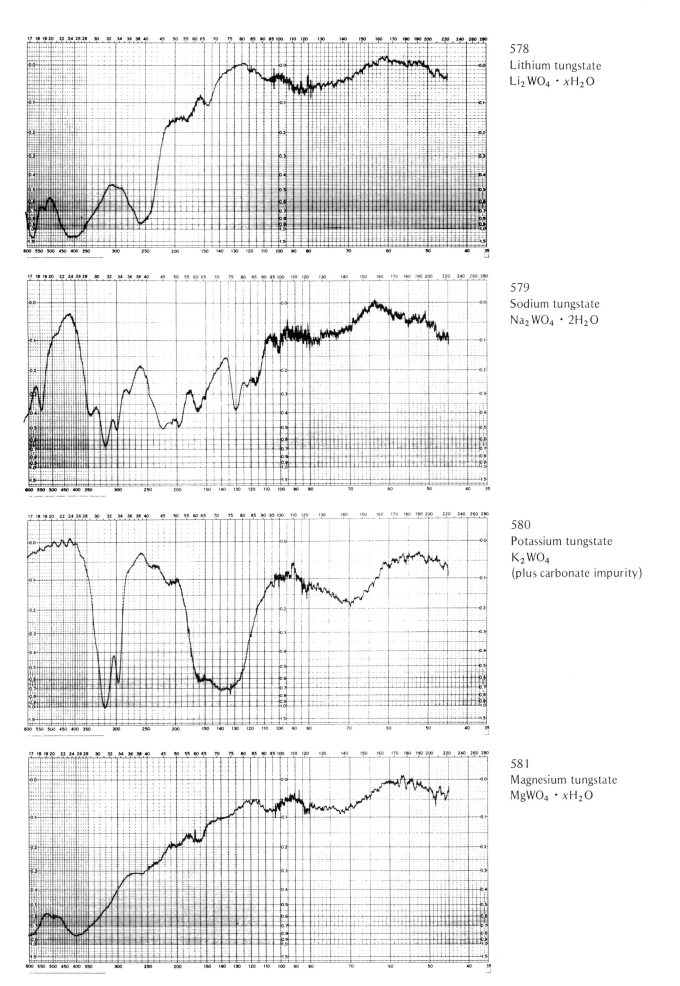

578
Lithium tungstate
Li$_2$WO$_4$ · xH$_2$O

579
Sodium tungstate
Na$_2$WO$_4$ · 2H$_2$O

580
Potassium tungstate
K$_2$WO$_4$
(plus carbonate impurity)

581
Magnesium tungstate
MgWO$_4$ · xH$_2$O

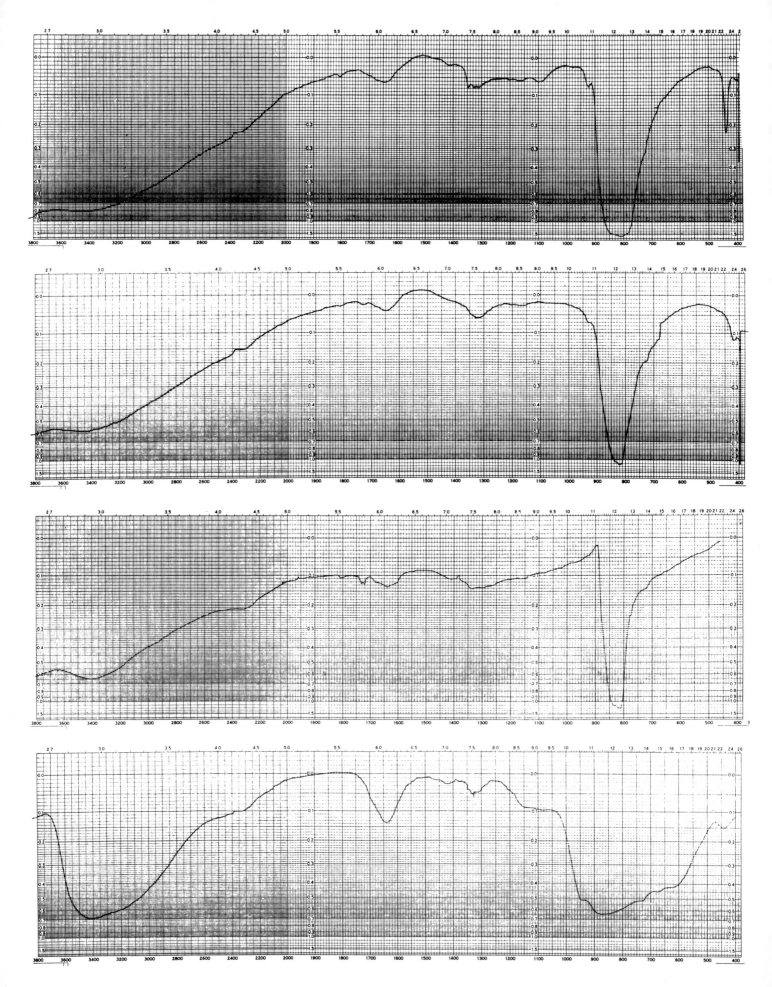

340

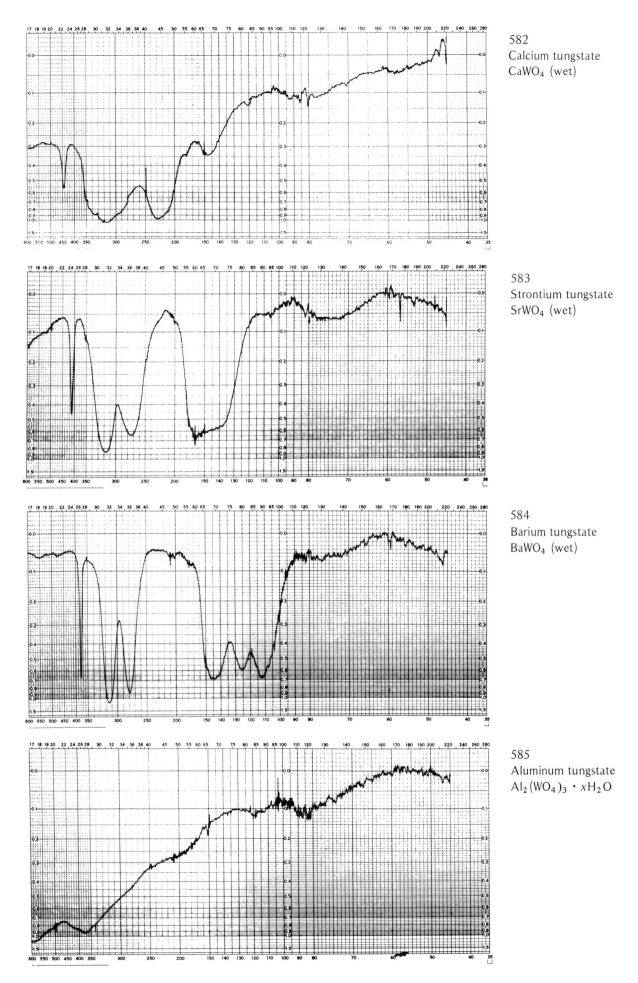

582
Calcium tungstate
CaWO$_4$ (wet)

583
Strontium tungstate
SrWO$_4$ (wet)

584
Barium tungstate
BaWO$_4$ (wet)

585
Aluminum tungstate
Al$_2$(WO$_4$)$_3$ · xH$_2$O

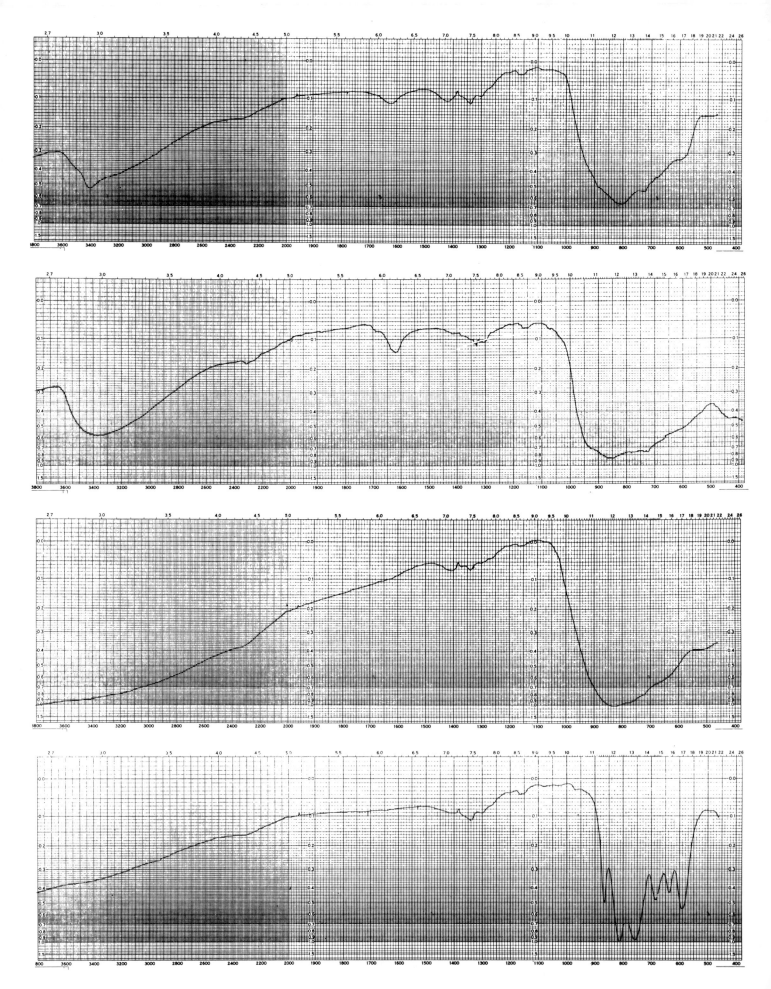

342

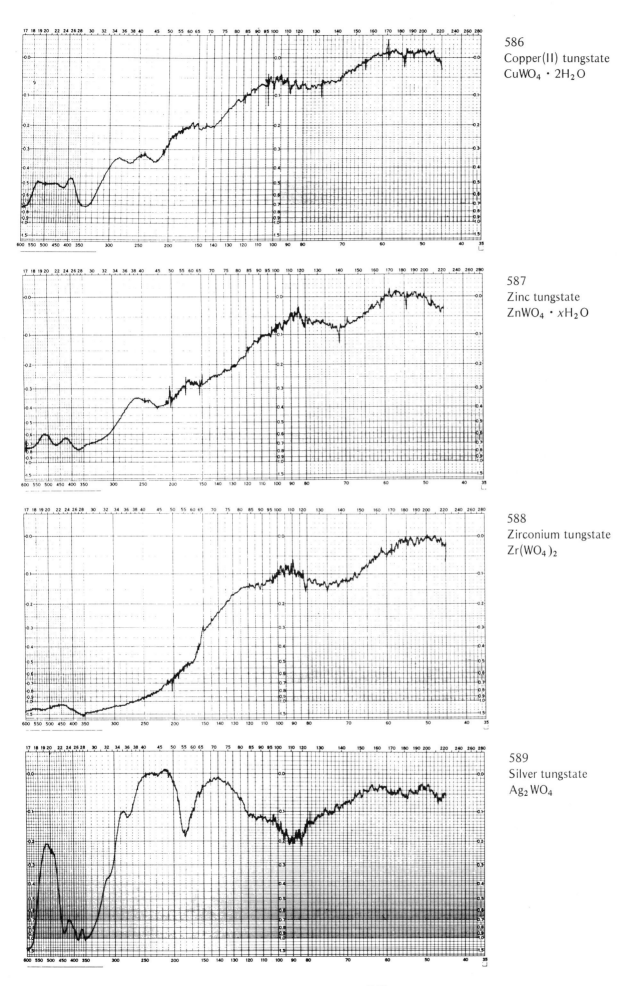

586
Copper(II) tungstate
$CuWO_4 \cdot 2H_2O$

587
Zinc tungstate
$ZnWO_4 \cdot xH_2O$

588
Zirconium tungstate
$Zr(WO_4)_2$

589
Silver tungstate
Ag_2WO_4

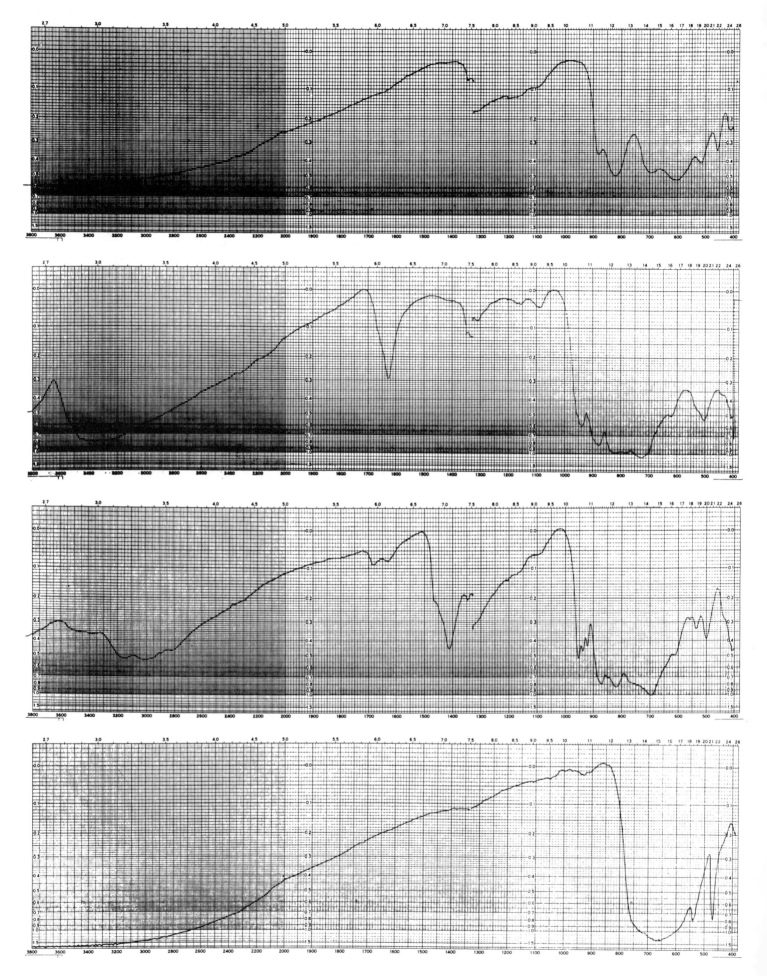

344

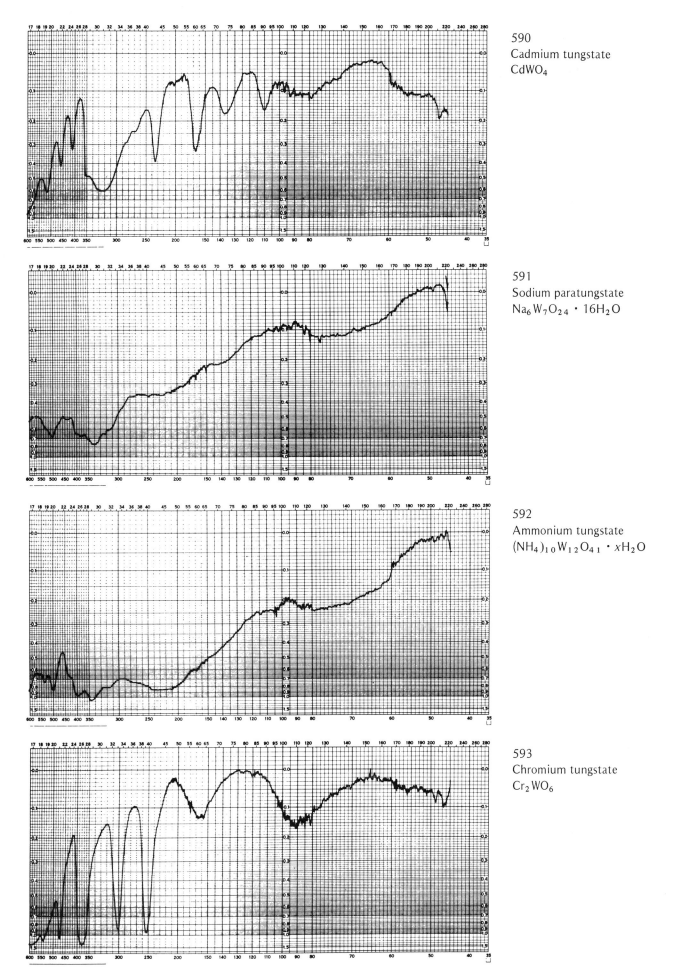

590
Cadmium tungstate
$CdWO_4$

591
Sodium paratungstate
$Na_6W_7O_{24} \cdot 16H_2O$

592
Ammonium tungstate
$(NH_4)_{10}W_{12}O_{41} \cdot xH_2O$

593
Chromium tungstate
Cr_2WO_6

345

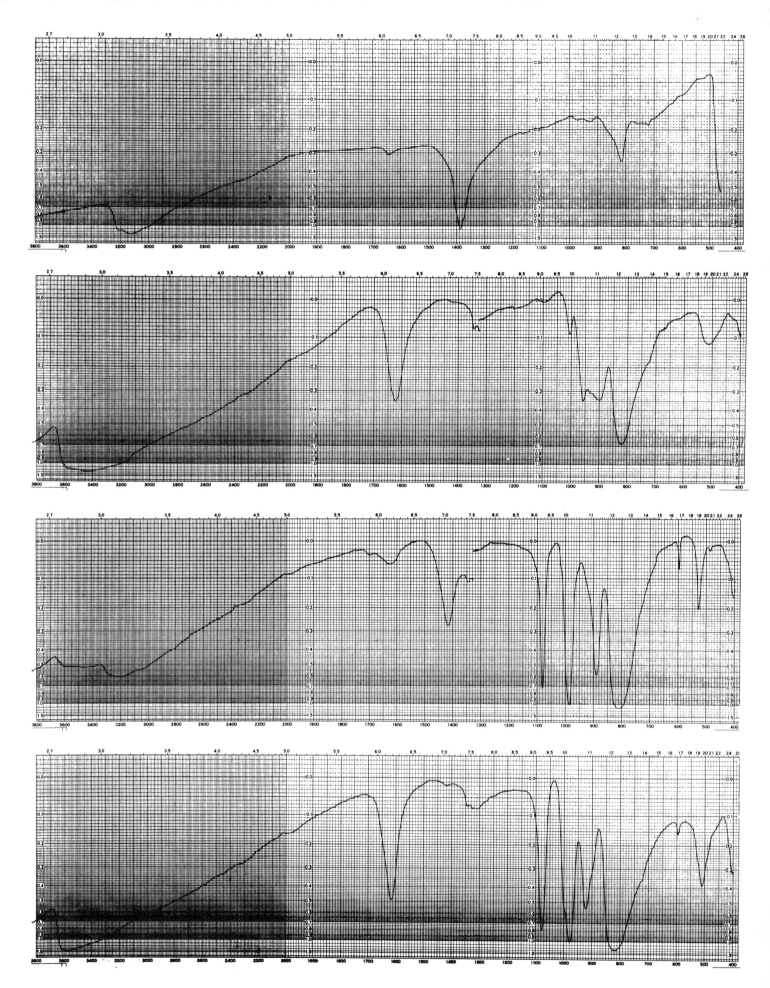

346

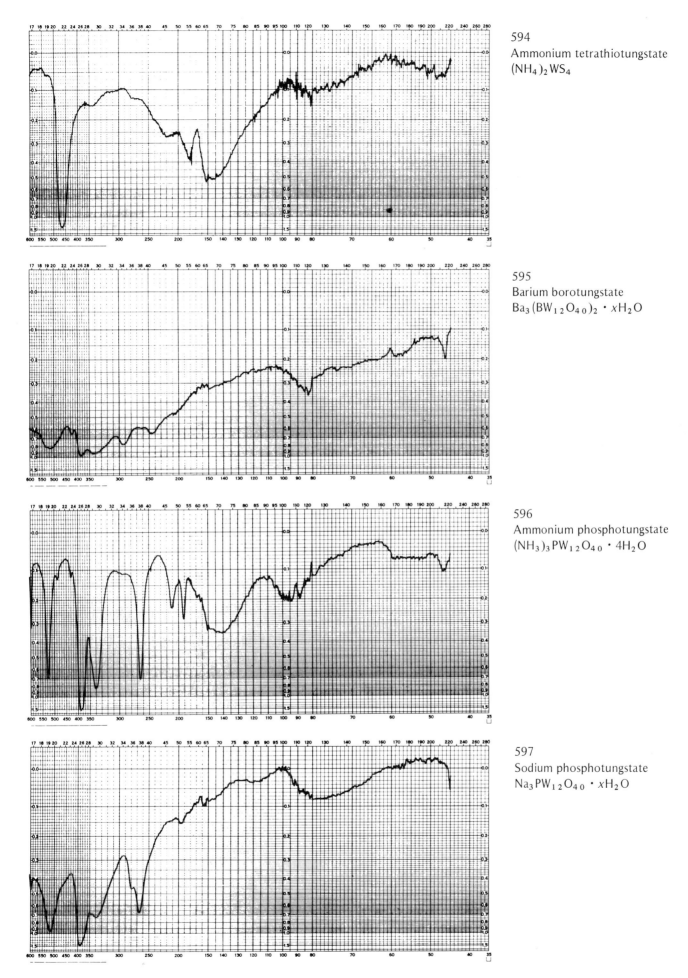

594
Ammonium tetrathiotungstate
$(NH_4)_2WS_4$

595
Barium borotungstate
$Ba_3(BW_{12}O_{40})_2 \cdot xH_2O$

596
Ammonium phosphotungstate
$(NH_3)_3PW_{12}O_{40} \cdot 4H_2O$

597
Sodium phosphotungstate
$Na_3PW_{12}O_{40} \cdot xH_2O$

347

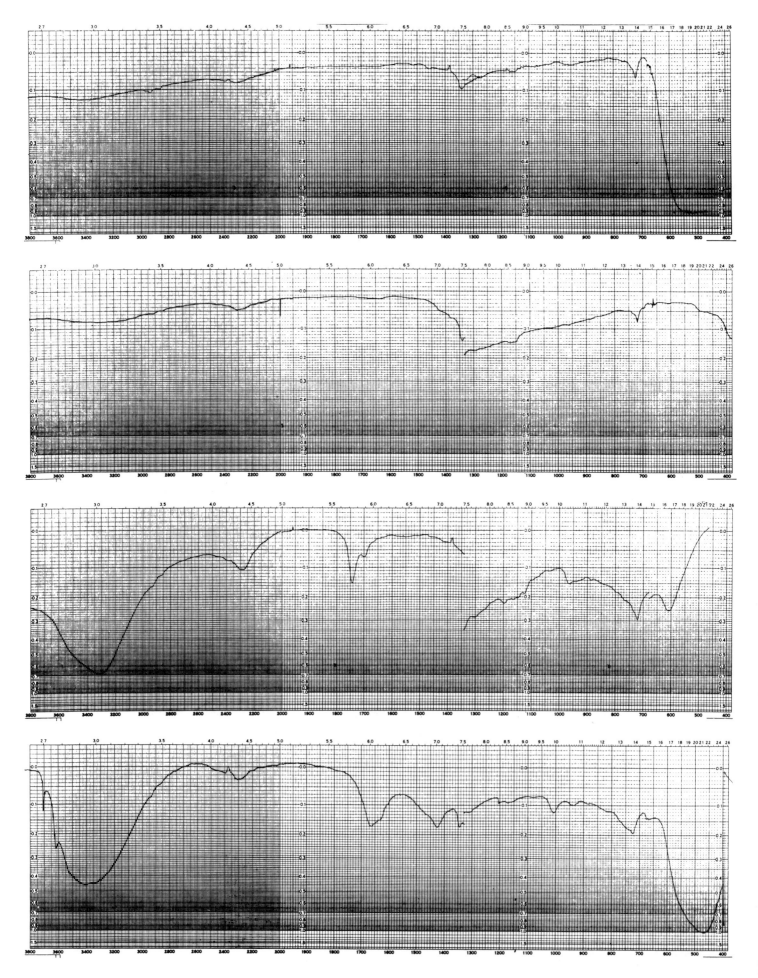

348

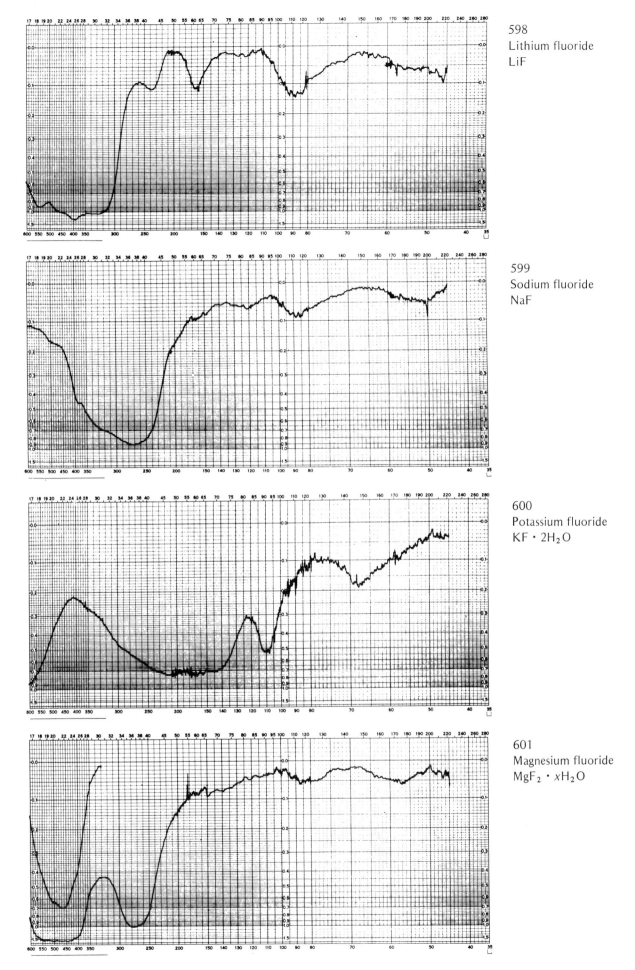

598
Lithium fluoride
LiF

599
Sodium fluoride
NaF

600
Potassium fluoride
KF · 2H$_2$O

601
Magnesium fluoride
MgF$_2$ · xH$_2$O

349

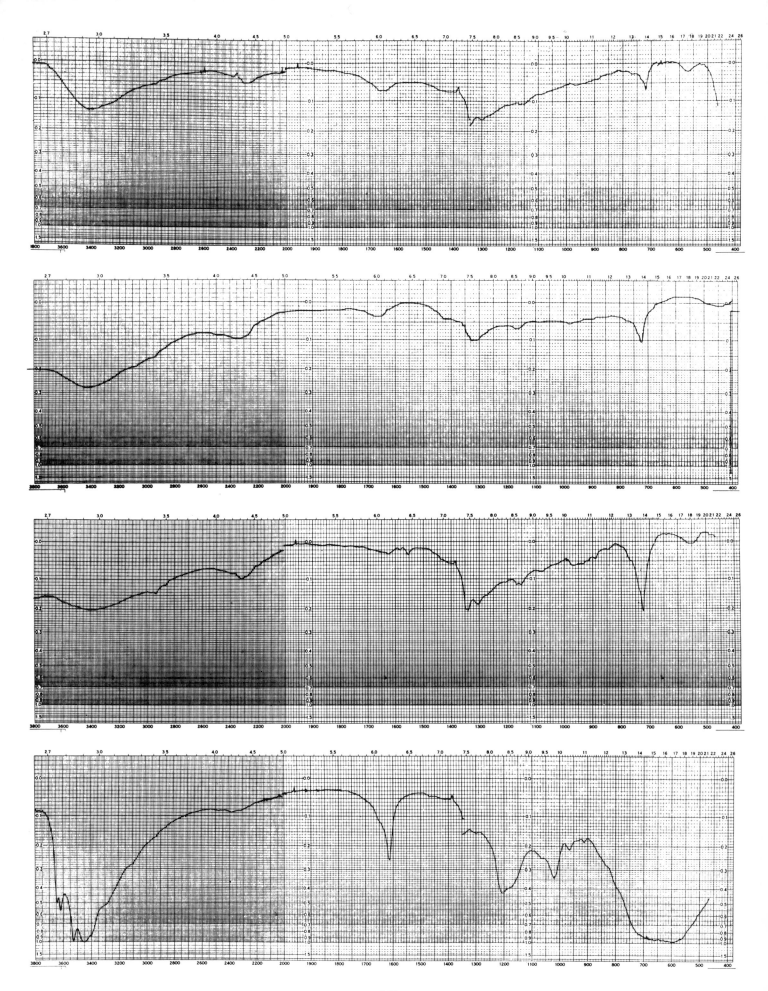

350

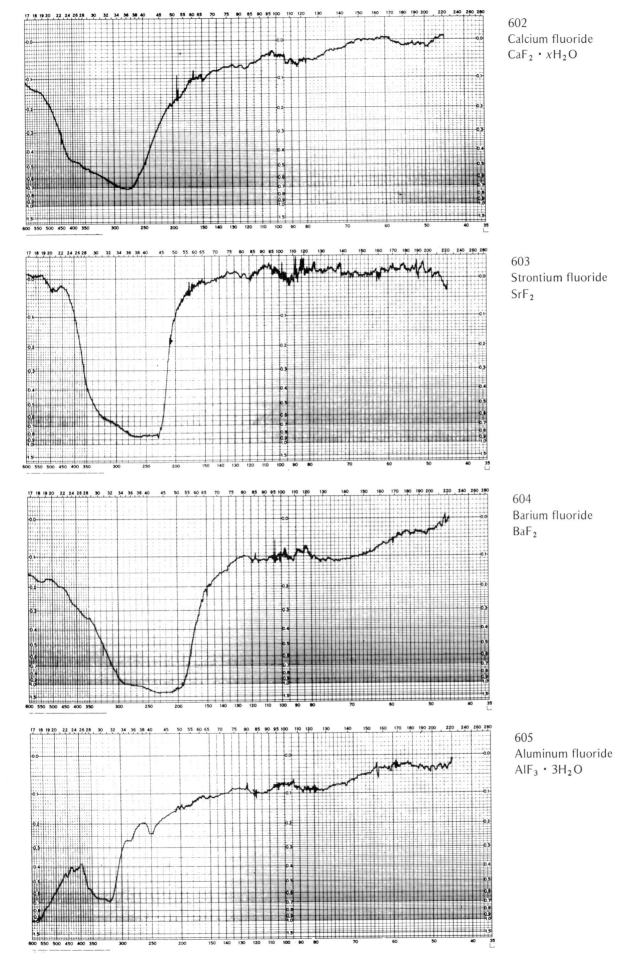

602
Calcium fluoride
$CaF_2 \cdot xH_2O$

603
Strontium fluoride
SrF_2

604
Barium fluoride
BaF_2

605
Aluminum fluoride
$AlF_3 \cdot 3H_2O$

351

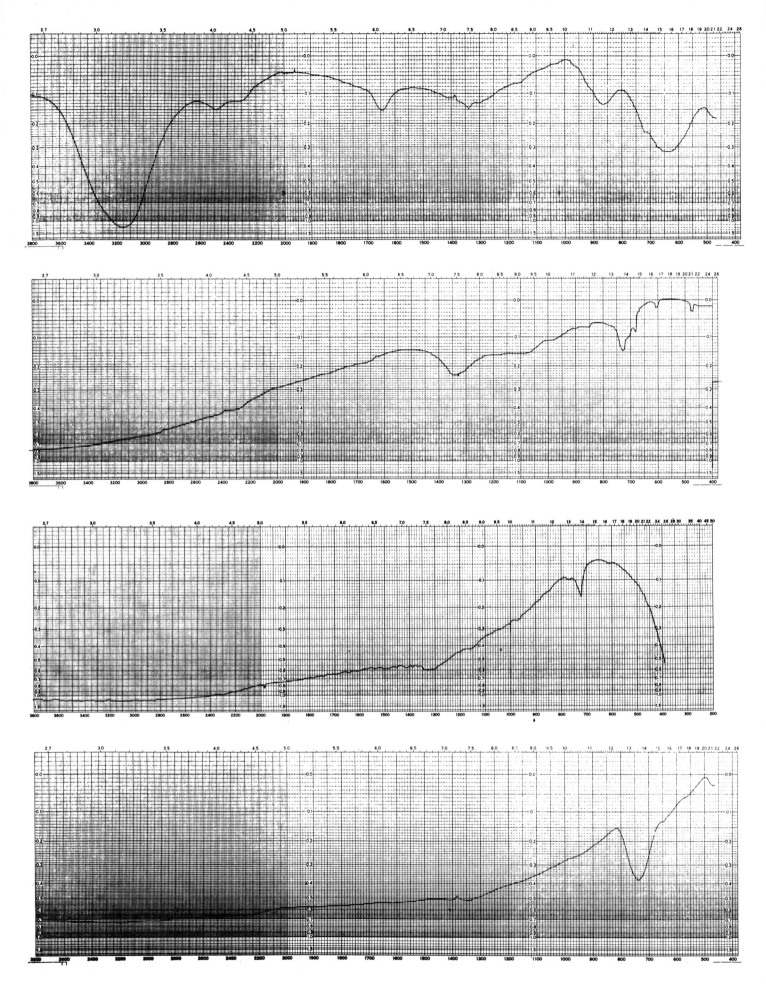

352

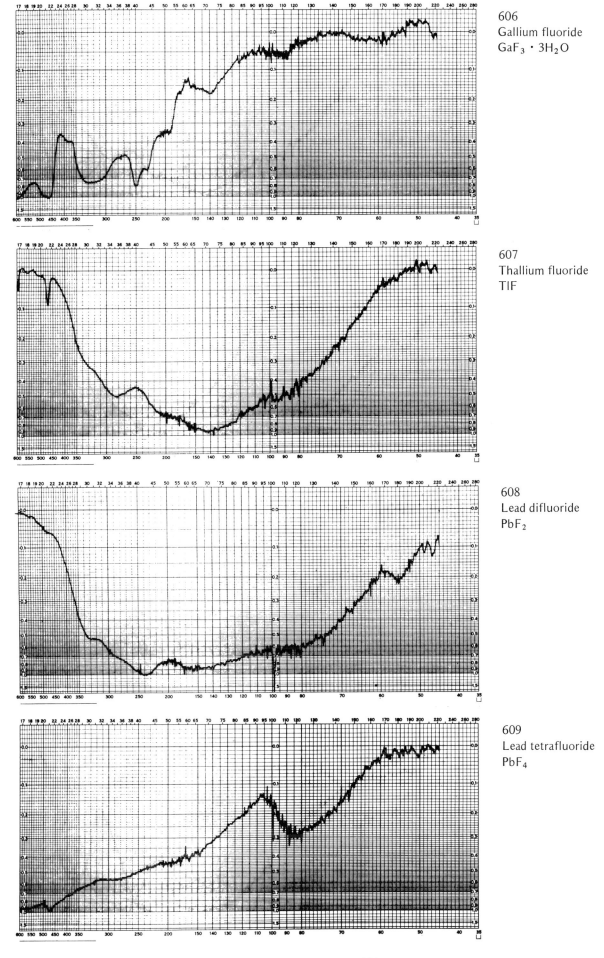

606
Gallium fluoride
GaF$_3$ · 3H$_2$O

607
Thallium fluoride
TlF

608
Lead difluoride
PbF$_2$

609
Lead tetrafluoride
PbF$_4$

353

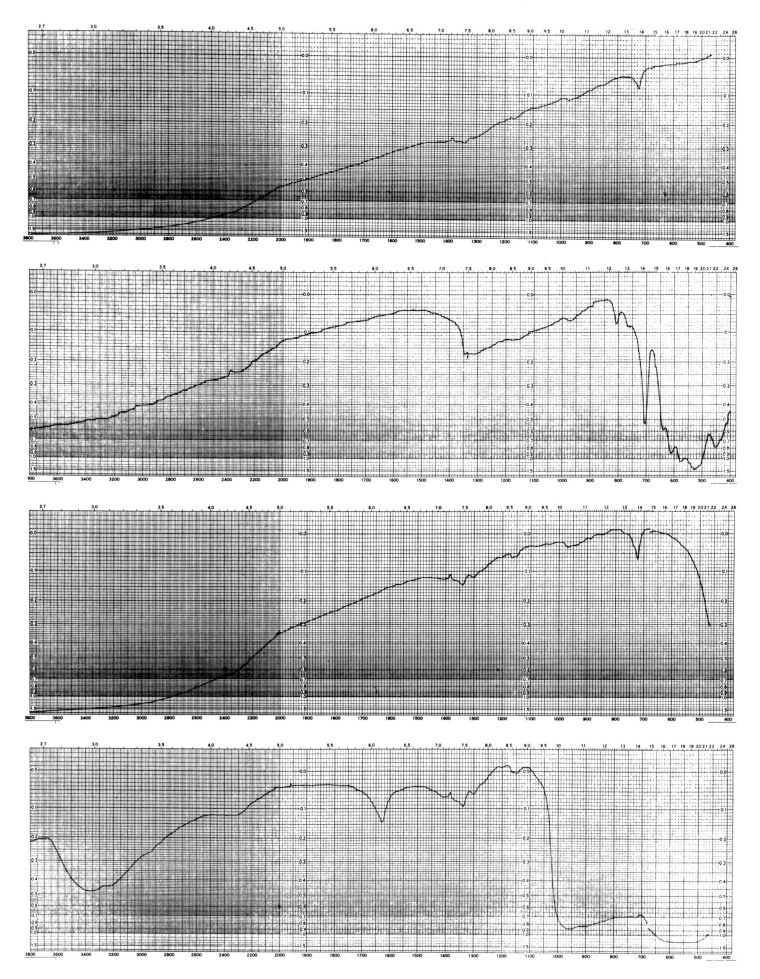

354

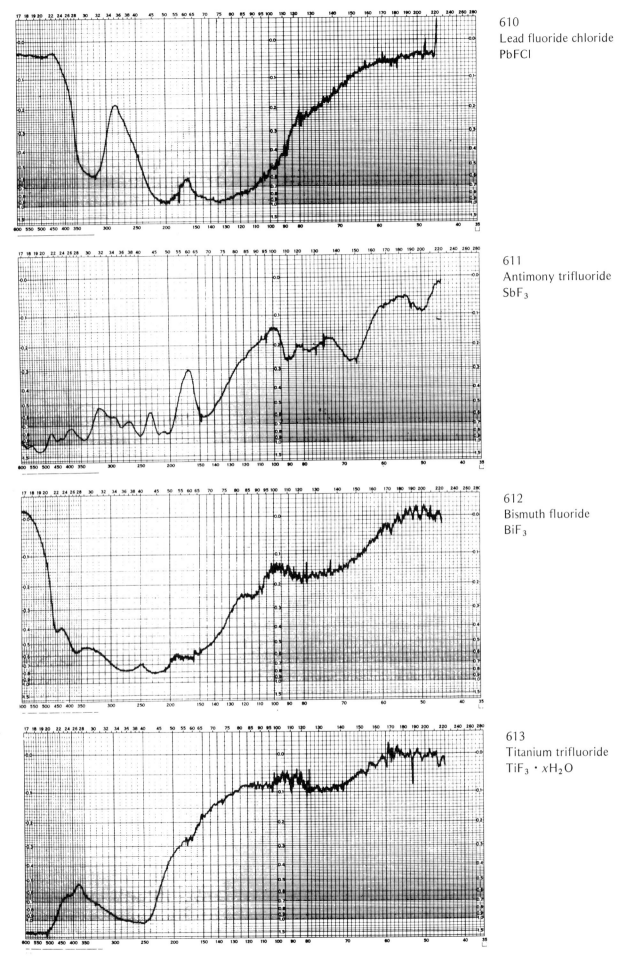

610
Lead fluoride chloride
PbFCl

611
Antimony trifluoride
SbF$_3$

612
Bismuth fluoride
BiF$_3$

613
Titanium trifluoride
TiF$_3$ · xH$_2$O

355

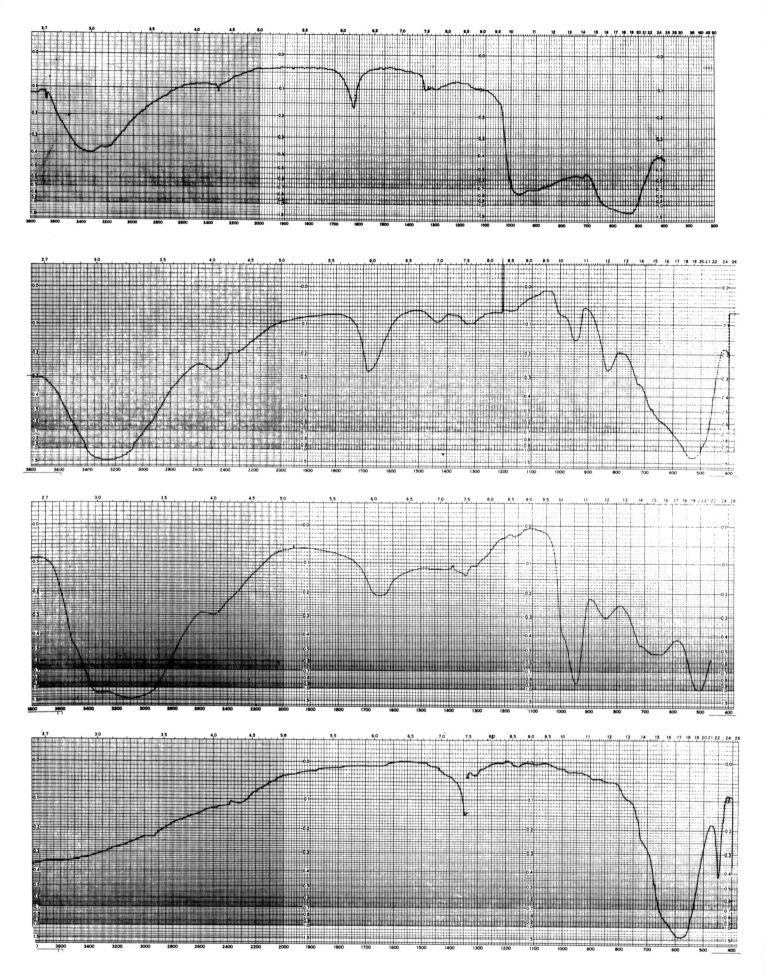

356

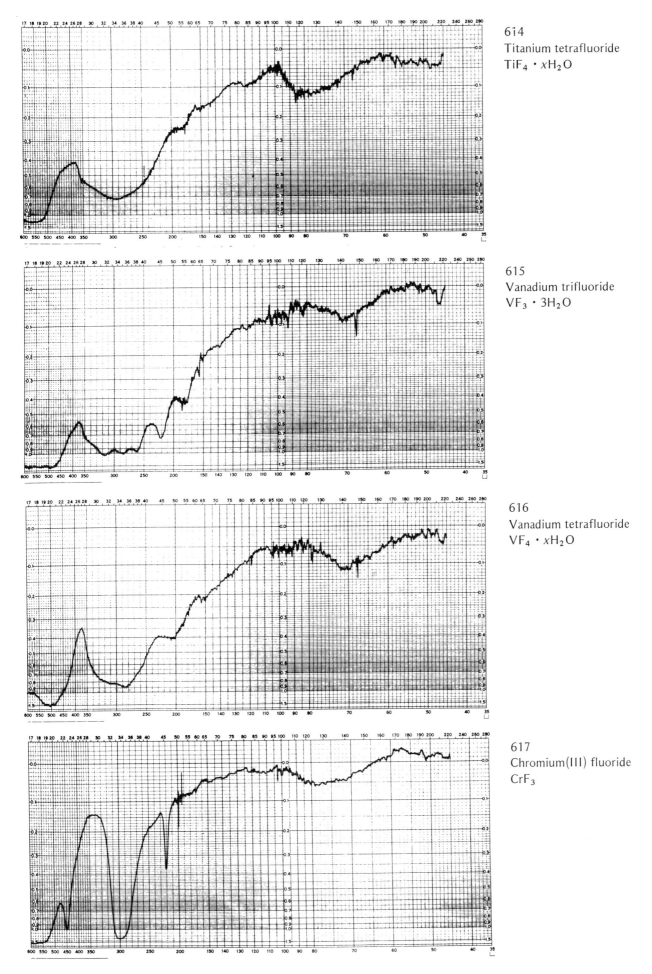

614
Titanium tetrafluoride
$TiF_4 \cdot xH_2O$

615
Vanadium trifluoride
$VF_3 \cdot 3H_2O$

616
Vanadium tetrafluoride
$VF_4 \cdot xH_2O$

617
Chromium(III) fluoride
CrF_3

357

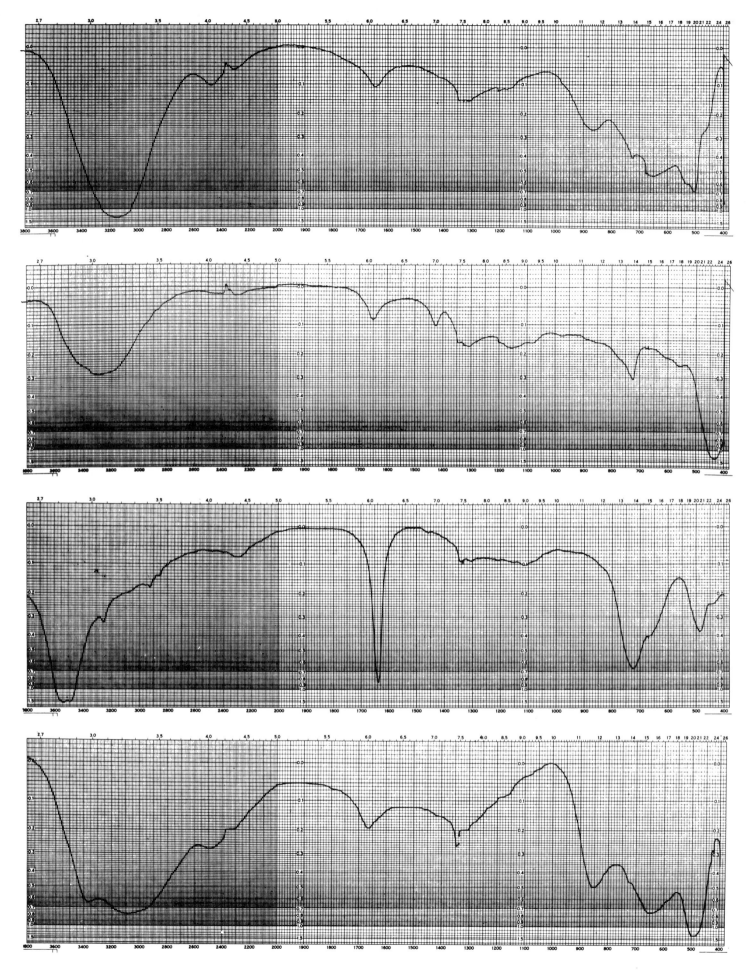

358

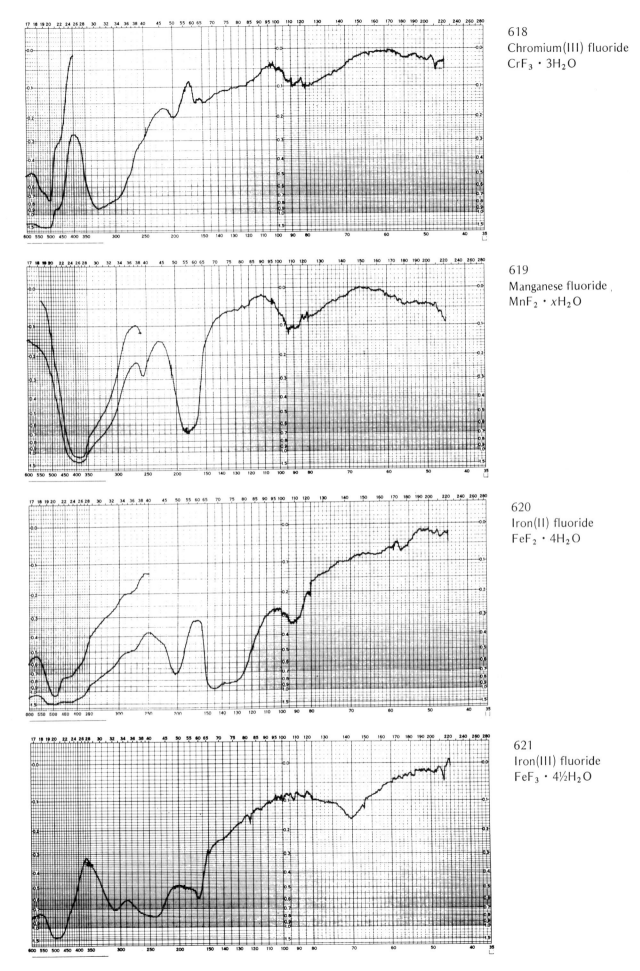

618
Chromium(III) fluoride
CrF$_3$ · 3H$_2$O

619
Manganese fluoride
MnF$_2$ · xH$_2$O

620
Iron(II) fluoride
FeF$_2$ · 4H$_2$O

621
Iron(III) fluoride
FeF$_3$ · 4½H$_2$O

359

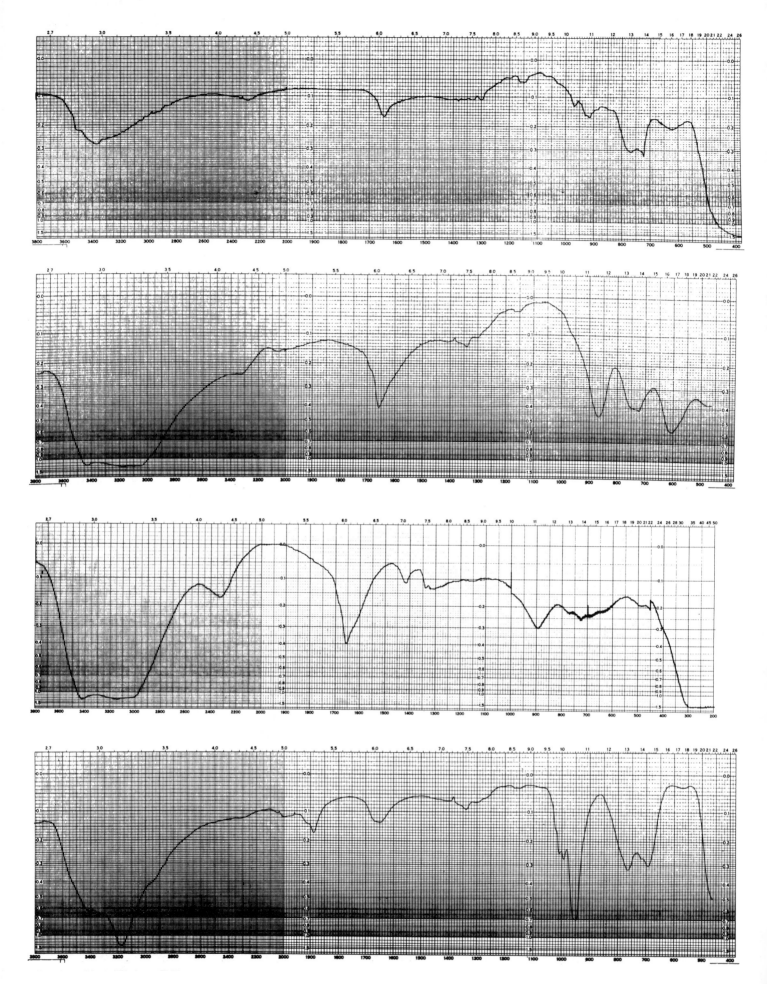

360

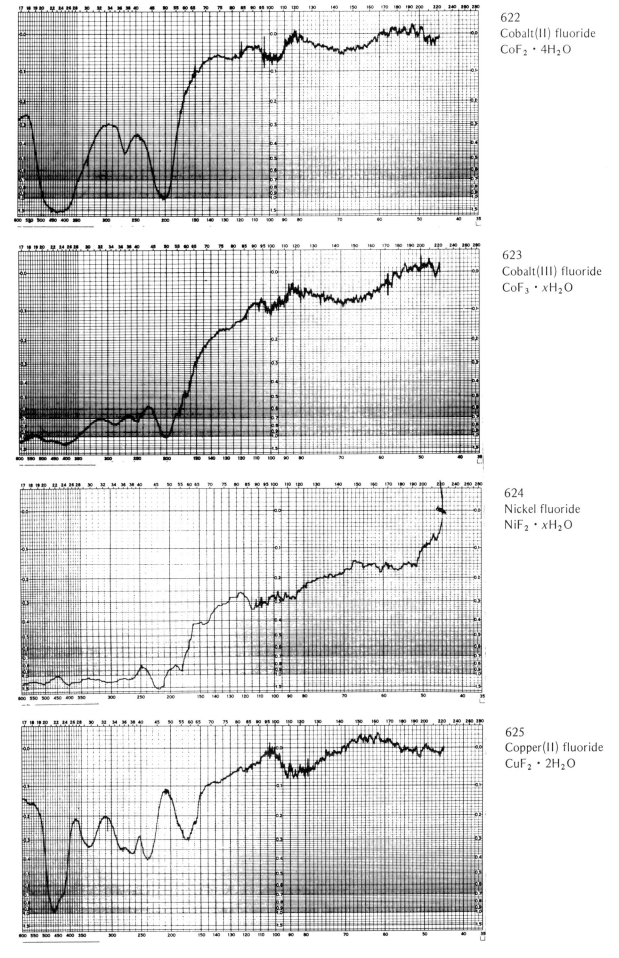

622
Cobalt(II) fluoride
$CoF_2 \cdot 4H_2O$

623
Cobalt(III) fluoride
$CoF_3 \cdot xH_2O$

624
Nickel fluoride
$NiF_2 \cdot xH_2O$

625
Copper(II) fluoride
$CuF_2 \cdot 2H_2O$

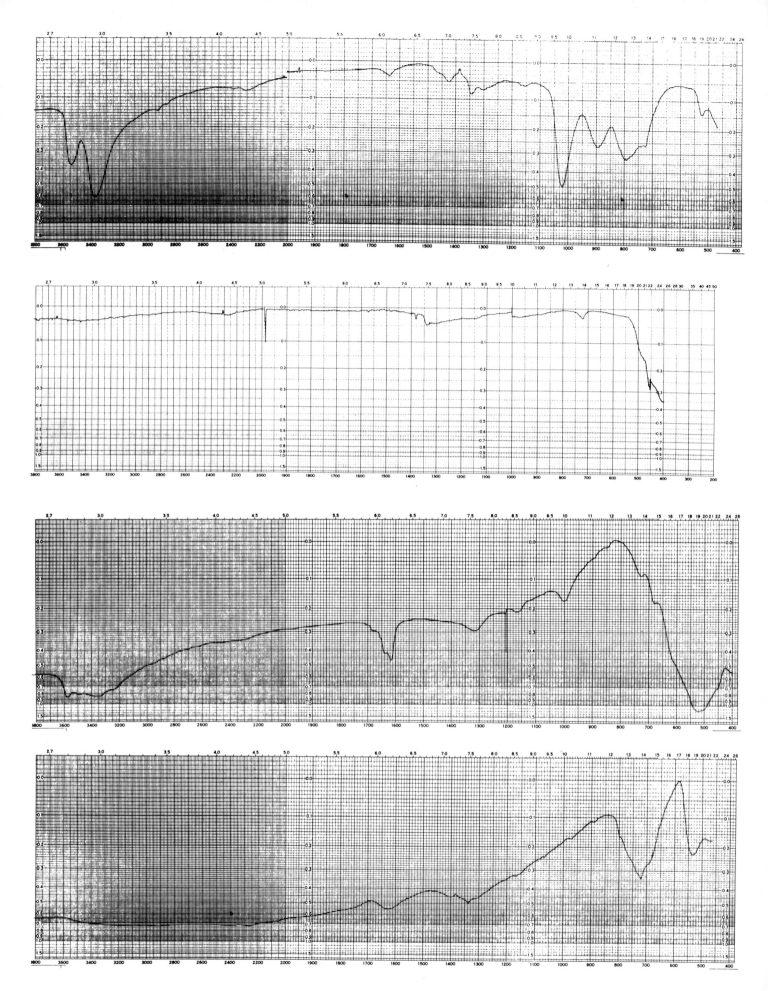

362

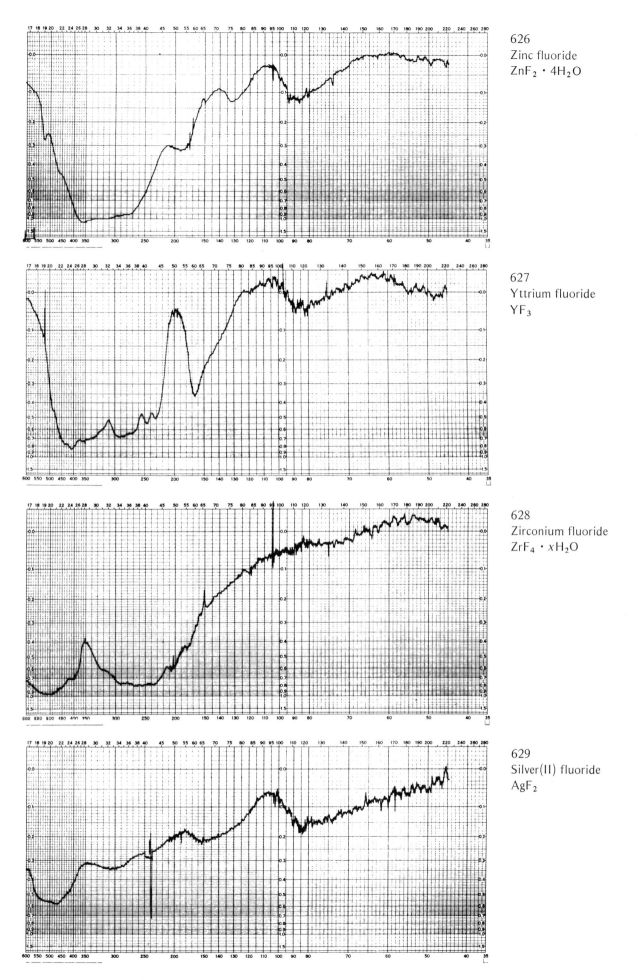

626
Zinc fluoride
$ZnF_2 \cdot 4H_2O$

627
Yttrium fluoride
YF_3

628
Zirconium fluoride
$ZrF_4 \cdot xH_2O$

629
Silver(II) fluoride
AgF_2

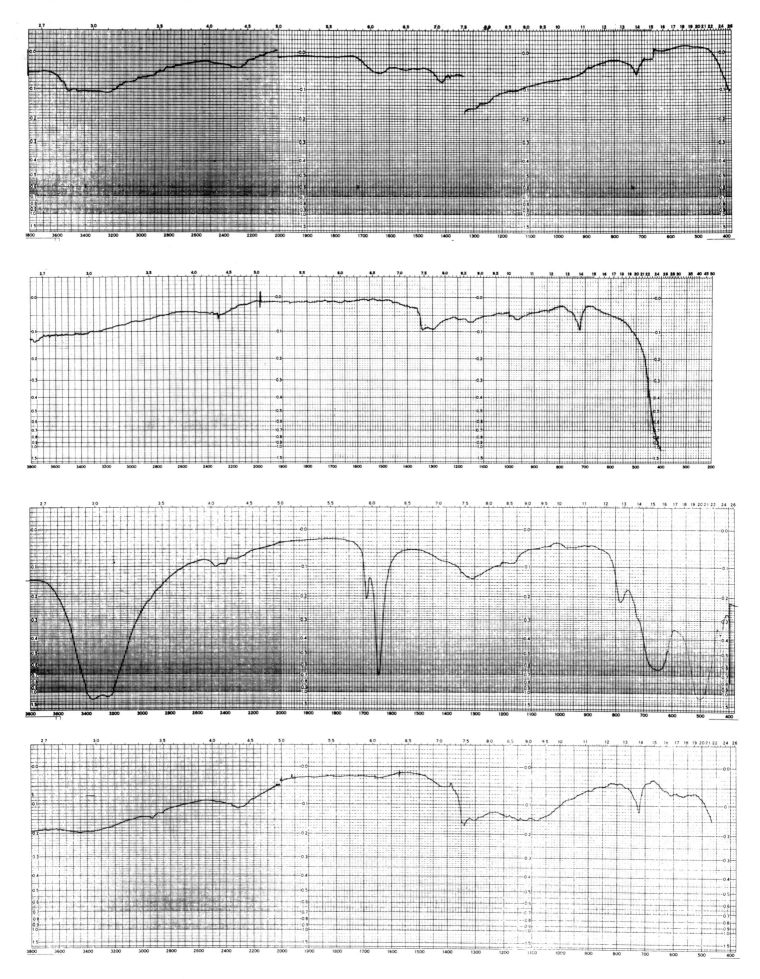

364

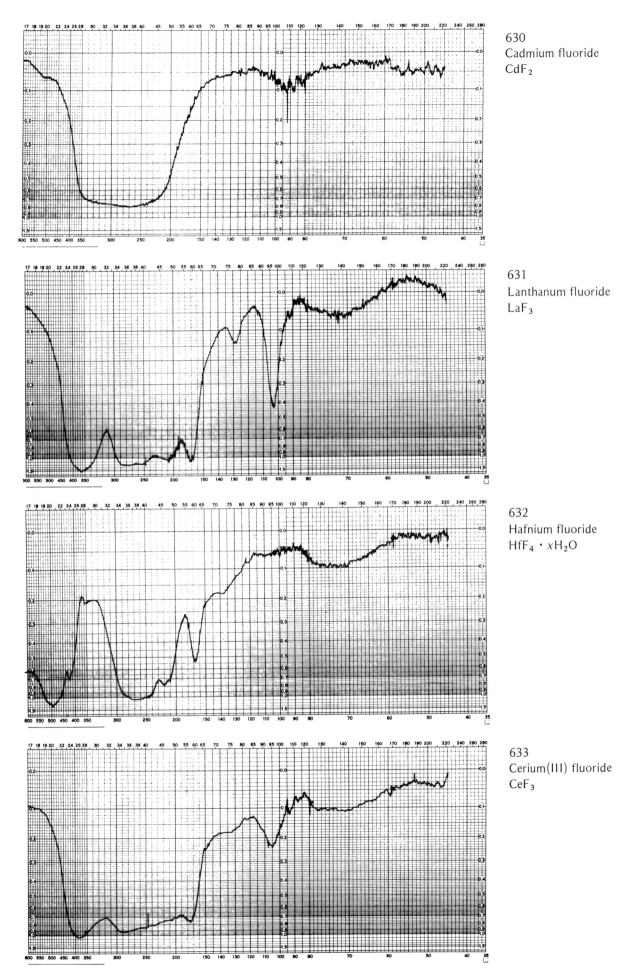

630
Cadmium fluoride
CdF$_2$

631
Lanthanum fluoride
LaF$_3$

632
Hafnium fluoride
HfF$_4$ · xH$_2$O

633
Cerium(III) fluoride
CeF$_3$

365

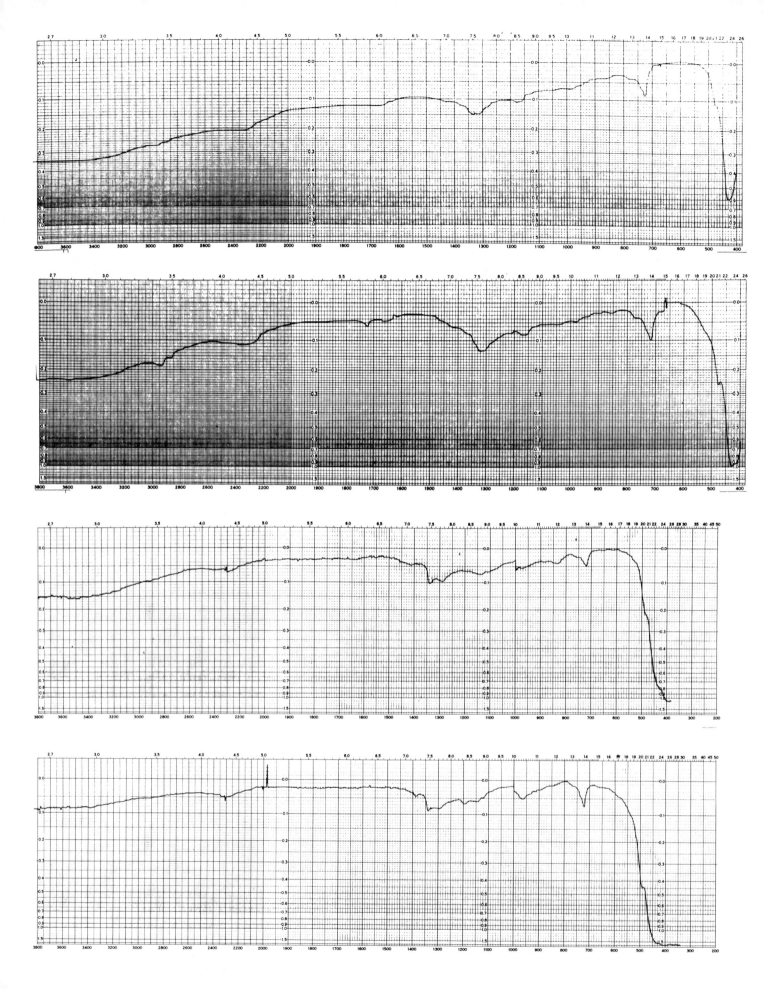

366

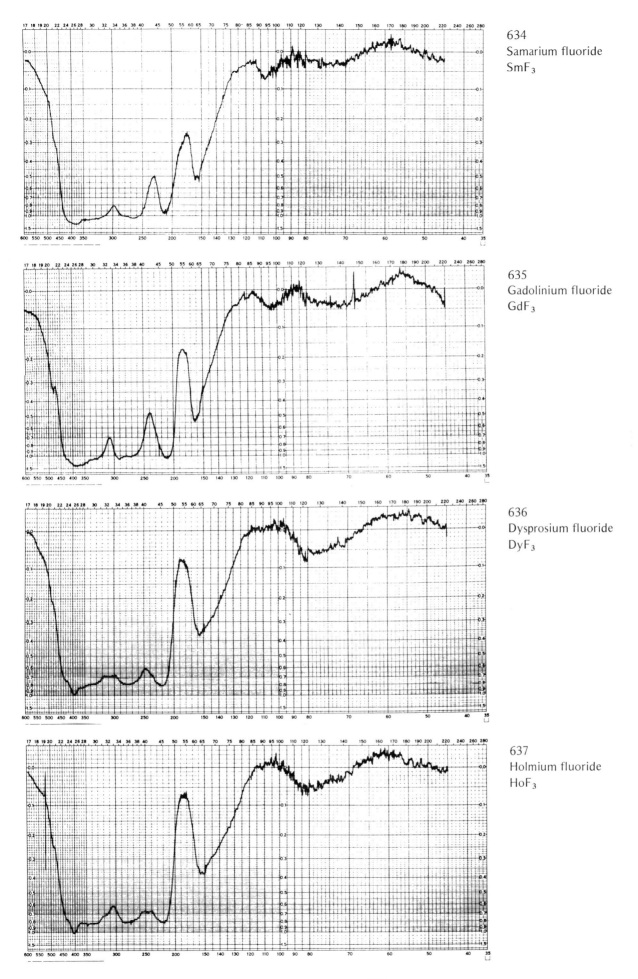

634
Samarium fluoride
SmF$_3$

635
Gadolinium fluoride
GdF$_3$

636
Dysprosium fluoride
DyF$_3$

637
Holmium fluoride
HoF$_3$

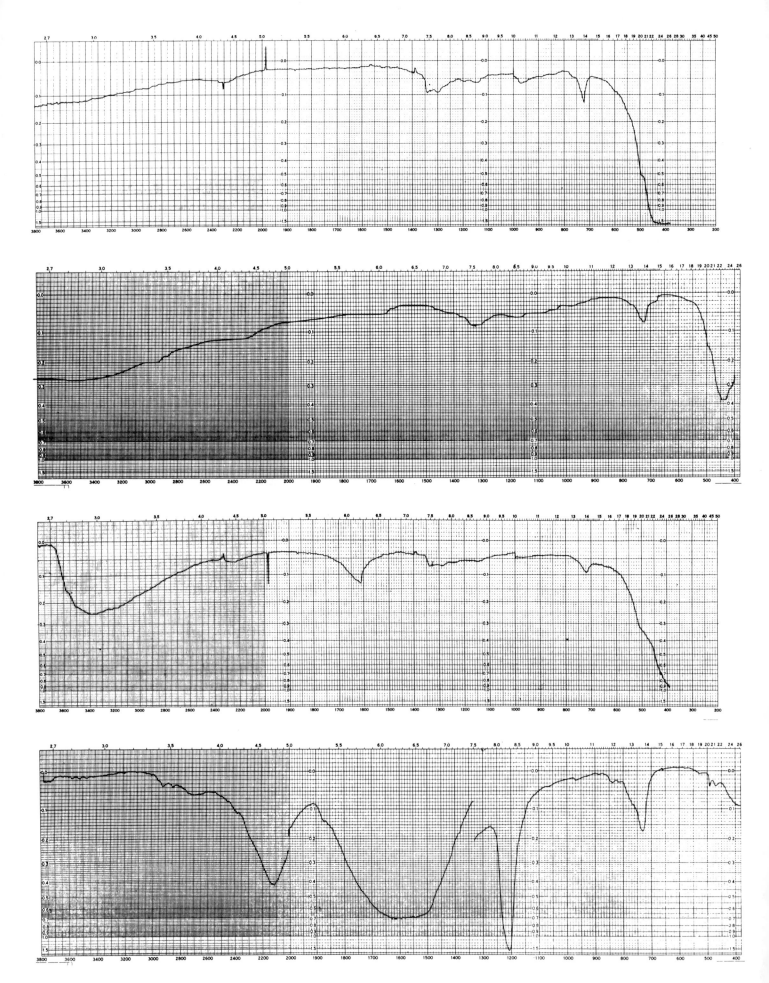

368

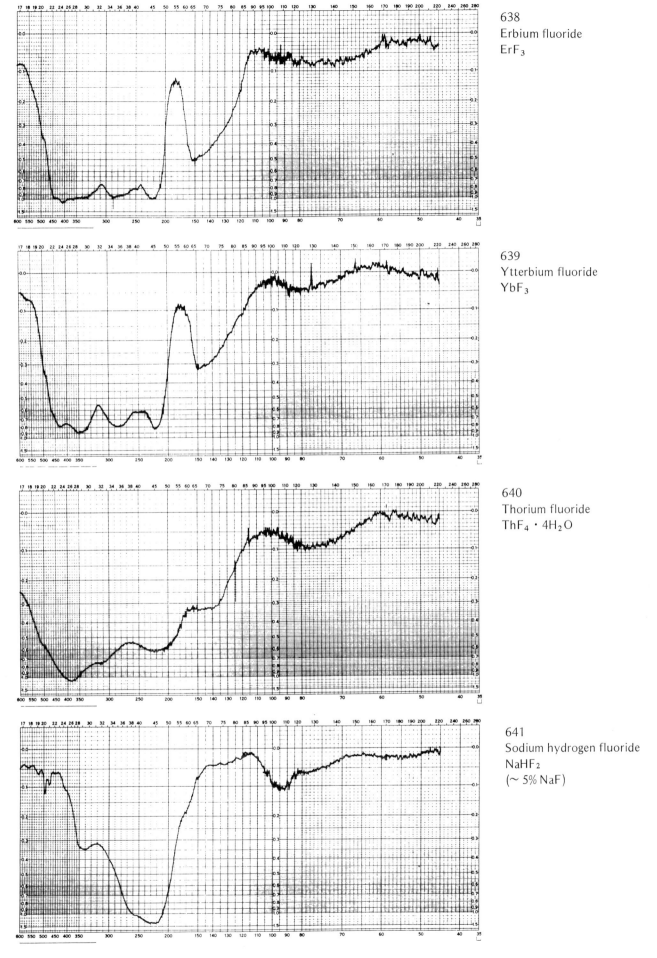

638
Erbium fluoride
ErF$_3$

639
Ytterbium fluoride
YbF$_3$

640
Thorium fluoride
ThF$_4 \cdot$ 4H$_2$O

641
Sodium hydrogen fluoride
NaHF$_2$
($\sim$ 5% NaF)

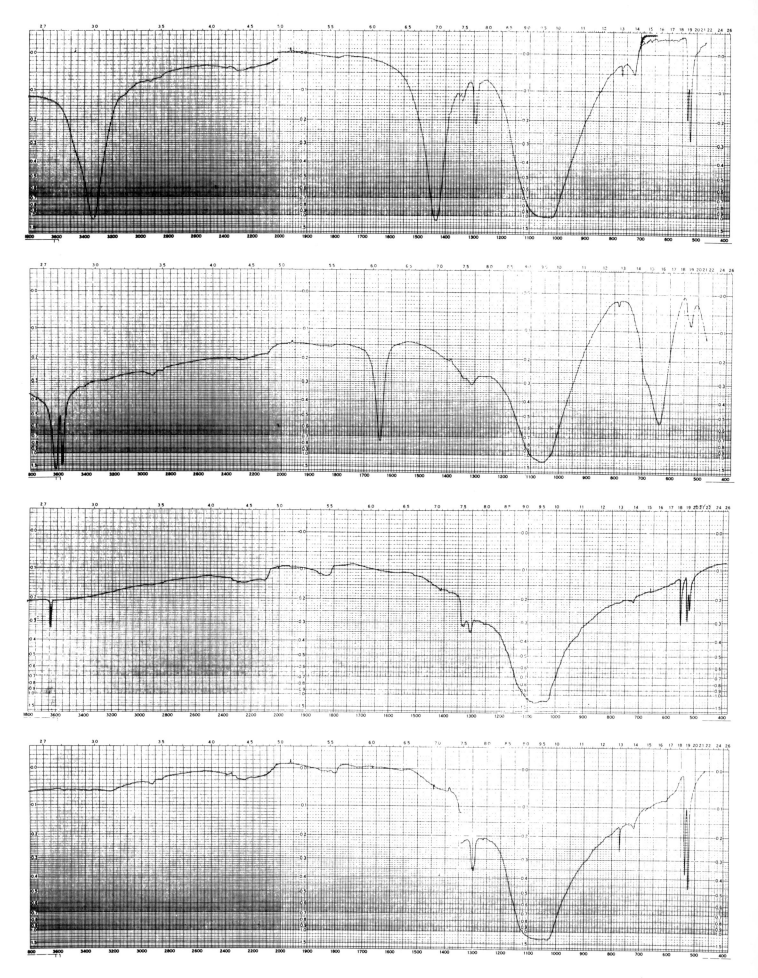

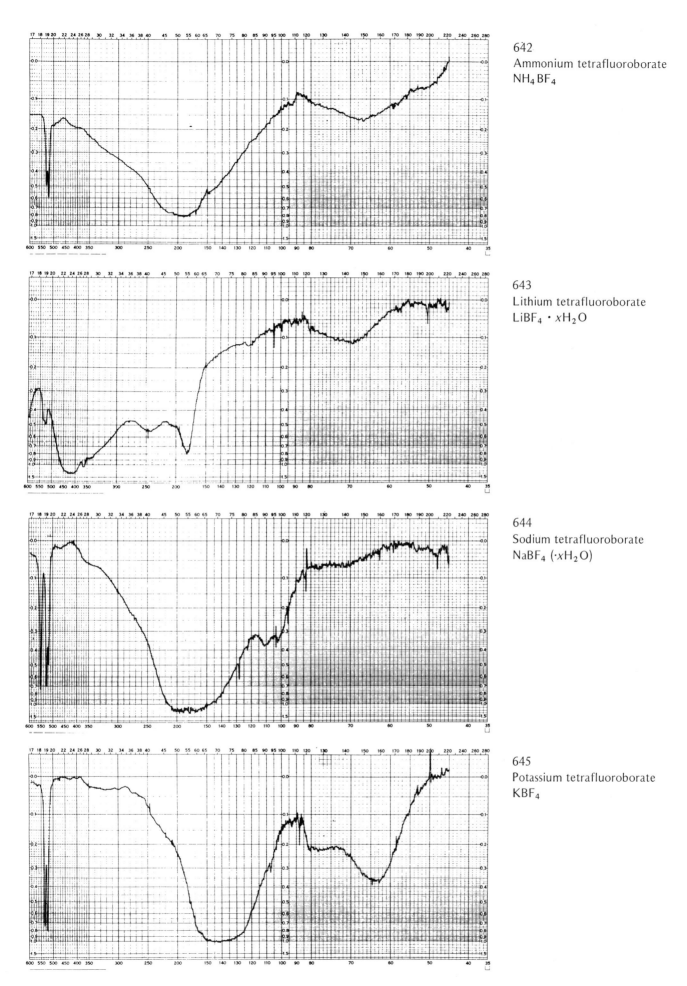

642
Ammonium tetrafluoroborate
NH_4BF_4

643
Lithium tetrafluoroborate
$LiBF_4 \cdot xH_2O$

644
Sodium tetrafluoroborate
$NaBF_4 \, (\cdot xH_2O)$

645
Potassium tetrafluoroborate
KBF_4

371

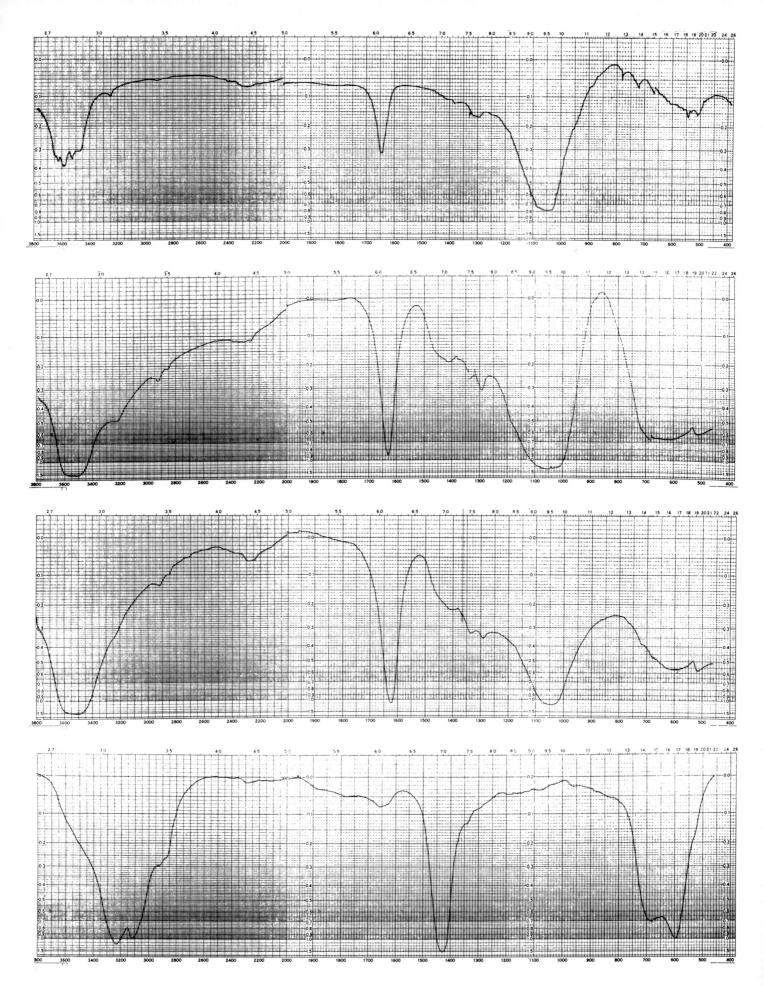

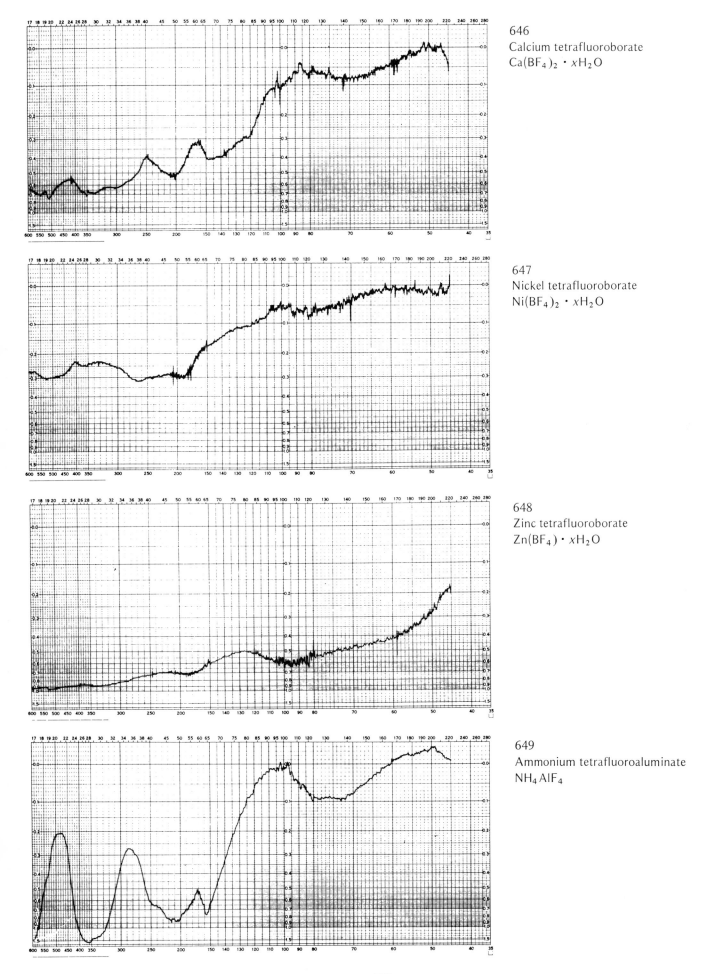

646
Calcium tetrafluoroborate
$Ca(BF_4)_2 \cdot xH_2O$

647
Nickel tetrafluoroborate
$Ni(BF_4)_2 \cdot xH_2O$

648
Zinc tetrafluoroborate
$Zn(BF_4) \cdot xH_2O$

649
Ammonium tetrafluoroaluminate
NH_4AlF_4

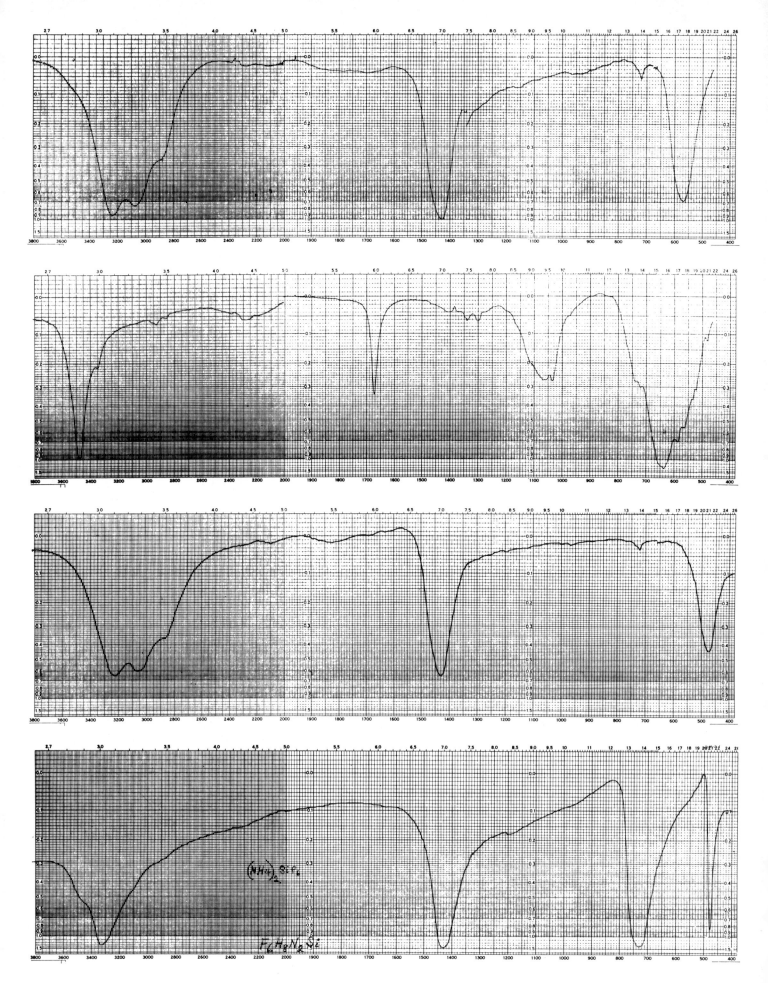

(NH₄)₂SiF₆

F₆H₈N₂Si

374

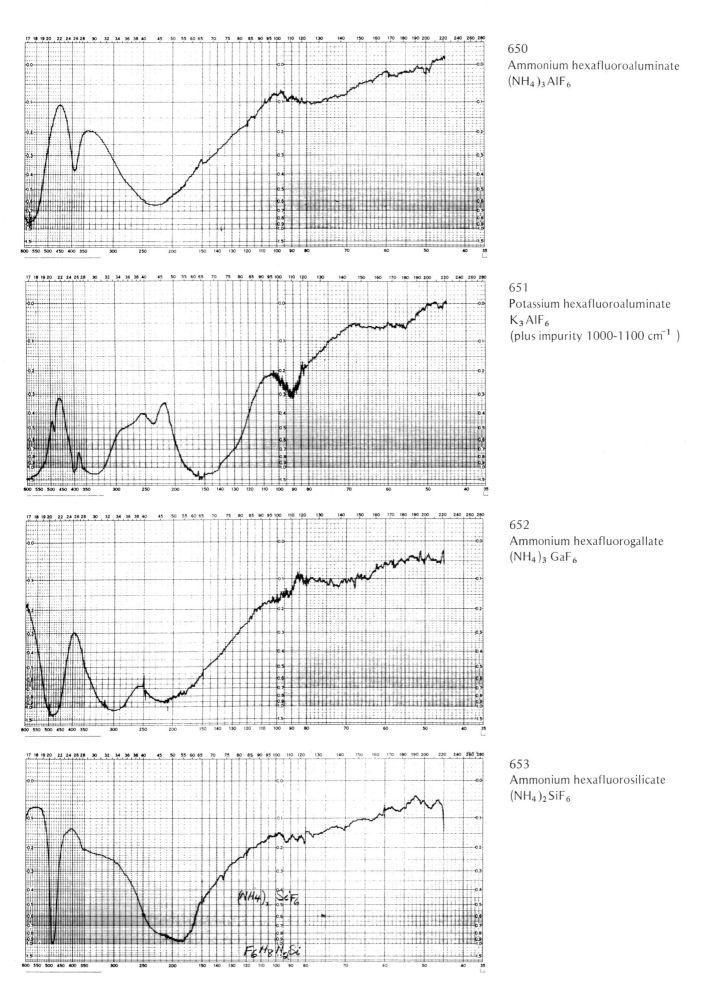

650
Ammonium hexafluoroaluminate
$(NH_4)_3 AlF_6$

651
Potassium hexafluoroaluminate
$K_3 AlF_6$
(plus impurity 1000-1100 cm^{-1})

652
Ammonium hexafluorogallate
$(NH_4)_3 GaF_6$

653
Ammonium hexafluorosilicate
$(NH_4)_2 SiF_6$

375

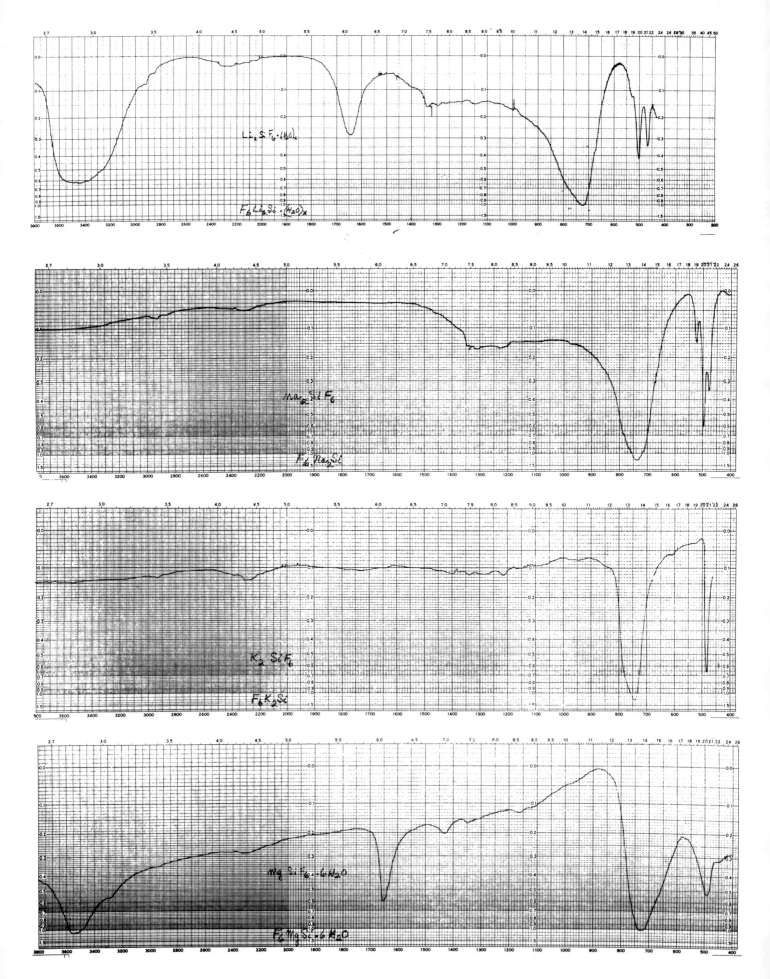

$Li_2 Si F_6 \cdot 6H_2 O)_x$

$F_6 Li_2 Si \cdot (H_2O)_x$

$Na_2 Si F_6$

$F_6 Na_2 Si$

$K_2 Si F_6$

$F_6 K_2 Si$

$Mg Si F_6 \cdot 6H_2O$

$F_6 Mg Si \cdot 6 H_2O$

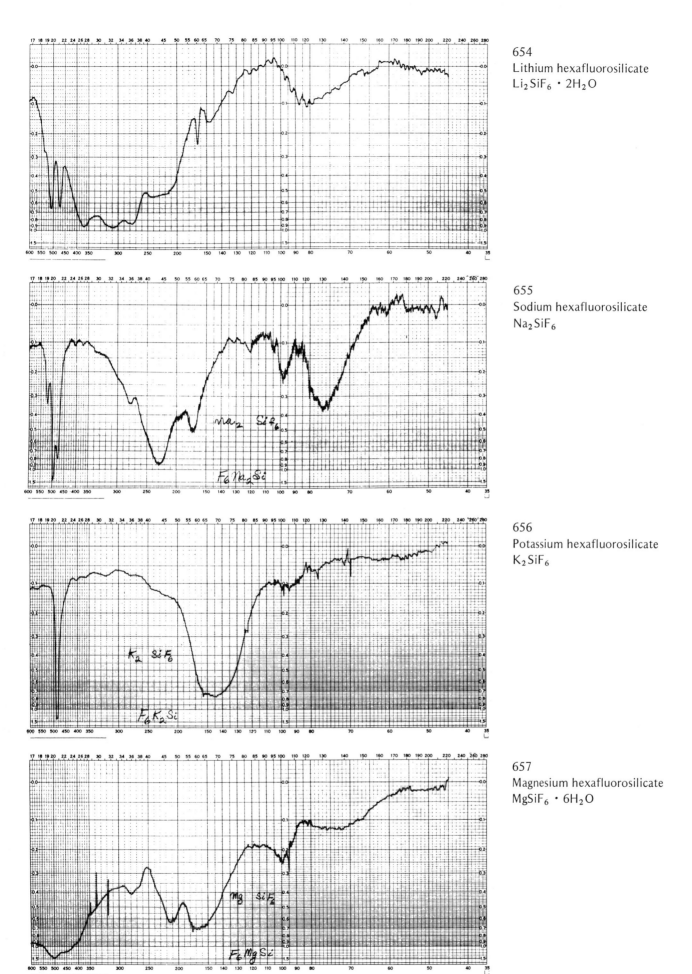

654
Lithium hexafluorosilicate
$Li_2SiF_6 \cdot 2H_2O$

655
Sodium hexafluorosilicate
Na_2SiF_6

656
Potassium hexafluorosilicate
K_2SiF_6

657
Magnesium hexafluorosilicate
$MgSiF_6 \cdot 6H_2O$

377

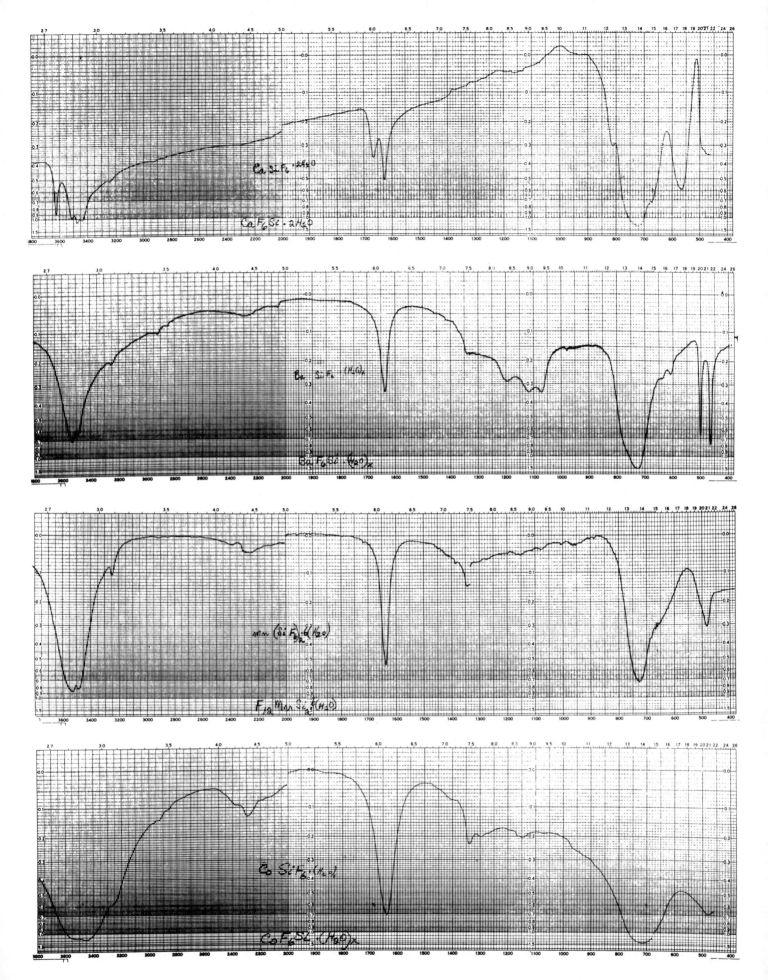

Ca Si F₆ · 2H₂O

CaF₆Si · 2H₂O

Ba Si F₆ (H₂O)ₓ

Ba F₆ Si (H₂O)ₓ

Mn (Si F₆)·(H₂O)

F₁₂ Mn Si₂ · (H₂O)

Co Si F₆ · (H₂O)ₓ

Co F₆ Si · (H₂O)ₓ

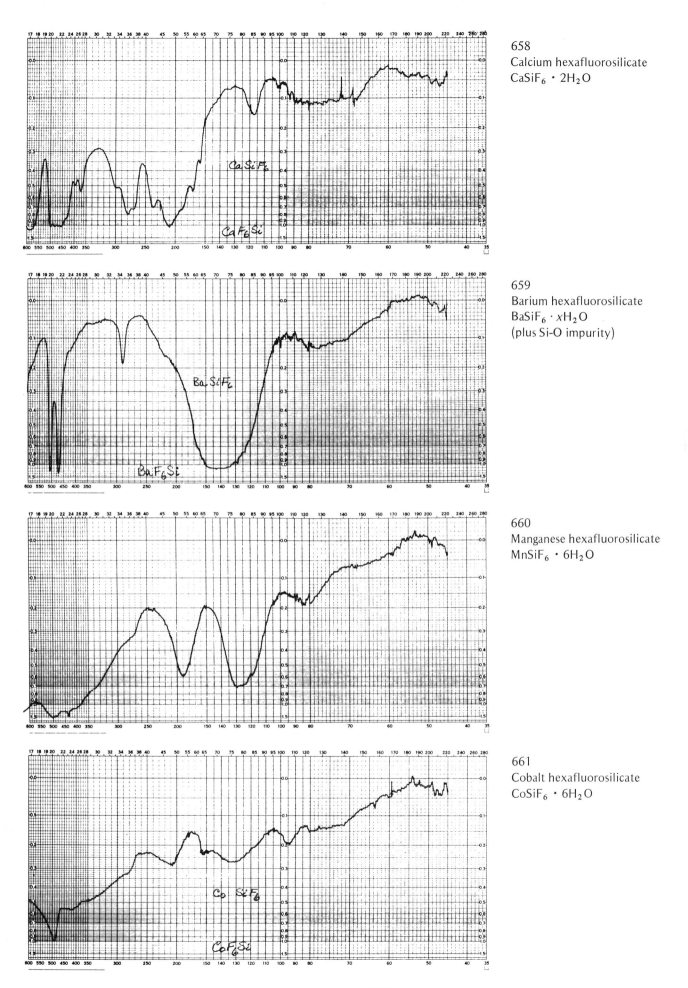

658
Calcium hexafluorosilicate
$CaSiF_6 \cdot 2H_2O$

659
Barium hexafluorosilicate
$BaSiF_6 \cdot xH_2O$
(plus Si-O impurity)

660
Manganese hexafluorosilicate
$MnSiF_6 \cdot 6H_2O$

661
Cobalt hexafluorosilicate
$CoSiF_6 \cdot 6H_2O$

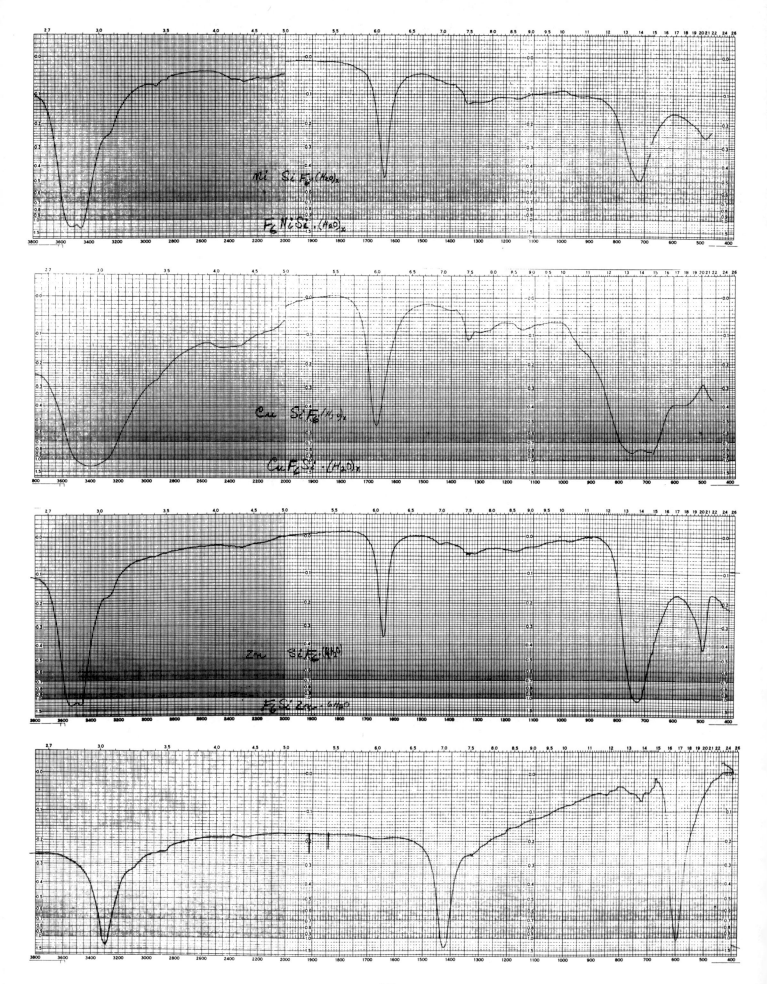

Ni SiF₆(H₂O)ₓ

F₆NiSi·(H₂O)ₓ

Cu SiF₆(H₂O)ₓ

CuF₆Si·(H₂O)ₓ

Zn SiF₆(H₂O)

F₆SiZn·6H₂O

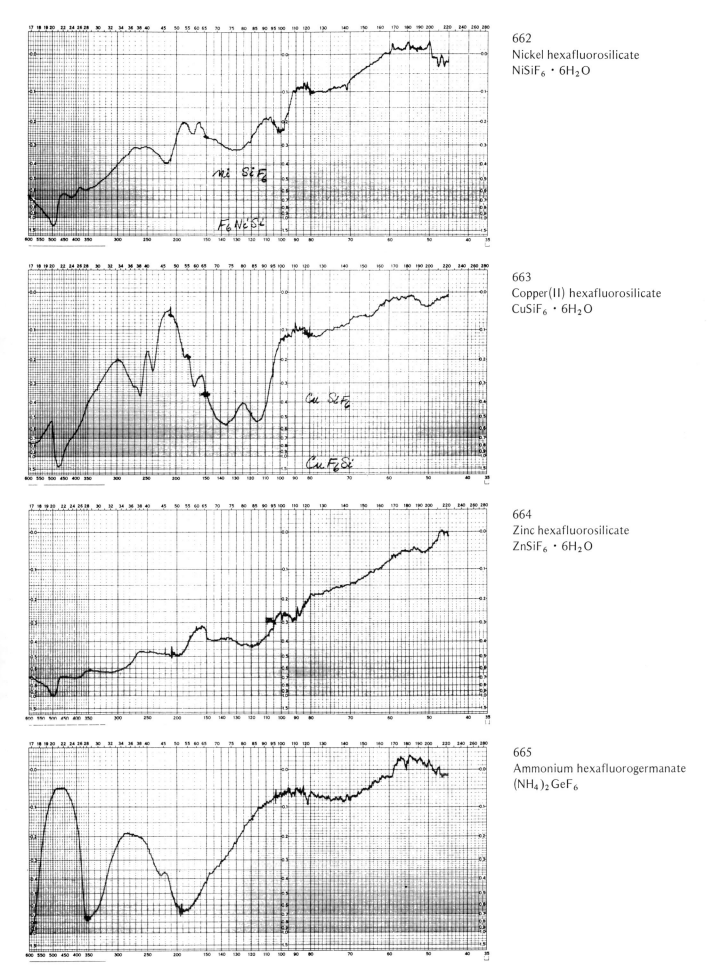

662
Nickel hexafluorosilicate
NiSiF$_6$ · 6H$_2$O

663
Copper(II) hexafluorosilicate
CuSiF$_6$ · 6H$_2$O

664
Zinc hexafluorosilicate
ZnSiF$_6$ · 6H$_2$O

665
Ammonium hexafluorogermanate
(NH$_4$)$_2$GeF$_6$

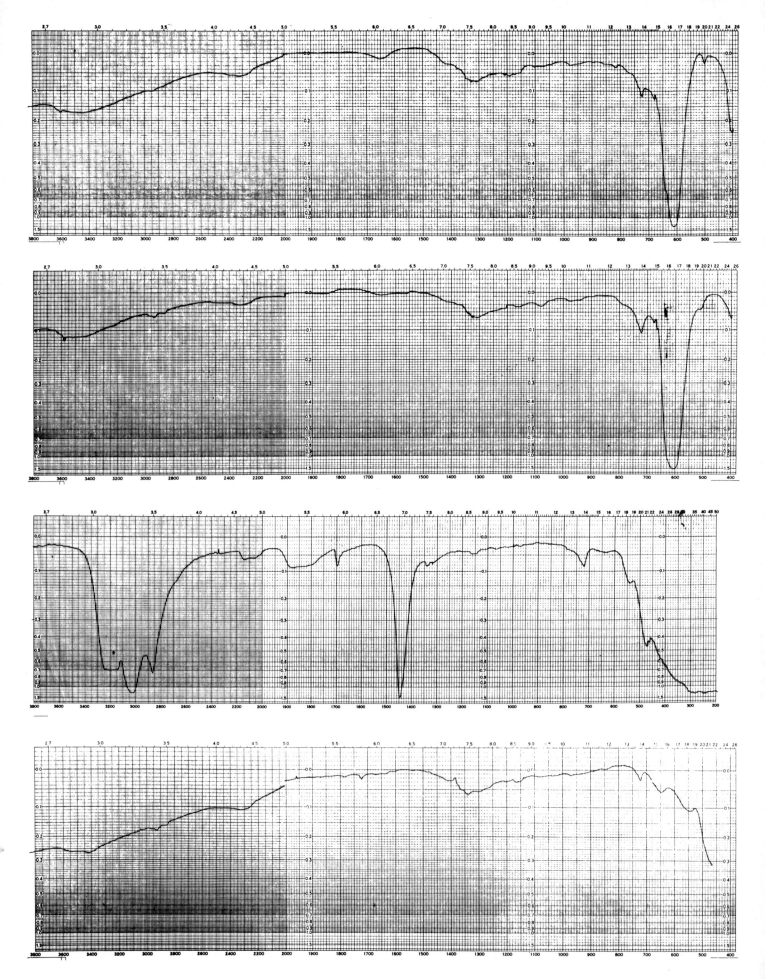

382

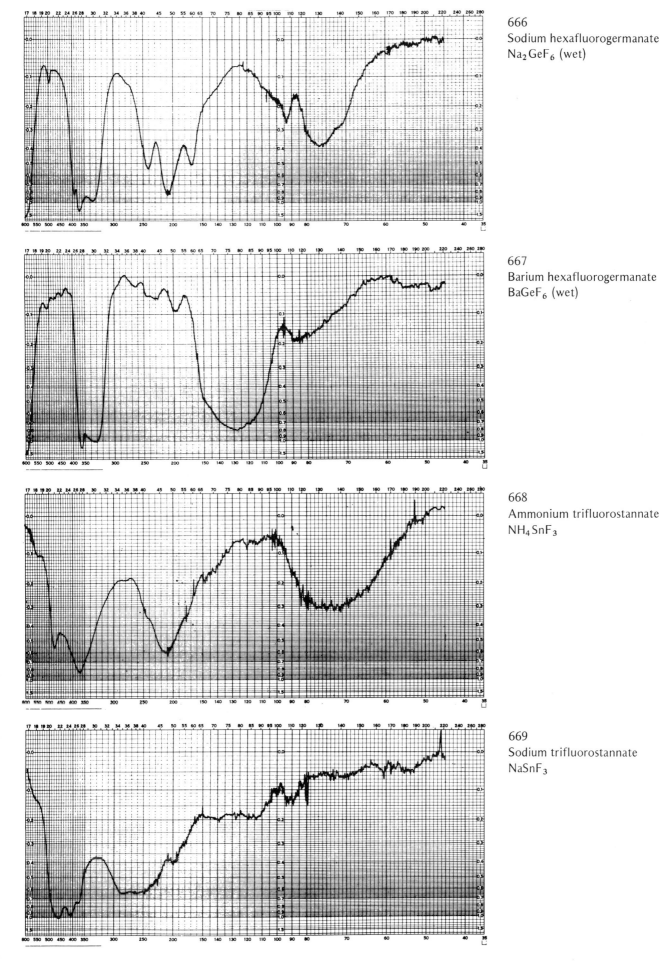

666
Sodium hexafluorogermanate
Na$_2$GeF$_6$ (wet)

667
Barium hexafluorogermanate
BaGeF$_6$ (wet)

668
Ammonium trifluorostannate
NH$_4$SnF$_3$

669
Sodium trifluorostannate
NaSnF$_3$

383

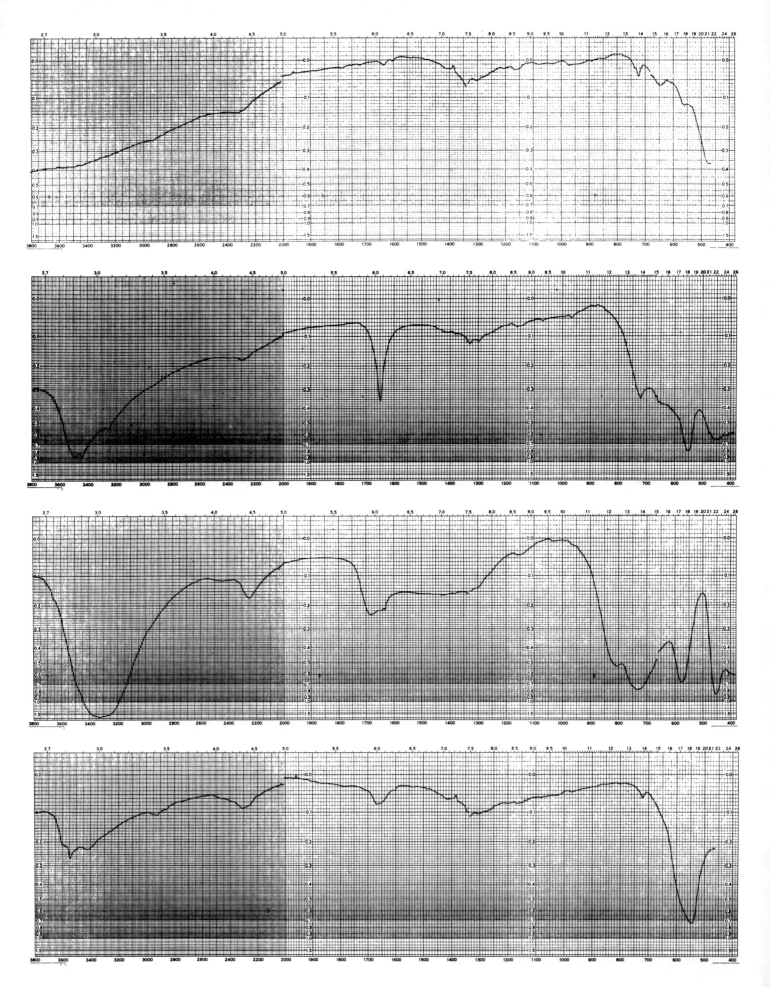

384

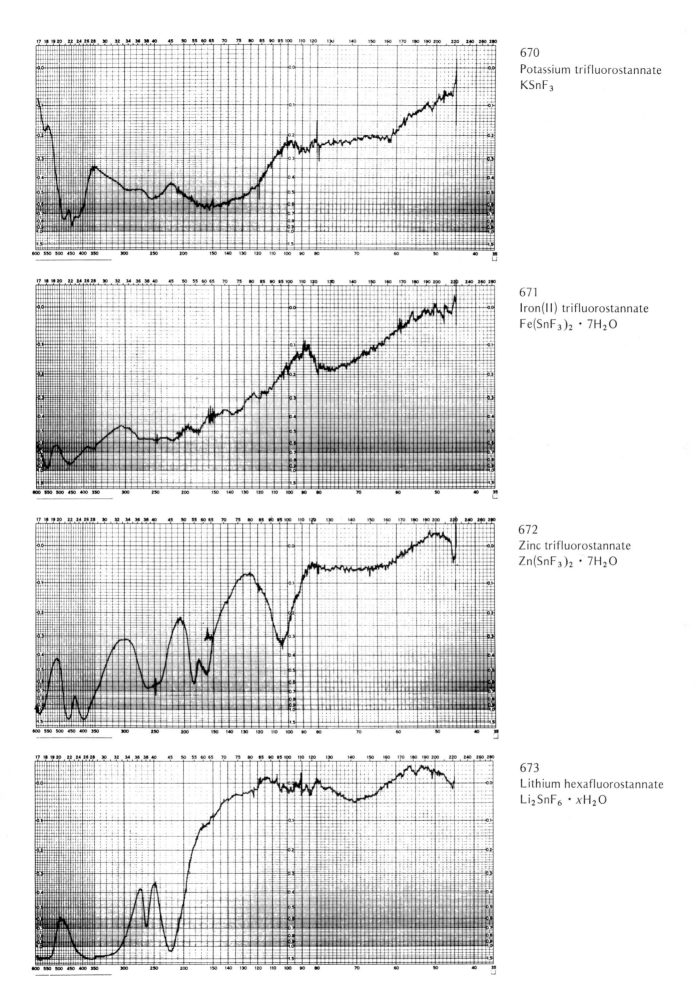

670
Potassium trifluorostannate
$KSnF_3$

671
Iron(II) trifluorostannate
$Fe(SnF_3)_2 \cdot 7H_2O$

672
Zinc trifluorostannate
$Zn(SnF_3)_2 \cdot 7H_2O$

673
Lithium hexafluorostannate
$Li_2SnF_6 \cdot xH_2O$

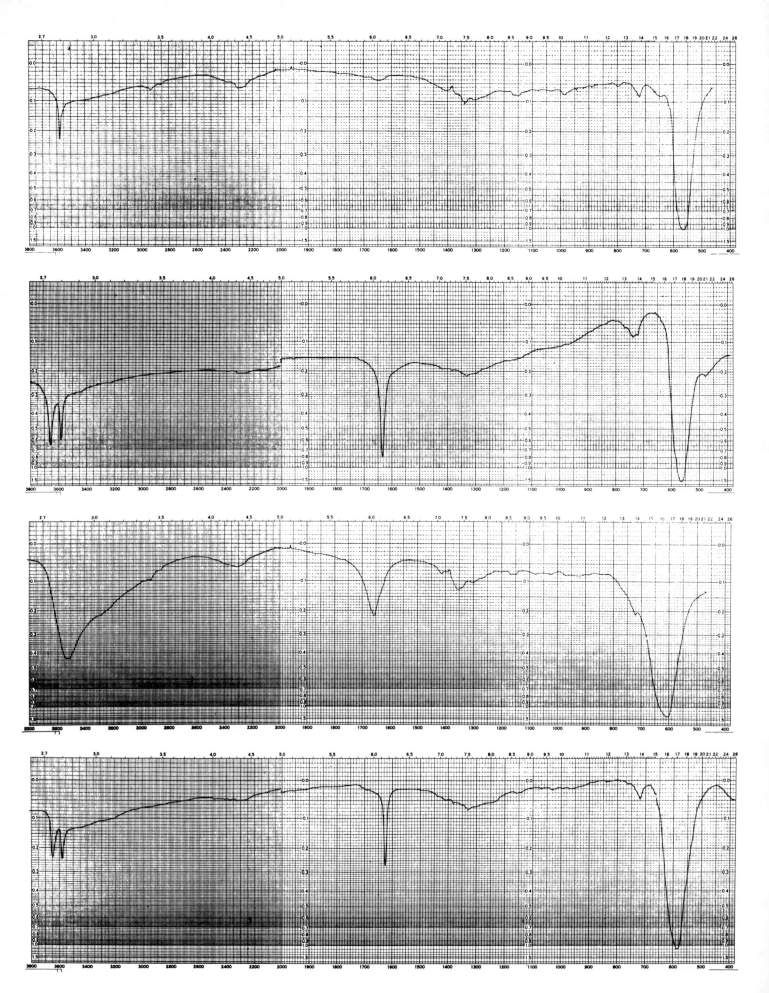

386

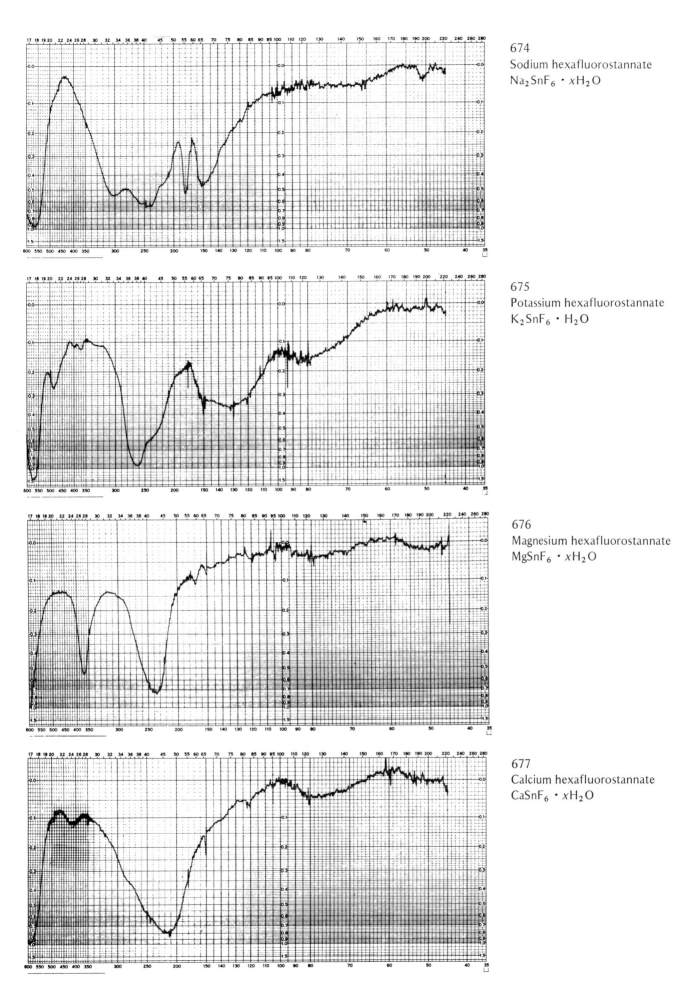

674
Sodium hexafluorostannate
$Na_2SnF_6 \cdot xH_2O$

675
Potassium hexafluorostannate
$K_2SnF_6 \cdot H_2O$

676
Magnesium hexafluorostannate
$MgSnF_6 \cdot xH_2O$

677
Calcium hexafluorostannate
$CaSnF_6 \cdot xH_2O$

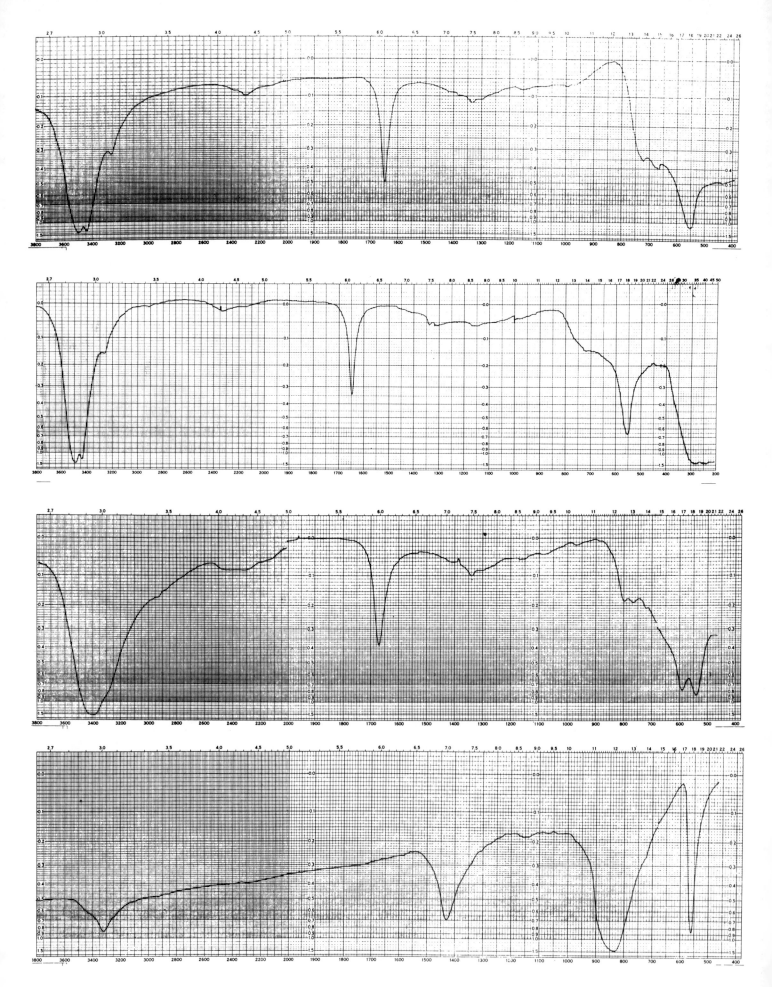

388

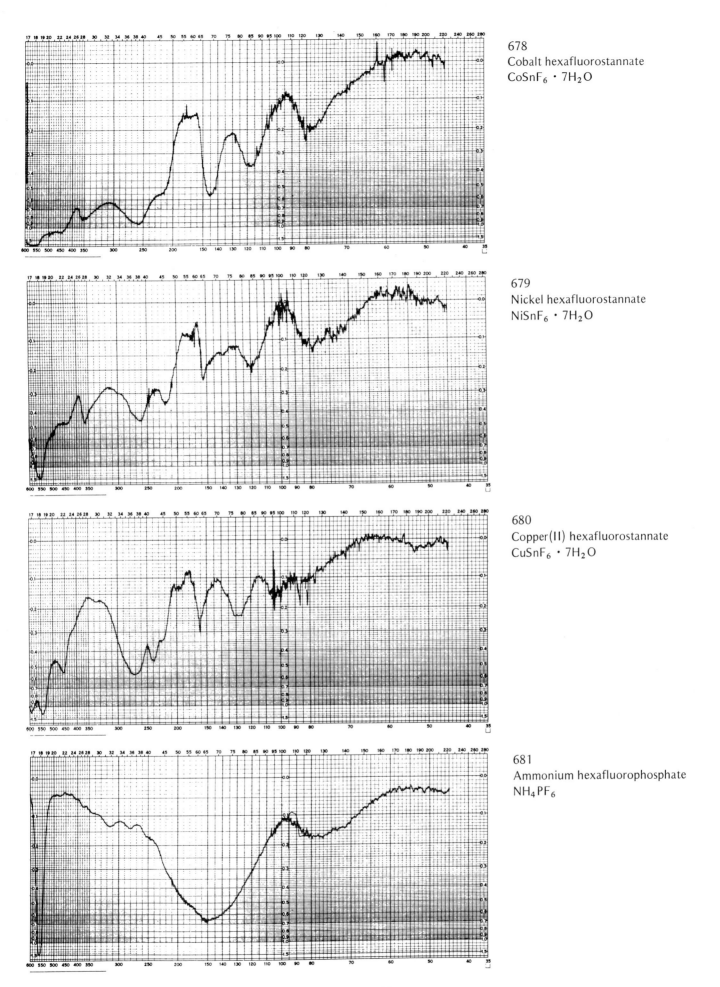

678
Cobalt hexafluorostannate
$CoSnF_6 \cdot 7H_2O$

679
Nickel hexafluorostannate
$NiSnF_6 \cdot 7H_2O$

680
Copper(II) hexafluorostannate
$CuSnF_6 \cdot 7H_2O$

681
Ammonium hexafluorophosphate
NH_4PF_6

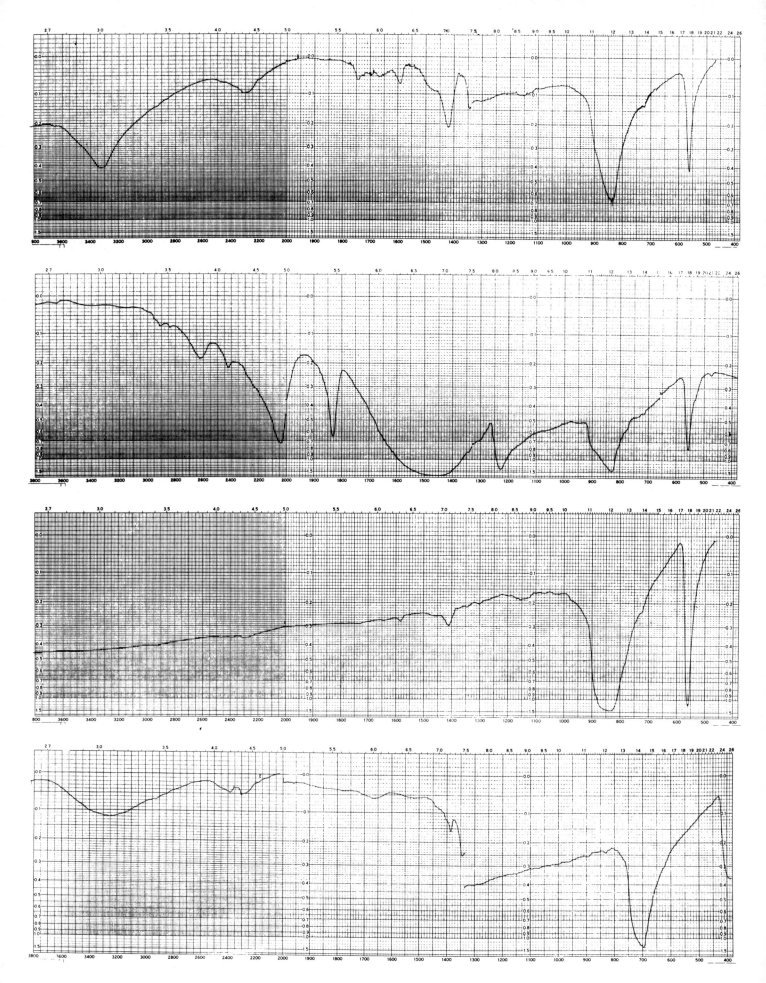

390

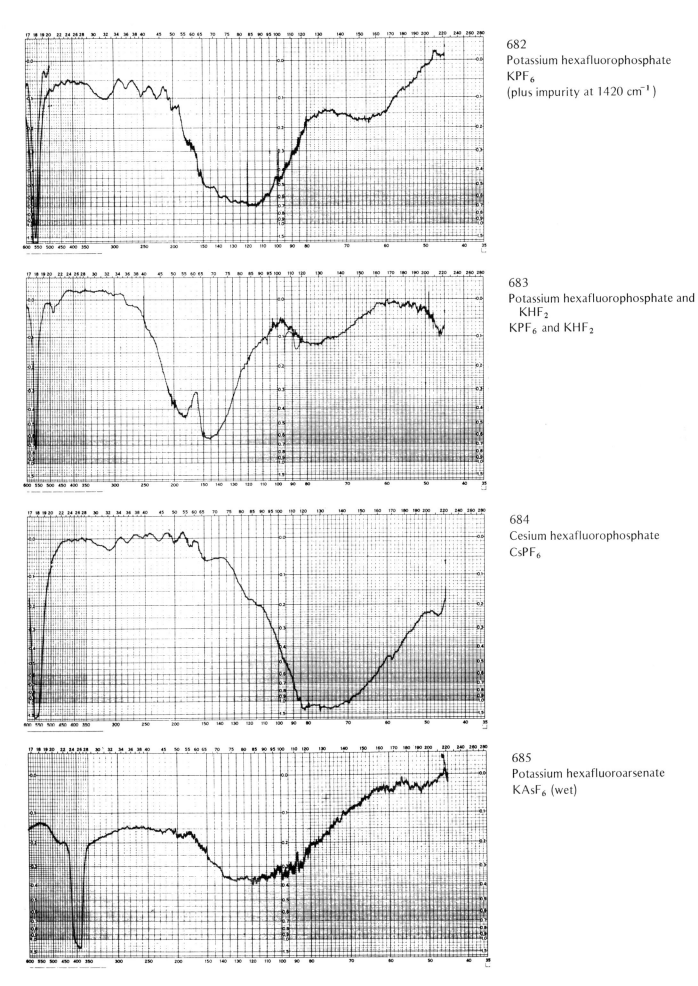

682
Potassium hexafluorophosphate
KPF₆
(plus impurity at 1420 cm⁻¹)

683
Potassium hexafluorophosphate and
KHF₂
KPF₆ and KHF₂

684
Cesium hexafluorophosphate
CsPF₆

685
Potassium hexafluoroarsenate
KAsF₆ (wet)

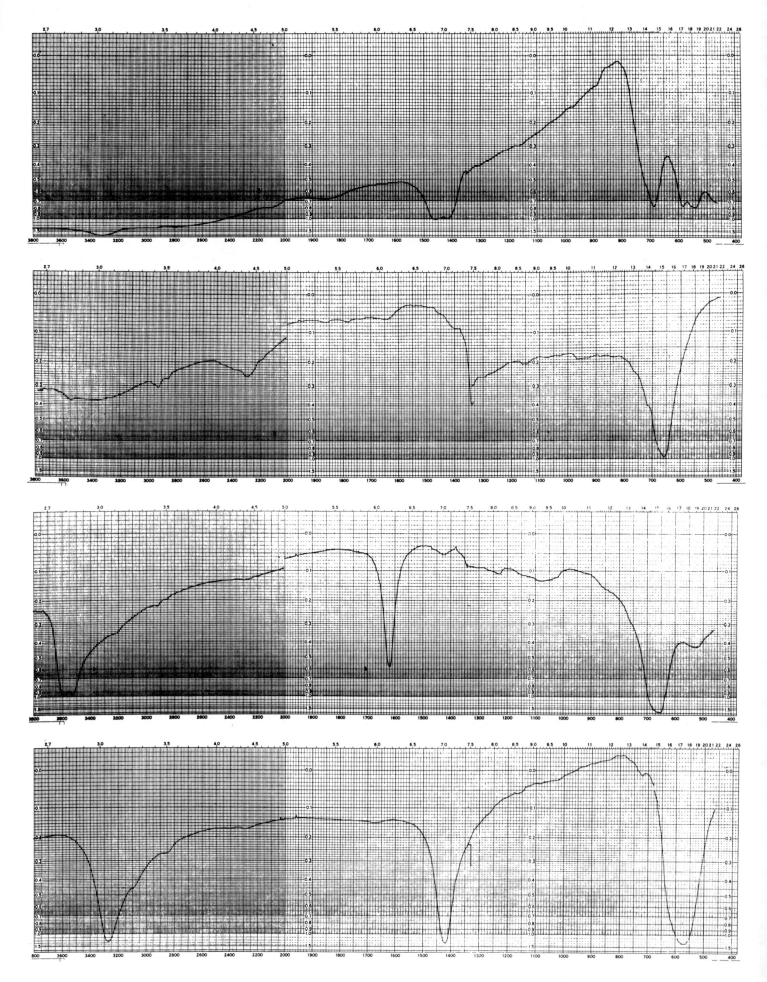

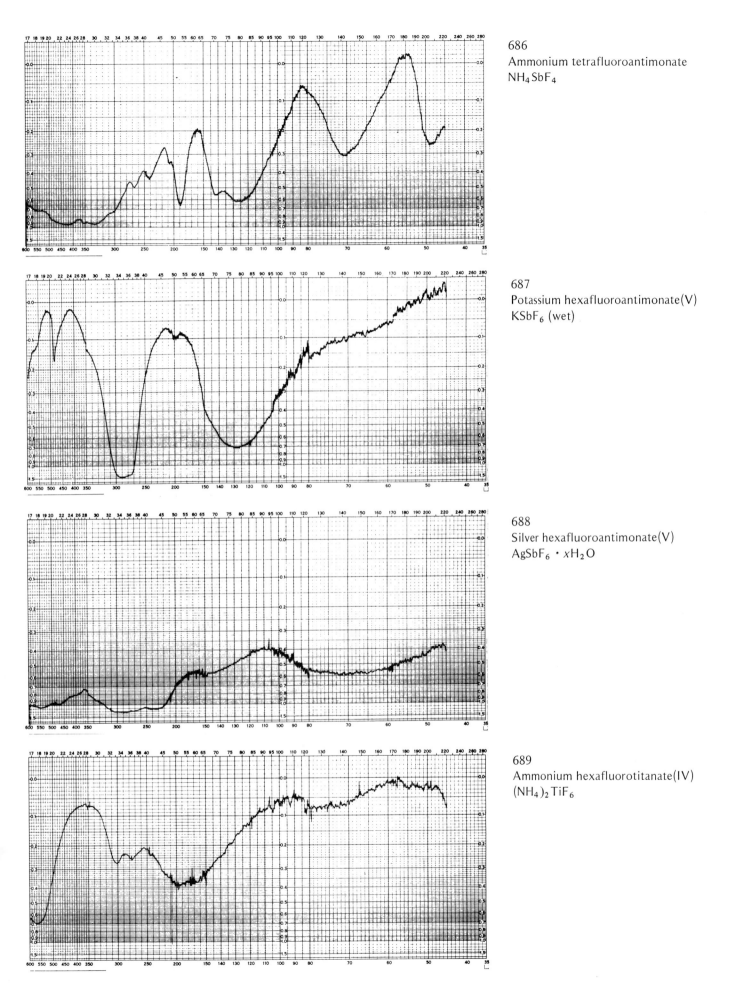

686
Ammonium tetrafluoroantimonate
NH_4SbF_4

687
Potassium hexafluoroantimonate(V)
$KSbF_6$ (wet)

688
Silver hexafluoroantimonate(V)
$AgSbF_6 \cdot xH_2O$

689
Ammonium hexafluorotitanate(IV)
$(NH_4)_2TiF_6$

393

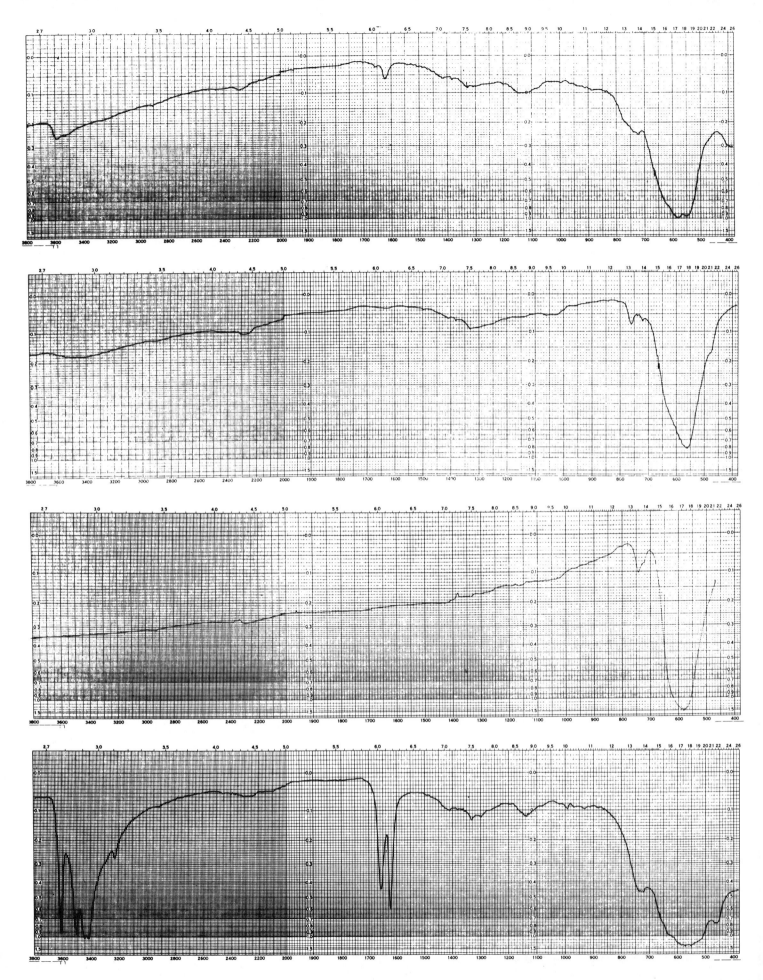

394

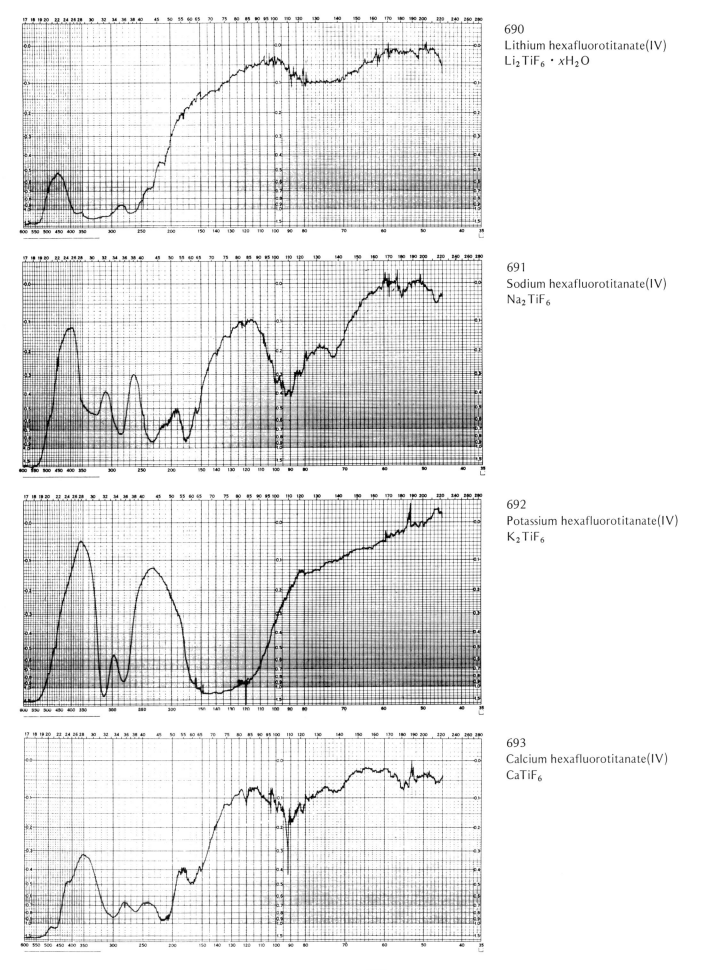

690
Lithium hexafluorotitanate(IV)
$Li_2TiF_6 \cdot xH_2O$

691
Sodium hexafluorotitanate(IV)
Na_2TiF_6

692
Potassium hexafluorotitanate(IV)
K_2TiF_6

693
Calcium hexafluorotitanate(IV)
$CaTiF_6$

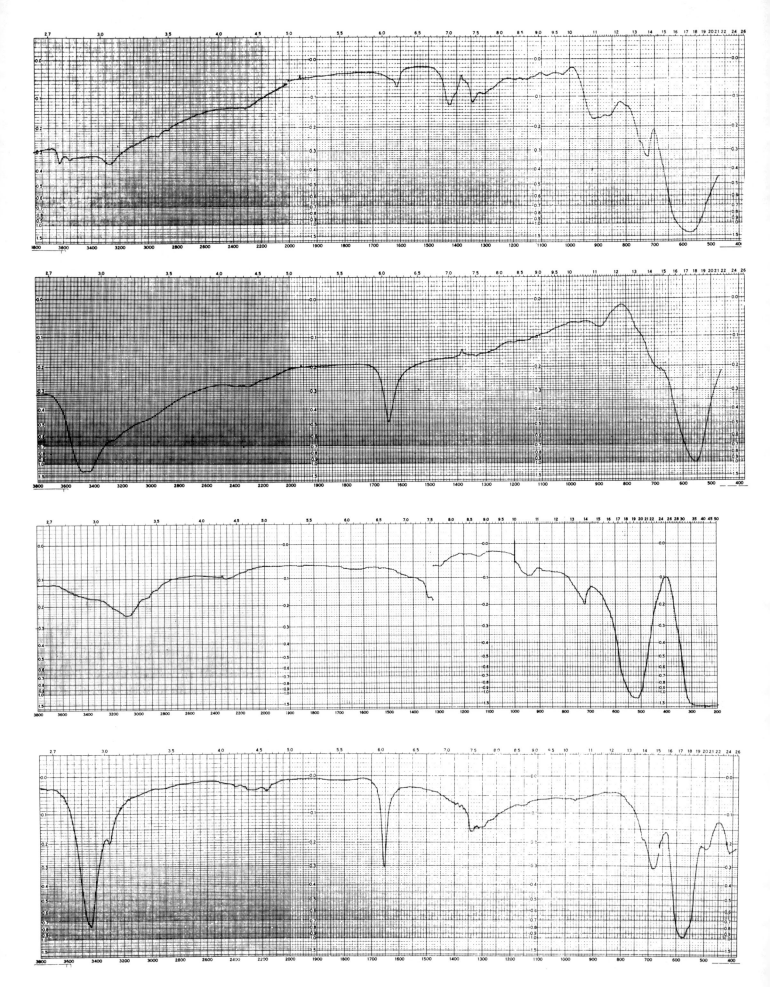

396

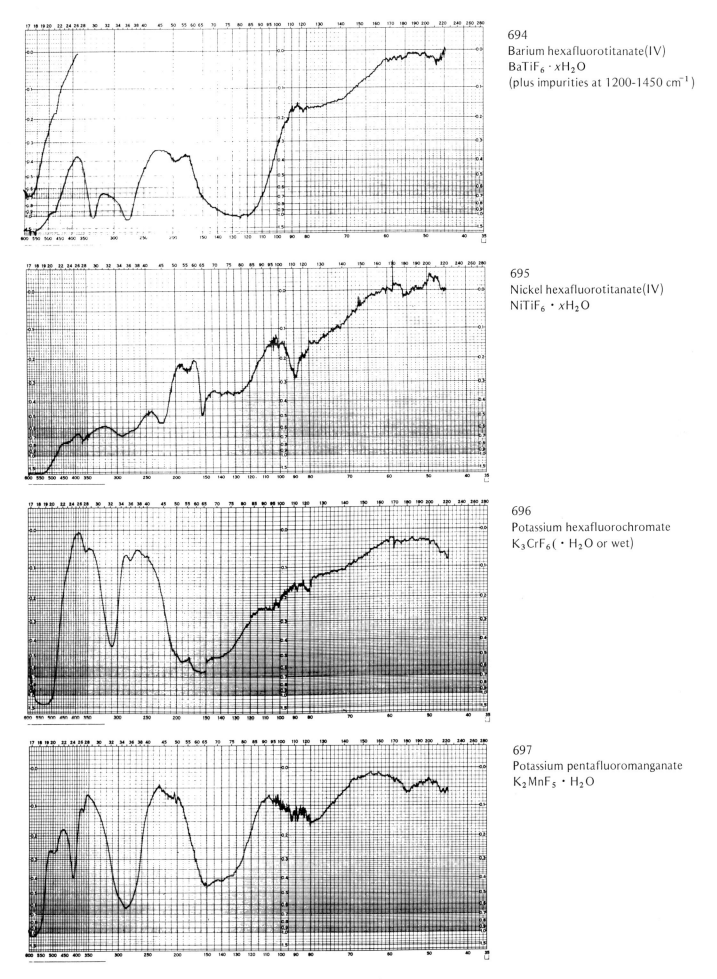

694
Barium hexafluorotitanate(IV)
BaTiF$_6 \cdot x$H$_2$O
(plus impurities at 1200-1450 cm^{-1})

695
Nickel hexafluorotitanate(IV)
NiTiF$_6 \cdot x$H$_2$O

696
Potassium hexafluorochromate
K$_3$CrF$_6$($\cdot$ H$_2$O or wet)

697
Potassium pentafluoromanganate
K$_2$MnF$_5 \cdot$ H$_2$O

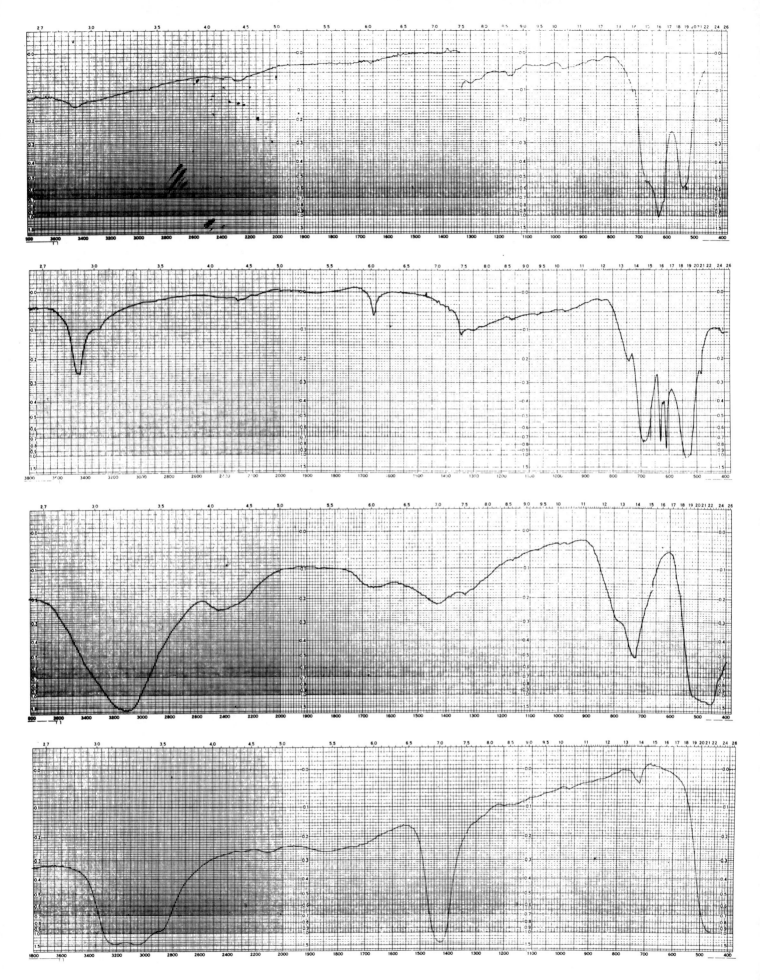

398

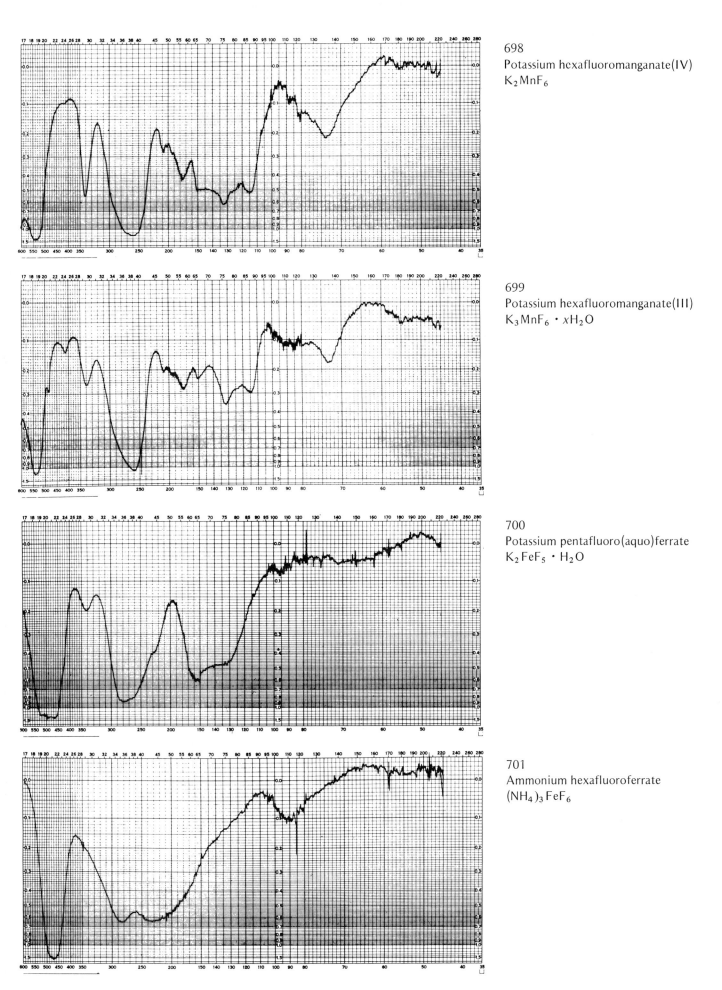

698
Potassium hexafluoromanganate(IV)
K_2MnF_6

699
Potassium hexafluoromanganate(III)
$K_3MnF_6 \cdot xH_2O$

700
Potassium pentafluoro(aquo)ferrate
$K_2FeF_5 \cdot H_2O$

701
Ammonium hexafluoroferrate
$(NH_4)_3FeF_6$

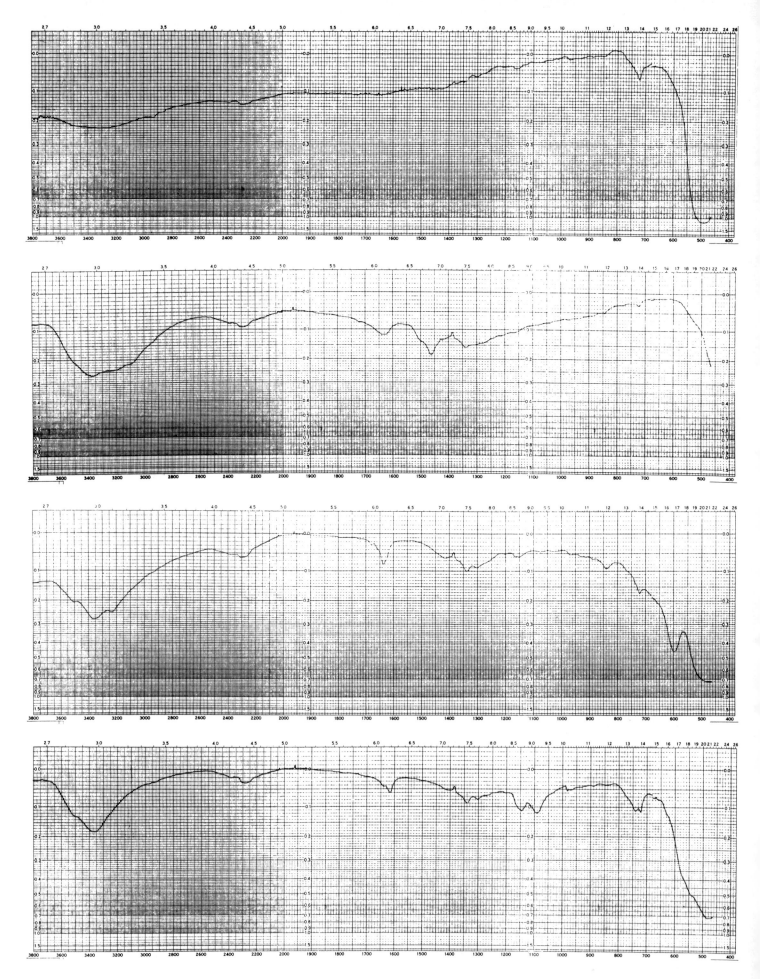

400

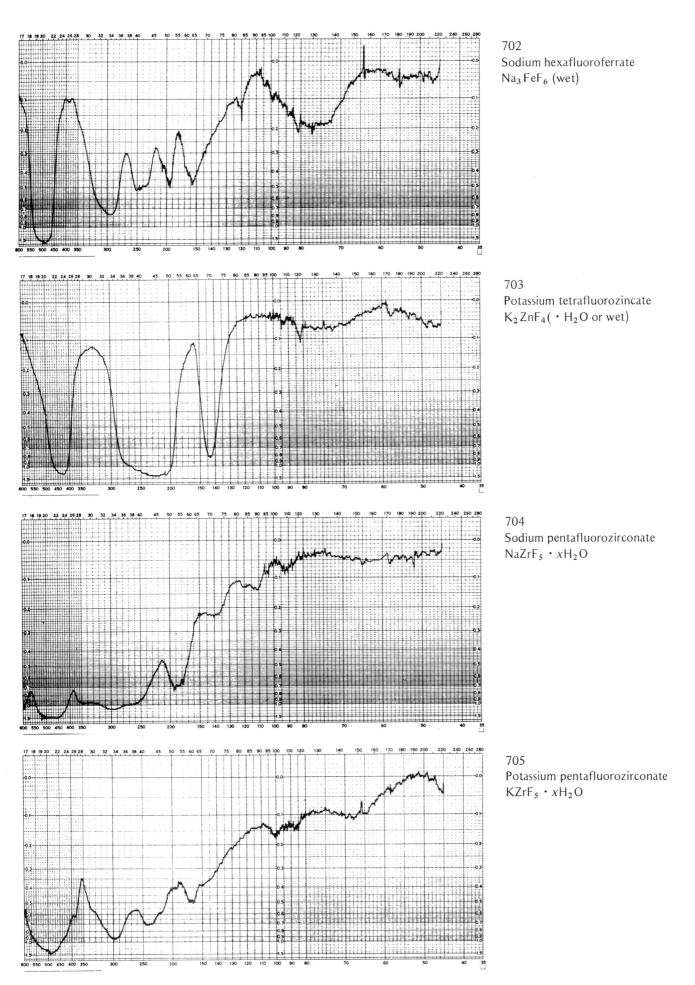

702
Sodium hexafluoroferrate
Na$_3$FeF$_6$ (wet)

703
Potassium tetrafluorozincate
K$_2$ZnF$_4$(· H$_2$O or wet)

704
Sodium pentafluorozirconate
NaZrF$_5$ · xH$_2$O

705
Potassium pentafluorozirconate
KZrF$_5$ · xH$_2$O

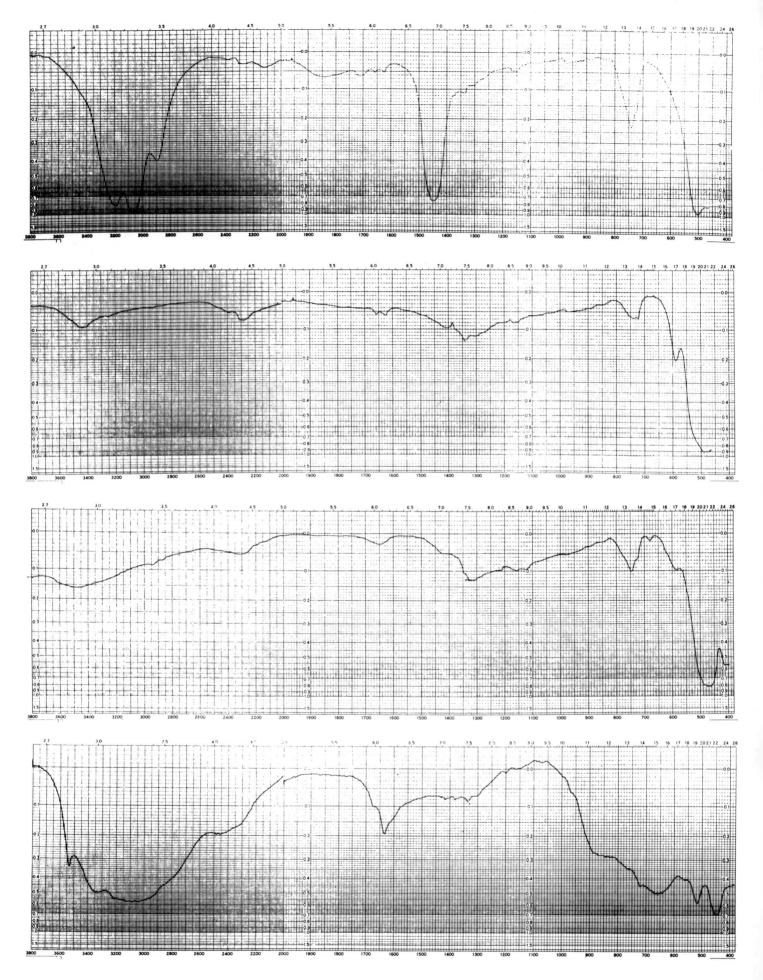

402

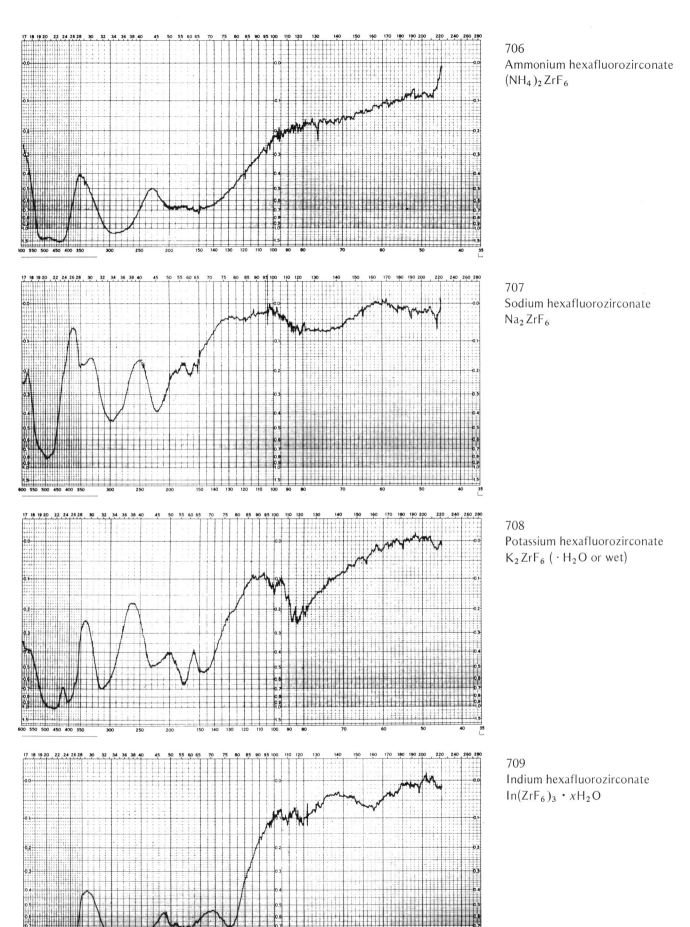

706
Ammonium hexafluorozirconate
$(NH_4)_2 ZrF_6$

707
Sodium hexafluorozirconate
$Na_2 ZrF_6$

708
Potassium hexafluorozirconate
$K_2 ZrF_6$ (· H_2O or wet)

709
Indium hexafluorozirconate
$In(ZrF_6)_3$ · $x H_2O$

403

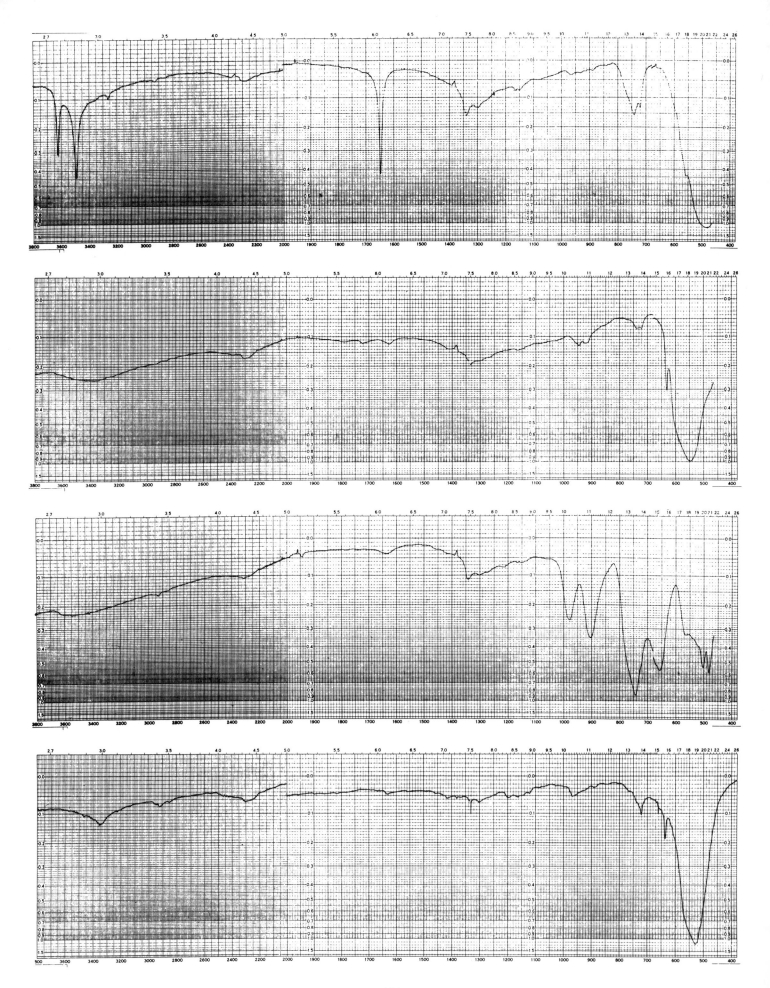

404

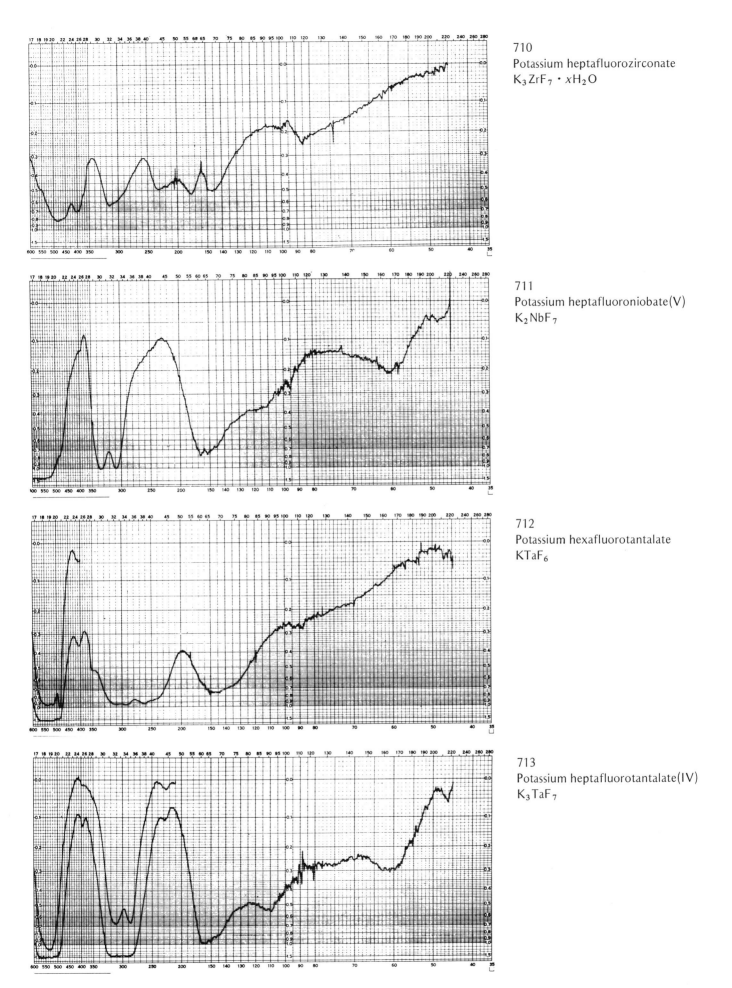

710
Potassium heptafluorozirconate
$K_3ZrF_7 \cdot xH_2O$

711
Potassium heptafluoroniobate(V)
K_2NbF_7

712
Potassium hexafluorotantalate
$KTaF_6$

713
Potassium heptafluorotantalate(IV)
K_3TaF_7

405

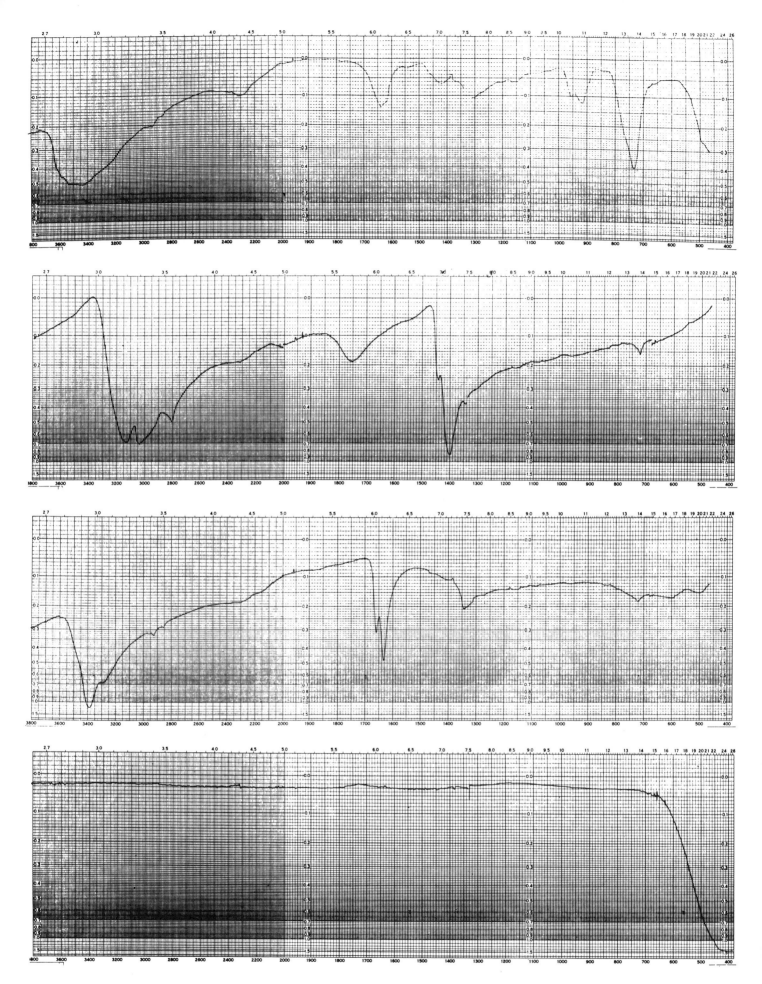

406

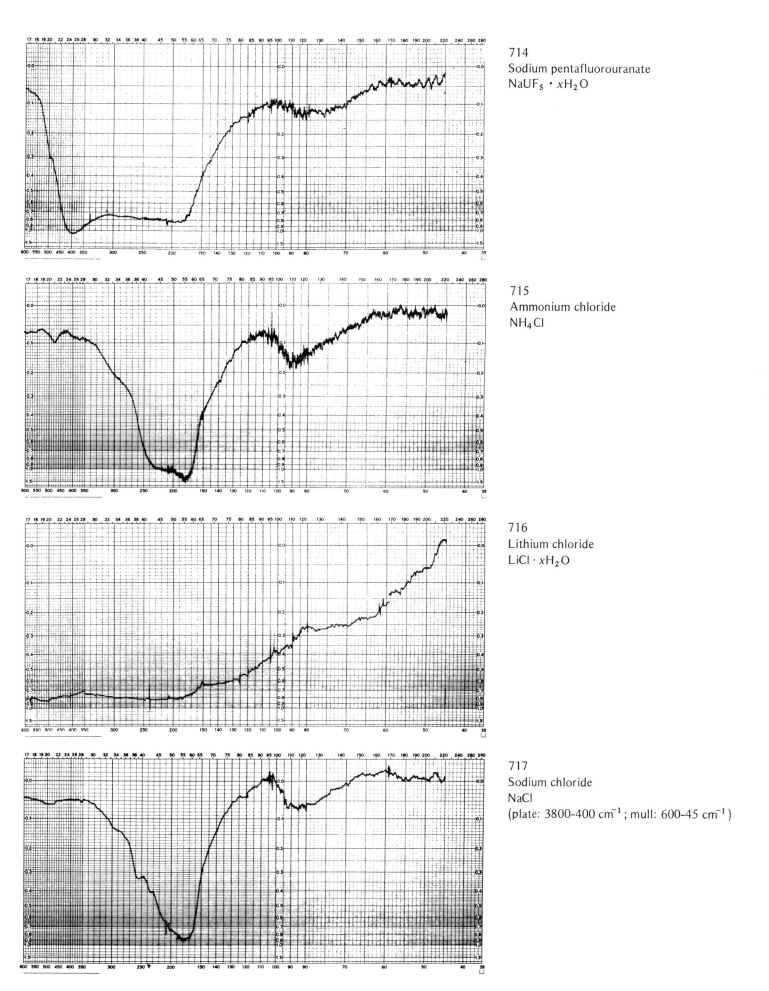

714
Sodium pentafluorouranate
NaUF$_5$ · xH$_2$O

715
Ammonium chloride
NH$_4$Cl

716
Lithium chloride
LiCl · xH$_2$O

717
Sodium chloride
NaCl
(plate: 3800-400 cm^{-1}; mull: 600-45 cm^{-1})

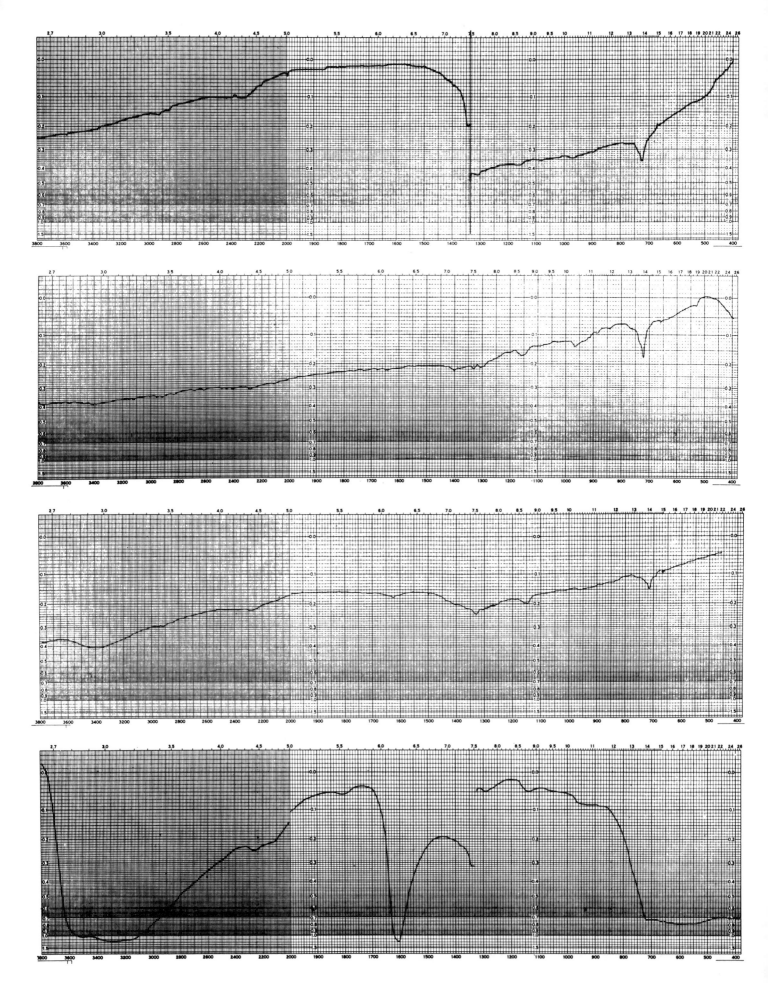

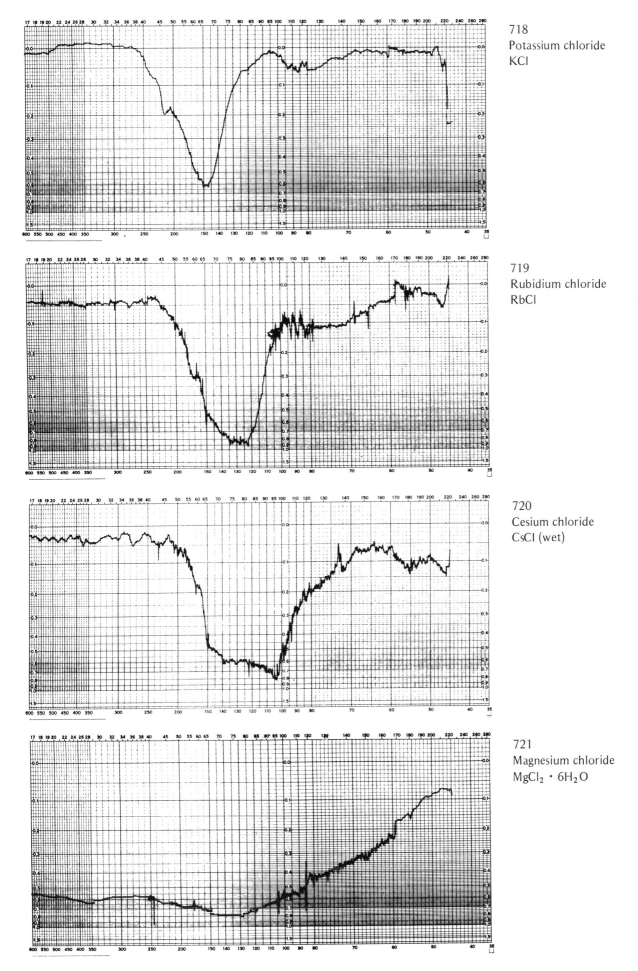

718
Potassium chloride
KCl

719
Rubidium chloride
RbCl

720
Cesium chloride
CsCl (wet)

721
Magnesium chloride
MgCl$_2$ · 6H$_2$O

409

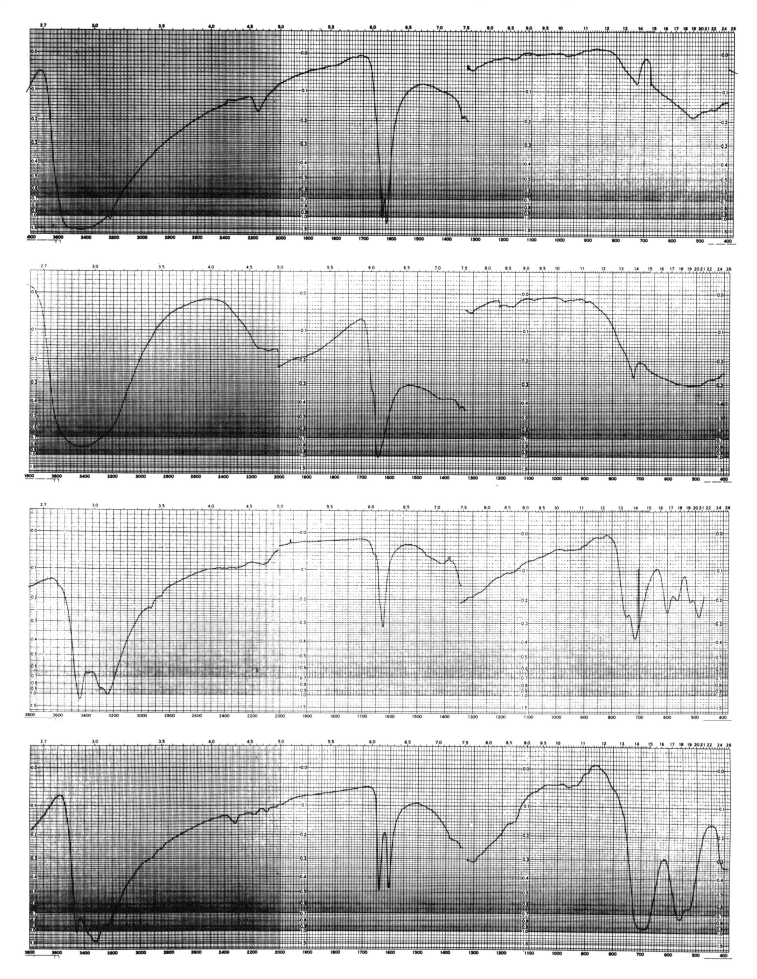

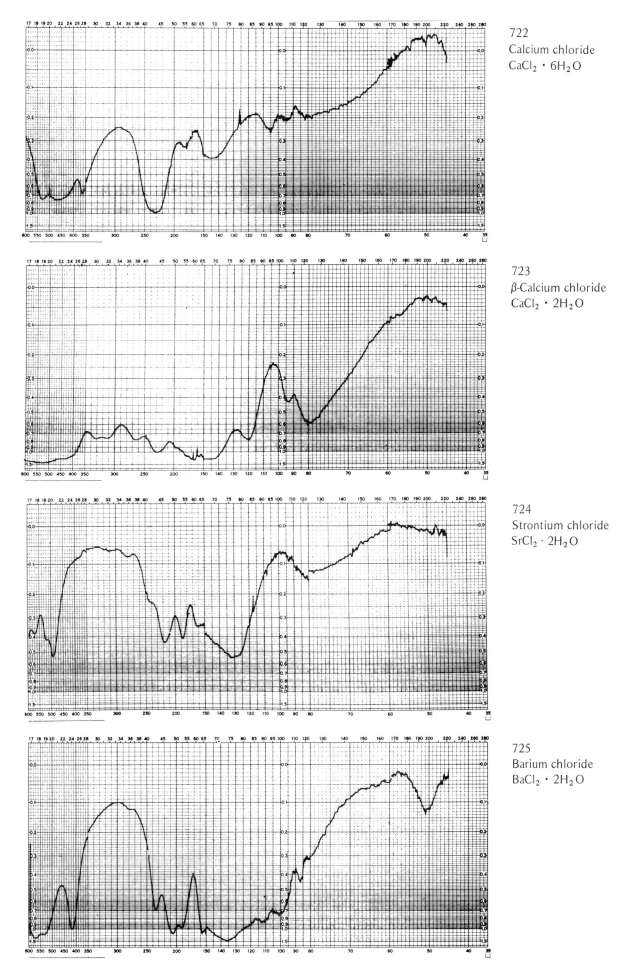

722
Calcium chloride
$CaCl_2 \cdot 6H_2O$

723
β-Calcium chloride
$CaCl_2 \cdot 2H_2O$

724
Strontium chloride
$SrCl_2 \cdot 2H_2O$

725
Barium chloride
$BaCl_2 \cdot 2H_2O$

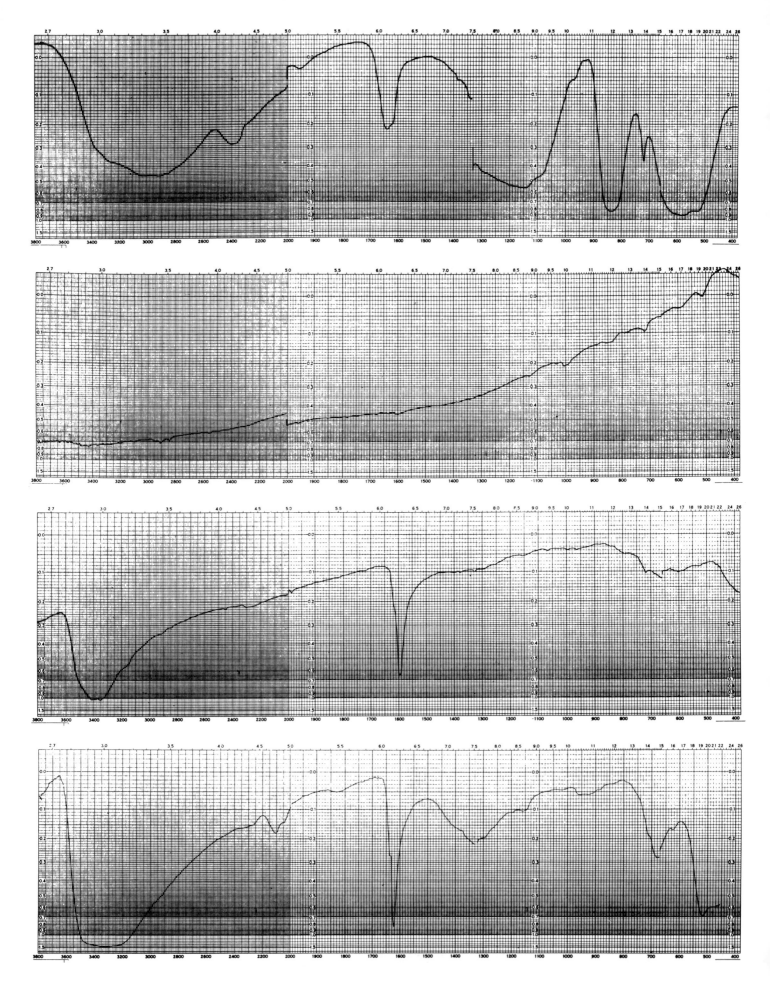

412

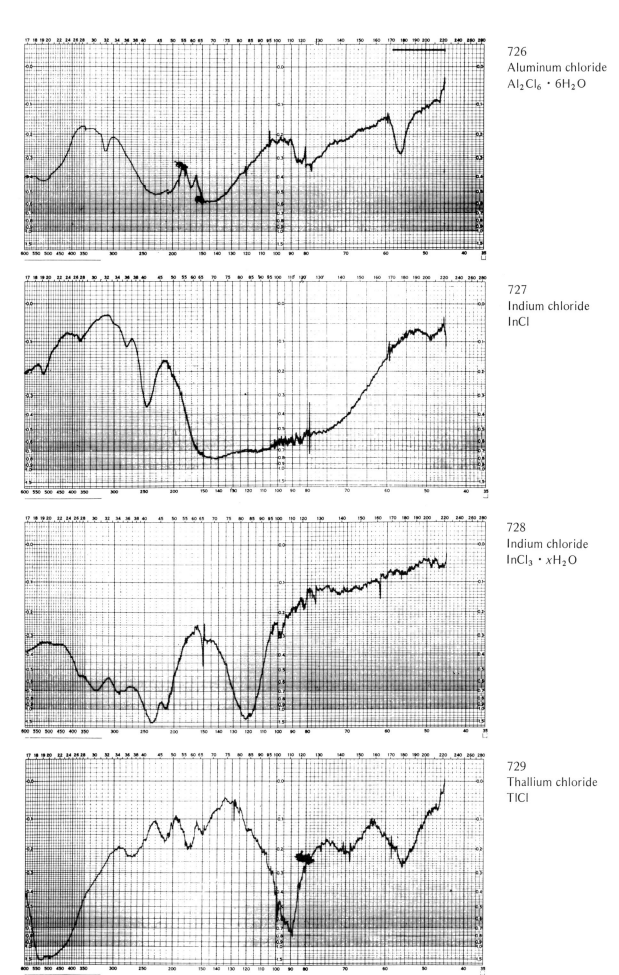

726
Aluminum chloride
$Al_2Cl_6 \cdot 6H_2O$

727
Indium chloride
InCl

728
Indium chloride
$InCl_3 \cdot xH_2O$

729
Thallium chloride
TlCl

413

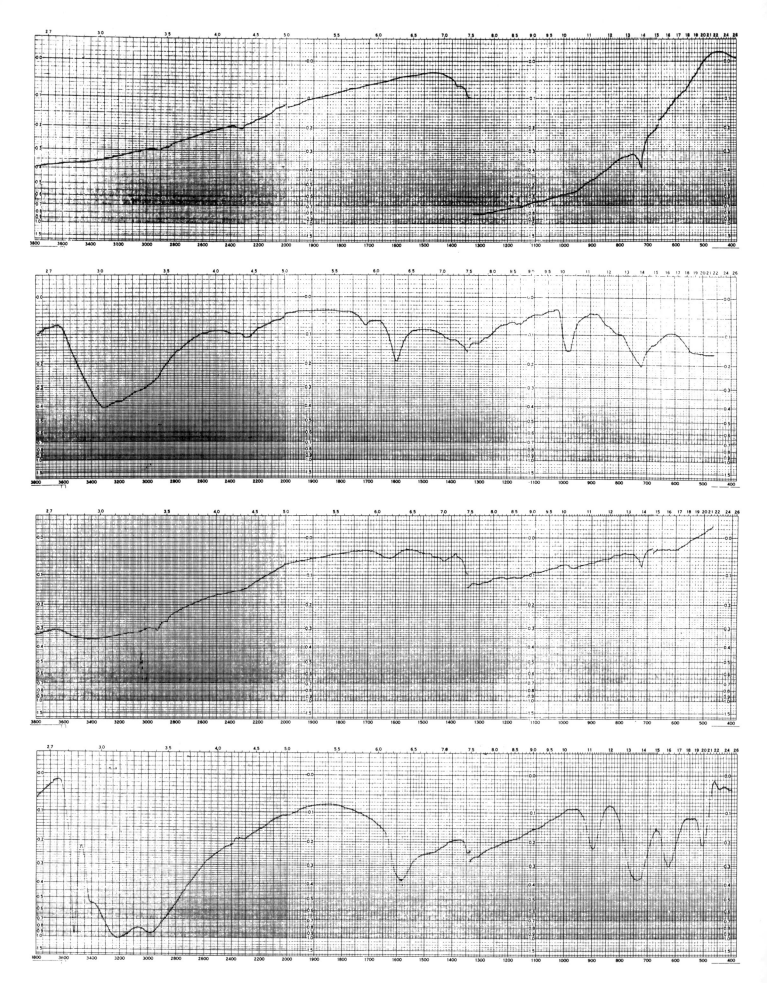

414

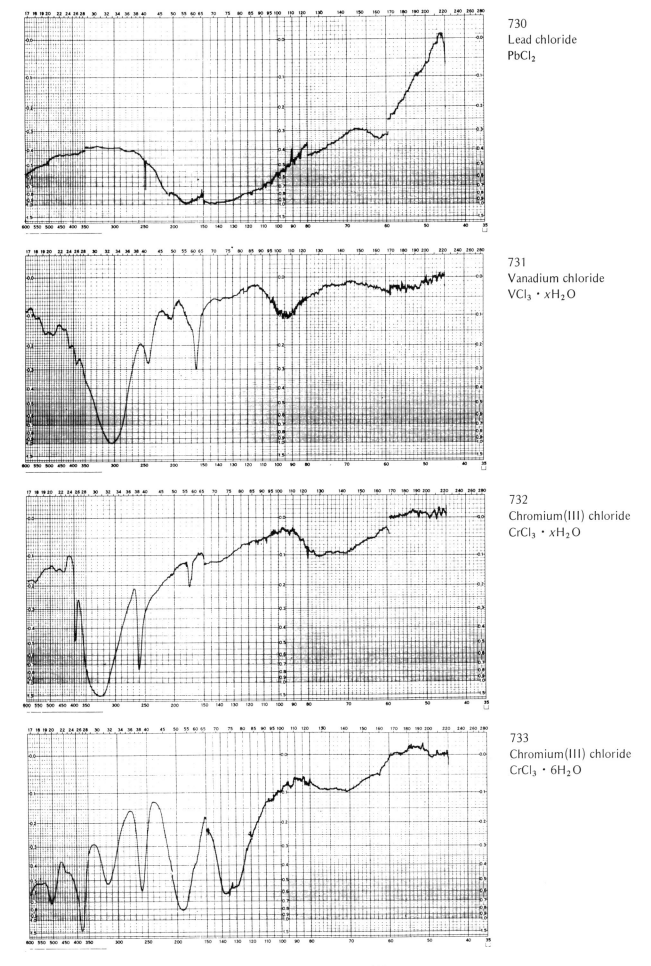

730
Lead chloride
PbCl$_2$

731
Vanadium chloride
VCl$_3$ · xH$_2$O

732
Chromium(III) chloride
CrCl$_3$ · xH$_2$O

733
Chromium(III) chloride
CrCl$_3$ · 6H$_2$O

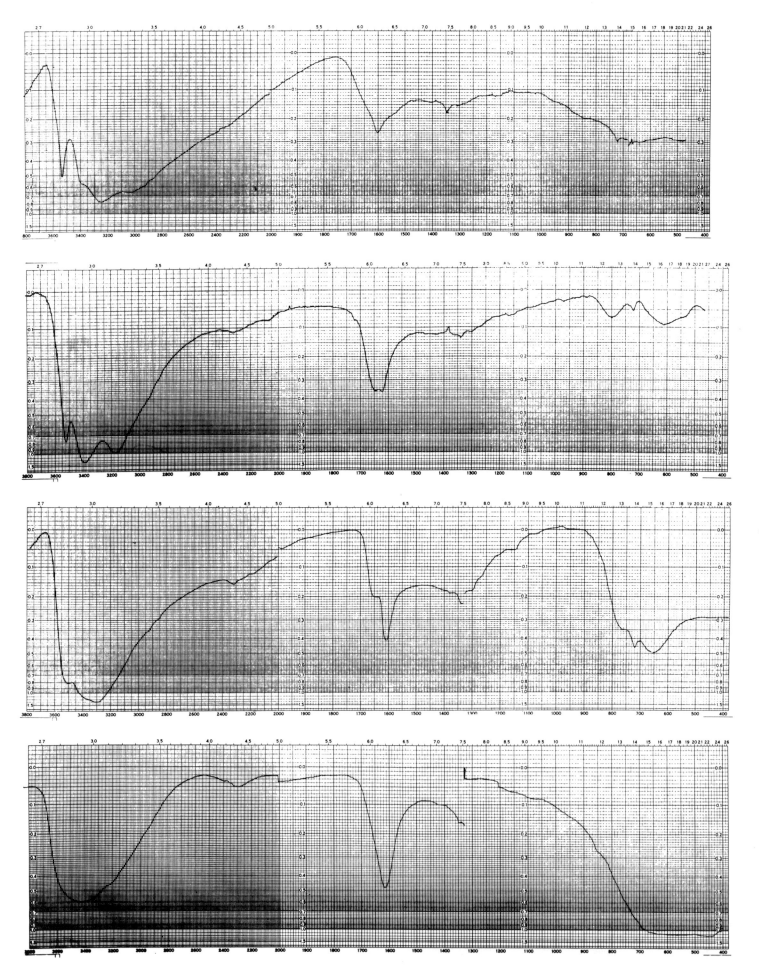

416

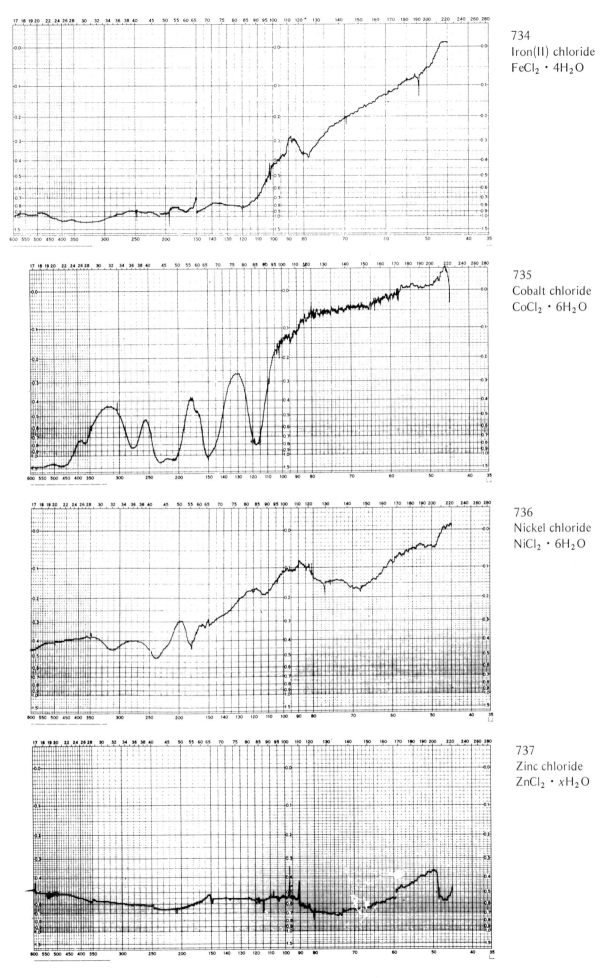

734
Iron(II) chloride
$FeCl_2 \cdot 4H_2O$

735
Cobalt chloride
$CoCl_2 \cdot 6H_2O$

736
Nickel chloride
$NiCl_2 \cdot 6H_2O$

737
Zinc chloride
$ZnCl_2 \cdot xH_2O$

417

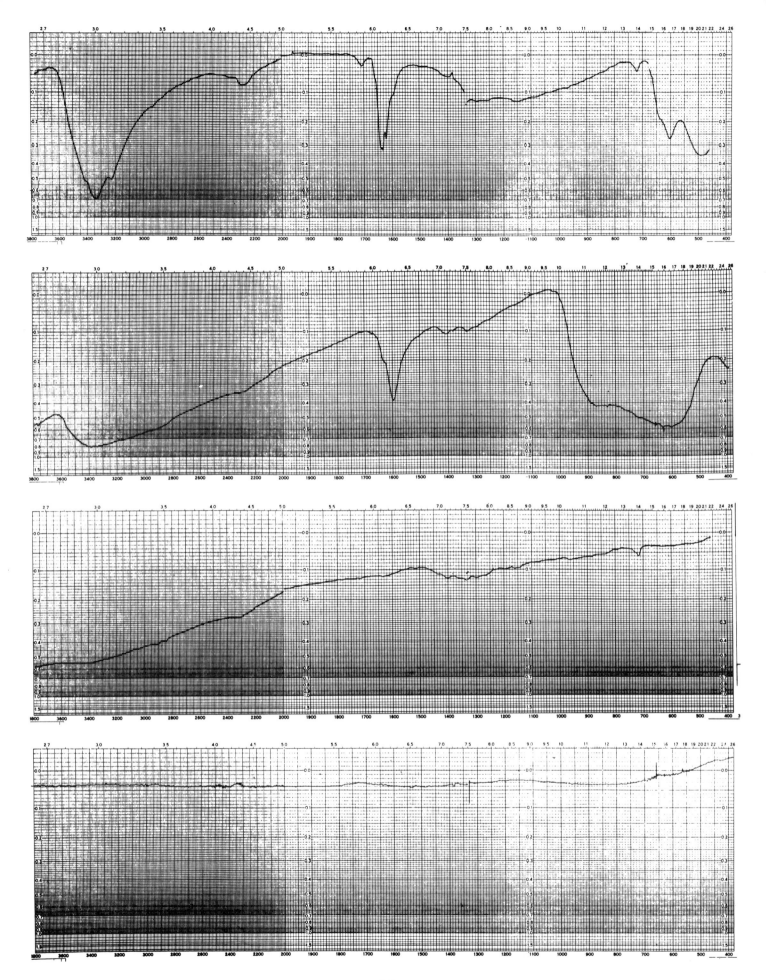

418

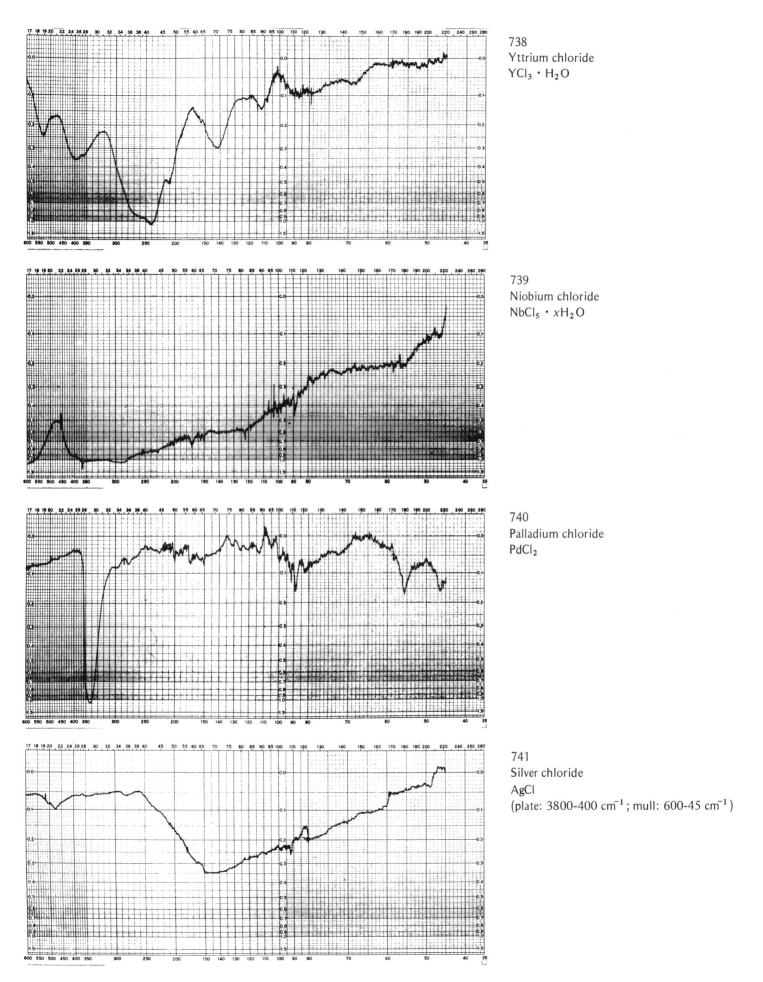

738
Yttrium chloride
$YCl_3 \cdot H_2O$

739
Niobium chloride
$NbCl_5 \cdot xH_2O$

740
Palladium chloride
$PdCl_2$

741
Silver chloride
AgCl
(plate: 3800-400 cm^{-1}; mull: 600-45 cm^{-1})

419

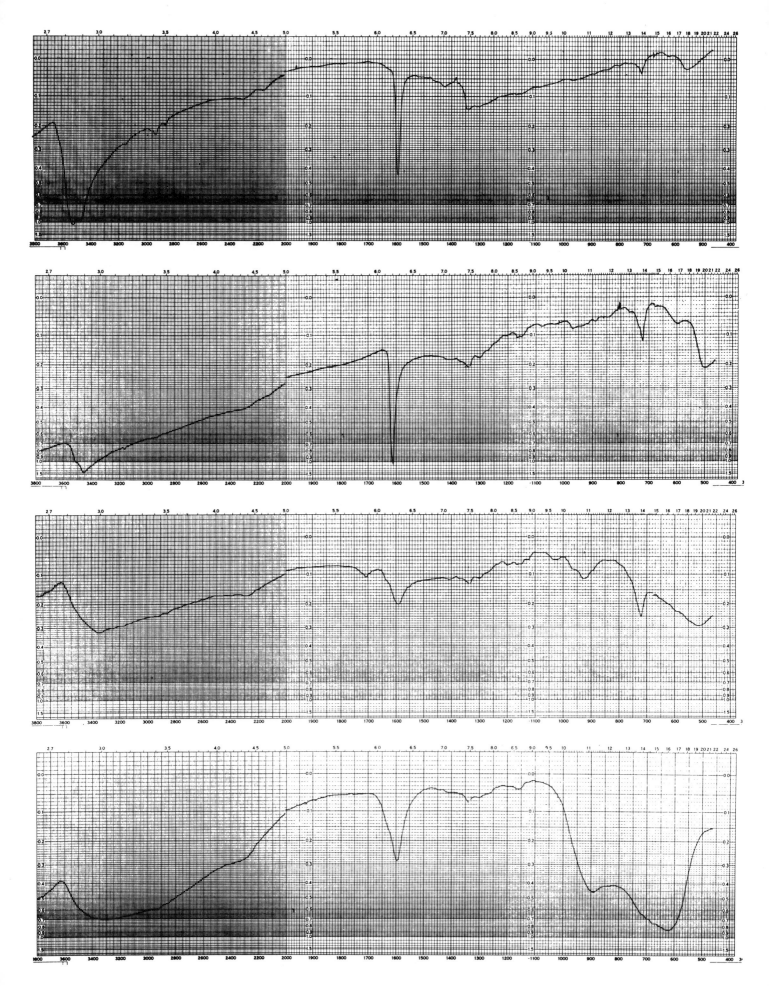

420

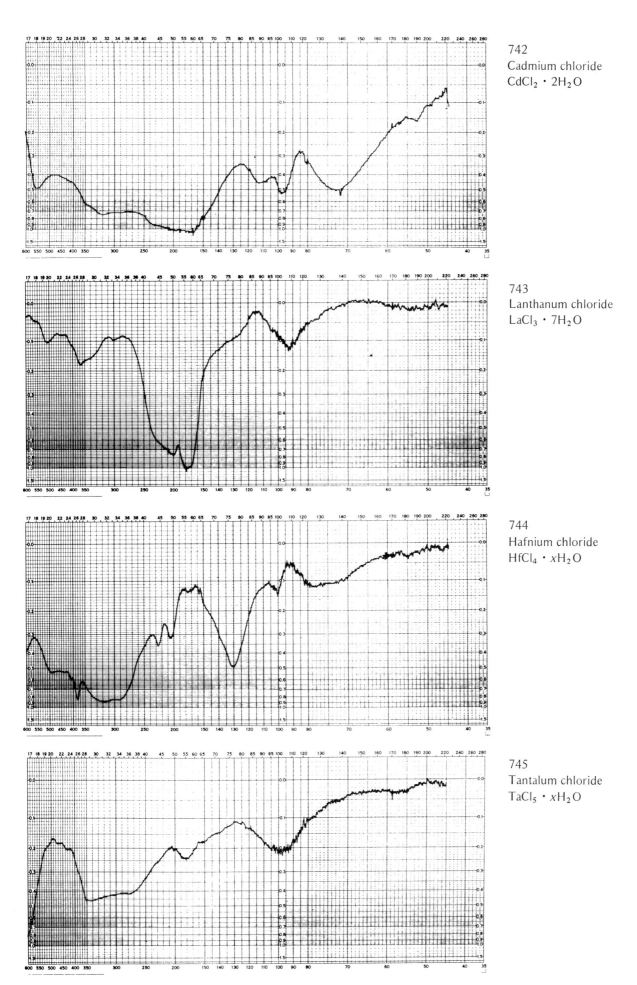

742
Cadmium chloride
$CdCl_2 \cdot 2H_2O$

743
Lanthanum chloride
$LaCl_3 \cdot 7H_2O$

744
Hafnium chloride
$HfCl_4 \cdot xH_2O$

745
Tantalum chloride
$TaCl_5 \cdot xH_2O$

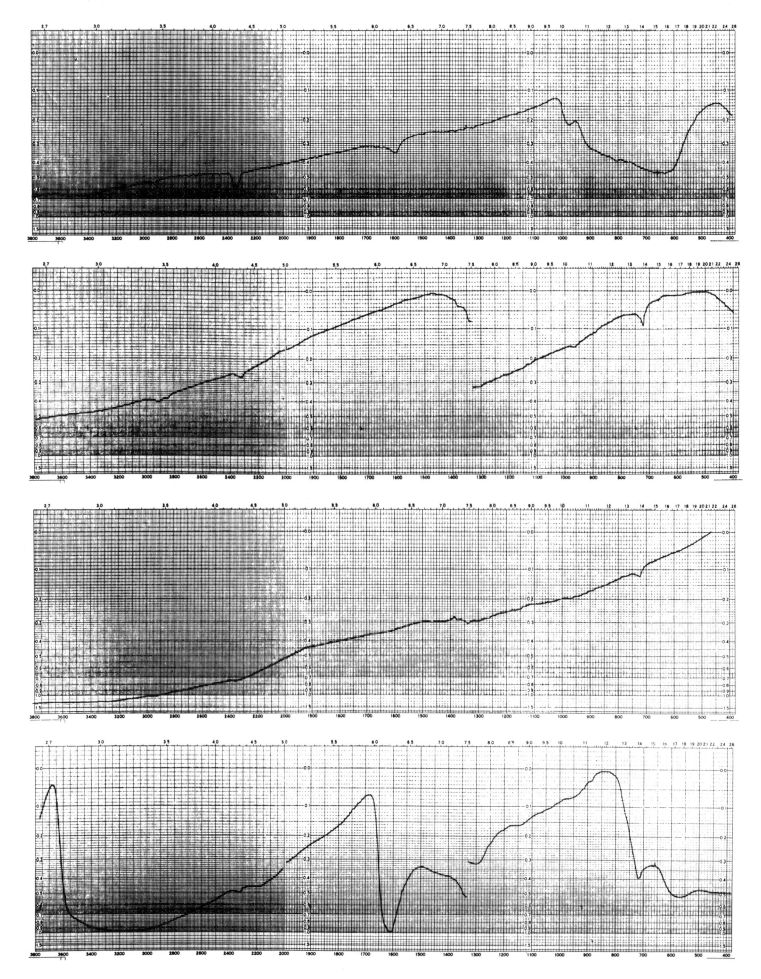

422

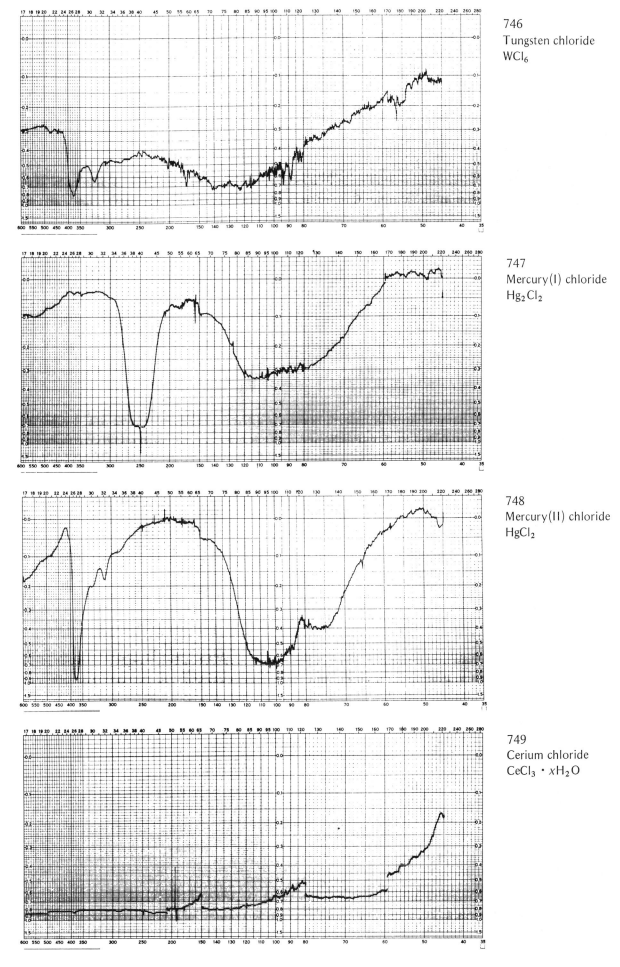

746
Tungsten chloride
WCl$_6$

747
Mercury(I) chloride
Hg$_2$Cl$_2$

748
Mercury(II) chloride
HgCl$_2$

749
Cerium chloride
CeCl$_3$ · xH$_2$O

423

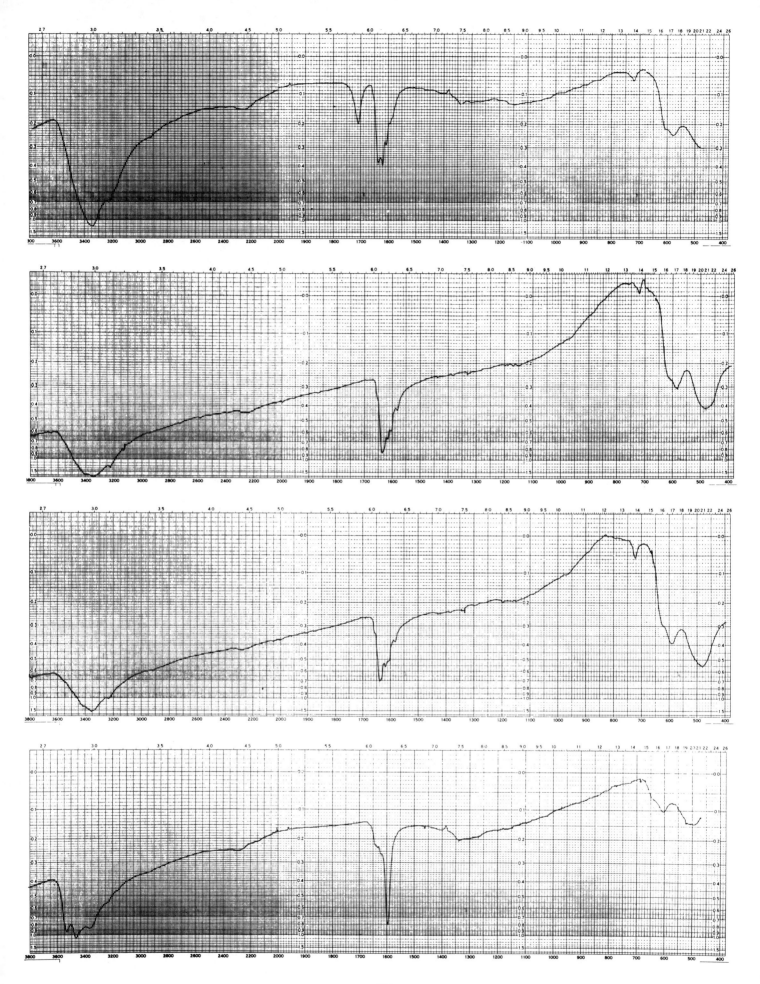

424

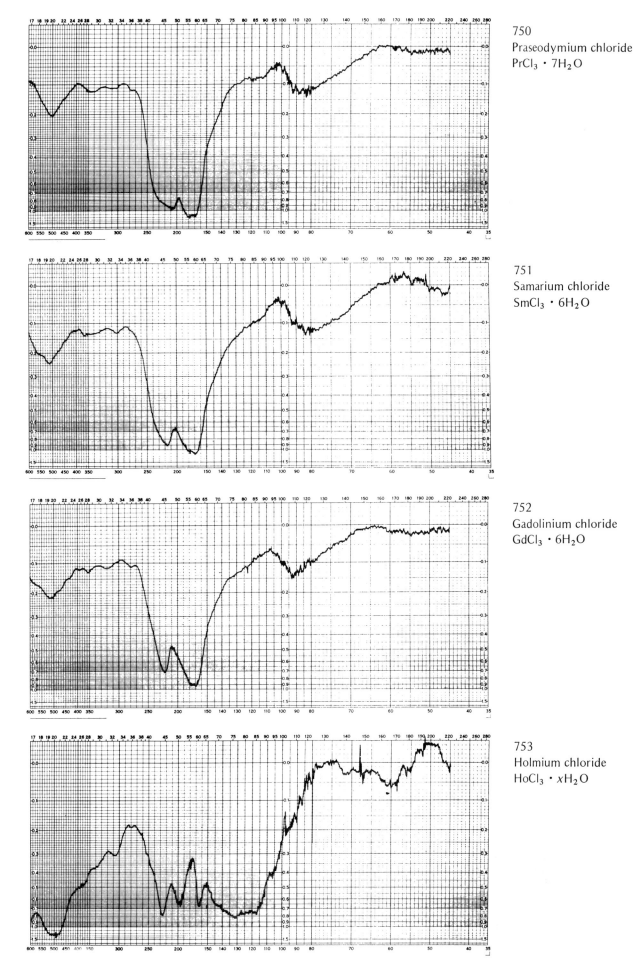

750
Praseodymium chloride
PrCl$_3$ · 7H$_2$O

751
Samarium chloride
SmCl$_3$ · 6H$_2$O

752
Gadolinium chloride
GdCl$_3$ · 6H$_2$O

753
Holmium chloride
HoCl$_3$ · xH$_2$O

425

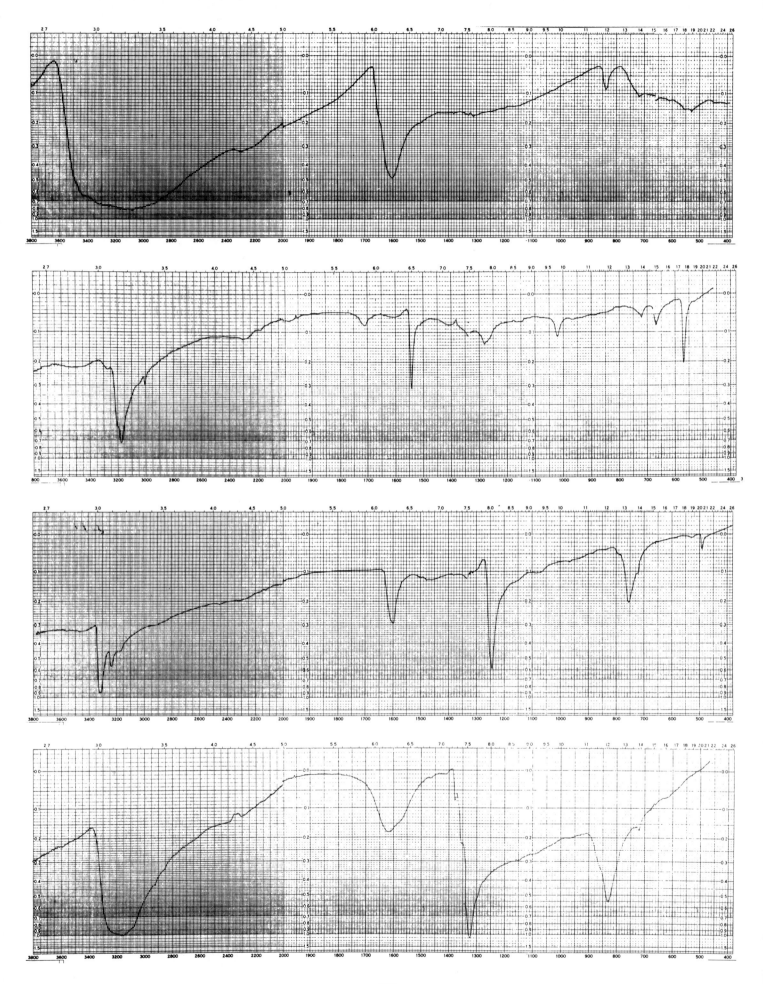

426

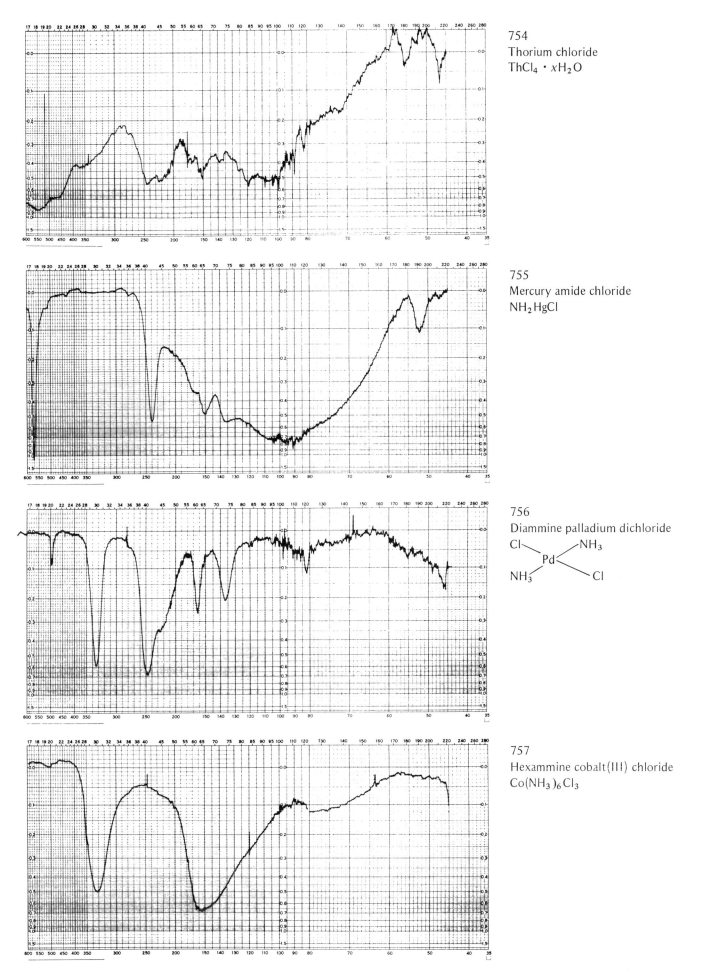

754
Thorium chloride
$ThCl_4 \cdot xH_2O$

755
Mercury amide chloride
NH_2HgCl

756
Diammine palladium dichloride

757
Hexammine cobalt(III) chloride
$Co(NH_3)_6Cl_3$

427

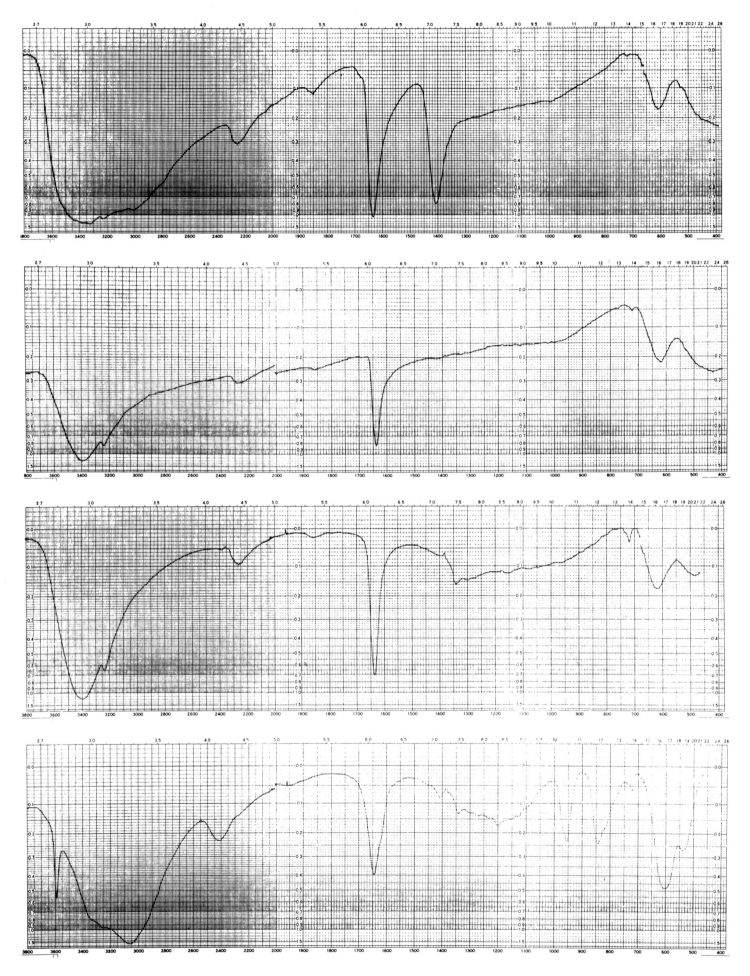

428

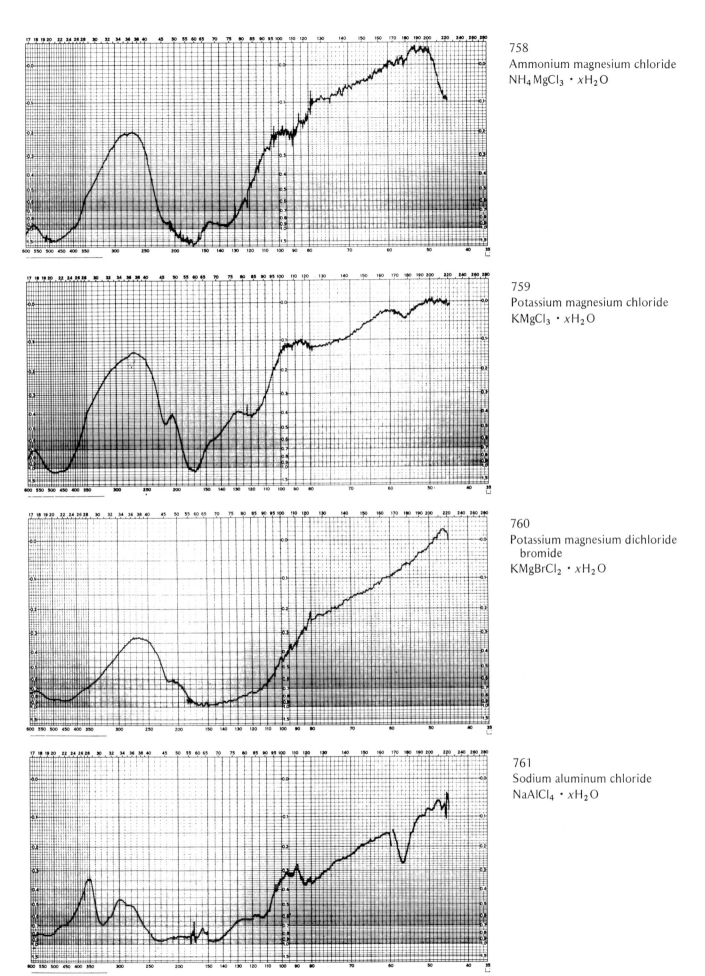

758
Ammonium magnesium chloride
$NH_4MgCl_3 \cdot xH_2O$

759
Potassium magnesium chloride
$KMgCl_3 \cdot xH_2O$

760
Potassium magnesium dichloride
bromide
$KMgBrCl_2 \cdot xH_2O$

761
Sodium aluminum chloride
$NaAlCl_4 \cdot xH_2O$

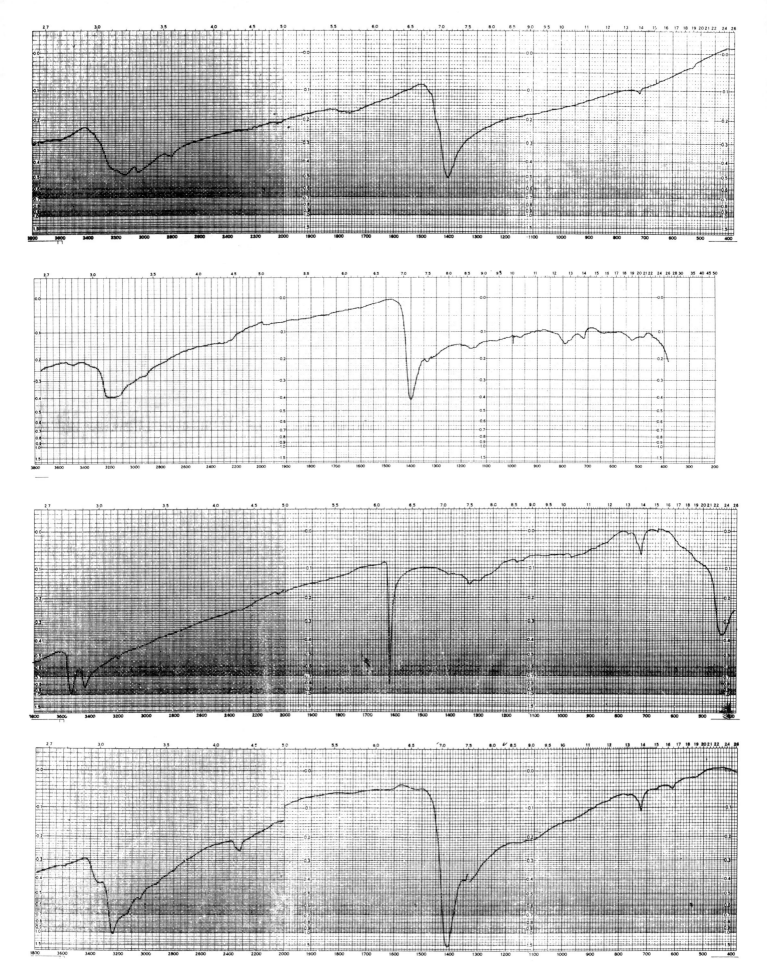

430

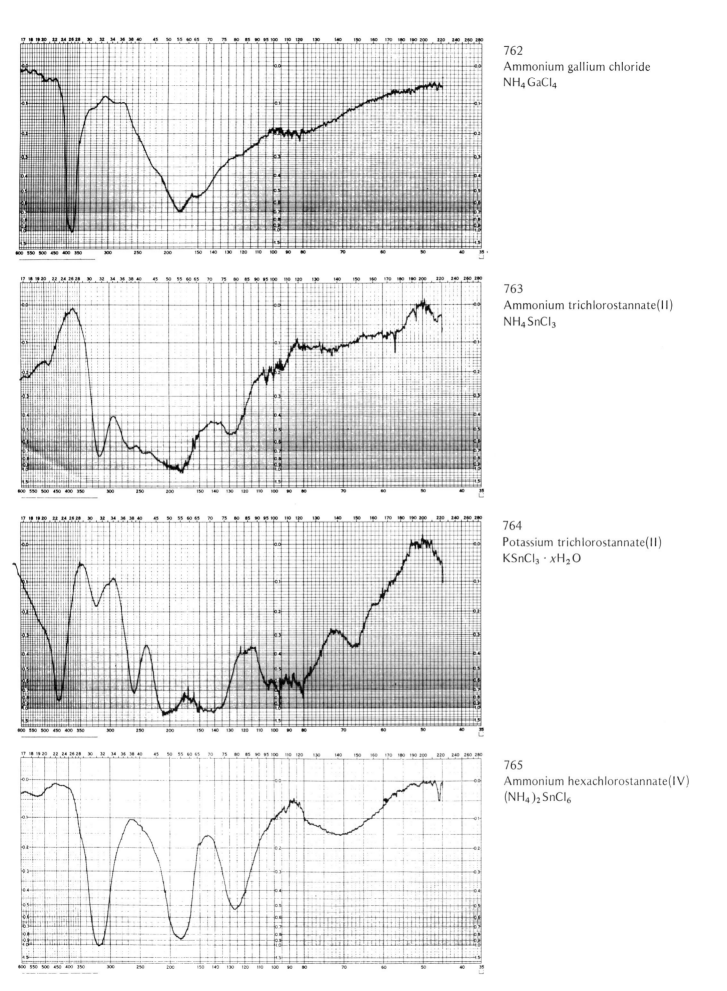

762
Ammonium gallium chloride
$NH_4 GaCl_4$

763
Ammonium trichlorostannate(II)
$NH_4 SnCl_3$

764
Potassium trichlorostannate(II)
$KSnCl_3 \cdot xH_2O$

765
Ammonium hexachlorostannate(IV)
$(NH_4)_2 SnCl_6$

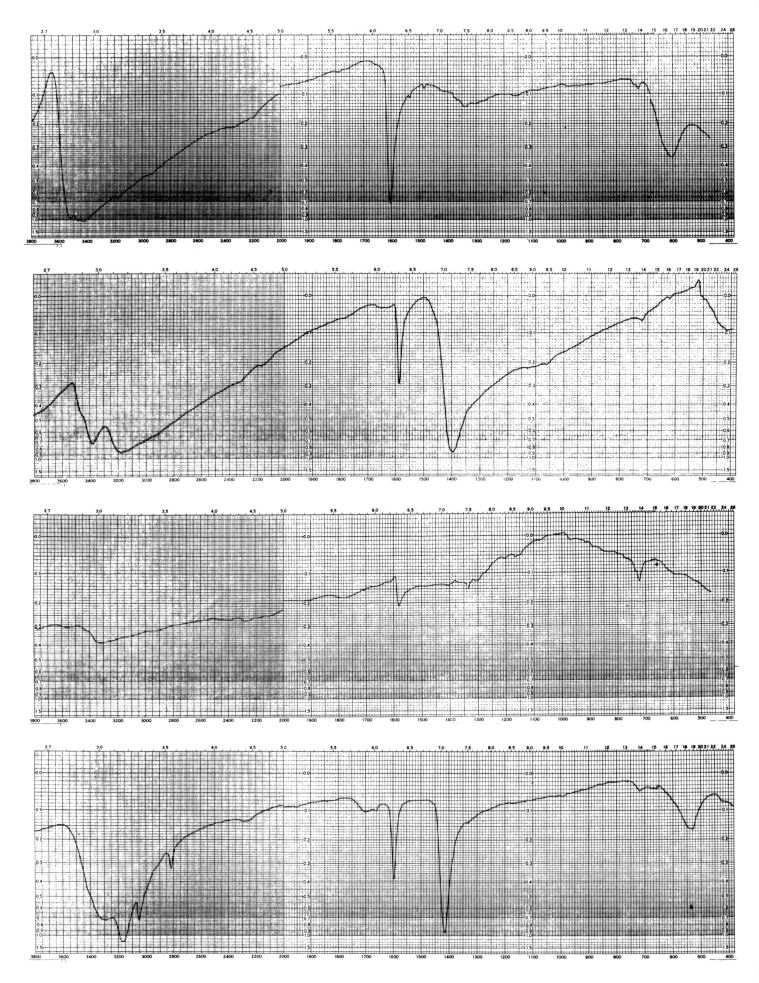

432

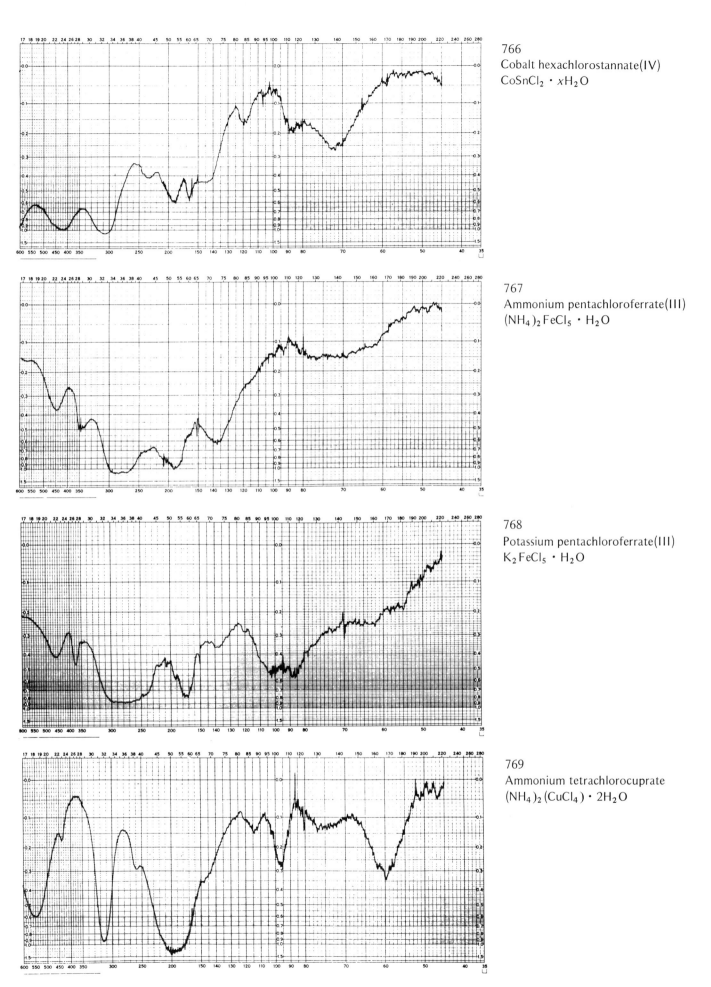

766
Cobalt hexachlorostannate(IV)
$CoSnCl_2 \cdot xH_2O$

767
Ammonium pentachloroferrate(III)
$(NH_4)_2FeCl_5 \cdot H_2O$

768
Potassium pentachloroferrate(III)
$K_2FeCl_5 \cdot H_2O$

769
Ammonium tetrachlorocuprate
$(NH_4)_2(CuCl_4) \cdot 2H_2O$

433

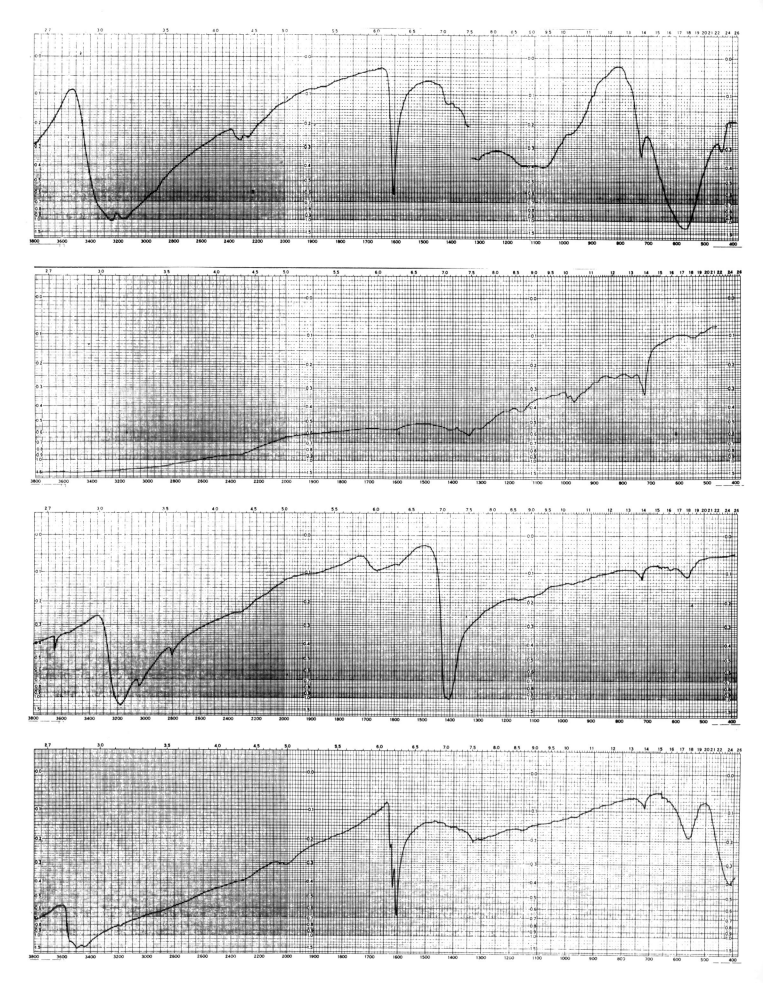

434

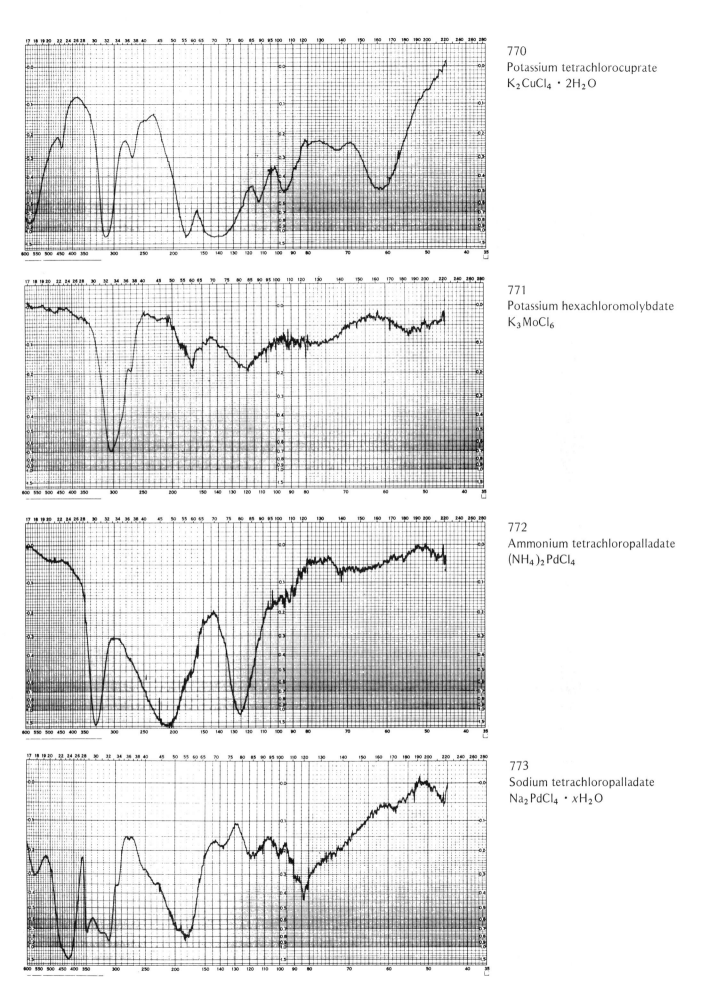

770
Potassium tetrachlorocuprate
$K_2CuCl_4 \cdot 2H_2O$

771
Potassium hexachloromolybdate
K_3MoCl_6

772
Ammonium tetrachloropalladate
$(NH_4)_2PdCl_4$

773
Sodium tetrachloropalladate
$Na_2PdCl_4 \cdot xH_2O$

435

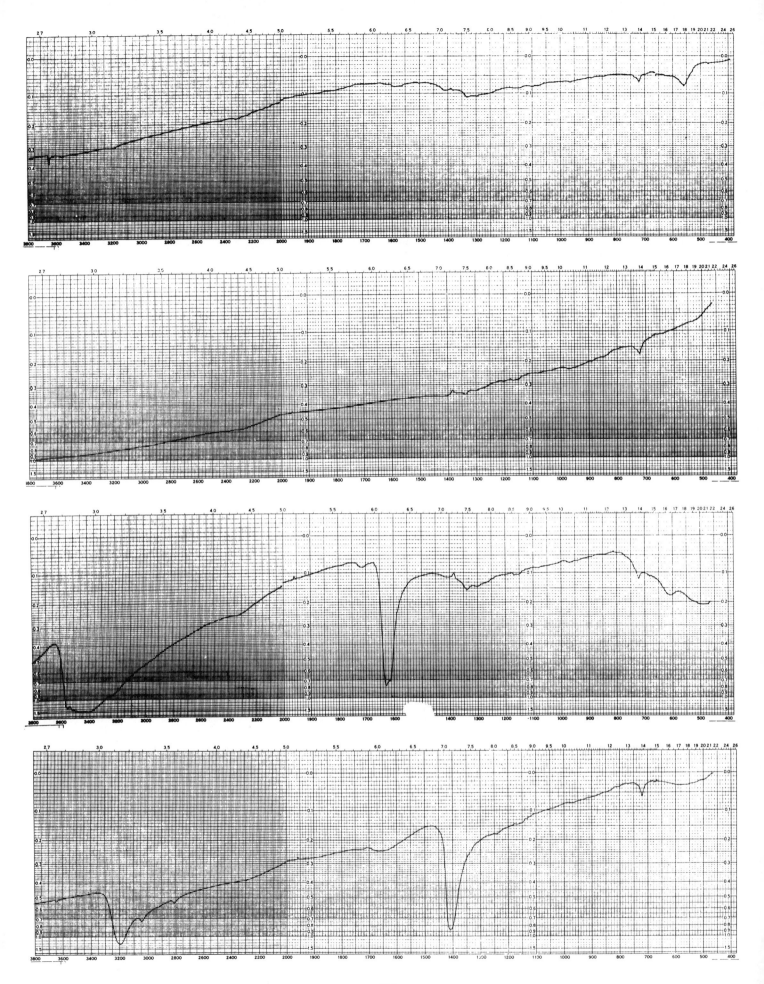

436

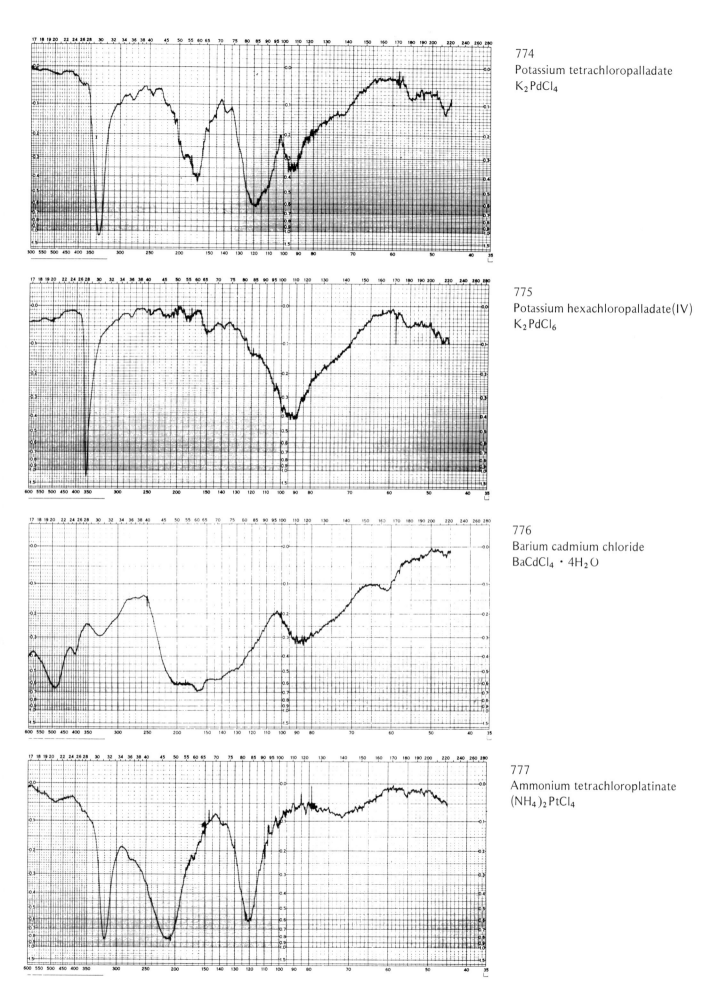

774
Potassium tetrachloropalladate
K₂PdCl₄

775
Potassium hexachloropalladate(IV)
K₂PdCl₆

776
Barium cadmium chloride
BaCdCl₄ · 4H₂O

777
Ammonium tetrachloroplatinate
(NH₄)₂PtCl₄

437

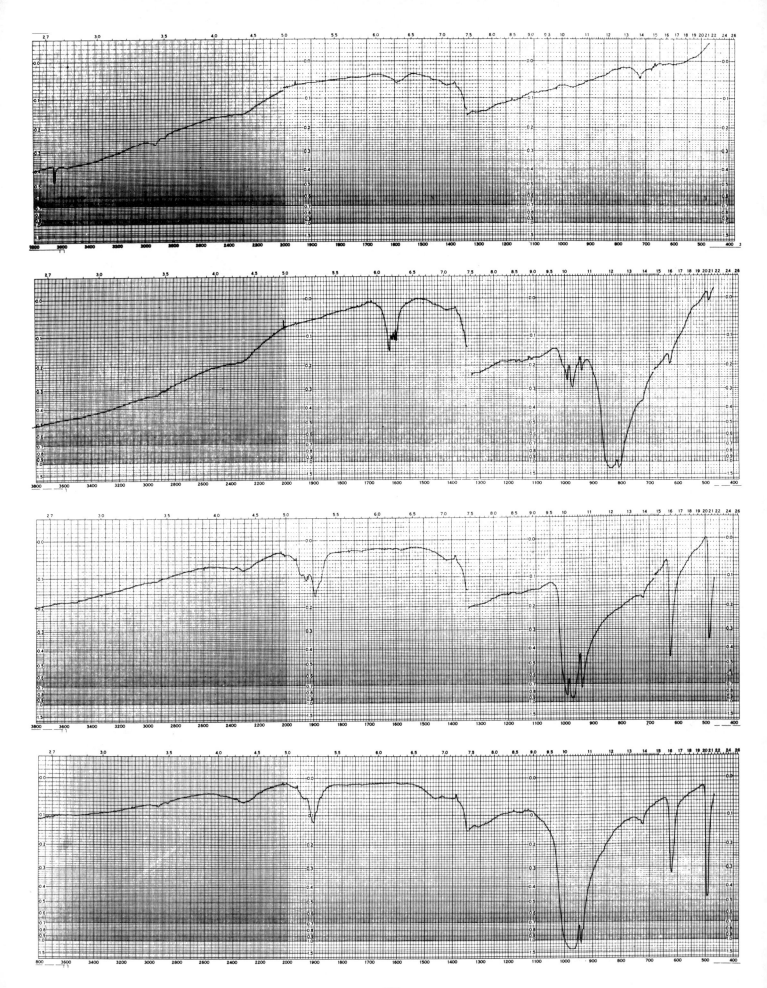

438

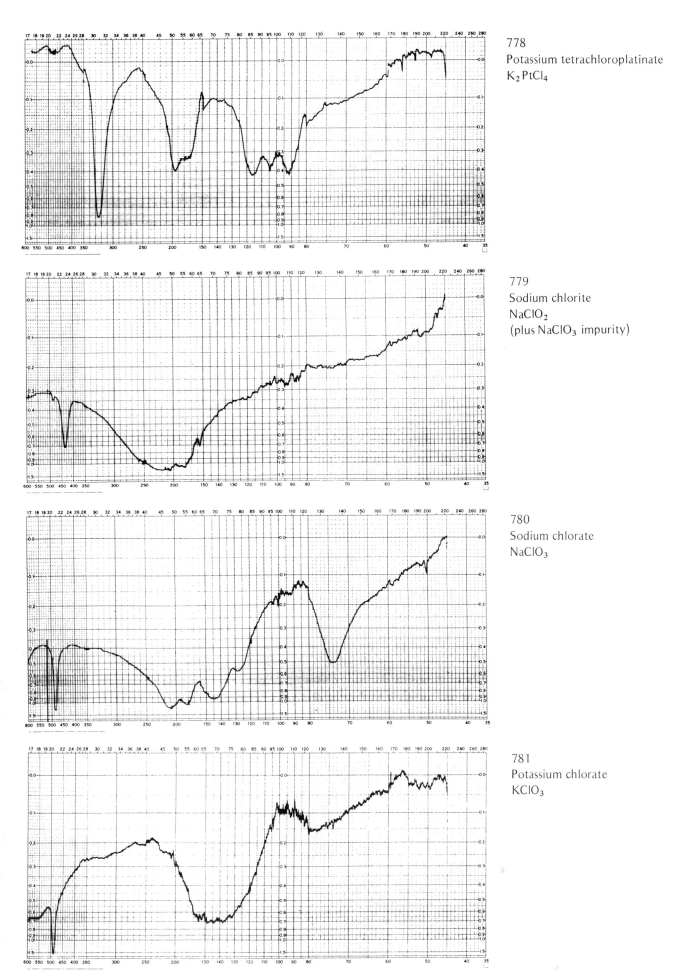

778
Potassium tetrachloroplatinate
K$_2$PtCl$_4$

779
Sodium chlorite
NaClO$_2$
(plus NaClO$_3$ impurity)

780
Sodium chlorate
NaClO$_3$

781
Potassium chlorate
KClO$_3$

439

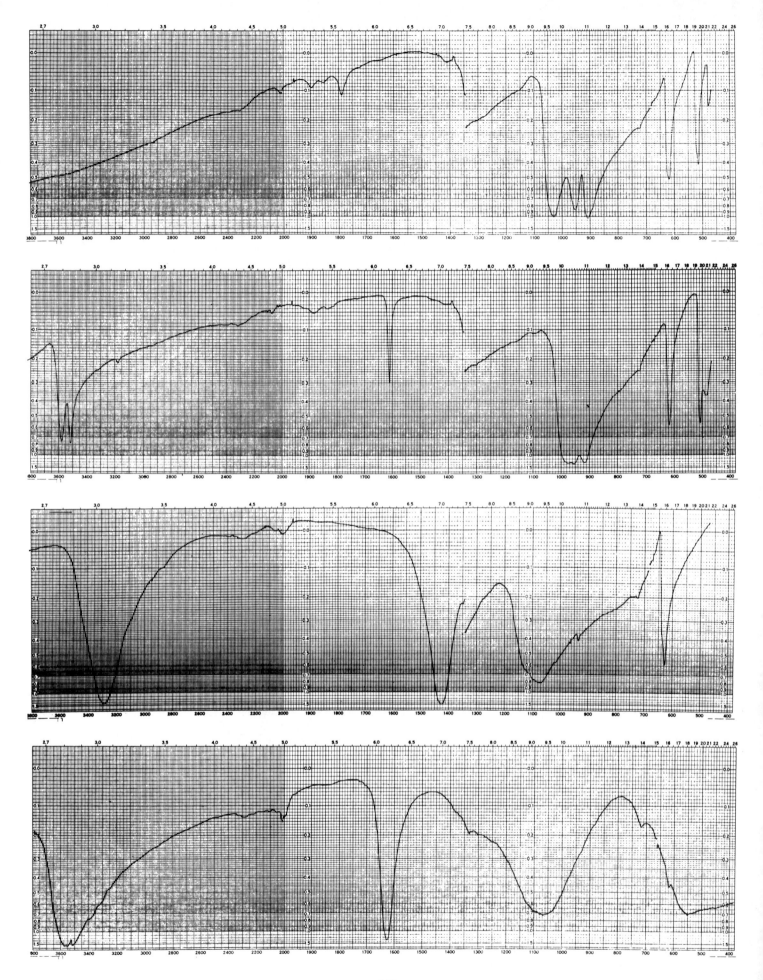

440

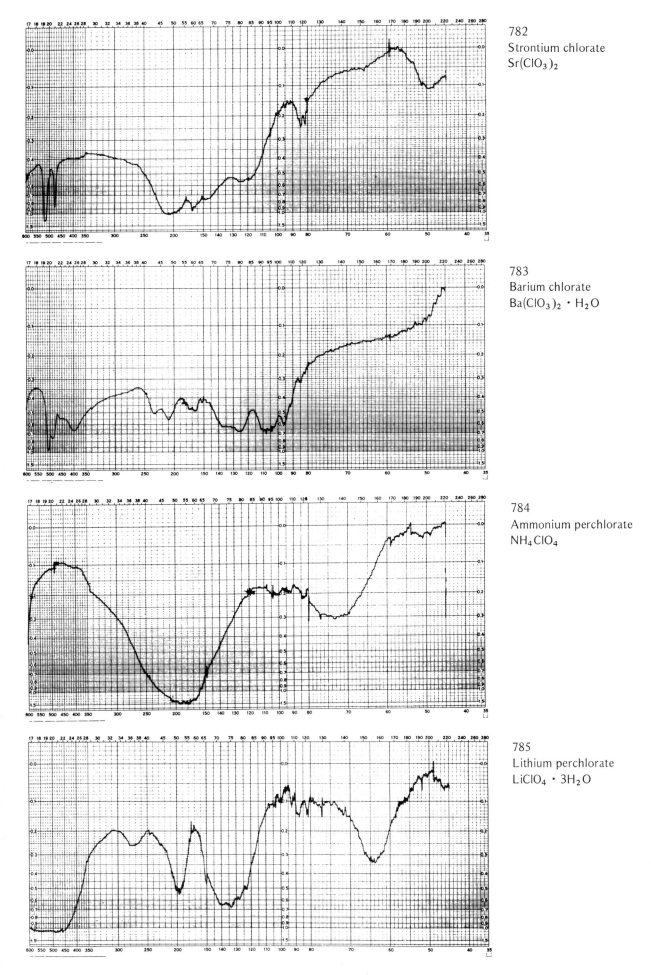

782
Strontium chlorate
Sr(ClO$_3$)$_2$

783
Barium chlorate
Ba(ClO$_3$)$_2$ · H$_2$O

784
Ammonium perchlorate
NH$_4$ClO$_4$

785
Lithium perchlorate
LiClO$_4$ · 3H$_2$O

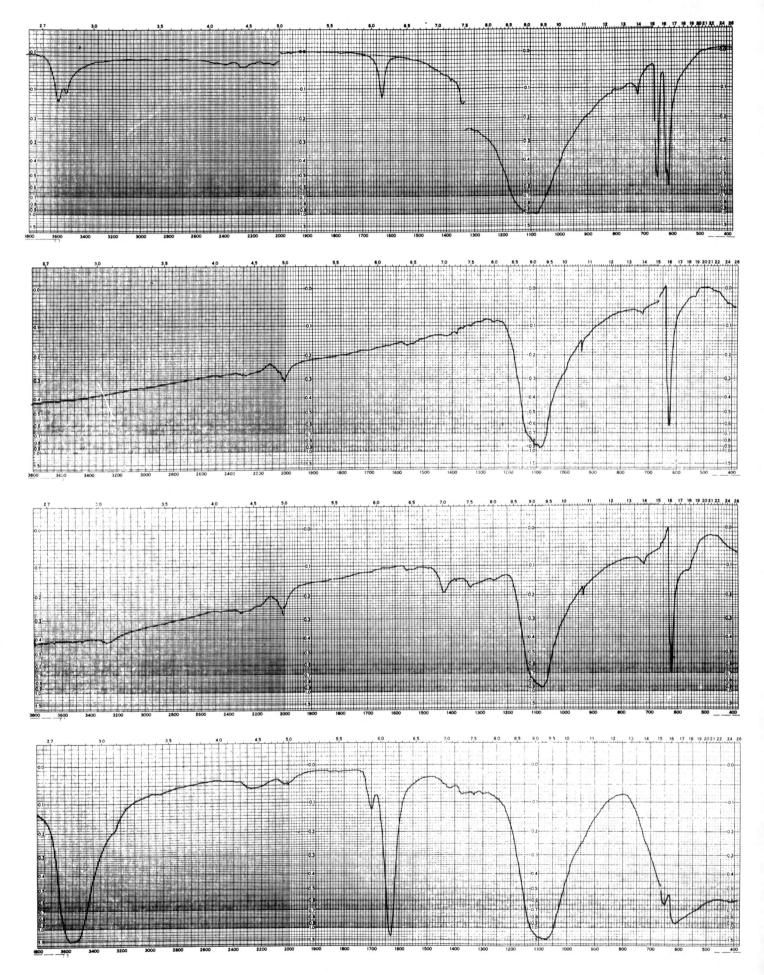

442

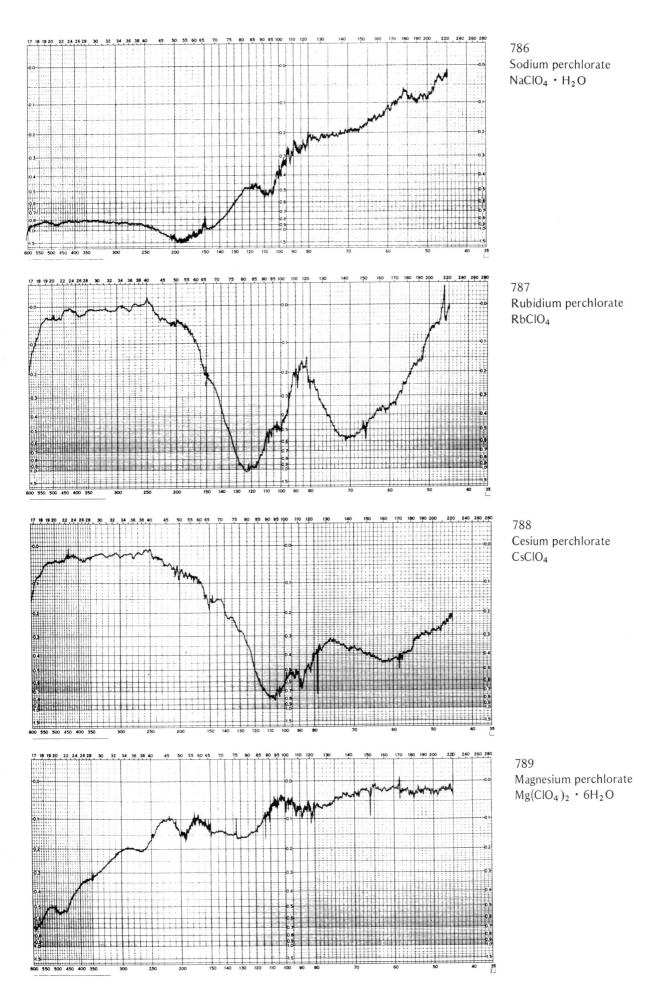

786
Sodium perchlorate
NaClO$_4$ · H$_2$O

787
Rubidium perchlorate
RbClO$_4$

788
Cesium perchlorate
CsClO$_4$

789
Magnesium perchlorate
Mg(ClO$_4$)$_2$ · 6H$_2$O

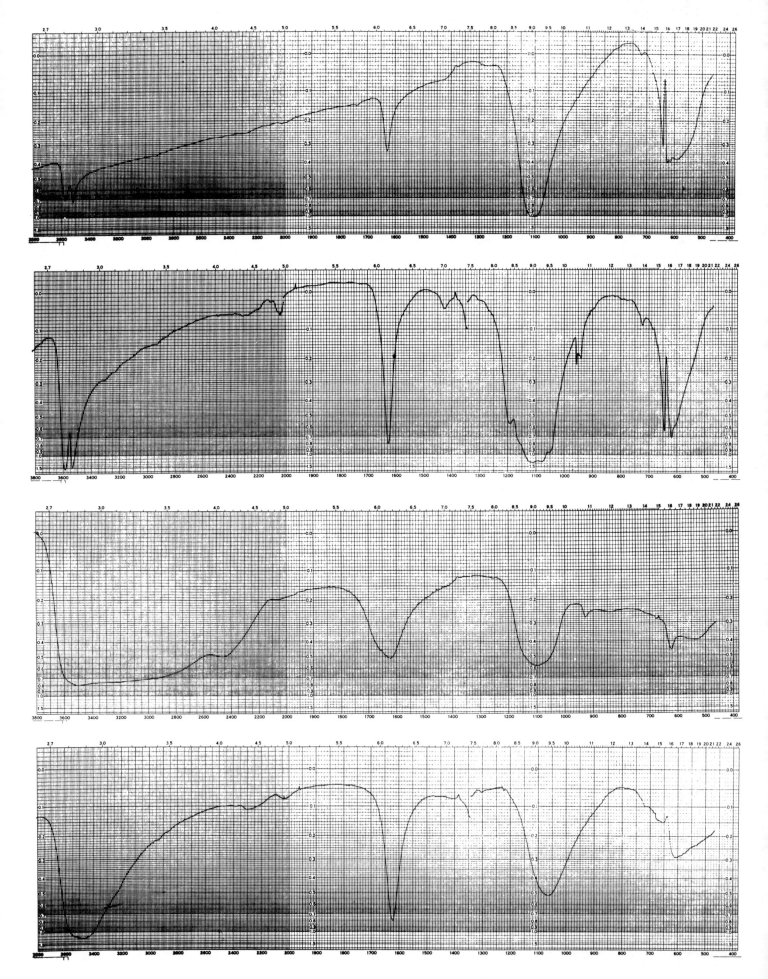

444

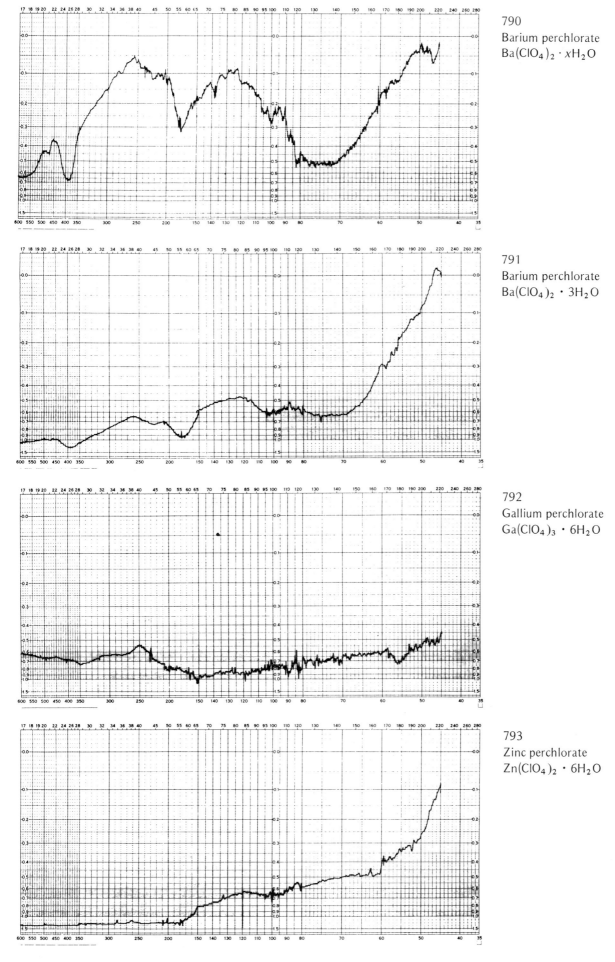

790
Barium perchlorate
$Ba(ClO_4)_2 \cdot xH_2O$

791
Barium perchlorate
$Ba(ClO_4)_2 \cdot 3H_2O$

792
Gallium perchlorate
$Ga(ClO_4)_3 \cdot 6H_2O$

793
Zinc perchlorate
$Zn(ClO_4)_2 \cdot 6H_2O$

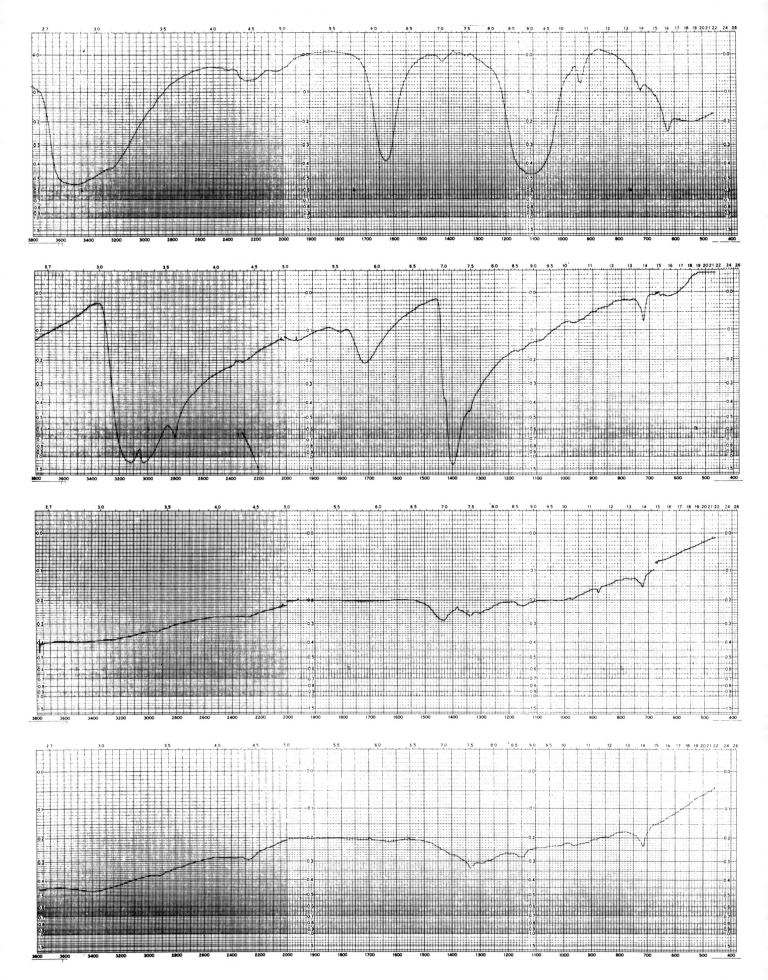

446

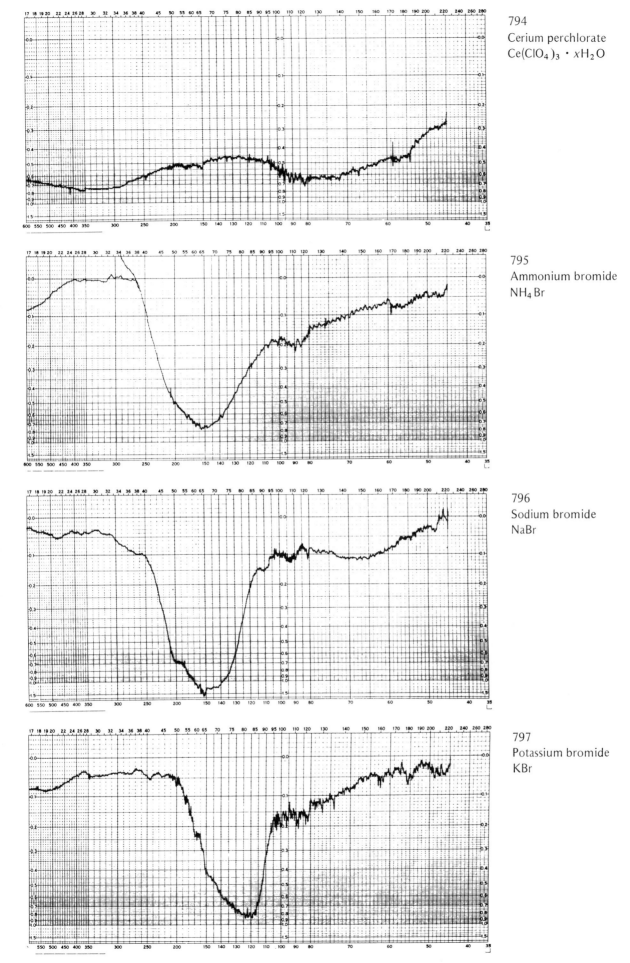

794
Cerium perchlorate
Ce(ClO$_4$)$_3$ · xH$_2$O

795
Ammonium bromide
NH$_4$Br

796
Sodium bromide
NaBr

797
Potassium bromide
KBr

447

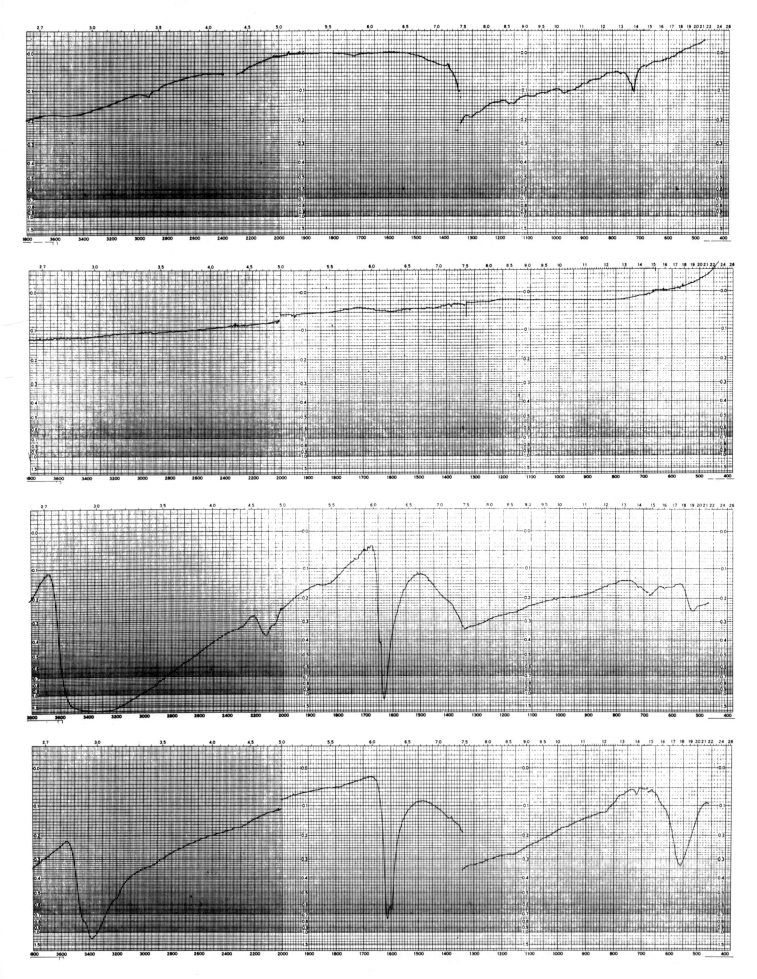

448

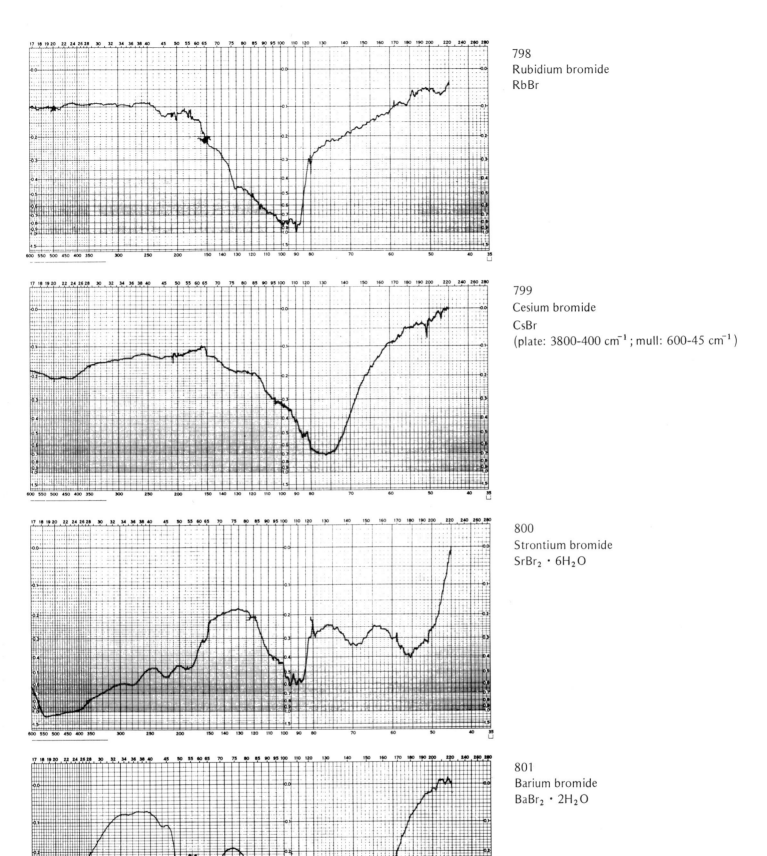

798
Rubidium bromide
RbBr

799
Cesium bromide
CsBr
(plate: 3800-400 cm^{-1}; mull: 600-45 cm^{-1})

800
Strontium bromide
SrBr$_2$ · 6H$_2$O

801
Barium bromide
BaBr$_2$ · 2H$_2$O

449

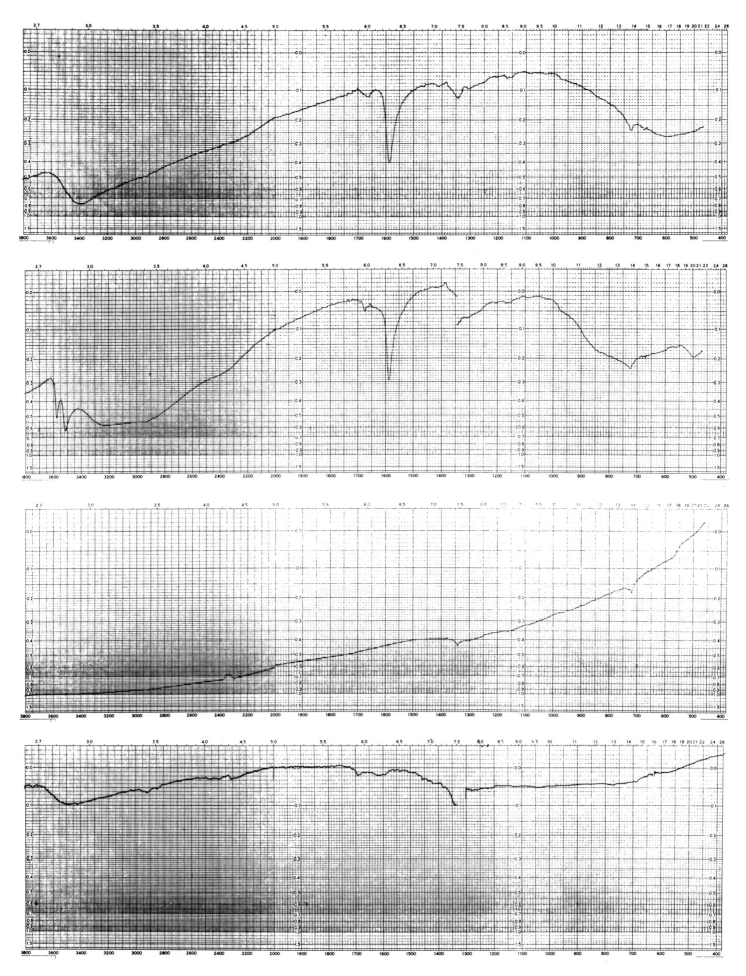

450

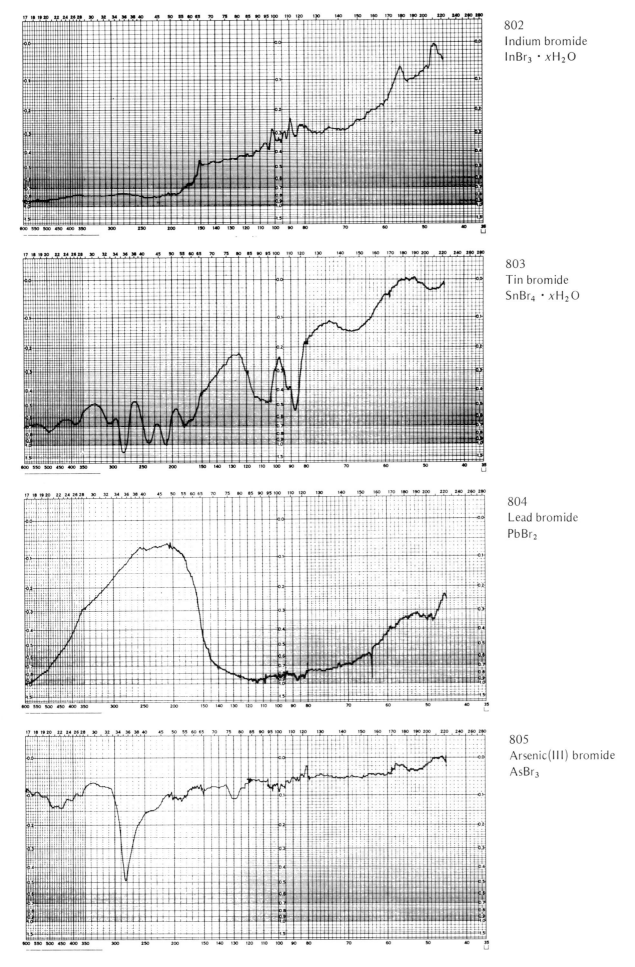

802
Indium bromide
InBr$_3 \cdot x$H$_2$O

803
Tin bromide
SnBr$_4 \cdot x$H$_2$O

804
Lead bromide
PbBr$_2$

805
Arsenic(III) bromide
AsBr$_3$

451

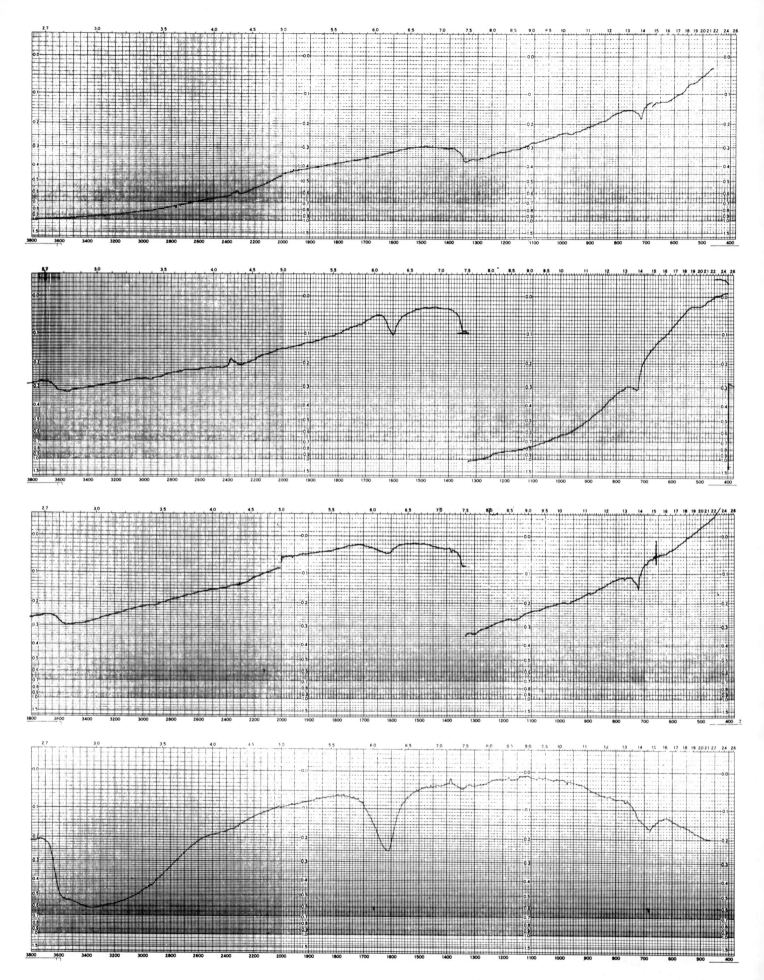

452

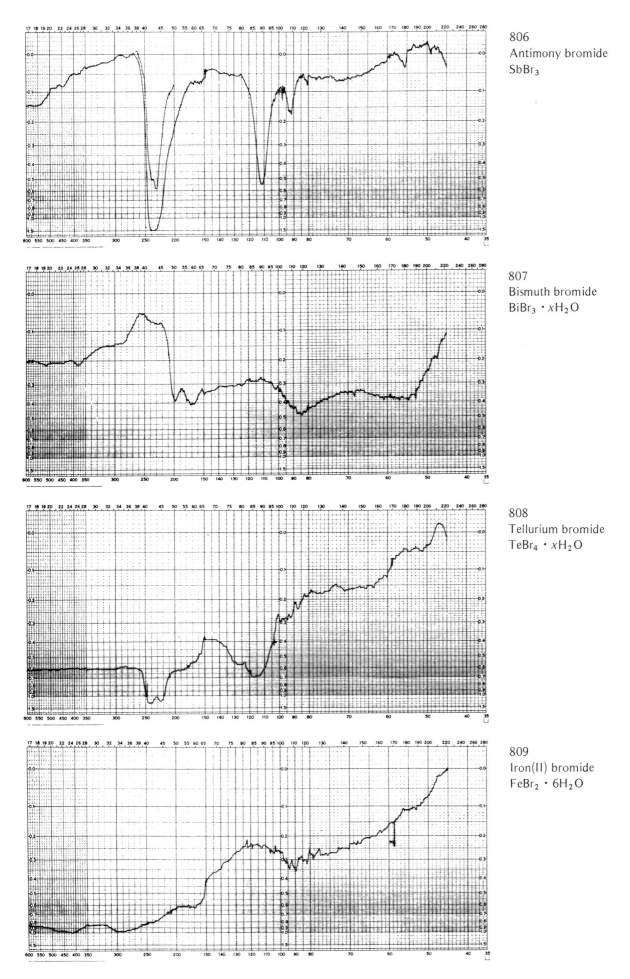

806
Antimony bromide
SbBr$_3$

807
Bismuth bromide
BiBr$_3 \cdot x$H$_2$O

808
Tellurium bromide
TeBr$_4 \cdot x$H$_2$O

809
Iron(II) bromide
FeBr$_2 \cdot 6$H$_2$O

453

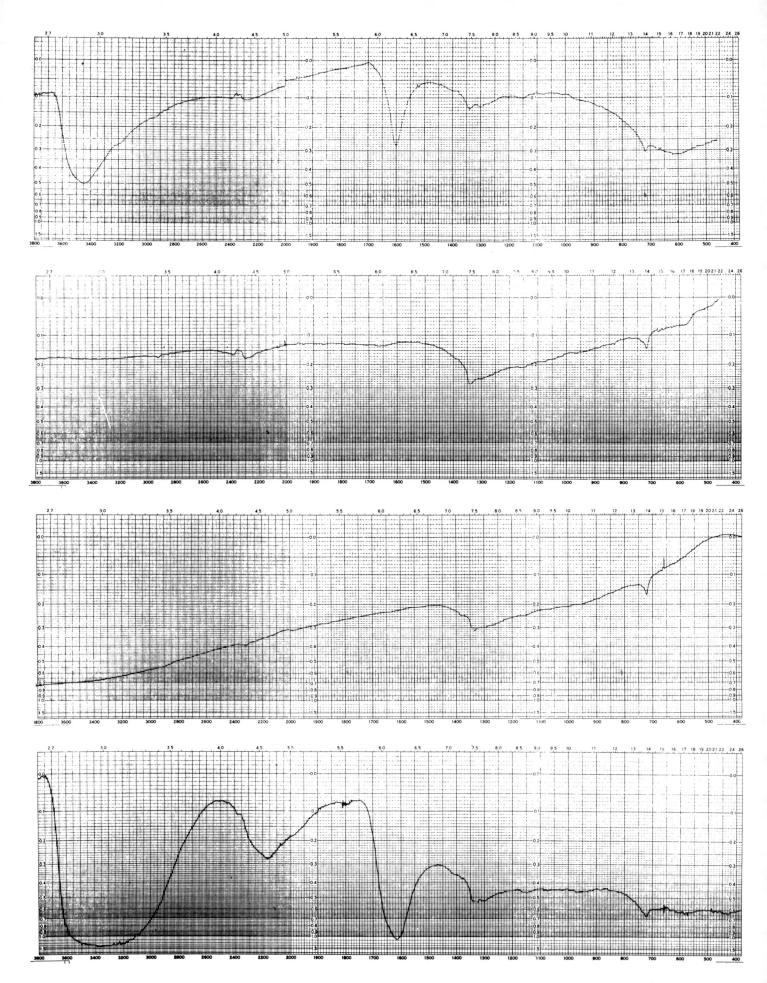

454

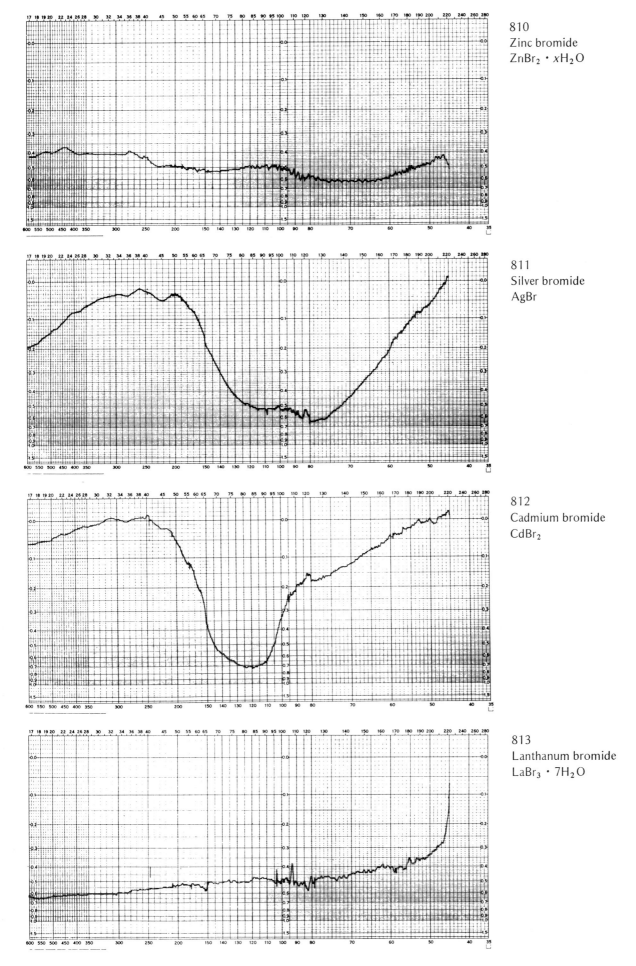

810
Zinc bromide
$ZnBr_2 \cdot xH_2O$

811
Silver bromide
AgBr

812
Cadmium bromide
$CdBr_2$

813
Lanthanum bromide
$LaBr_3 \cdot 7H_2O$

455

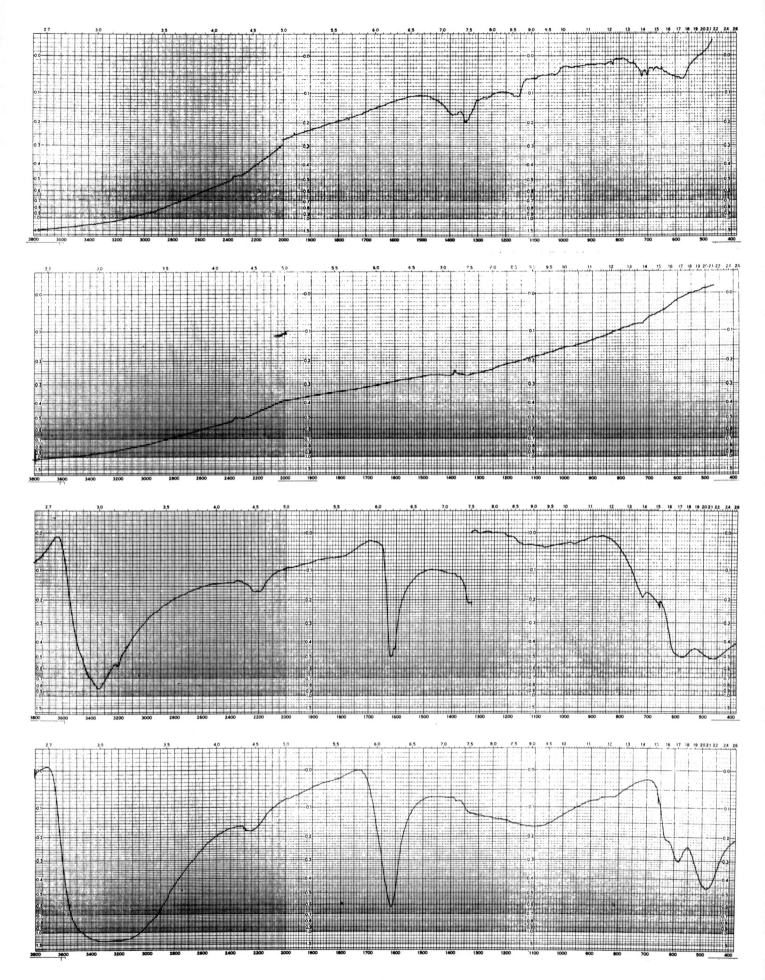

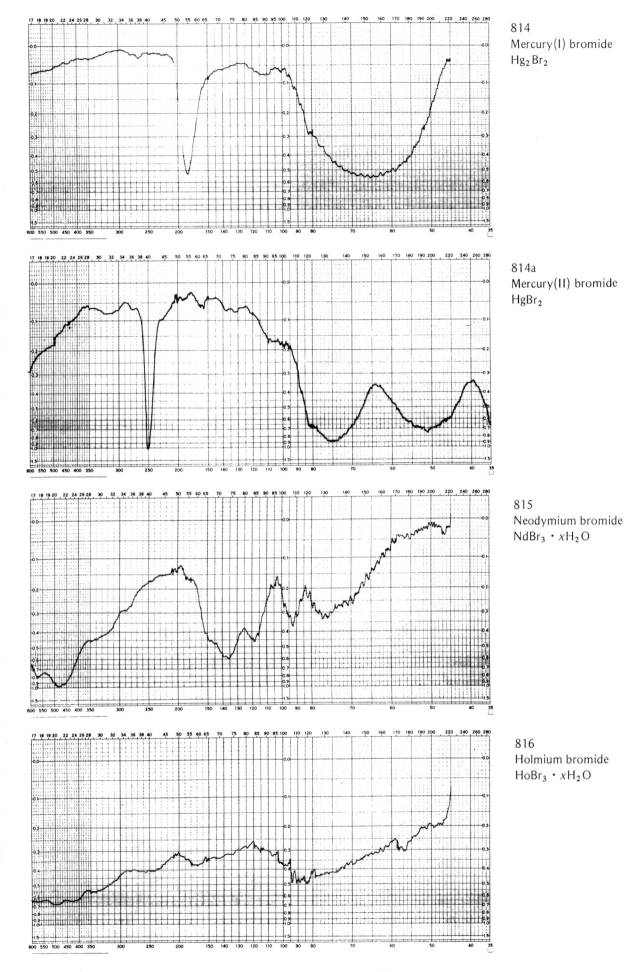

814
Mercury(I) bromide
Hg$_2$Br$_2$

814a
Mercury(II) bromide
HgBr$_2$

815
Neodymium bromide
NdBr$_3 \cdot x$H$_2$O

816
Holmium bromide
HoBr$_3 \cdot x$H$_2$O

457

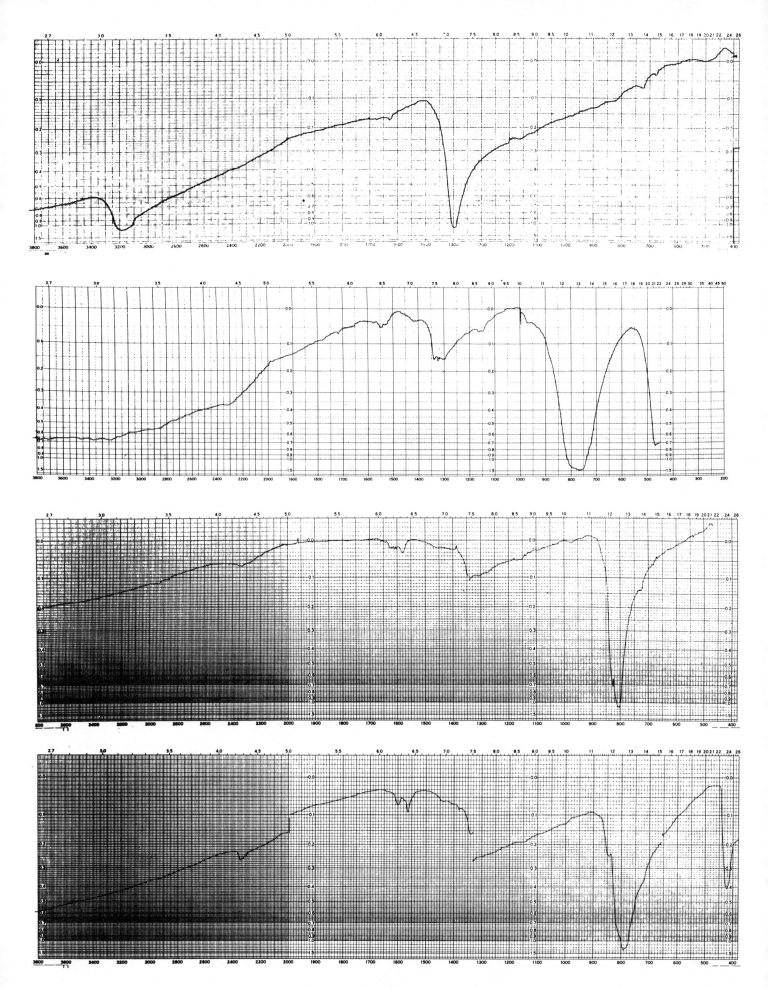

458

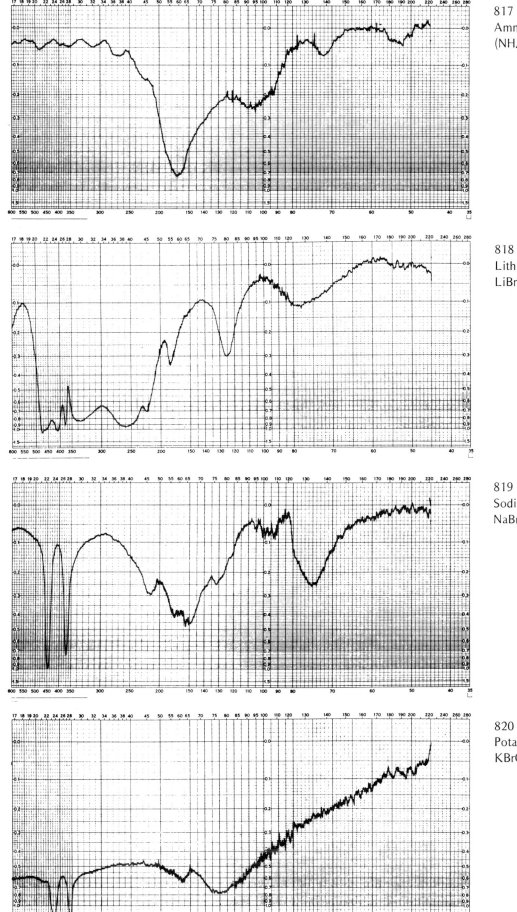

817
Ammonium cadmium bromide
$(NH_4)_2CdBr_4$

818
Lithium bromate
$LiBrO_3$

819
Sodium bromate
$NaBrO_3$

820
Potassium bromate
$KBrO_3$

459

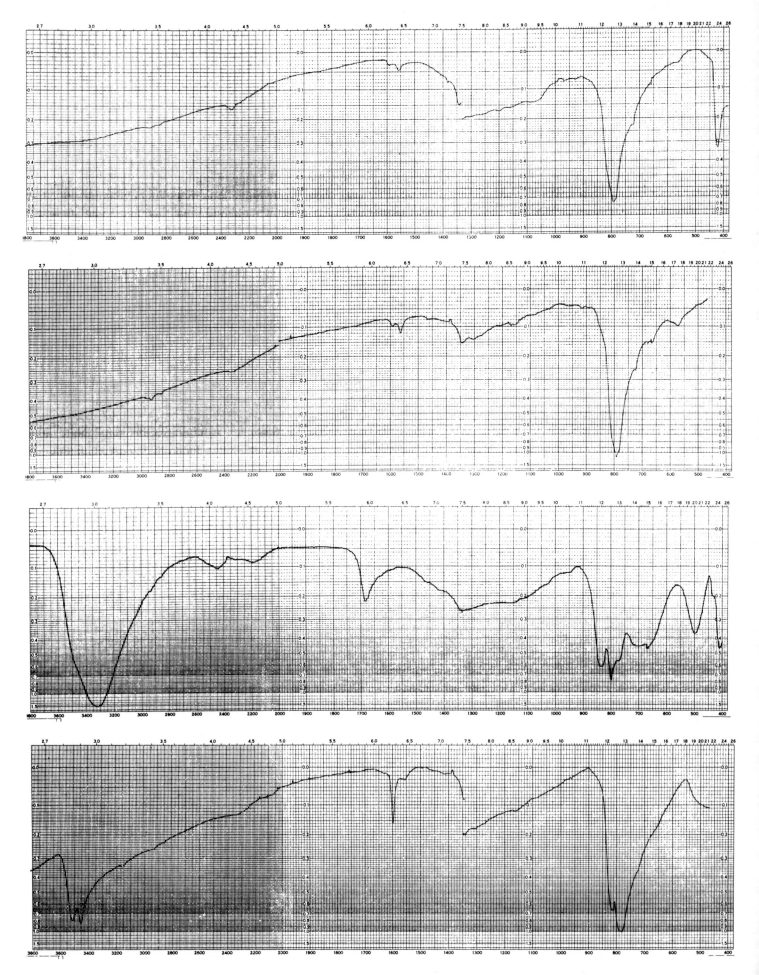

460

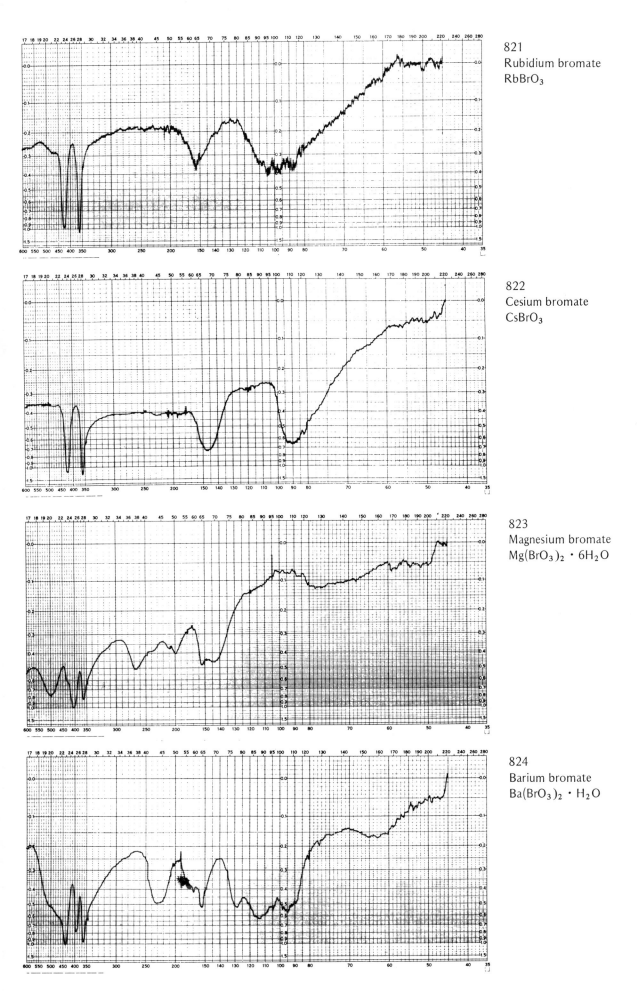

821
Rubidium bromate
RbBrO$_3$

822
Cesium bromate
CsBrO$_3$

823
Magnesium bromate
Mg(BrO$_3$)$_2$ · 6H$_2$O

824
Barium bromate
Ba(BrO$_3$)$_2$ · H$_2$O

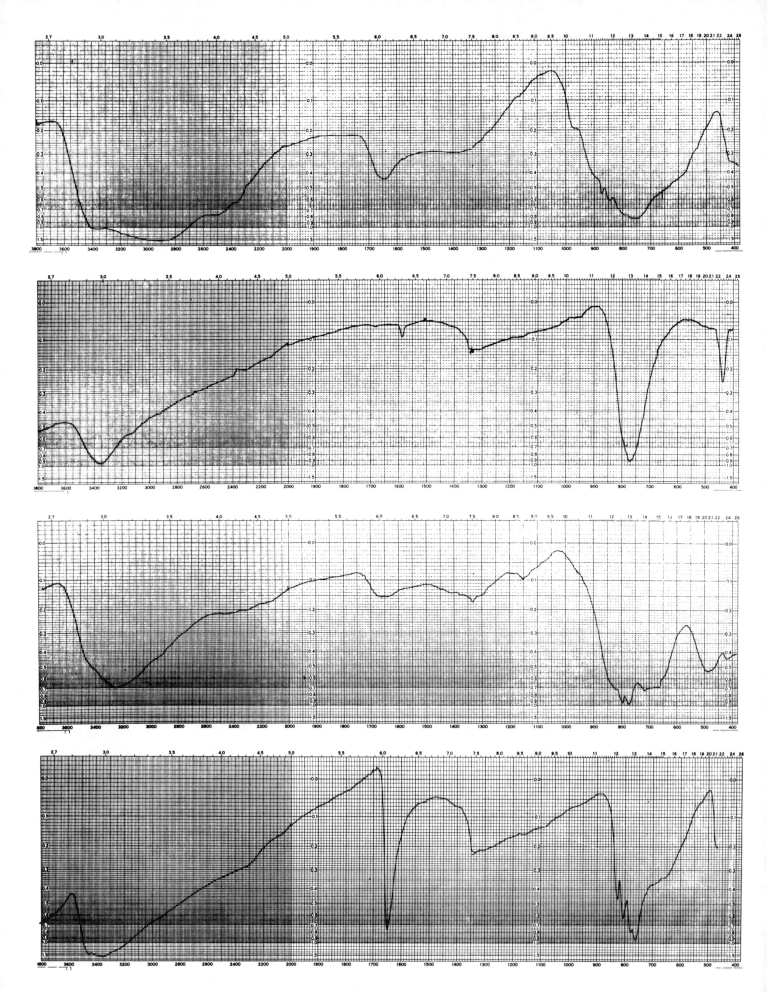

462

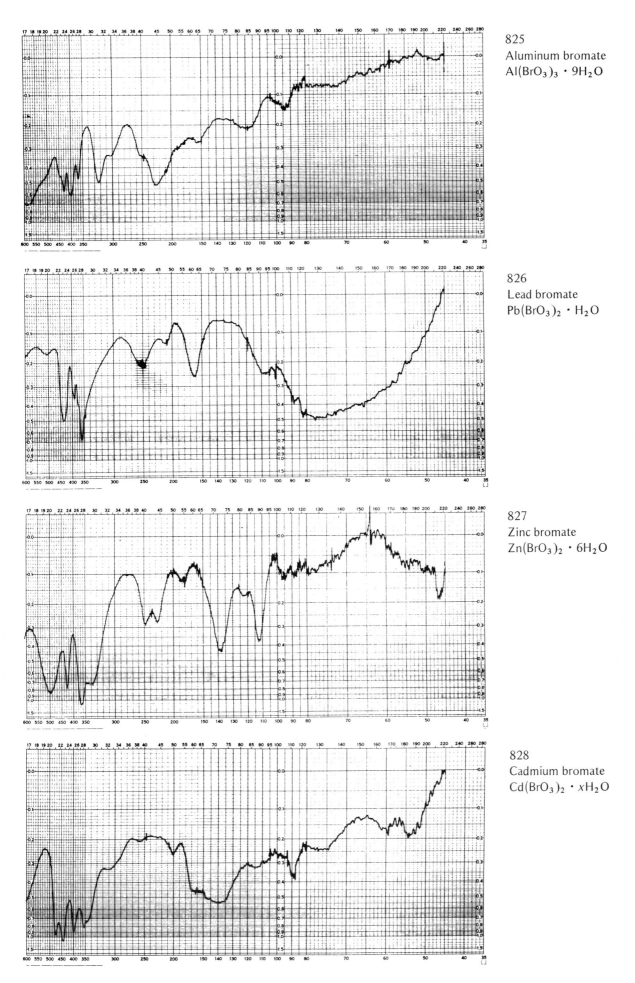

825
Aluminum bromate
Al(BrO$_3$)$_3$ · 9H$_2$O

826
Lead bromate
Pb(BrO$_3$)$_2$ · H$_2$O

827
Zinc bromate
Zn(BrO$_3$)$_2$ · 6H$_2$O

828
Cadmium bromate
Cd(BrO$_3$)$_2$ · xH$_2$O

463

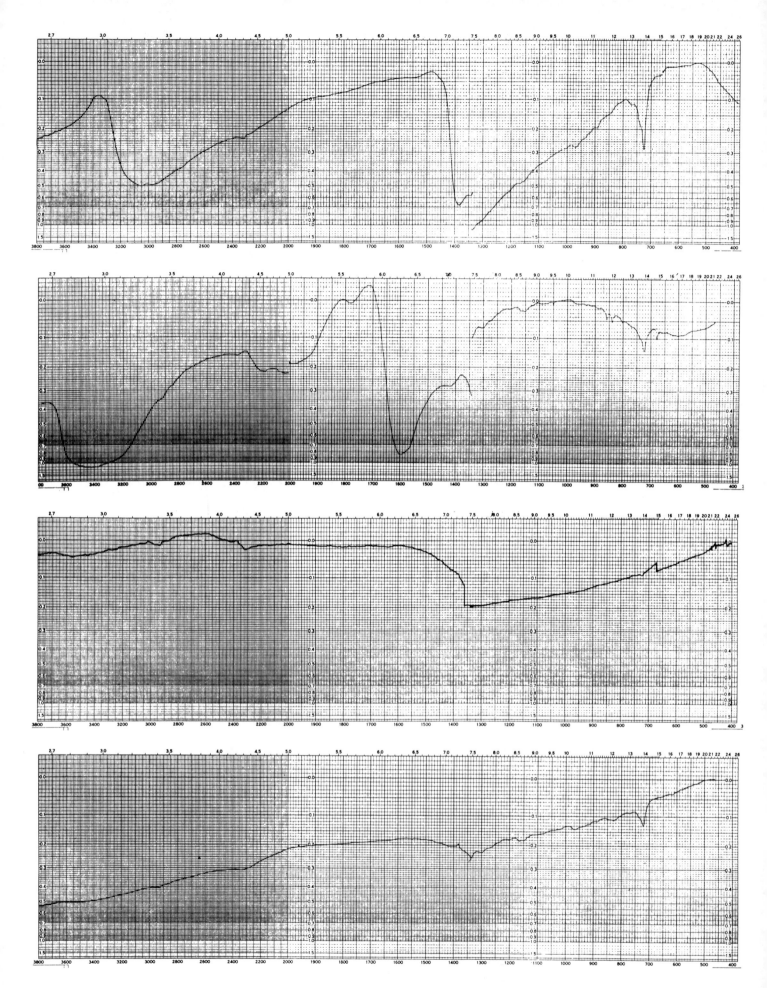

464

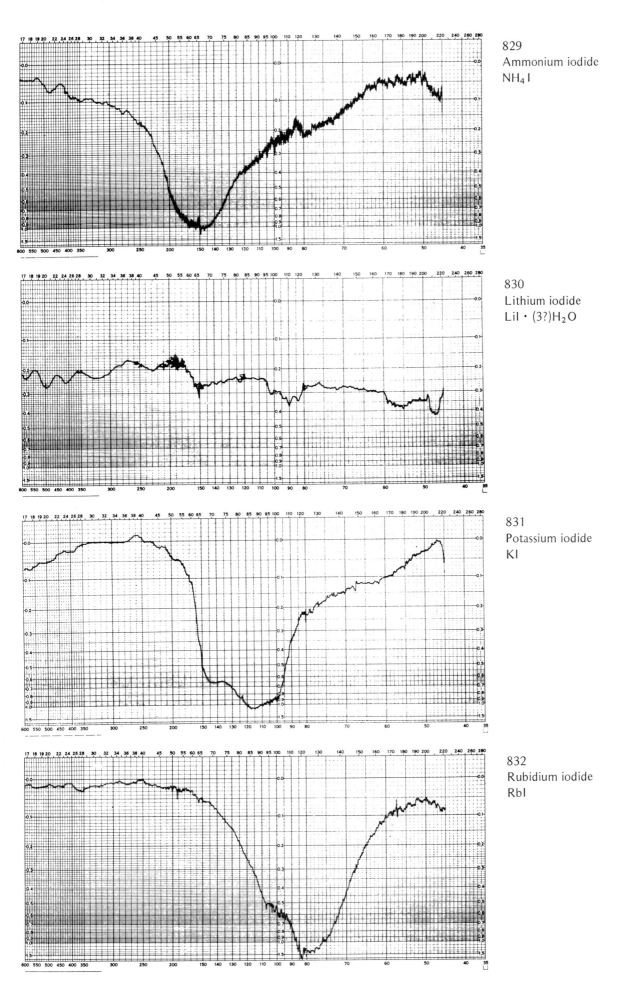

829
Ammonium iodide
NH₄I

830
Lithium iodide
LiI · (3?)H₂O

831
Potassium iodide
KI

832
Rubidium iodide
RbI

465

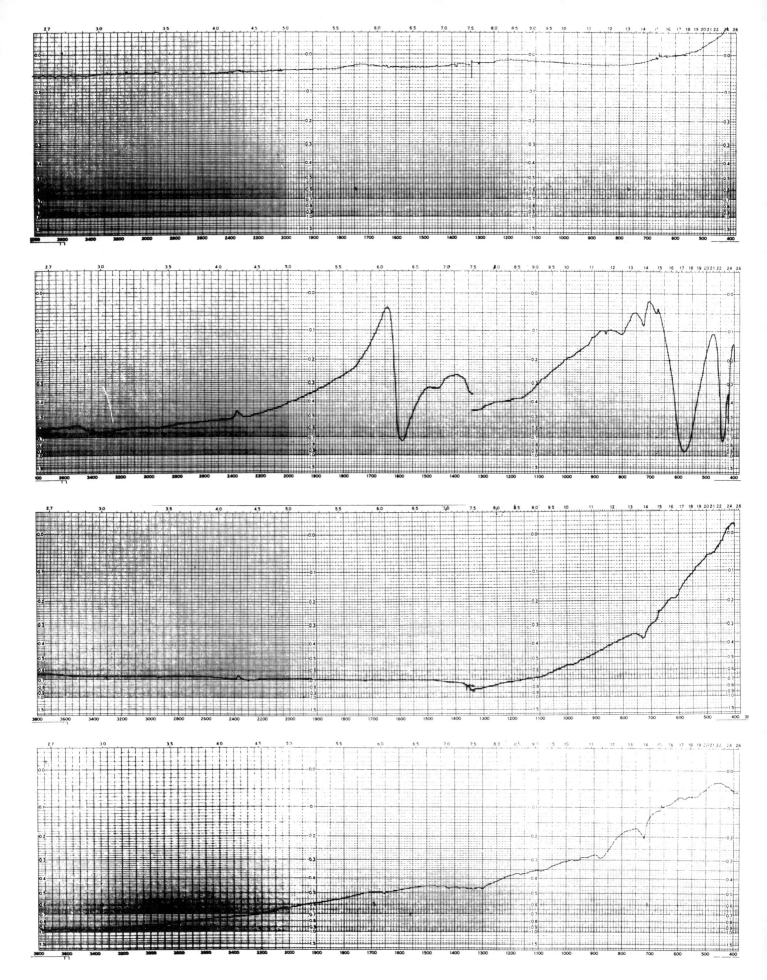

466

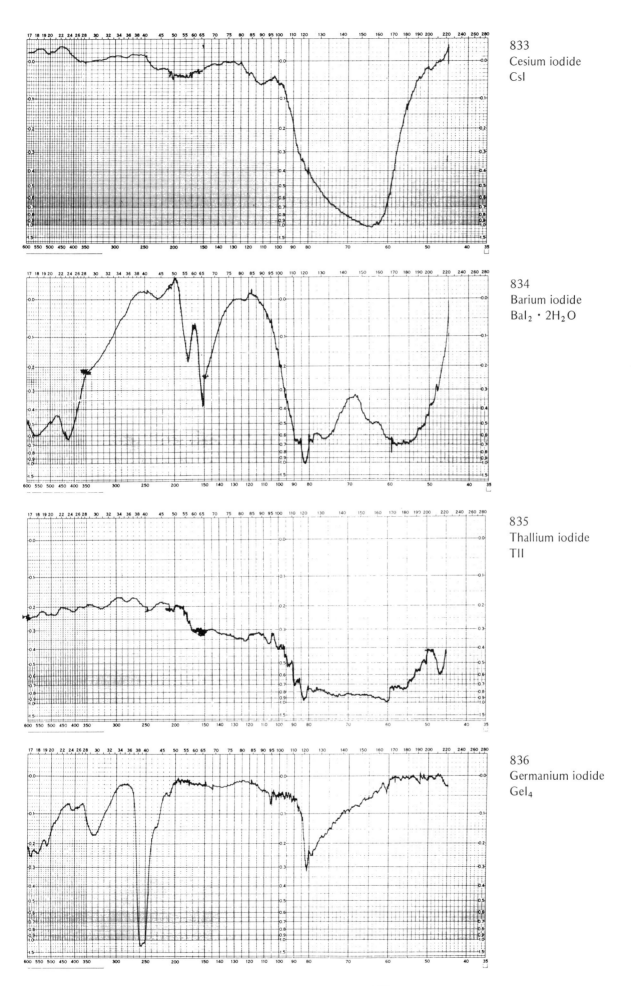

833
Cesium iodide
CsI

834
Barium iodide
BaI$_2$ · 2H$_2$O

835
Thallium iodide
TlI

836
Germanium iodide
GeI$_4$

467

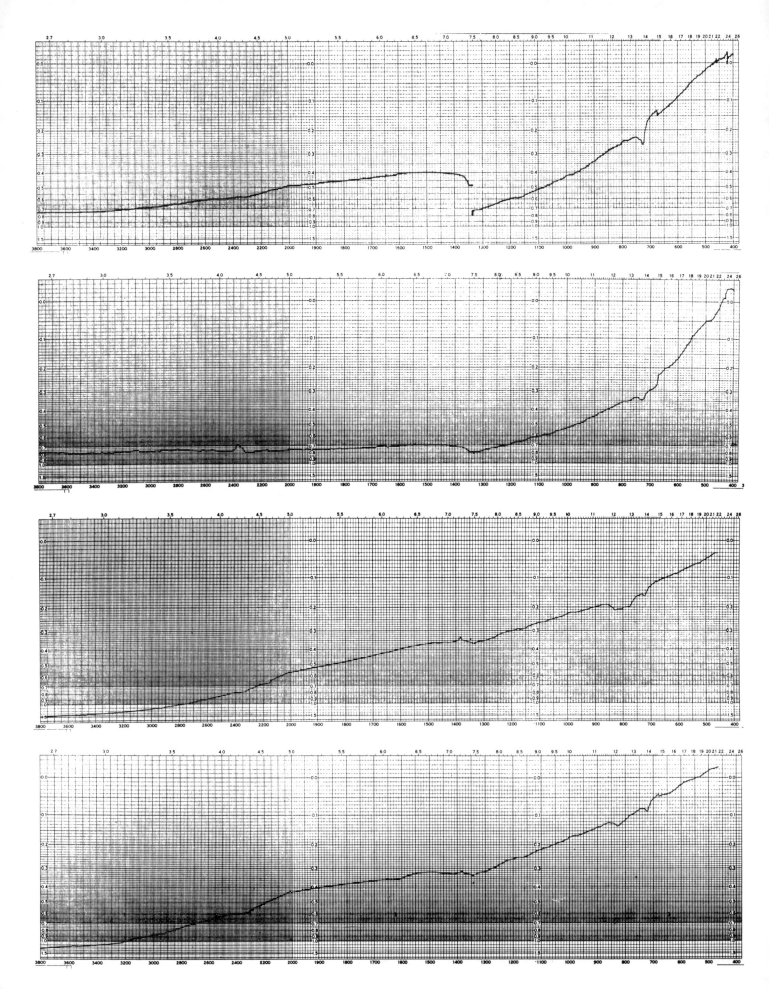

468

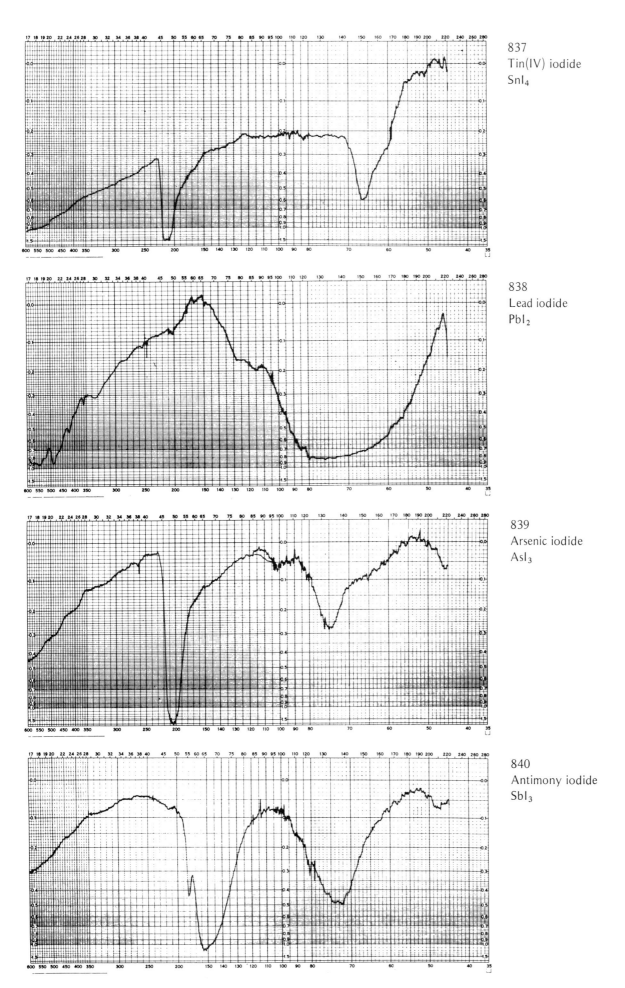

837
Tin(IV) iodide
SnI_4

838
Lead iodide
PbI_2

839
Arsenic iodide
AsI_3

840
Antimony iodide
SbI_3

469

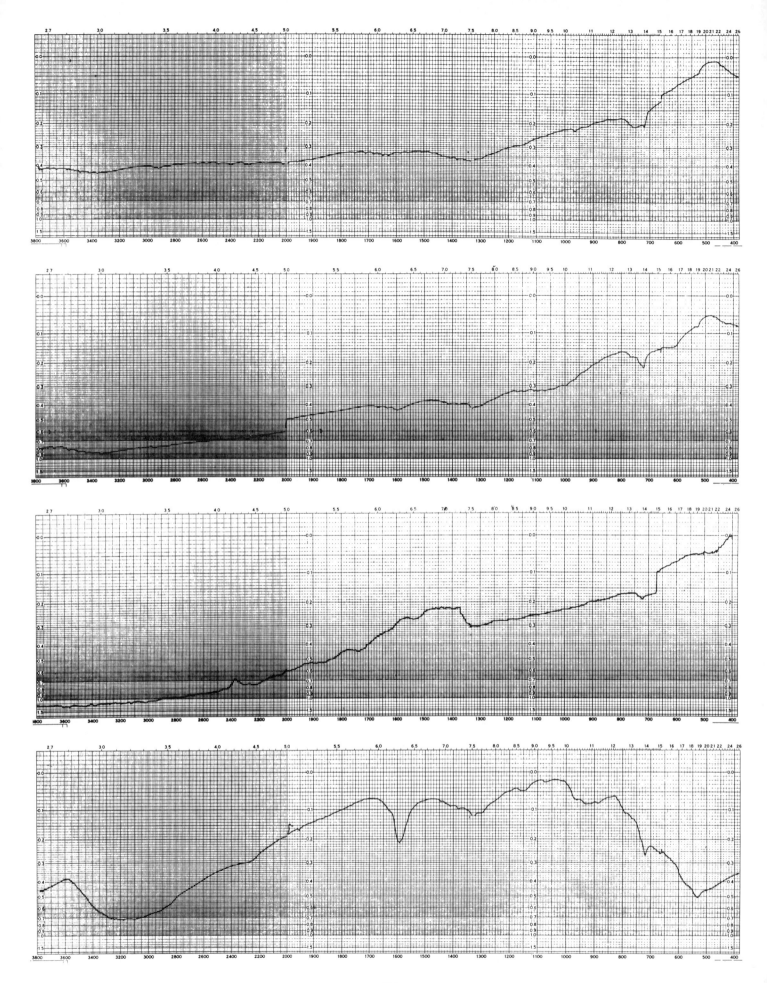

470

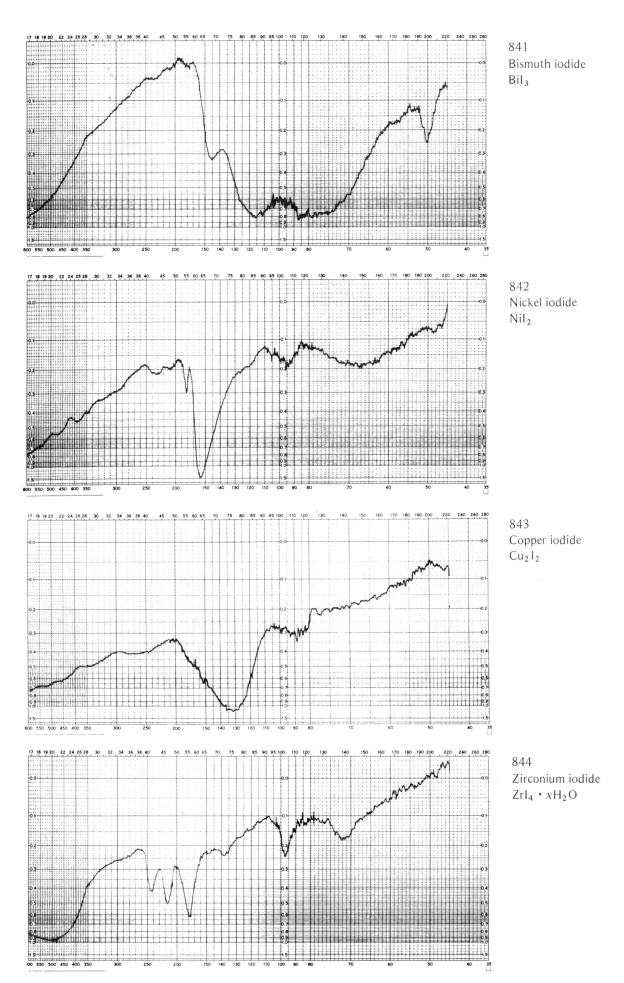

841
Bismuth iodide
BiI$_3$

842
Nickel iodide
NiI$_2$

843
Copper iodide
Cu$_2$I$_2$

844
Zirconium iodide
ZrI$_4$ · xH$_2$O

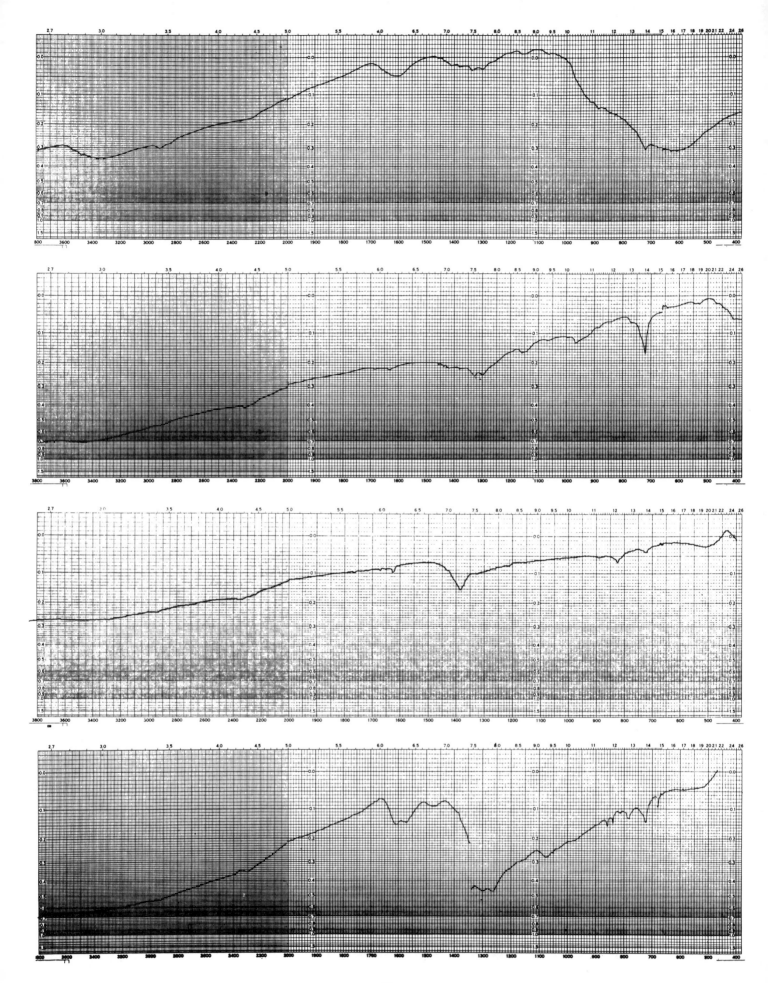

472

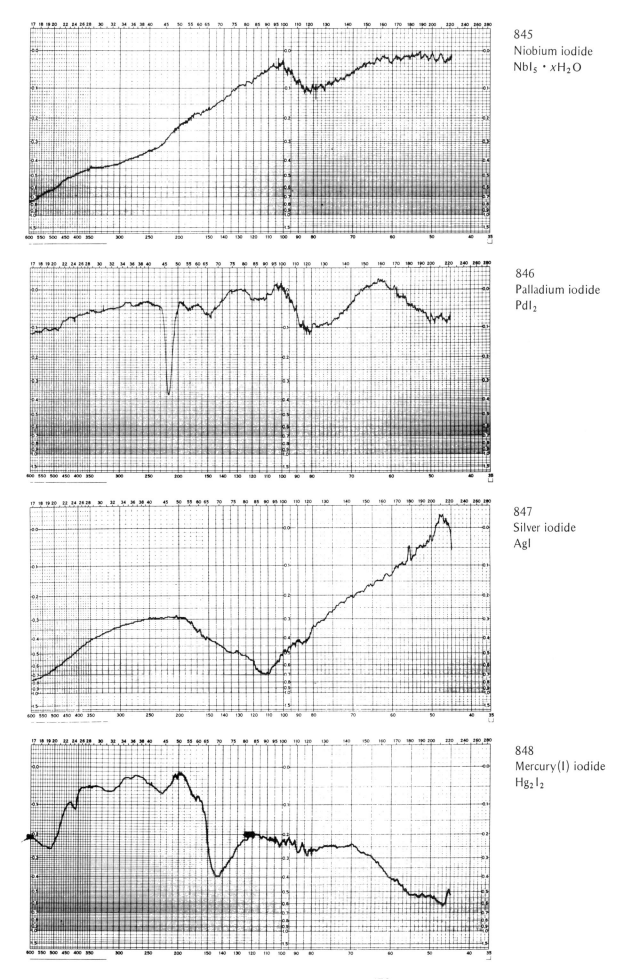

845
Niobium iodide
NbI$_5$ · xH$_2$O

846
Palladium iodide
PdI$_2$

847
Silver iodide
AgI

848
Mercury(I) iodide
Hg$_2$I$_2$

473

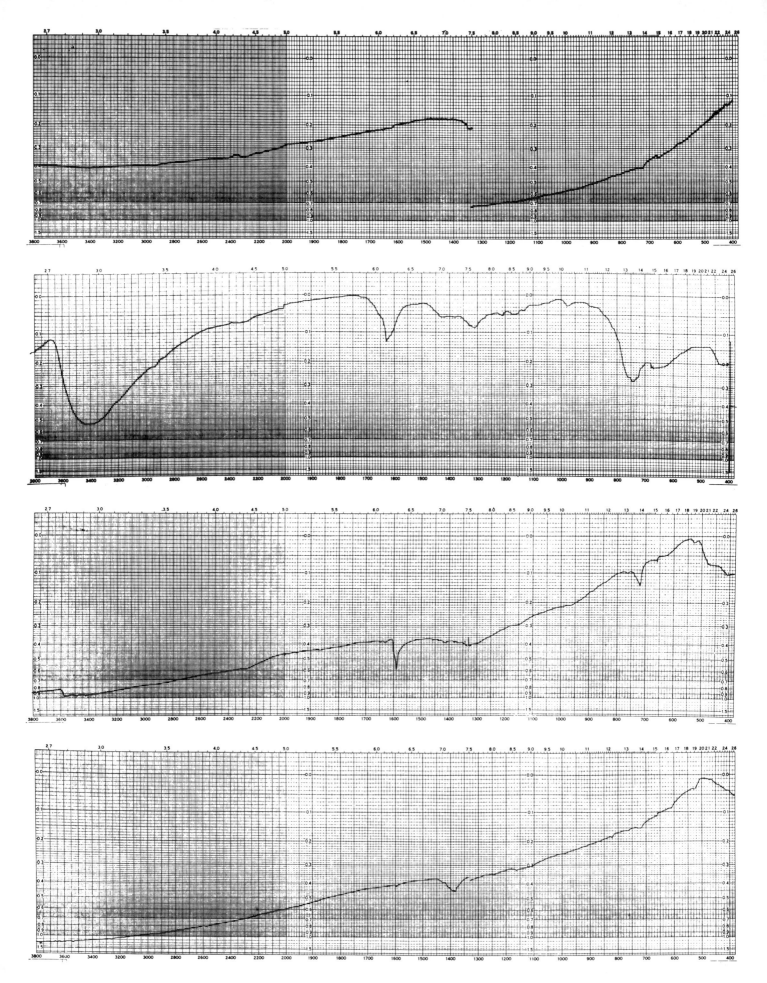

474

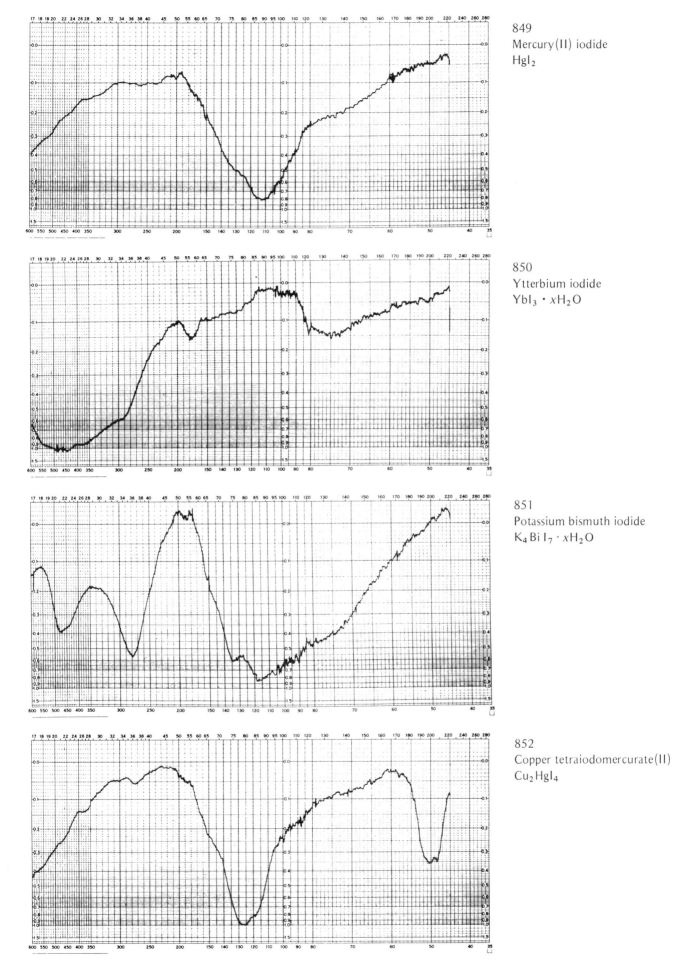

849
Mercury(II) iodide
HgI$_2$

850
Ytterbium iodide
YbI$_3 \cdot x$H$_2$O

851
Potassium bismuth iodide
K$_4$BiI$_7 \cdot x$H$_2$O

852
Copper tetraiodomercurate(II)
Cu$_2$HgI$_4$

475

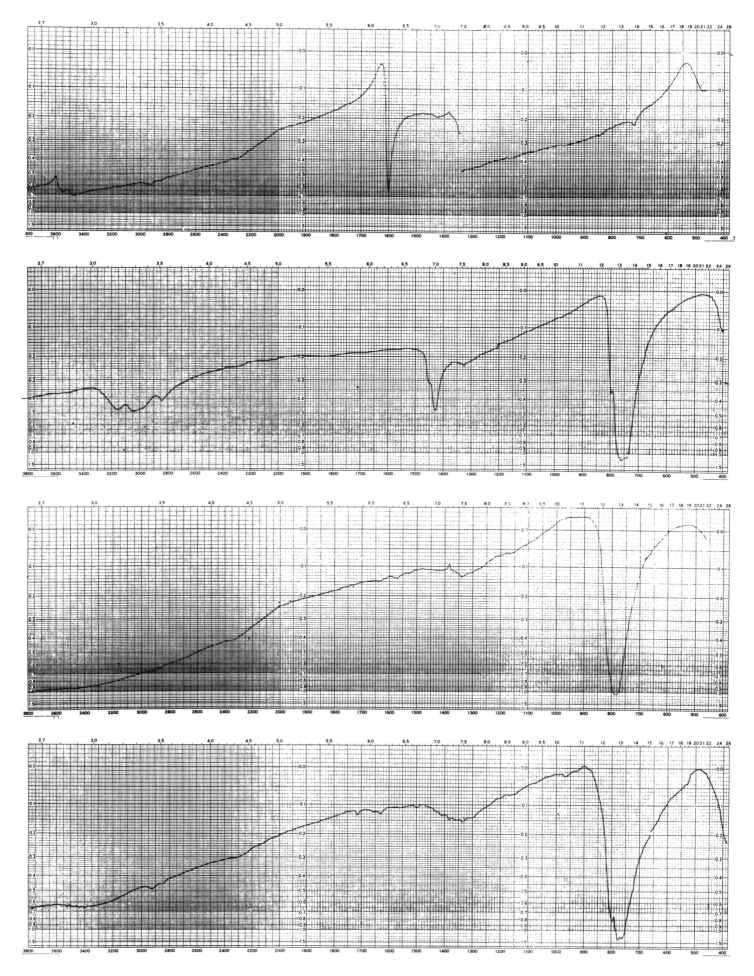

476

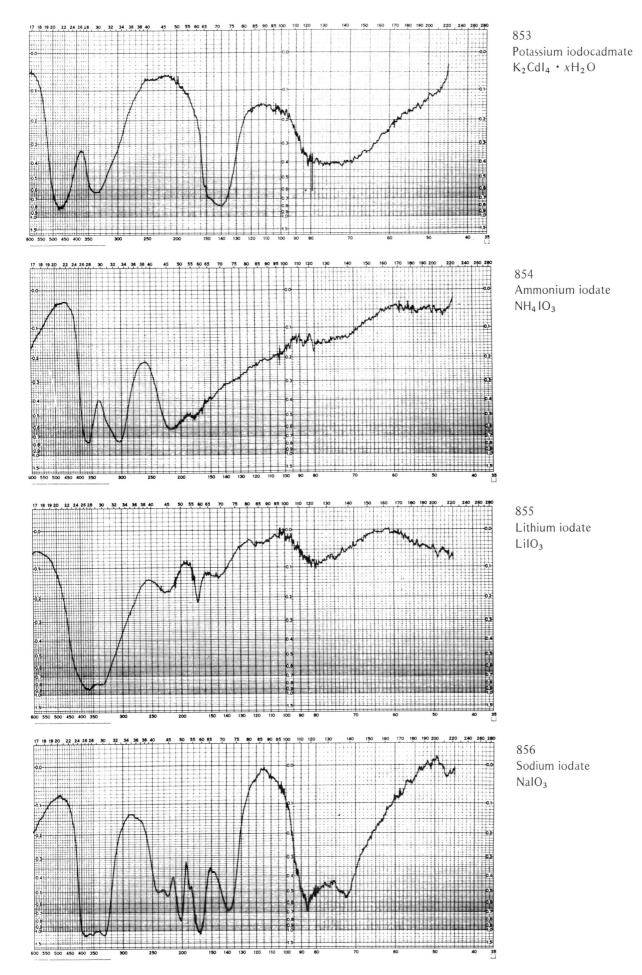

853
Potassium iodocadmate
$K_2CdI_4 \cdot xH_2O$

854
Ammonium iodate
NH_4IO_3

855
Lithium iodate
$LiIO_3$

856
Sodium iodate
$NaIO_3$

477

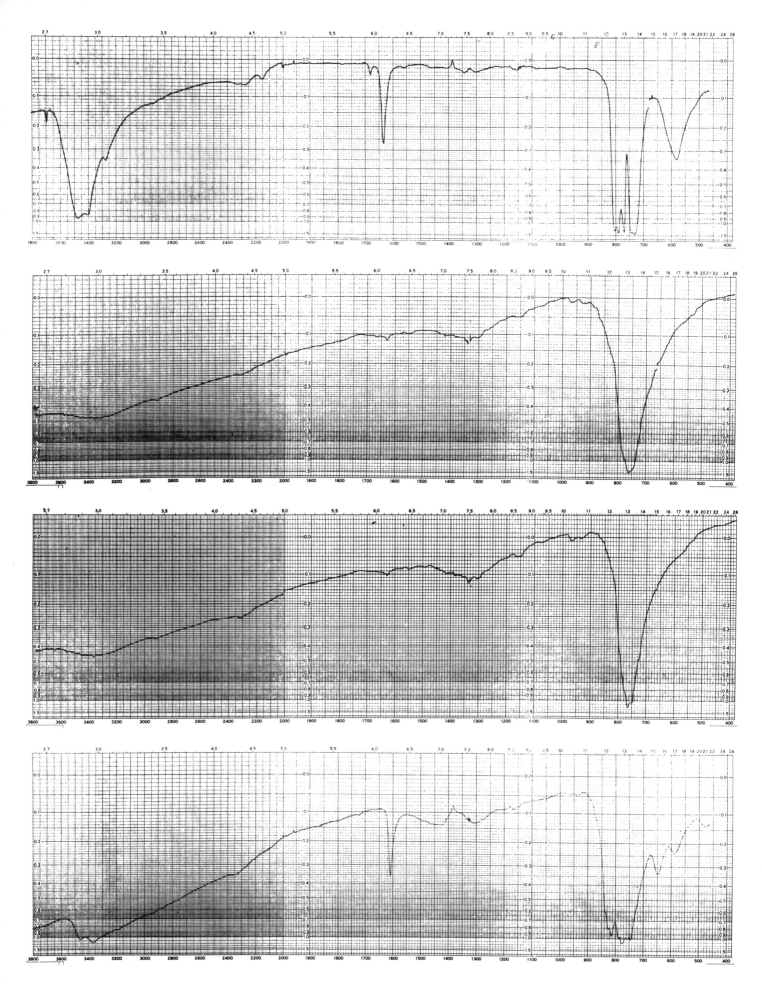

478

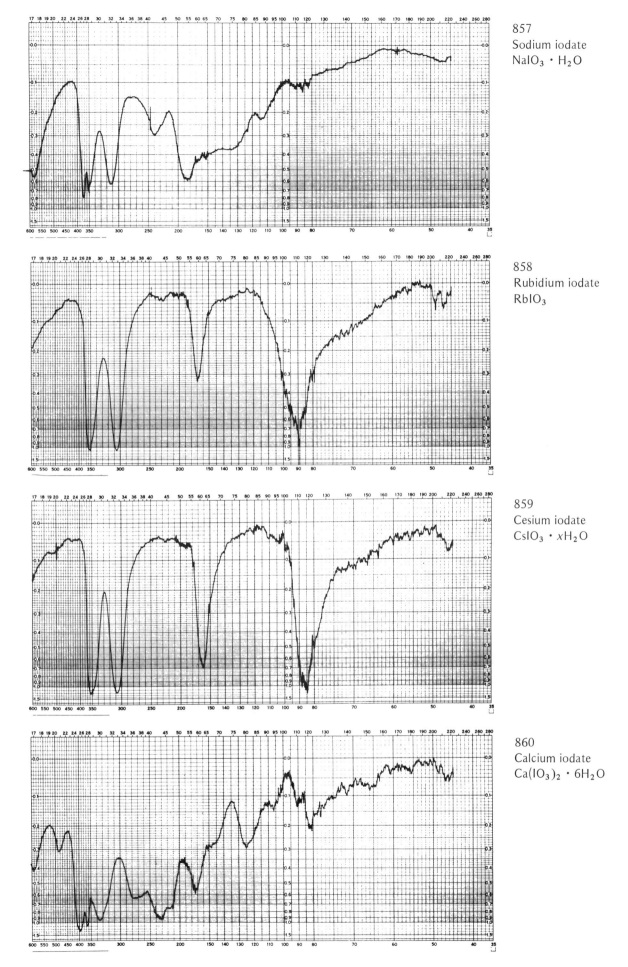

857
Sodium iodate
$NaIO_3 \cdot H_2O$

858
Rubidium iodate
$RbIO_3$

859
Cesium iodate
$CsIO_3 \cdot xH_2O$

860
Calcium iodate
$Ca(IO_3)_2 \cdot 6H_2O$

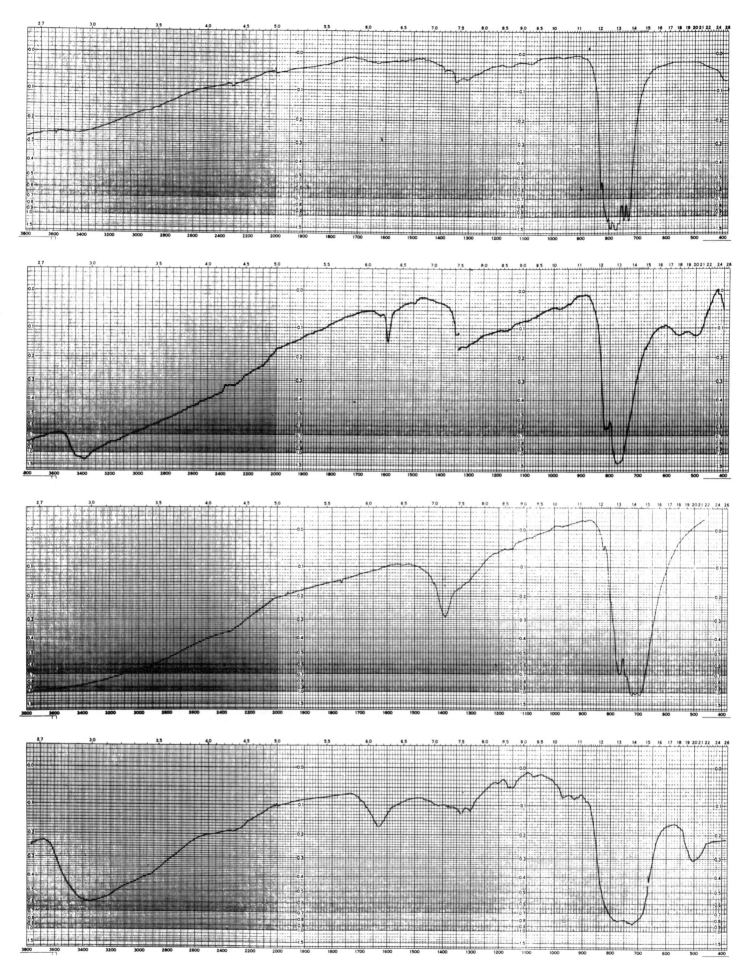

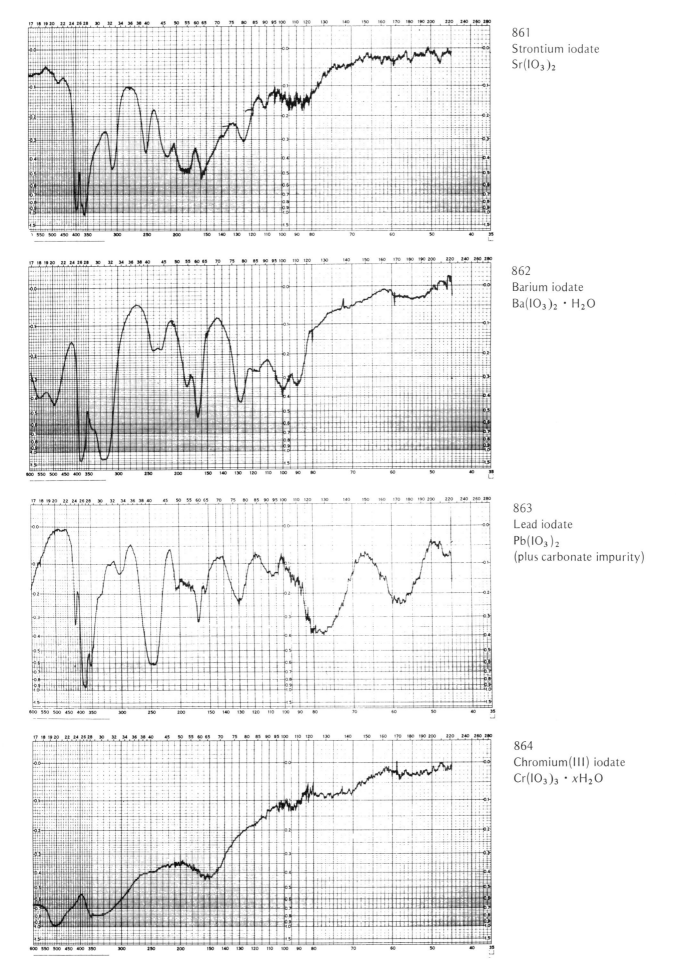

861
Strontium iodate
Sr(IO₃)₂

862
Barium iodate
Ba(IO₃)₂ · H₂O

863
Lead iodate
Pb(IO₃)₂
(plus carbonate impurity)

864
Chromium(III) iodate
Cr(IO₃)₃ · xH₂O

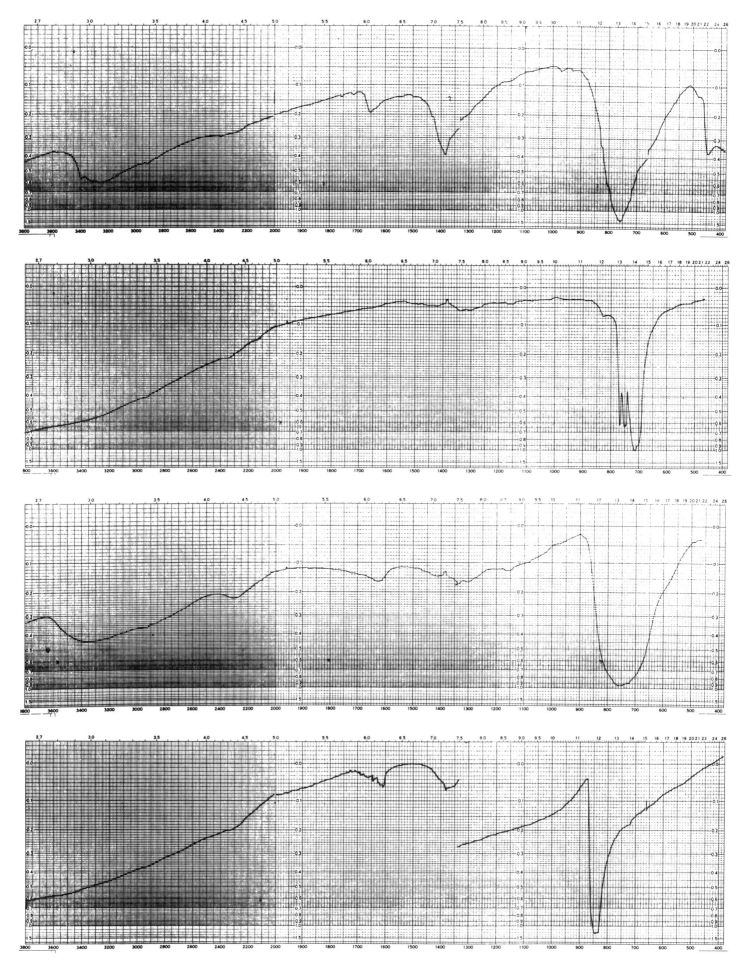

482

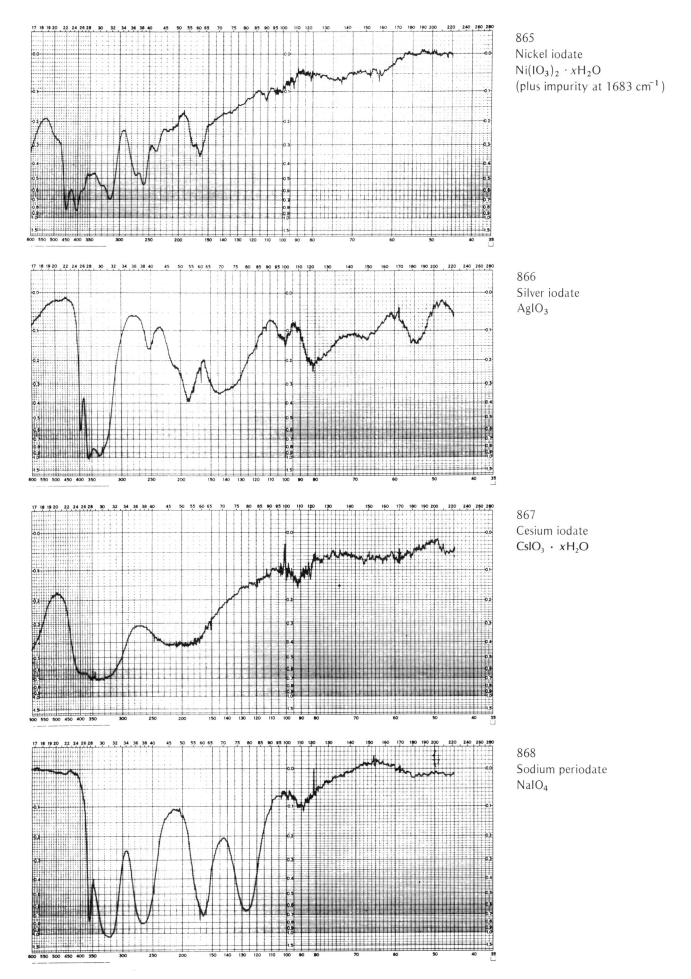

865
Nickel iodate
$Ni(IO_3)_2 \cdot xH_2O$
(plus impurity at 1683 cm^{-1})

866
Silver iodate
$AgIO_3$

867
Cesium iodate
$CsIO_3 \cdot xH_2O$

868
Sodium periodate
$NaIO_4$

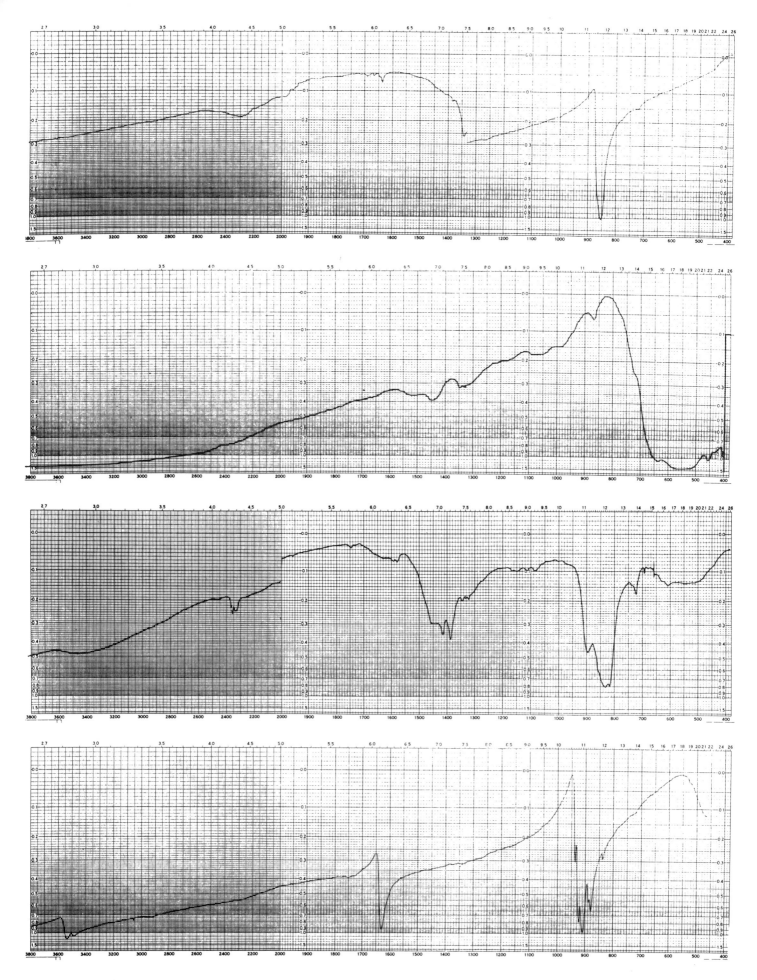

484

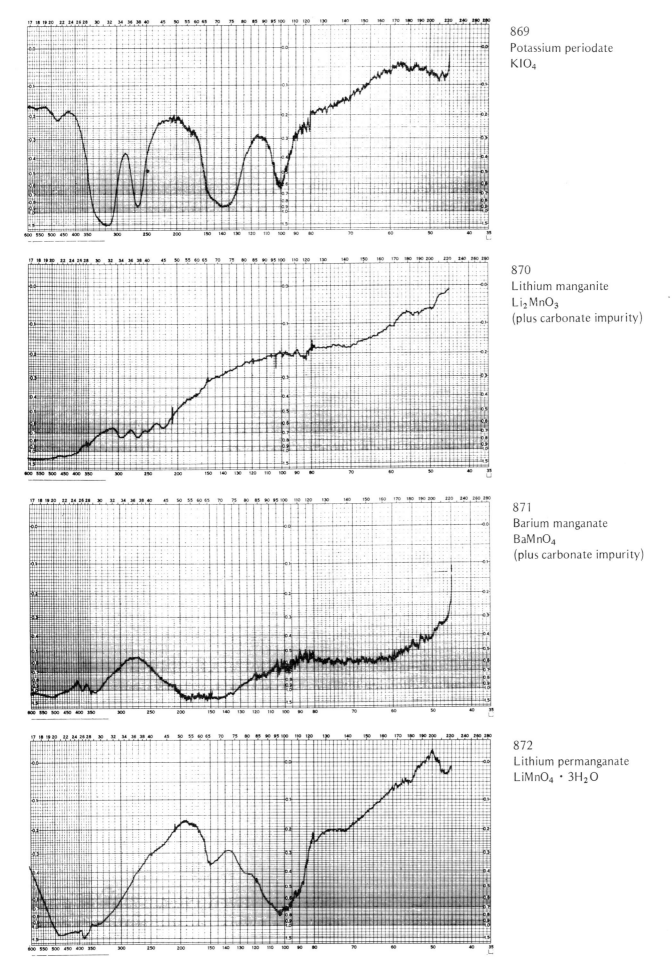

869
Potassium periodate
KIO_4

870
Lithium manganite
Li_2MnO_3
(plus carbonate impurity)

871
Barium manganate
$BaMnO_4$
(plus carbonate impurity)

872
Lithium permanganate
$LiMnO_4 \cdot 3H_2O$

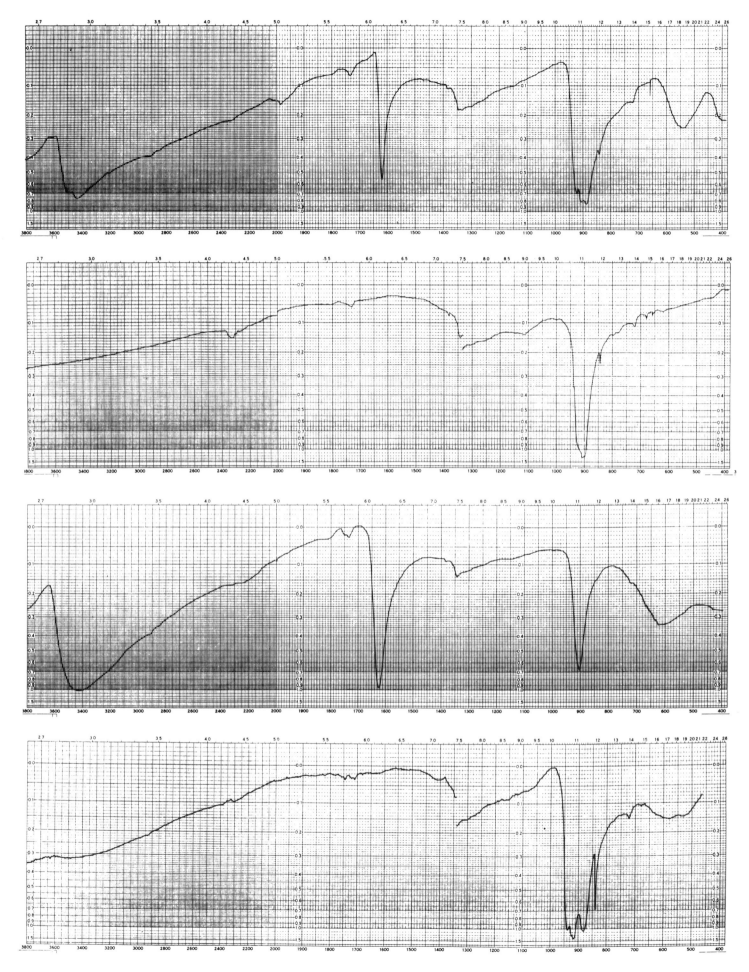

486

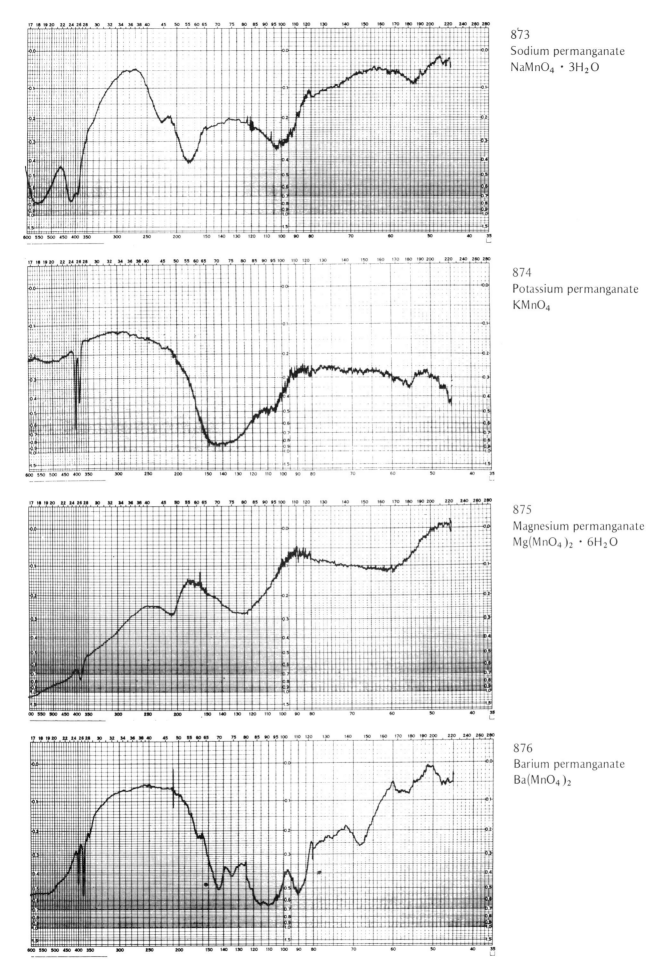

873
Sodium permanganate
NaMnO$_4$ · 3H$_2$O

874
Potassium permanganate
KMnO$_4$

875
Magnesium permanganate
Mg(MnO$_4$)$_2$ · 6H$_2$O

876
Barium permanganate
Ba(MnO$_4$)$_2$

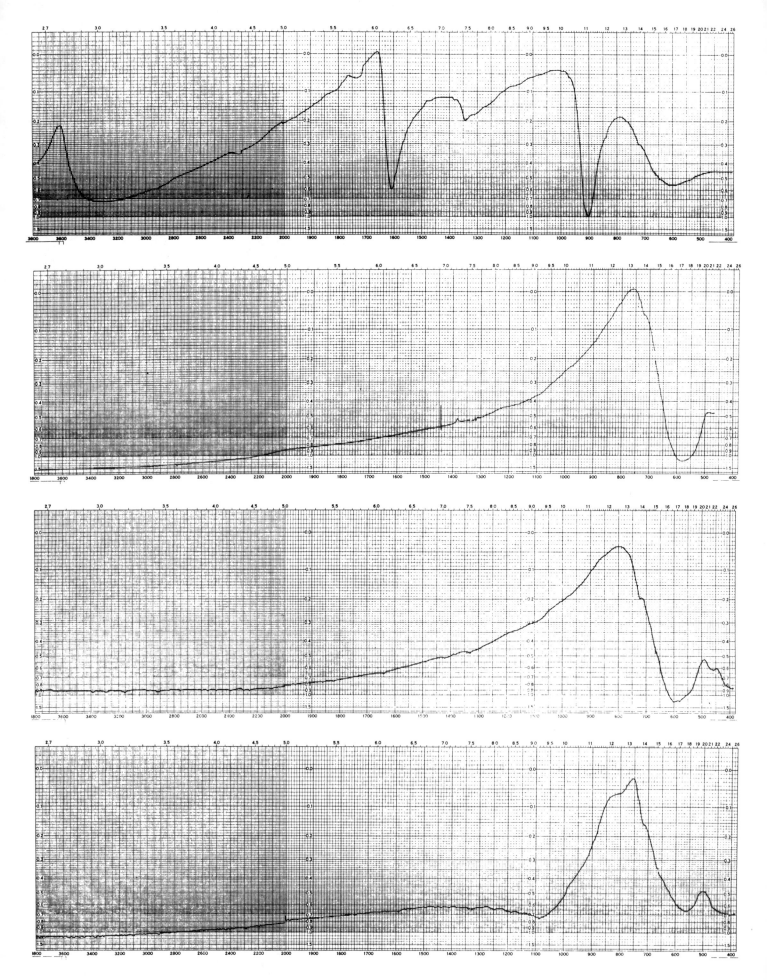

488

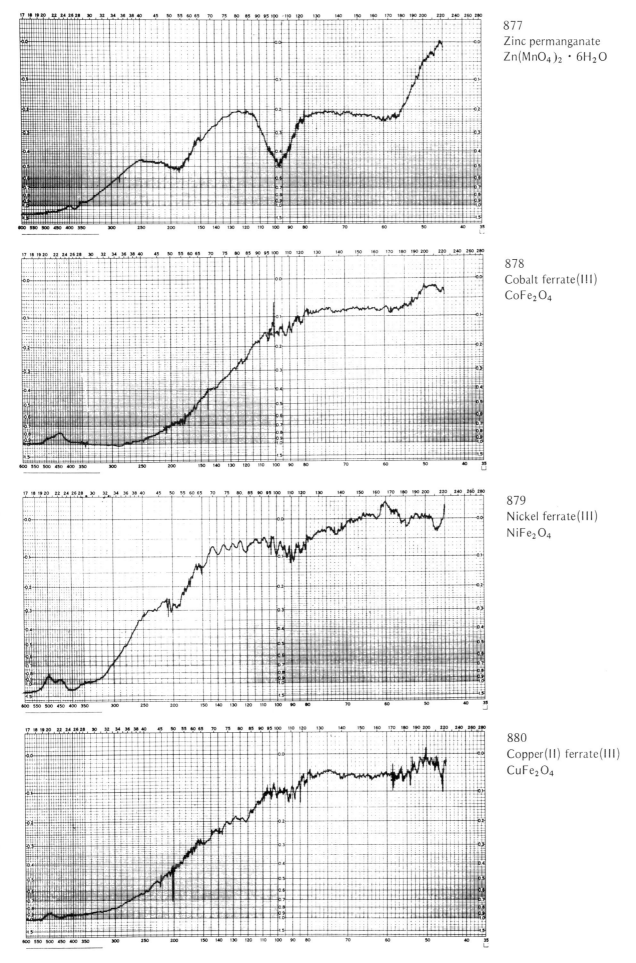

877
Zinc permanganate
$Zn(MnO_4)_2 \cdot 6H_2O$

878
Cobalt ferrate(III)
$CoFe_2O_4$

879
Nickel ferrate(III)
$NiFe_2O_4$

880
Copper(II) ferrate(III)
$CuFe_2O_4$

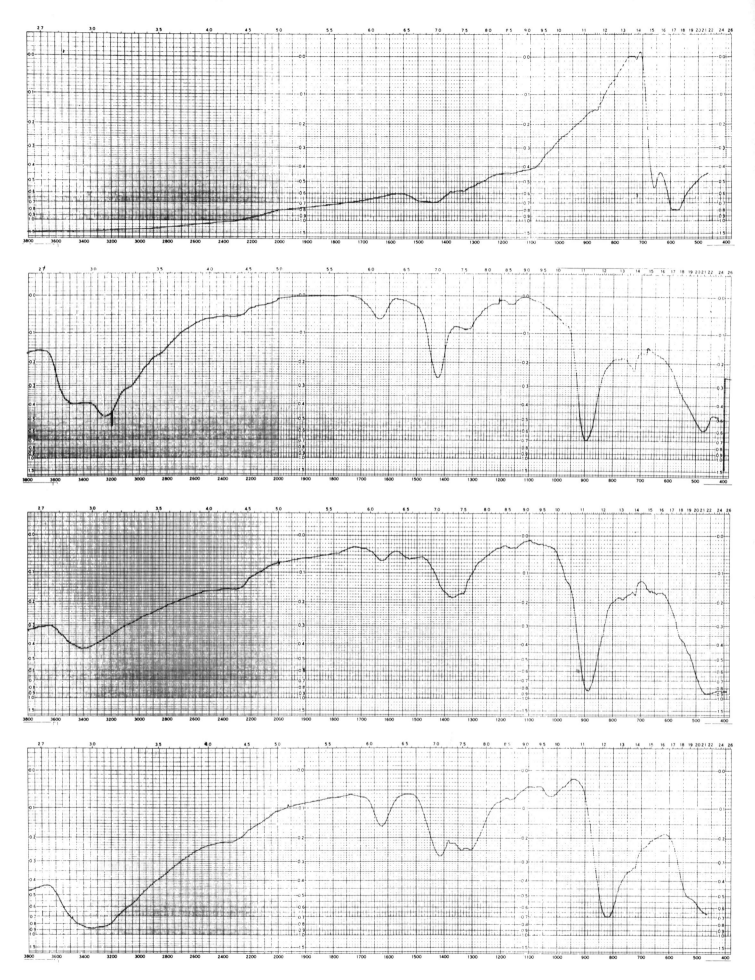

490

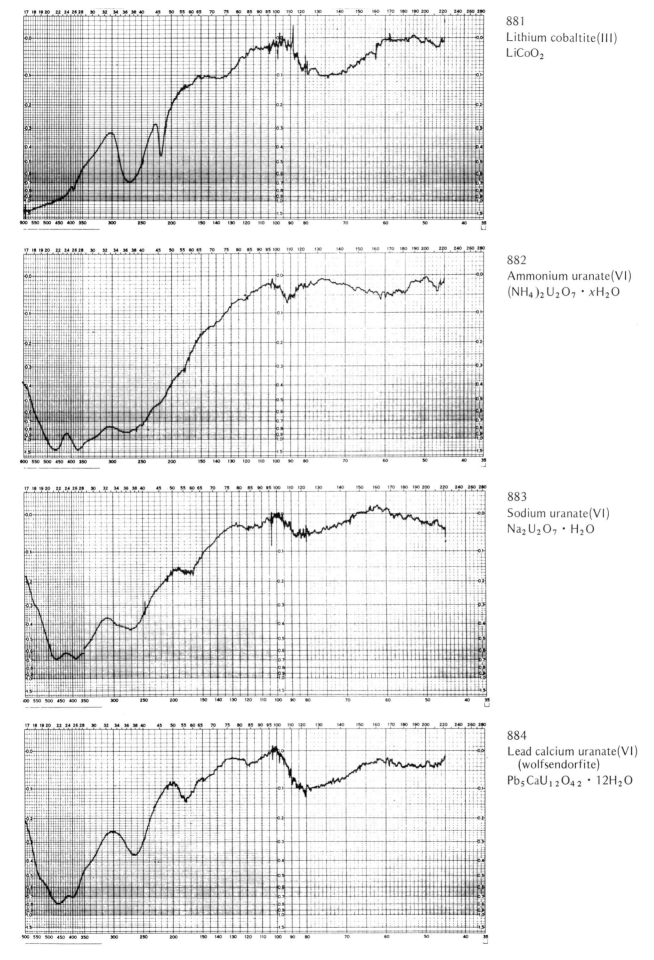

881
Lithium cobaltite(III)
$LiCoO_2$

882
Ammonium uranate(VI)
$(NH_4)_2U_2O_7 \cdot xH_2O$

883
Sodium uranate(VI)
$Na_2U_2O_7 \cdot H_2O$

884
Lead calcium uranate(VI)
(wolfsendorfite)
$Pb_5CaU_{12}O_{42} \cdot 12H_2O$

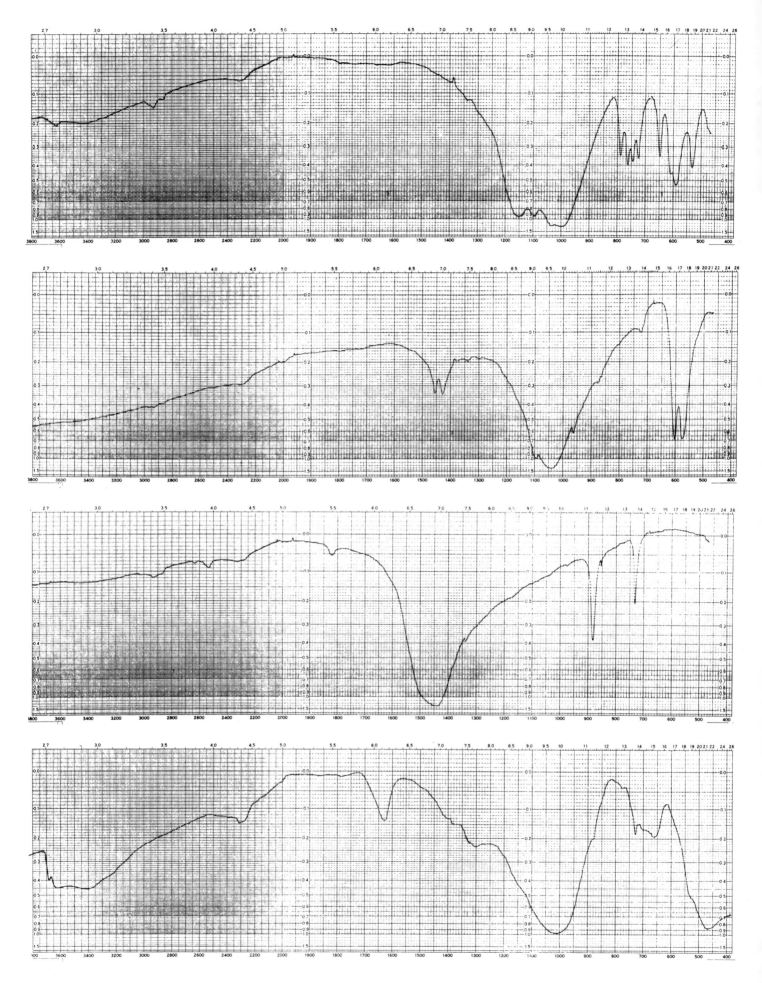

492

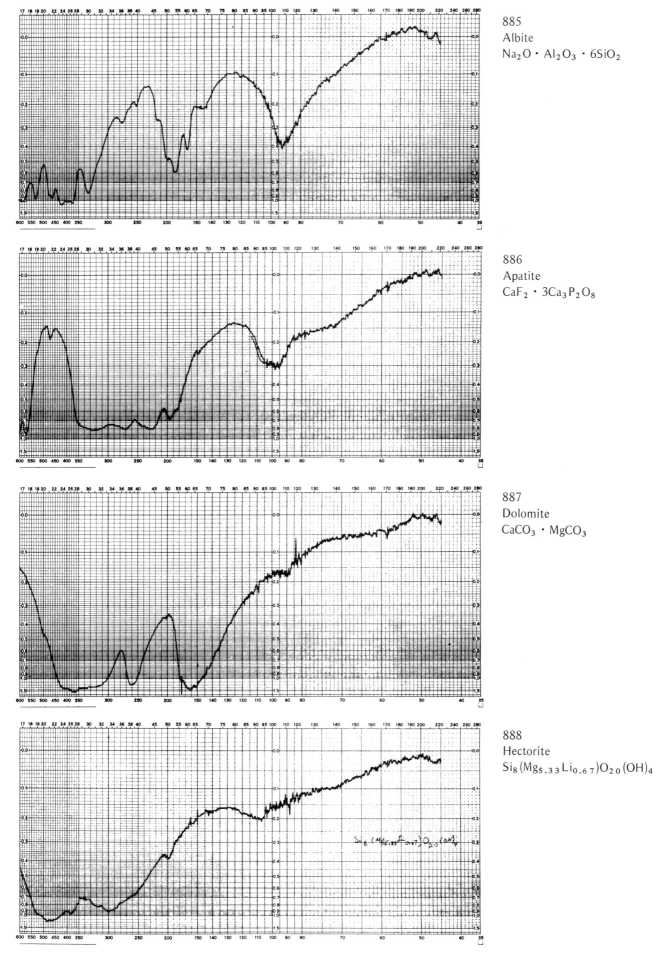

885
Albite
$Na_2O \cdot Al_2O_3 \cdot 6SiO_2$

886
Apatite
$CaF_2 \cdot 3Ca_3P_2O_8$

887
Dolomite
$CaCO_3 \cdot MgCO_3$

888
Hectorite
$Si_8(Mg_{5.33}Li_{0.67})O_{20}(OH)_4$

493

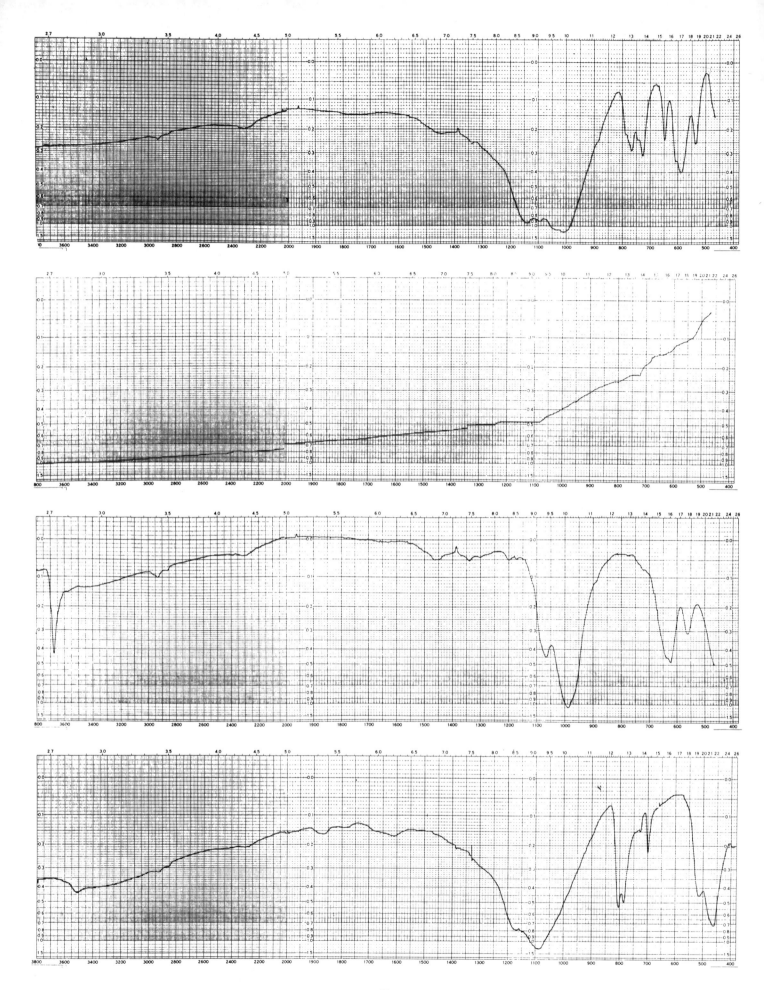

494

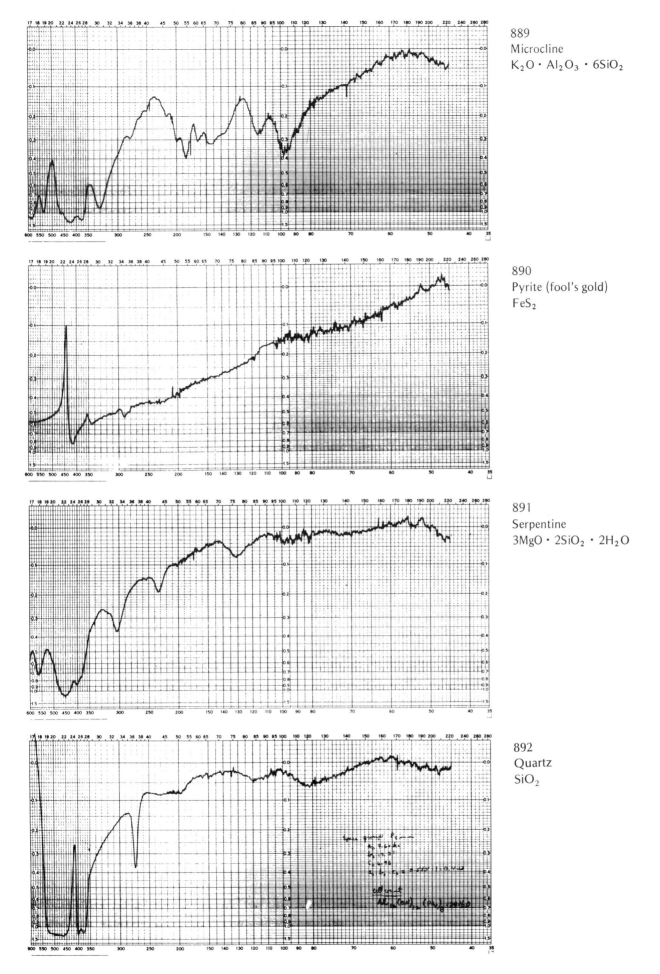

889
Microcline
$K_2O \cdot Al_2O_3 \cdot 6SiO_2$

890
Pyrite (fool's gold)
FeS_2

891
Serpentine
$3MgO \cdot 2SiO_2 \cdot 2H_2O$

892
Quartz
SiO_2

495

ISBN 0-12-523450-3